Lecture Notes in Computer Science 9155

Commenced Publication in 1973
Founding and Former Series Editors:
Gerhard Goos, Juris Hartmanis, and Jan van Leeuwen

Editorial Board

More information about this series at http://www.springer.com/series/7407

Osvaldo Gervasi · Beniamino Murgante
Sanjay Misra · Marina L. Gavrilova
Ana Maria Alves Coutinho Rocha · Carmelo Torre
David Taniar · Bernady O. Apduhan (Eds.)

Computational Science and Its Applications – ICCSA 2015

15th International Conference
Banff, AB, Canada, June 22–25, 2015
Proceedings, Part I

 Springer

Editors
Osvaldo Gervasi
University of Perugia
Perugia
Italy

Beniamino Murgante
University of Basilicata
Potenza
Italy

Sanjay Misra
Covenant University
Canaanland
Nigeria

Marina L. Gavrilova
University of Calgary
Calgary, AB
Canada

Ana Maria Alves Coutinho Rocha
University of Minho
Braga
Portugal

Carmelo Torre
Polytechnic University
Bari
Italy

David Taniar
Monash University
Clayton, VIC
Australia

Bernady O. Apduhan
Kyushu Sangyo University
Fukuoka
Japan

ISSN 0302-9743   ISSN 1611-3349   (electronic)
Lecture Notes in Computer Science
ISBN 978-3-319-21403-0   ISBN 978-3-319-21404-7   (eBook)
DOI 10.1007/978-3-319-21404-7

Library of Congress Control Number: 2015943360

LNCS Sublibrary: SL1 – Theoretical Computer Science and General Issues

Springer Cham Heidelberg New York Dordrecht London

Printed on acid-free paper

Springer International Publishing AG Switzerland is part of Springer Science+Business Media
(www.springer.com)

Preface

The year 2015 is a memorable year for the International Conference on Computational Science and Its Applications. In 2003, the First International Conference on Computational Science and Its Applications (chaired by C.J.K. Tan and M. Gavrilova) took place in Montreal, Canada (2003), and the following year it was hosted by A. Laganà and O. Gervasi in Assisi, Italy (2004). It then moved to Singapore (2005), Glasgow, UK (2006), Kuala-Lumpur, Malaysia (2007), Perugia, Italy (2008), Seoul, Korea (2009), Fukuoka, Japan (2010), Santander, Spain (2011), Salvador de Bahia, Brazil (2012), Ho Chi Minh City, Vietnam (2013), and Guimarães, Portugal (2014). The current installment of ICCSA 2015 took place in majestic Banff National Park, Banff, Alberta, Canada, during June 22–25, 2015.

The event received approximately 780 submissions from over 45 countries, evaluated by over 600 reviewers worldwide.

Its main track acceptance rate was approximately 29.7 % for full papers. In addition to full papers, published by Springer, the event accepted short papers, poster papers, and PhD student showcase works that are published in the IEEE CPS proceedings.

It also runs a number of parallel workshops, some for over 10 years, with new ones appearing for the first time this year. The success of ICCSA is largely contributed to the continuous support of the computational sciences community as well as researchers working in the applied relevant fields, such as graphics, image processing, biometrics, optimization, computer modeling, information systems, geographical sciences, physics, biology, astronomy, biometrics, virtual reality, and robotics, to name a few.

Over the past decade, the vibrant and promising area focusing on performance-driven computing and big data has became one of the key points of research enhancing the performance of information systems and supported processes. In addition to high-quality research at the frontier of these fields, consistently presented at ICCSA, a number of special journal issues are being planned following ICCSA 2015, including TCS Springer (*Transactions on Computational Sciences,* LNCS).

The contribution of the International Steering Committee and the International Program Committee are invaluable in the conference success. The dedication of members of these committees, the majority of whom have fulfilled this difficult role for the last 10 years, is astounding. Our warm appreciation also goes to the invited speakers, all event sponsors, supporting organizations, and volunteers. Finally, we thank all the authors for their submissions making the ICCSA conference series a well recognized and a highly successful event year after year.

June 2015

Marina L. Gavrilova
Osvaldo Gervasi
Bernady O. Apduhan

Organization

ICCSA 2015 was organized by the University of Calgary (Canada), the University of Perugia (Italy), the University of Basilicata (Italy), Monash University (Australia), Kyushu Sangyo University (Japan), and the University of Minho, (Portugal)

Honorary General Chairs

Antonio Laganà	University of Perugia, Italy
Norio Shiratori	Tohoku University, Japan
Kenneth C.J. Tan	Sardina Systems, Estonia

General Chairs

Marina L. Gavrilova	University of Calgary, Canada
Osvaldo Gervasi	University of Perugia, Italy
Bernady O. Apduhan	Kyushu Sangyo University, Japan

Program Committee Chairs

Beniamino Murgante	University of Basilicata, Italy
Ana Maria A.C. Rocha	University of Minho, Portugal
David Taniar	Monash University, Australia

International Advisory Committee

Jemal Abawajy	Deakin University, Australia
Dharma P. Agrawal	University of Cincinnati, USA
Claudia Bauzer Medeiros	University of Campinas, Brazil
Manfred M. Fisher	Vienna University of Economics and Business, Austria
Yee Leung	Chinese University of Hong Kong, SAR China

International Liaison Chairs

Ana Carla P. Bitencourt	Universidade Federal do Reconcavo da Bahia, Brazil
Alfredo Cuzzocrea	ICAR-CNR and University of Calabria, Italy
Maria Irene Falcão	University of Minho, Portugal
Marina L. Gavrilova	University of Calgary, Canada
Robert C.H. Hsu	Chung Hua University, Taiwan
Andrés Iglesias	University of Cantabria, Spain
Tai-Hoon Kim	Hannam University, Korea
Sanjay Misra	University of Minna, Nigeria
Takashi Naka	Kyushu Sangyo University, Japan

Rafael D.C. Santos   Brazilian National Institute for Space Research, Brazil
Maribel Yasmina Santos   University of Minho, Portugal

Workshop and Session Organizing Chairs

Beniamino Murgante   University of Basilicata, Italy
Jorge Gustavo Rocha   University of Minho, Portugal

Local Arrangement Chairs

Marina Gavrilova   University of Calgary, Canada (Chair)
Madeena Sultana   University of Calgary, Canada
Padma Polash Paul   University of Calgary, Canada
Faisal Ahmed   University of Calgary, Canada
Hossein Talebi   University of Calgary, Canada
Camille Sinanan   University of Calgary, Canada

Venue

ICCSA 2015 took place in the Banff Park Lodge Conference Center, Alberta (Canada).

Workshop Organizers

Agricultural and Environment Information and Decision Support Systems (AEIDSS 2015)

Sandro Bimonte   IRSTEA, France
André Miralles   IRSTEA, France
Frederic Hubert   University of Laval, Canada
François Pinet   IRSTEA, France

Approaches or Methods of Security Engineering (AMSE 2015)

TaiHoon Kim   Sungshin W. University, Korea

Advances in Information Systems and Technologies for Emergency Preparedness and Risk Assessment (ASTER 2015)

Maurizio Pollino   ENEA, Italy
Marco Vona   University of Basilicata, Italy
Beniamino Murgante   University of Basilicata, Italy

Advances in Web-Based Learning (AWBL 2015)

Mustafa Murat Inceoglu   Ege University, Turkey

Bio-inspired Computing and Applications (BIOCA 2015)

Nadia Nedjah   State University of Rio de Janeiro, Brazil
Luiza de Macedo   State University of Rio de Janeiro, Brazil
  Mourell

Computer-Aided Modeling, Simulation, and Analysis (CAMSA 2015)

Jie Shen   University of Michigan, USA, and Jilin University, China
Hao Chen   Shanghai University of Engineering Science, China
Xiaoqiang Liun   Donghua University, China
Weichun Shi   Shanghai Maritime University, China

Computational and Applied Statistics (CAS 2015)

Ana Cristina Braga   University of Minho, Portugal
Ana Paula Costa   University of Minho, Portugal
  Conceicao Amorim

Computational Geometry and Security Applications (CGSA 2015)

Marina L. Gavrilova   University of Calgary, Canada

Computational Algorithms and Sustainable Assessment (CLASS 2015)

Antonino Marvuglia   Public Research Centre Henri Tudor, Luxembourg
Beniamino Murgante   University of Basilicata, Italy

Chemistry and Materials Sciences and Technologies (CMST 2015)

Antonio Laganà   University of Perugia, Italy
Alessandro Costantini   INFN, Italy
Noelia Faginas Lago   University of Perugia, Italy
Leonardo Pacifici   University of Perugia, Italy

Computational Optimization and Applications (COA 2015)

Ana Maria Rocha   University of Minho, Portugal
Humberto Rocha   University of Coimbra, Portugal

Cities, Technologies and Planning (CTP 2015)

Giuseppe Borruso   University of Trieste, Italy
Beniamino Murgante   University of Basilicata, Italy

Econometrics and Multidimensional Evaluation in the Urban Environment (EMEUE 2015)

Carmelo M. Torre   Polytechnic of Bari, Italy
Maria Cerreta   University of Naples Federico II, Italy
Paola Perchinunno   University of Bari, Italy

Simona Panaro	University of Naples Federico II, Italy
Raffaele Attardi	University of Naples Federico II, Italy
Claudia Ceppi	Polytechnic of Bari, Italy

Future Computing Systems, Technologies, and Applications (FISTA 2015)

Bernady O. Apduhan	Kyushu Sangyo University, Japan
Rafael Santos	Brazilian National Institute for Space Research, Brazil
Jianhua Ma	Hosei University, Japan
Qun Jin	Waseda University, Japan

Geographical Analysis, Urban Modeling, Spatial Statistics (GEOGAN-MOD 2015)

Giuseppe Borruso	University of Trieste, Italy
Beniamino Murgante	University of Basilicata, Italy
Hartmut Asche	University of Potsdam, Germany

Land Use Monitoring for Soil Consumption Reduction (LUMS 2015)

Carmelo M. Torre	Polytechnic of Bari, Italy
Alessandro Bonifazi	Polytechnic of Bari, Italy
Valentina Sannicandro	University Federico II of Naples, Italy
Massimiliano Bencardino	University of Salerno, Italy
Gianluca di Cugno	Polytechnic of Bari, Italy
Beniamino Murgante	University of Basilicata, Italy

Mobile Communications (MC 2015)

| Hyunseung Choo | Sungkyunkwan University, Korea |

Mobile Computing, Sensing, and Actuation for Cyber Physical Systems (MSA4CPS 2015)

| Saad Qaisar | NUST School of Electrical Engineering and Computer Science, Pakistan |
| Moonseong Kim | Korean Intellectual Property Office, Korea |

Quantum Mechanics: Computational Strategies and Applications (QMCSA 2015)

Mirco Ragni	Universidad Federal de Bahia, Brazil
Ana Carla Peixoto Bitencourt	Universidade Estadual de Feira de Santana, Brazil
Roger Anderson	University of California, USA
Vincenzo Aquilanti	University of Perugia, Italy
Frederico Vasconcellos Prudente	Universidad Federal de Bahia, Brazil

Remote Sensing Data Analysis, Modeling, Interpretation and Applications: From a Global View to a Local Analysis (RS2015)

Rosa Lasaponara   Institute of Methodologies for Environmental Analysis,
                         National Research Council, Italy

Scientific Computing Infrastructure (SCI 2015)

Alexander Bodganov   St. Petersburg State University, Russia
Elena Stankova   St. Petersburg State University, Russia

Software Engineering Processes and Applications (SEPA 2015)

Sanjay Misra   Covenant University, Nigeria

Software Quality (SQ 2015)

Sanjay Misra   Covenant University, Nigeria

Advances in Spatio-Temporal Analytics (ST-Analytics 2015)

Joao Moura Pires   New University of Lisbon, Portugal
Maribel Yasmina Santos   New University of Lisbon, Portugal

Tools and Techniques in Software Development Processes (TTSDP 2015)

Sanjay Misra   Covenant University, Nigeria

Virtual Reality and Its Applications (VRA 2015)

Osvaldo Gervasi   University of Perugia, Italy
Lucio Depaolis   University of Salento, Italy

Program Committee

Jemal Abawajy   Deakin University, Australia
Kenny Adamson   University of Ulster, UK
Filipe Alvelos   University of Minho, Portugal
Paula Amaral   Universidade Nova de Lisboa, Portugal
Hartmut Asche   University of Potsdam, Germany
Md. Abul Kalam Azad   University of Minho, Portugal
Michela Bertolotto   University College Dublin, Ireland
Sandro Bimonte   CEMAGREF, TSCF, France
Rod Blais   University of Calgary, Canada
Ivan Blecic   University of Sassari, Italy
Giuseppe Borruso   University of Trieste, Italy
Yves Caniou   Lyon University, France
José A. Cardoso e   Universidade Nova de Lisboa, Portugal
   Cunha
Leocadio G. Casado   University of Almeria, Spain

Carlo Cattani	University of Salerno, Italy
Mete Celik	Erciyes University, Turkey
Alexander Chemeris	National Technical University of Ukraine KPI, Ukraine
Min Young Chung	Sungkyunkwan University, Korea
Gilberto Corso Pereira	Federal University of Bahia, Brazil
M. Fernanda Costa	University of Minho, Portugal
Gaspar Cunha	University of Minho, Portugal
Alfredo Cuzzocrea	ICAR-CNR and University of Calabria, Italy
Carla Dal Sasso Freitas	Universidade Federal do Rio Grande do Sul, Brazil
Pradesh Debba	The Council for Scientific and Industrial Research (CSIR), South Africa
Hendrik Decker	Instituto Tecnológico de Informática, Spain
Frank Devai	London South Bank University, UK
Rodolphe Devillers	Memorial University of Newfoundland, Canada
Prabu Dorairaj	NetApp, India/USA
M. Irene Falcao	University of Minho, Portugal
Cherry Liu Fang	U.S. DOE Ames Laboratory, USA
Edite M.G.P. Fernandes	University of Minho, Portugal
Jose-Jesus Fernandez	National Centre for Biotechnology, CSIS, Spain
Maria Antonia Forjaz	University of Minho, Portugal
Maria Celia Furtado Rocha	PRODEB/UFBA, Brazil
Akemi Galvez	University of Cantabria, Spain
Paulino Jose Garcia Nieto	University of Oviedo, Spain
Marina Gavrilova	University of Calgary, Canada
Jerome Gensel	LSR-IMAG, France
Maria Giaoutzi	National Technical University, Athens, Greece
Andrzej M. Goscinski	Deakin University, Australia
Alex Hagen-Zanker	University of Cambridge, UK
Malgorzata Hanzl	Technical University of Lodz, Poland
Shanmugasundaram Hariharan	B.S. Abdur Rahman University, India
Eligius M.T. Hendrix	University of Malaga/Wageningen University, Spain/The Netherlands
Tutut Herawan	Universitas Teknologi Yogyakarta, Indonesia
Hisamoto Hiyoshi	Gunma University, Japan
Fermin Huarte	University of Barcelona, Spain
Andres Iglesias	University of Cantabria, Spain
Mustafa Inceoglu	EGE University, Turkey
Peter Jimack	University of Leeds, UK
Qun Jin	Waseda University, Japan
Farid Karimipour	Vienna University of Technology, Austria
Baris Kazar	Oracle Corp., USA
DongSeong Kim	University of Canterbury, New Zealand
Taihoon Kim	Hannam University, Korea

Ivana Kolingerova   University of West Bohemia, Czech Republic
Dieter Kranzlmueller   LMU and LRZ Munich, Germany
Antonio Laganà   University of Perugia, Italy
Rosa Lasaponara   National Research Council, Italy
Maurizio Lazzari   National Research Council, Italy
Cheng Siong Lee   Monash University, Australia
Sangyoun Lee   Yonsei University, Korea
Jongchan Lee   Kunsan National University, Korea
Clement Leung   Hong Kong Baptist University, Hong Kong, SAR China
Chendong Li   University of Connecticut, USA
Gang Li   Deakin University, Australia
Ming Li   East China Normal University, China
Fang Liu   AMES Laboratories, USA
Xin Liu   University of Calgary, Canada
Savino Longo   University of Bari, Italy
Tinghuai Ma   NanJing University of Information Science and Technology, China

Sergio Maffioletti   University of Zurich, Switzerland
Ernesto Marcheggiani   Katholieke Universiteit Leuven, Belgium
Antonino Marvuglia   Research Centre Henri Tudor, Luxembourg
Nicola Masini   National Research Council, Italy
Nirvana Meratnia   University of Twente, The Netherlands
Alfredo Milani   University of Perugia, Italy
Sanjay Misra   Federal University of Technology Minna, Nigeria
Giuseppe Modica   University of Reggio Calabria, Italy
José Luis Montaña   University of Cantabria, Spain
Beniamino Murgante   University of Basilicata, Italy
Jiri Nedoma   Academy of Sciences of the Czech Republic, Czech Republic
Laszlo Neumann   University of Girona, Spain
Kok-Leong Ong   Deakin University, Australia
Belen Palop   Universidad de Valladolid, Spain
Marcin Paprzycki   Polish Academy of Sciences, Poland
Eric Pardede   La Trobe University, Australia
Kwangjin Park   Wonkwang University, Korea
Ana Isabel Pereira   Polytechnic Institute of Braganca, Portugal
Maurizio Pollino   Italian National Agency for New Technologies, Energy and Sustainable Economic Development, Italy
Alenka Poplin   University of Hamburg, Germany
Vidyasagar Potdar   Curtin University of Technology, Australia
David C. Prosperi   Florida Atlantic University, USA
Wenny Rahayu   La Trobe University, Australia
Jerzy Respondek   Silesian University of Technology Poland
Ana Maria A.C. Rocha   University of Minho, Portugal

Humberto Rocha	INESC-Coimbra, Portugal
Alexey Rodionov	Institute of Computational Mathematics and Mathematical Geophysics, Russia
Cristina S. Rodrigues	University of Minho, Portugal
Octavio Roncero	CSIC, Spain
Maytham Safar	Kuwait University, Kuwait
Chiara Saracino	A.O. Ospedale Niguarda Ca' Granda - Milano, Italy
Haiduke Sarafian	The Pennsylvania State University, USA
Jie Shen	University of Michigan, USA
Qi Shi	Liverpool John Moores University, UK
Dale Shires	U.S. Army Research Laboratory, USA
Takuo Suganuma	Tohoku University, Japan
Sergio Tasso	University of Perugia, Italy
Ana Paula Teixeira	University of Tras-os-Montes and Alto Douro, Portugal
Senhorinha Teixeira	University of Minho, Portugal
Parimala Thulasiraman	University of Manitoba, Canada
Carmelo Torre	Polytechnic of Bari, Italy
Javier Martinez Torres	Centro Universitario de la Defensa Zaragoza, Spain
Giuseppe A. Trunfio	University of Sassari, Italy
Unal Ufuktepe	Izmir University of Economics, Turkey
Toshihiro Uchibayashi	Kyushu Sangyo University, Japan
Mario Valle	Swiss National Supercomputing Centre, Switzerland
Pablo Vanegas	University of Cuenca, Equador
Piero Giorgio Verdini	INFN Pisa and CERN, Italy
Marco Vizzari	University of Perugia, Italy
Koichi Wada	University of Tsukuba, Japan
Krzysztof Walkowiak	Wroclaw University of Technology, Poland
Robert Weibel	University of Zurich, Switzerland
Roland Wismüller	Universität Siegen, Germany
Mudasser Wyne	SOET National University, USA
Chung-Huang Yang	National Kaohsiung Normal University, Taiwan
Xin-She Yang	National Physical Laboratory, UK
Salim Zabir	France Telecom Japan Co., Japan
Haifeng Zhao	University of California, Davis, USA
Kewen Zhao	University of Qiongzhou, China
Albert Y. Zomaya	University of Sydney, Australia

Reviewers

Abawajy Jemal	Deakin University, Australia
Abdi Samane	University College Cork, Ireland
Aceto Lidia	University of Pisa, Italy
Acharjee Shukla	Dibrugarh University, India
Adriano Elias	Universidade Nova de Lisboa, Portugal
Afreixo Vera	University of Aveiro, Portugal
Aguiar Ademar	Universidade do Porto, Portugal

Aguilar Antonio	University of Barcelona, Spain
Aguilar José Alfonso	Universidad Autónoma de Sinaloa, Mexico
Ahmed Faisal	University of Calgary, Canada
Aktas Mehmet	Yildiz Technical University, Turkey
Al-Juboori AliAlwan	International Islamic University Malaysia, Malaysia
Alarcon Vladimir	Universidad Diego Portales, Chile
Alberti Margarita	University of Barcelona, Spain
Ali Salman	NUST, Pakistan
Alkazemi Basem Qassim	University, Saudi Arabia
Alvanides Seraphim	Northumbria University, UK
Alvelos Filipe	University of Minho, Portugal
Alves Cláudio	University of Minho, Portugal
Alves José Luis	University of Minho, Portugal
Alves Maria Joo	Universidade de Coimbra, Portugal
Amin Benatia Mohamed	Groupe Cesi, France
Amorim Ana Paula	University of Minho, Portugal
Amorim Paulo	Federal University of Rio de Janeiro, Brazil
Andrade Wilkerson	Federal University of Campina Grande, Brazil
Andrianov Serge	Yandex, Russia
Aniche Mauricio	University of São Paulo, Brazil
Andrienko Gennady	Fraunhofer Institute for Intelligent Analysis and Informations Systems, Germany
Apduhan Bernady	Kyushu Sangyo University, Japan
Aquilanti Vincenzo	University of Perugia, Italy
Aquino Gibeon	UFRN, Brazil
Argiolas Michele	University of Cagliari, Italy
Asche Hartmut	Potsdam University, Germany
Athayde Maria Emilia Feijão Queiroz	University of Minho, Portugal
Attardi Raffaele	University of Napoli Federico II, Italy
Azad Md. Abdul	Indian Institute of Technology Kanpur, India
Azad Md. Abul Kalam	University of Minho, Portugal
Bao Fernando	Universidade Nova de Lisboa, Portugal
Badard Thierry	Laval University, Canada
Bae Ihn-Han	Catholic University of Daegu, South Korea
Baioletti Marco	University of Perugia, Italy
Balena Pasquale	Polytechnic of Bari, Italy
Banerjee Mahua	Xavier Institute of Social Sciences, India
Barroca Filho Itamir	UFRN, Brazil
Bartoli Daniele	University of Perugia, Italy
Bastanfard Azam	Islamic Azad University, Iran
Belanzoni Paola	University of Perugia, Italy
Bencardino Massimiliano	University of Salerno, Italy
Benigni Gladys	University of Oriente, Venezuela

Bertolotto Michela	University College Dublin, Ireland
Bilancia Massimo	Università di Bari, Italy
Blanquer Ignacio	Universitat Politècnica de València, Spain
Bodini Olivier	Université Pierre et Marie Curie Paris and CNRS, France
Bogdanov Alexander	Saint-Petersburg State University, Russia
Bollini Letizia	University of Milano, Italy
Bonifazi Alessandro	Polytechnic of Bari, Italy
Borruso Giuseppe	University of Trieste, Italy
Bostenaru Maria	"Ion Mincu" University of Architecture and Urbanism, Romania
Boucelma Omar	University of Marseille, France
Braga Ana Cristina	University of Minho, Portugal
Branquinho Amilcar	University of Coimbra, Portugal
Brás Carmo	Universidade Nova de Lisboa, Portugal
Cacao Isabel	University of Aveiro, Portugal
Cadarso-Suárez Carmen	University of Santiago de Compostela, Spain
Caiaffa Emanuela	ENEA, Italy
Calamita Giuseppe	National Research Council, Italy
Campagna Michele	University of Cagliari, Italy
Campobasso Francesco	University of Bari, Italy
Campos José	University of Minho, Portugal
Caniato Renhe Marcelo	Universidade Federal de Juiz de Fora, Brazil
Cannatella Daniele	University of Napoli Federico II, Italy
Canora Filomena	University of Basilicata, Italy
Cannatella Daniele	University of Napoli Federico II, Italy
Canora Filomena	University of Basilicata, Italy
Carbonara Sebastiano	University of Chieti, Italy
Carlini Maurizio	University of Tuscia, Italy
Carneiro Claudio	École Polytechnique Fédérale de Lausanne, Switzerland
Ceppi Claudia	Polytechnic of Bari, Italy
Cerreta Maria	University Federico II of Naples, Italy
Chen Hao	Shanghai University of Engineering Science, China
Choi Joonsoo	Kookmin University, South Korea
Choo Hyunseung	Sungkyunkwan University, South Korea
Chung Min Young	Sungkyunkwan University, South Korea
Chung Myoungbeom	Sungkyunkwan University, South Korea
Chung Tai-Myoung	Sungkyunkwan University, South Korea
Cirrincione Maurizio	Université de Technologie Belfort-Montbeliard, France
Clementini Eliseo	University of L'Aquila, Italy
Coelho Leandro dos Santos	PUC-PR, Brazil
Coletti Cecilia	University of Chieti, Italy
Conceicao Ana	Universidade do Algarve, Portugal
Correia Elisete	University of Trás-Os-Montes e Alto Douro, Portugal
Correia Filipe	FEUP, Portugal

Correia Florbela Maria da Cruz Domingues	Instituto Politécnico de Viana do Castelo, Portugal
Corso Pereira Gilberto	UFPA, Brazil
Cortés Ana	Universitat Autònoma de Barcelona, Spain
Cosido Oscar	Ayuntamiento de Santander, Spain
Costa Carlos	Faculdade Engenharia U. Porto, Portugal
Costa Fernanda	University of Minho, Portugal
Costantini Alessandro	INFN, Italy
Crasso Marco	National Scientific and Technical Research Council, Argentina
Crawford Broderick	Universidad Catolica de Valparaiso, Chile
Crestaz Ezio	GiScience, Italia
Cristia Maximiliano	CIFASIS and UNR, Argentina
Cunha Gaspar	University of Minho, Portugal
Cutini Valerio	University of Pisa, Italy
Danese Maria	IBAM, CNR, Italy
Daneshpajouh Shervin	University of Western Ontario, Canada
De Almeida Regina	University of Trás-os-Montes e Alto Douro, Portugal
de Doncker Elise	University of Michgan, USA
De Fino Mariella	Polytechnic of Bari, Italy
De Paolis Lucio Tommaso	University of Salento, Italy
de Rezende Pedro J.	Universidade Estadual de Campinas, Brazil
De Rosa Fortuna	University of Napoli Federico II, Italy
De Toro Pasquale	University of Napoli Federico II, Italy
Decker Hendrik	Instituto Tecnológico de Informática, Spain
Degtyarev Alexander	Saint-Petersburg State University, Russia
Deiana Andrea	Geoinfolab, Italia
Deniz Berkhan	Aselsan Electronics Inc., Turkey
Desjardin Eric	University of Reims, France
Devai Frank	London South Bank University, UK
Dwivedi Sanjay Kumar	Babasaheb Bhimrao Ambedkar University, India
Dhawale Chitra	PR Pote College, Amravati, India
Di Cugno Gianluca	Polytechnic of Bari, Italy
Di Gangi Massimo	University of Messina, Italy
Di Leo Margherita	JRC, European Commission, Belgium
Dias Joana	University of Coimbra, Portugal
Dias d'Almeida Filomena	University of Porto, Portugal
Diez Teresa	Universidad de Alcalá, Spain
Dilo Arta	University of Twente, The Netherlands
Dixit Veersain	Delhi University, India
Doan Anh Vu	Université Libre de Bruxelles, Belgium
Durrieu Sylvie	Maison de la Teledetection Montpellier, France
Dutra Inês	University of Porto, Portugal
Dyskin Arcady	The University of Western Australia, Australia

Eichelberger Hanno	University of Tübingen, Germany
El-Zawawy Mohamed A.	Cairo University, Egypt
Escalona Maria-Jose	University of Seville, Spain
Falcão M. Irene	University of Minho, Portugal
Farantos Stavros	University of Crete and FORTH, Greece
Faria Susana	University of Minho, Portugal
Fernandes Edite	University of Minho, Portugal
Fernandes Rosário	University of Minho, Portugal
Fernandez Joao P.	Universidade da Beira Interior, Portugal
Ferrão Maria	University of Beira Interior and CEMAPRE, Portugal
Ferreira Fátima	University of Trás-Os-Montes e Alto Douro, Portugal
Figueiredo Manuel Carlos	University of Minho, Portugal
Filipe Ana	University of Minho, Portugal
Flouvat Frederic	University New Caledonia, New Caledonia
Forjaz Maria Antónia	University of Minho, Portugal
Formosa Saviour	University of Malta, Malta
Fort Marta	University of Girona, Spain
Franciosa Alfredo	University of Napoli Federico II, Italy
Freitas Adelaide de Fátima Baptista Valente	University of Aveiro, Portugal
Frydman Claudia	Laboratoire des Sciences de l'Information et des Systèmes, France
Fusco Giovanni	CNRS - UMR ESPACE, France
Gabrani Goldie	University of Delhi, India Galleguillos Cristian, Pontificia Universidad Catlica de Valparaso, Chile
Gao Shang	Zhongnan University of Economics and Law, China
Garau Chiara	University of Cagliari, Italy
Garcia Ernesto	University of the Basque Country, Spain
Garca Omar Vicente	Universidad Autònoma de Sinaloa, Mexico
Garcia Tobio Javier	Centro de Supercomputación de Galicia, CESGA, Spain
Gavrilova Marina	University of Calgary, Canada
Gazzea Nicoletta	ISPRA, Italy
Gensel Jerome	IMAG, France
Geraldi Edoardo	National Research Council, Italy
Gervasi Osvaldo	University of Perugia, Italy
Giaoutzi Maria	National Technical University Athens, Greece
Gil Artur	University of the Azores, Portugal
Gizzi Fabrizio	National Research Council, Italy
Gomes Abel	Universidad de Beira Interior, Portugal
Gomes Maria Cecilia	Universidade Nova de Lisboa, Portugal
Gomes dos Anjos Eudisley	Federal University of Paraba, Brazil
Gonçalves Alexandre	Instituto Superior Tecnico Lisboa, Portugal

Gonçalves Arminda Manuela	University of Minho, Portugal
Gonzaga de Oliveira Sanderson Lincohn	Universidade Do Estado De Santa Catarina, Brazil
Gonzalez-Aguilera Diego	Universidad de Salamanca, Spain
Gorbachev Yuriy	Geolink Technologies, Russia
Govani Kishan	Darshan Institute of Engineering Technology, India
Grandison Tyrone	Proficiency Labs International, USA
Gravagnuolo Antonia	University of Napoli Federico II, Italy
Grilli Luca	University of Perugia, Italy
Guerra Eduardo	National Institute for Space Research, Brazil
Guo Hua	Carleton University, Canada
Hanazumi Simone	University of São Paulo, Brazil
Hanif Mohammad Abu	Chonbuk National University, South Korea
Hansen Henning Sten	Aalborg University, Denmark
Hanzl Malgorzata	University of Lodz, Poland
Hegedus Peter	University of Szeged, Hungary
Heijungs Reinout	VU University Amsterdam, The Netherlands
Hendrix Eligius M.T.	University of Malaga/Wageningen University, Spain/The Netherlands
Henriques Carla	Escola Superior de Tecnologia e Gestão, Portugal
Herawan Tutut	University of Malaya, Malaysia
Hiyoshi Hisamoto	Gunma University, Japan
Hodorog Madalina	Austria Academy of Science, Austria
Hong Choong Seon	Kyung Hee University, South Korea
Hsu Ching-Hsien	Chung Hua University, Taiwan
Hsu Hui-Huang	Tamkang University, Taiwan
Hu Hong	The Honk Kong Polytechnic University, China
Huang Jen-Fa	National Cheng Kung University, Taiwan
Hubert Frederic	Université Laval, Canada
Iglesias Andres	University of Cantabria, Spain
Jamal Amna	National University of Singapore, Singapore
Jank Gerhard	Aachen University, Germany
Jeong Jongpil	Sungkyunkwan University, South Korea
Jiang Bin	University of Gävle, Sweden
Johnson Franklin	Universidad de Playa Ancha, Chile
Kalogirou Stamatis	Harokopio University of Athens, Greece
Kamoun Farouk	Université de la Manouba, Tunisia
Kanchi Saroja	Kettering University, USA
Kanevski Mikhail	University of Lausanne, Switzerland
Kang Myoung-Ah	ISIMA Blaise Pascal University, France
Karandikar Varsha	Devi Ahilya University, Indore, India
Karimipour Farid	Vienna University of Technology, Austria
Kavouras Marinos	University of Lausanne, Switzerland
Kazar Baris	Oracle Corp., USA

Keramat Alireza	Jundi-Shapur Univ. of Technology, Iran
Khan Murtaza	NUST, Pakistan
Khattak Asad Masood	Kyung Hee University, Korea
Khazaei Hamzeh	Ryerson University, Canada
Khurshid Khawar	NUST, Pakistan
Kim Dongsoo	Indiana University-Purdue University Indianapolis, USA
Kim Mihui	Hankyong National University, South Korea
Koo Bonhyun	Samsung, South Korea
Korkhov Vladimir	St. Petersburg State University, Russia
Kotzinos Dimitrios	Université de Cergy-Pontoise, France
Kumar Dileep	SR Engineering College, India
Kurdia Anastasia	Buknell University, USA
Lachance-Bernard Nicolas	École Polytechnique Fédérale de Lausanne, Switzerland
Laganà Antonio	University of Perugia, Italy
Lai Sabrina	University of Cagliari, Italy
Lanorte Antonio	CNR-IMAA, Italy
Lanza Viviana	Lombardy Regional Institute for Research, Italy
Lasaponara Rosa	National Research Council, Italy
Lassoued Yassine	University College Cork, Ireland
Lazzari Maurizio	CNR IBAM, Italy
Le Duc Tai	Sungkyunkwan University, South Korea
Le Duc Thang	Sungkyunkwan University, South Korea
Le-Thi Kim-Tuyen	Sungkyunkwan University, South Korea
Ledoux Hugo	Delft University of Technology, The Netherlands
Lee Dong-Wook	INHA University, South Korea
Lee Hongseok	Sungkyunkwan University, South Korea
Lee Ickjai	James Cook University, Australia
Lee Junghoon	Jeju National University, South Korea
Lee KangWoo	Sungkyunkwan University, South Korea
Legatiuk Dmitrii	Bauhaus University Weimar, Germany
Lendvay Gyorgy	Hungarian Academy of Science, Hungary
Leonard Kathryn	California State University, USA
Li Ming	East China Normal University, China
Libourel Thrse	LIRMM, France
Lin Calvin	University of Texas at Austin, USA
Liu Xin	University of Calgary, Canada
Loconte Pierangela	Technical University of Bari, Italy
Lombardi Andrea	University of Perugia, Italy
Longo Savino	University of Bari, Italy
Lopes Cristina	University of California Irvine, USA
Lopez Cabido Ignacio	Centro de Supercomputación de Galicia, CESGA
Lourenço Vanda Marisa	University Nova de Lisboa, Portugal
Luaces Miguel	University of A Coruña, Spain
Lucertini Giulia	IUAV, Italy
Luna Esteban Robles	Universidad Nacional de la Plata, Argentina

M.M.H. Gregori Rodrigo	Universidade Tecnológica Federal do Paraná, Brazil
Machado Gaspar	University of Minho, Portugal
Machado Jose	University of Minho, Portugal
Mahinderjit Singh Manmeet	University Sains Malaysia, Malaysia
Malonek Helmuth	University of Aveiro, Portugal
Manfreda Salvatore	University of Basilicata, Italy
Manns Mary Lynn	University of North Carolina Asheville, USA
Manso Callejo Miguel Angel	Universidad Politécnica de Madrid, Spain
Marechal Bernard	Universidade Federal de Rio de Janeiro, Brazil
Marechal Franois	École Polytechnique Fédérale de Lausanne, Switzerland
Margalef Tomas	Universitat Autònoma de Barcelona, Spain
Marghany Maged	Universiti Teknologi Malaysia, Malaysia
Marsal-Llacuna Maria-Llusa	Universitat de Girona, Spain
Marsh Steven	University of Ontario, Canada
Martins Ana Mafalda	Universidade de Aveiro, Portugal
Martins Pedro	Universidade do Minho, Portugal
Marvuglia Antonino	Public Research Centre Henri Tudor, Luxembourg
Mateos Cristian	Universidad Nacional del Centro, Argentina
Matos Inés	Universidade de Aveiro, Portugal
Matos Jose	Instituto Politecnico do Porto, Portugal
Matos João	ISEP, Portugal
Mauro Giovanni	University of Trieste, Italy
Mauw Sjouke	University of Luxembourg, Luxembourg
Medeiros Pedro	Universidade Nova de Lisboa, Portugal
Melle Franco Manuel	University of Minho, Portugal
Melo Ana	Universidade de São Paulo, Brazil
Michikawa Takashi	University of Tokio, Japan
Milani Alfredo	University of Perugia, Italy
Millo Giovanni	Generali Assicurazioni, Italy
Min-Woo Park	SungKyunKwan University, South Korea
Miranda Fernando	University of Minho, Portugal
Misra Sanjay	Covenant University, Nigeria
Mo Otilia	Universidad Autonoma de Madrid, Spain
Modica Giuseppe	Università Mediterranea di Reggio Calabria, Italy
Mohd Nawi Nazri	Universiti Tun Hussein Onn Malaysia, Malaysia
Morais João	University of Aveiro, Portugal
Moreira Adriano	University of Minho, Portugal
Moerig Marc	University of Magdeburg, Germany
Morzy Mikolaj	University of Poznan, Poland
Mota Alexandre	Universidade Federal de Pernambuco, Brazil
Moura Pires João	Universidade Nova de Lisboa - FCT, Portugal
Mourão Maria	Polytechnic Institute of Viana do Castelo, Portugal

Mourelle Luiza de Macedo	UERJ, Brazil
Mukhopadhyay Asish	University of Windsor, Canada
Mulay Preeti	Bharti Vidyapeeth University, India
Murgante Beniamino	University of Basilicata, Italy
Naghizadeh Majid Reza	Qazvin Islamic Azad University, Iran
Nagy Csaba	University of Szeged, Hungary
Nandy Subhas	Indian Statistical Institute, India
Nash Andrew	Vienna Transport Strategies, Austria
Natário Isabel Cristina Maciel	University Nova de Lisboa, Portugal
Navarrete Gutierrez Tomas	Luxembourg Institute of Science and Technology, Luxembourg
Nedjah Nadia	State University of Rio de Janeiro, Brazil
Nguyen Hong-Quang	Ho Chi Minh City University, Vietnam
Nguyen Tien Dzung	Sungkyunkwan University, South Korea
Nickerson Bradford	University of New Brunswick, Canada
Nielsen Frank	Université Paris Saclay CNRS, France
NM Tuan	Ho Chi Minh City University of Technology, Vietnam
Nogueira Fernando	University of Coimbra, Portugal
Nole Gabriele	IRMAA National Research Council, Italy
Nourollah Ali	Amirkabir University of Technology, Iran
Olivares Rodrigo	UCV, Chile
Oliveira Irene	University of Trás-Os-Montes e Alto Douro, Portugal
Oliveira José A.	University of Minho, Portugal
Oliveira e Silva Luis	University of Lisboa, Portugal
Osaragi Toshihiro	Tokyo Institute of Technology, Japan
Ottomanelli Michele	Polytechnic of Bari, Italy
Ozturk Savas	TUBITAK, Turkey
Pagliara Francesca	University of Naples, Italy
Painho Marco	New University of Lisbon, Portugal
Pantazis Dimos	Technological Educational Institute of Athens, Greece
Paolotti Luisa	University of Perugia, Italy
Papa Enrica	University of Amsterdam, The Netherlands
Papathanasiou Jason	University of Macedonia, Greece
Pardede Eric	La Trobe University, Australia
Parissis Ioannis	Grenoble INP - LCIS, France
Park Gyung-Leen	Jeju National University, South Korea
Park Sooyeon	Korea Polytechnic University, South Korea
Pascale Stefania	University of Basilicata, Italy
Parker Gregory	University of Oklahoma, USA
Parvin Hamid	Iran University of Science and Technology, Iran
Passaro Pierluigi	University of Bari Aldo Moro, Italy
Pathan Al-Sakib Khan	International Islamic University Malaysia, Malaysia
Paul Padma Polash	University of Calgary, Canada

Peixoto Bitencourt Ana Carla	Universidade Estadual de Feira de Santana, Brazil
Peraza Juan Francisco	Autonomous University of Sinaloa, Mexico
Perchinunno Paola	University of Bari, Italy
Pereira Ana	Polytechnic Institute of Bragança, Portugal
Pereira Francisco	Instituto Superior de Engenharia, Portugal
Pereira Paulo	University of Minho, Portugal
Pereira Javier	Diego Portales University, Chile
Pereira Oscar	Universidade de Aveiro, Portugal
Pereira Ricardo	Portugal Telecom Inovacao, Portugal
Perez Gregorio	Universidad de Murcia, Spain
Pesantes Mery	CIMAT, Mexico
Pham Quoc Trung	HCMC University of Technology, Vietnam
Pietrantuono Roberto	University of Napoli "Federico II", Italy
Pimentel Carina	University of Aveiro, Portugal
Pina Antonio	University of Minho, Portugal
Piñar Miguel	Universidad de Granada, Spain
Pinciu Val	Southern Connecticut State University, USA
Pinet Francois	IRSTEA, France
Piscitelli Claudia	Polytechnic University of Bari, Italy
Pollino Maurizio	ENEA, Italy
Poplin Alenka	University of Hamburg, Germany
Porschen Stefan	University of Köln, Germany
Potena Pasqualina	University of Bergamo, Italy
Prata Paula	University of Beira Interior, Portugal
Previtali Mattia	Polytechnic of Milan, Italy
Prosperi David	Florida Atlantic University, USA
Protheroe Dave	London South Bank University, UK
Pusatli Tolga	Cankaya University, Turkey
Qaisar Saad	NURST, Pakistan
Qi Yu	Mesh Capital LLC, USA
Quan Tho	Ho Chi Minh City University of Technology, Vietnam
Raffaeta Alessandra	University of Venice, Italy
Ragni Mirco	Universidade Estadual de Feira de Santana, Brazil
Rahayu Wenny	La Trobe University, Australia
Rautenberg Carlos	University of Graz, Austria
Ravat Franck	IRIT, France
Raza Syed Muhammad	Sungkyunkwan University, South Korea
Rinaldi Antonio	DIETI - UNINA, Italy
Rinzivillo Salvatore	University of Pisa, Italy
Rios Gordon	University College Dublin, Ireland
Riva Sanseverino Eleonora	University of Palermo, Italy
Roanes-Lozano Eugenio	Universidad Complutense de Madrid, Spain
Rocca Lorena	University of Padova, Italy
Roccatello Eduard	3DGIS, Italy

Rocha Ana Maria	University of Minho, Portugal
Rocha Humberto	University of Coimbra, Portugal
Rocha Jorge	University of Minho, Portugal
Rocha Maria Clara	ESTES Coimbra, Portugal
Rocha Miguel	University of Minho, Portugal
Rodrigues Armanda	Universidade Nova de Lisboa, Portugal
Rodrigues Cristina	DPS, University of Minho, Portugal
Rodrigues Joel	University of Minho, Portugal
Rodriguez Daniel	University of Alcala, Spain
Rodrguez Gonzlez Alejandro	Universidad Carlos III Madrid, Spain
Roh Yongwan	Korean IP, South Korea
Romano Bernardino	University of l'Aquila, Italy
Roncaratti Luiz	Instituto de Física, University of Brasilia, Brazil
Roshannejad Ali	University of Calgary, Canada
Rosi Marzio	University of Perugia, Italy
Rossi Gianfranco	University of Parma, Italy
Rotondo Francesco	Polytechnic of Bari, Italy
Roussey Catherine	IRSTEA, France
Ruj Sushmita	Indian Statistical Institute, India
S. Esteves Jorge	University of Aveiro, Portugal
Saeed Husnain	NUST, Pakistan
Sahore Mani	Lovely Professional University, India
Saini Jatinder Singh	Baba Banda Singh Bahadur Engineering College, India
Salzer Reiner	Technical University Dresden, Germany
Sameh Ahmed	The American University in Cairo, Egypt
Sampaio Alcinia Zita	Instituto Superior Tecnico Lisboa, Portugal
Sannicandro Valentina	Polytechnic of Bari, Italy
Santiago Jnior Valdivino	Instituto Nacional de Pesquisas Espaciais, Brazil
Santos Josué	UFABC, Brazil
Santos Rafael	INPE, Brazil
Santos Viviane	Universidade de São Paulo, Brazil
Santucci Valentino	University of Perugia, Italy
Saracino Gloria	University of Milano-Bicocca, Italy
Sarafian Haiduke	Pennsylvania State University, USA
Saraiva João	University of Minho, Portugal
Sarrazin Renaud	Université Libre de Bruxelles, Belgium
Schirone Dario Antonio	University of Bari, Italy
Schneider Michel	ISIMA, France
Schoier Gabriella	University of Trieste, Italy
Schuhmacher Marta	Universitat Rovira i Virgili, Spain
Scorza Francesco	University of Basilicata, Italy
Seara Carlos	Universitat Politècnica de Catalunya, Spain
Sellares J. Antoni	Universitat de Girona, Spain
Selmaoui Nazha	University of New Caledonia, New Caledonia
Severino Ricardo Jose	University of Minho, Portugal

Shaik Mahaboob Hussain	JNTUK Vizianagaram, A.P., India
Sheikho Kamel	KACST, Saudi Arabia
Shen Jie	University of Michigan, USA
Shi Xuefei	University of Science Technology Beijing, China
Shin Dong Hee	Sungkyunkwan University, South Korea
Shojaeipour Shahed	Universiti Kebangsaan Malaysia, Malaysia
Shon Minhan	Sungkyunkwan University, South Korea
Shukla Ruchi	University of Johannesburg, South Africa
Silva Carlos	University of Minho, Portugal
Silva J.C.	IPCA, Portugal
Silva de Souza Laudson	Federal University of Rio Grande do Norte, Brazil
Silva-Fortes Carina	ESTeSL-IPL, Portugal
Simão Adenilso	Universidade de São Paulo, Brazil
Singh R.K.	Delhi University, India
Singh V.B.	University of Delhi, India
Singhal Shweta	GGSIPU, India
Sipos Gergely	European Grid Infrastructure, The Netherlands
Smolik Michal	University of West Bohemia, Czech Republic
Soares Inês	INESC Porto, Portugal
Soares Michel	Federal University of Sergipe, Brazil
Sobral Joao	University of Minho, Portugal
Son Changhwan	Sungkyunkwan University, South Korea
Song Kexing	Henan University of Science and Technology, China
Sosnin Petr	Ulyanovsk State Technical University, Russia
Souza Eric	Universidade Nova de Lisboa, Portugal
Sproessig Wolfgang	Technical University Bergakademie Freiberg, Germany
Sreenan Cormac	University College Cork, Ireland
Stankova Elena	Saint-Petersburg State University, Russia
Starczewski Janusz	Institute of Computational Intelligence, Poland
Stehn Fabian	University of Bayreuth, Germany
Sultana Madeena	University of Calgary, Canada
Swarup Das	Ananda Kalinga Institute of Industrial Technology, India
Tahar Sofiène	Concordia University, Canada
Takato Setsuo	Toho University, Japan
Talebi Hossein	University of Calgary, Canada
Tanaka Kazuaki	Kyushu Institute of Technology, Japan
Taniar David	Monash University, Australia
Taramelli Andrea	Columbia University, USA
Tarantino Eufemia	Polytechnic of Bari, Italy
Tariq Haroon	Connekt Lab, Pakistan
Tasso Sergio	University of Perugia, Italy
Teixeira Ana Paula	University of Trás-Os-Montes e Alto Douro, Portugal
Tesseire Maguelonne	IRSTEA, France
Thi Thanh Huyen Phan	Japan Advanced Institute of Science and Technology, Japan

Thorat Pankaj	Sungkyunkwan University, South Korea
Tilio Lucia	University of Basilicata, Italy
Tiwari Rupa	University of Minnesota, USA
Toma Cristian	Polytechnic University of Bucarest, Romania
Tomaz Graça	Polytechnic Institute of Guarda, Portugal
Tortosa Leandro	University of Alicante, Spain
Tran Nguyen	Kyung Hee University, South Korea
Tripp Barba, Carolina	Universidad Autnoma de Sinaloa, Mexico
Trunfio Giuseppe A.	University of Sassari, Italy
Uchibayashi Toshihiro	Kyushu Sangyo University, Japan
Ugalde Jesus	Universidad del Pais Vasco, Spain
Urbano Joana	LIACC University of Porto, Portugal
Van de Weghe Nico	Ghent University, Belgium
Varella Evangelia	Aristotle University of Thessaloniki, Greece
Vasconcelos Paulo	University of Porto, Portugal
Vella Flavio	University of Rome La Sapienza, Italy
Velloso Pedro	Universidade Federal Fluminense, Brazil
Viana Ana	INESC Porto, Portugal
Vidacs Laszlo	MTA-SZTE, Hungary
Vieira Ramadas Gisela	Polytechnic of Porto, Portugal
Vijay NLankalapalli	National Institute for Space Research, Brazil
Vijaykumar Nandamudi	INPE, Brazil
Viqueira José R.R.	University of Santiago de Compostela, Spain
Vitellio Ilaria	University of Naples, Italy
Vizzari Marco	University of Perugia, Italy
Wachowicz Monica	University of New Brunswick, Canada
Walentynski Ryszard	Silesian University of Technology, Poland
Walkowiak Krzysztof	Wroclav University of Technology, Poland
Wallace Richard J.	University College Cork, Ireland
Waluyo Agustinus Borgy	Monash University, Australia
Wanderley Fernando	FCT/UNL, Portugal
Wang Chao	University of Science and Technology of China, China
Wang Yanghui	Beijing Jiaotong University, China
Wei Hoo Chong	Motorola, USA
Won Dongho	Sungkyunkwan University, South Korea
Wu Jian-Da	National Changhua University of Education, Taiwan
Xin Liu	École Polytechnique Fédérale de Lausanne, Switzerland
Yadav Nikita	Delhi Universty, India
Yamauchi Toshihiro	Okayama University, Japan
Yao Fenghui	Tennessee State University, USA
Yatskevich Mikalai	Assioma, Italy
Yeoum Sanggil	Sungkyunkwan University, South Korea
Yoder Joseph	Refactory Inc., USA
Zalyubovskiy Vyacheslav	Russian Academy of Sciences, Russia

Zeile Peter	Technische Universität Kaiserslautern, Germany
Zemek Michael	University of West Bohemia, Czech Republic
Zemlika Michal	Charles University, Czech Republic
Zolotarev Valeriy	Saint-Petersburg State University, Russia
Zunino Alejandro	Universidad Nacional del Centro, Argentina
Zurita Cruz Carlos Eduardo	Autonomous University of Sinaloa, Mexico

Sponsoring Organizations

ICCSA 2015 would not have been possible without the tremendous support of many organizations and institutions, for which all organizers and participants of ICCSA 2015 express their sincere gratitude:

University of Calgary, Canada (http://www.ucalgary.ca)

University of Perugia, Italy (http://www.unipg.it)

University of Basilicata, Italy (http://www.unibas.it)

Monash University, Australia (http://monash.edu)

Kyushu Sangyo University, Japan (www.kyusan-u.ac.jp)

Universidade do Minho, Portugal (http://www.uminho.pt)

Contents – Part I

Workshop on Advances in Information Systems and Technologies for Emergency Preparedness and Risk Assessment (ASTER 2015)

Workshop on Advances in Web Based Learning (AWBL 2015)

General Tracks

Workshop on Agricultural and Environmental Information and Decision Support Systems (AEIDSS 2015)

Adapting to Climate Change - An Open Data Platform for Cumulative Environmental Analysis and Management

Donald Cowan[1]([✉]), Paulo Alencar[1], Fred McGarry[2], and R. Mark Palmer[3]

[1] David R. Cheriton School of Computer Science, University of Waterloo, Waterloo, ON N2L 3G1, Canada
{dcowan,palencar}@uwaterloo.ca
[2] Centre for Community Mapping, Waterloo, ON N2L 2R5, Canada
mcgarry@comap.ca
[3] Greenland International Consulting Ltd., Collingwood, ON L9Y 1V5, Canada
mpalmer@grnland.com

Abstract. The frequency of extreme weather events has accelerated, an apparent outcome of progressive climate change. Excess water is a significant consequence of these events and is now the leading cause of insurance claims for infrastructure and property damage.

Governments recognize that plans for growth must reflect communities' needs, strengths and opportunities while balancing the *cumulative effects* of economic growth with environmental concerns. Legislation must incorporate the cumulative effects of economic growth with adaptation to weather events to protect the environment and citizens, while ensuring that products of growth such as buildings and infrastructure are resilient. For such a process to be effective it will be necessary for the private sector to develop and operate cumulative effect decision support software (CEDSS) tools and to work closely with all levels of government including watershed management authorities (WMAs) that supply environmental data. Such cooperation and sharing will require a new *Open Data* information-sharing platform managed by the private sector. This paper outlines that platform, its operation and possible governance model.

Keywords: Climate change · Open data · Environmental analysis · Cumulative effects · Software platform

1 Introduction

The frequency of extreme weather events has accelerated over the last decade. Although excess water is one of the significant impacts of these weather events and is the leading cause of insurance claims for infrastructure and property damage [10,11], lack of water can be just as problematic. Prolonged periods of drought can limit our access to drinking water and impair agricultural production, while also changing the landscape ecosystem, encouraging invasive species and putting terrestrial species at risk.

© Springer International Publishing Switzerland 2015
O. Gervasi et al. (Eds.): ICCSA 2015, Part I, LNCS 9155, pp. 3–15, 2015.
DOI: 10.1007/978-3-319-21404-7_1

Economic development such as construction of buildings and infrastructure, and resource development modifies the landscape and its response to extreme weather. Economic development is *cumulative*, as changes in one area will at a minimum impact adjacent areas if not further afield. Thus, planning for adaptation to weather events must recognize these *cumulative effects* [4,13].

Many governments acknowledge that plans for growth must reflect the needs, strengths and opportunities of the communities involved while balancing economic needs with environmental concerns [5,6]. The essential element of balancing the economy with environmental issues must be embodied in legislation that governs how the land is developed. Such statutes must reconcile environmental concerns and the *cumulative effects* of economic development to:

- Protect the environment; and
- Ensure that the products of growth such as buildings, infrastructure and resource development
  - are durable and resilient to the impacts of climate change; and
  - have sustainable environmental impact.

For such a process to be effective it will be necessary for the private sector to develop and operate cumulative effects decision support software (CEDSS) tools and to work closely with all levels of government, including watershed management organizations (WMAs) and other parties that supply environmental data. Such cooperation and sharing will require a new **Open Data** information-sharing platform managed by the private sector but with the full cooperation of the public sector and associated NGOs in supplying and maintaining appropriate **Open Data**. This paper outlines such an information platform, its operation, a governance model and benefits to both the public and private sectors.

2   Why an Open Data Platform?

Currently the private sector works with data from government and WMAs, so why is there a need for an Open Data platform to assess environmental impacts and propose remediation solutions? To answer this question we should first examine the process that is currently followed and discover its deficiencies. Then we can illustrate a new form of cooperation between the public and private sector that will be far more effective in addressing issues related to the interactions between the cumulative effects of economic growth and the environment.

2.1   Current Processes for Environmental Impact Analysis

How is environmental impact analysis conducted currently? The following steps outline the process for a specific geographic area.

1. Identify the modelling procedures that are required.
2. Find or develop the modelling tools to implement the procedures.
3. Acquire the necessary mapping and environmental data from all the different government agencies and WMAs for the analysis and simulation.

4. Do the analysis and simulation and prepare a report.

What happens to the tools and data that have been gathered and generated for the analysis and simulation relating to a specific land use change, development or infrastructure plan? Typically a report is produced and the data used to generate the report are stored in a computer system or a desk drawer. The next time a similar analysis or simulation is required the process is repeated, all the data are assembled again and the raw results of the original analysis and simulation are not accessible. Thinking of the cost of repeating all these operations, it is clear there must be a better way. In addition, because the results of previous analysis and simulations are not available it is impossible to show the cumulative effects when projects changing the environmental landscape build upon one another.

2.2   New Ways to Gather, Maintain and Analyze Environmental Data

The issues just described give rise to a number of questions including:

- Can mapping and environmental data from the different government agencies and WMAs be assembled and stored in an accessible Open Data platform?
- Can an accessible suite of analysis and simulation software tools often called cumulative effects decision support software (CEDSS) be provided?
- Can results from previous analyses be captured so that future analyses can build upon previous results showing cumulative effects?

If we could answer these questions positively, then we could ensure that current cumulative data and the tools representing the latest available science and engineering practice are being used in environmental analysis and modelling. Current government policies and existing information technologies can support this wide accessibility of existing data, cumulative data and CEDSS tools. In fact answering the first question alone would reduce the cost of any analysis and simulation exercise by eliminating repetitive data gathering.

Governments and agencies such as WMAs are adopting Open Data policies. Open Data is based on the concept that certain data should be freely available to everyone to use and republish as they wish, without restrictions from copyright, patents or other mechanisms of control. However, republishing does imply citing the original source, not only to give credit, but to ensure that the data has not been modified or the results misrepresented [2].

Current technologies such as web-based and mobile cloud computing services provide the ability to access both Open Data and the latest CEDSS tools. Instead of each organization assembling data and acquiring software tools such as CEDSSs, organizations can access a common pool of data and tools. Users of a web-based cloud system will also be able to store the results of running a CEDSS in the cloud system. Thus, the next uses based on the same geographic area will have access to the latest results and will be able to show the cumulative effects of economic development.

2.3   Challenges to Governance and Operation

Building such a web-based cloud system is technically feasible and would provide substantial benefits to any organization that needs to model and understand specific environmental behaviour. However, this is a new model of operation and requires answers to several key questions including:

- How will the web-based and mobile cloud system be operated?
- How will policies be set for the operation of this cloud system?
- Who will supply and maintain the Open Data?
- Who will supply and maintain the CEDSS software tools?
- What is the pricing model for the services provided?
- How does a web-based and mobile cloud system scale so as to become operational for a significant portion of a geographic area?

The answers to many of these questions require different modes of operation involving both the private and public sectors. Business is likely to supply the software tools and build and operate the web-based and mobile cloud infrastructure, while governments and NGOs such as WMAs, will supply the data and ensure that the results are properly applied by the parties involved.

The next sections of this paper explore how this web-based and mobile cloud system might be started and grow into a viable, sustainable operation. The paper starts by describing the architecture of the CEDSS web-based and mobile cloud and then its initial operation. Then the paper presents a partial road map for growth of the operation into wider jurisdictions.

3   Making Environmental Data Accessible

Making data easy to store and maintain while making it accessible is critical. Could we find a way to:

- assemble and store accessible mapping and environmental data; and
- accumulate results from previous simulations to show cumulative effects?

Such an approach would reduce the cost of any analysis and simulation with CEDSS tools and make it easier for organizations with fewer financial resources to participate.

Fortunately, the Internet and Web provide connectivity for access and the new cloud structures provide accessible, scalable and inexpensive processing and storage. Mapping and environmental data can be stored so that it is accessible through an Open Data storage mechanism. Not only maps and data collected from direct measurement of the environment, but also the results of running CEDSS tools would be accessible. Future studies could cumulatively build on the results of current work.

Instead of each organization assembling data and acquiring software tools such as CEDSSs, organizations can access a common pool of data and tools. Members of each organization will have to understand the data, keep the data current, and be able to use the tools, but will no longer have to maintain the database and the software. This maintenance will be performed for the cloud system by the "cloud staff."

The remainder of this section provides a detailed description of how such an Open Data platform could be structured. However, this is a work in progress and it is too soon to commit to a detailed architecture.

3.1 The Structure of the Cloud

The authors' organizations are cooperating to design, implement and deploy an initial version of a web-based and mobile cloud for environmental impact analysis called an Integrated Science and Watershed Management System (ISWMS™). A version of the system is shown as three layers (Open Data & Other Data Sources; Science & Engineering; and Human Interface) in Figure 1.

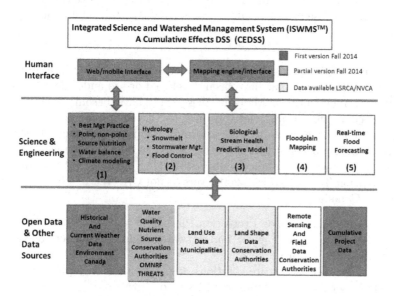

Fig. 1. ISWMS™- architectural layers and components

Open Data and Other Data Sources Layer. The Open Data layer indicates the sources of data, much of which will come from governments and NGOs such as WMAs. Current weather data comes from the Federal government, while a significant amount of water quality data comes from provincial or state governments while gathered by local WMAs. Present and future land use data is typically provided by municipalities, while drainage characteristics such as catchment and stream delineation are derived from digital elevation maps. Remote sensing and field data may come from WMAs, university researchers, professional consultants or government and cumulative data from previous studies.

An Open Data layer will provide multiple advantages:

- Data will <u>not</u> have to be gathered every time there is a requirement to analyze or model aspects of the land or water such as a watershed, it will be accessible through the Open Data layer;
- Projects will be able to build upon the results of previous analysis and modelling runs, thus able to anticipate cumulative effects of development; and
- Costs of "understanding" and managing interactions with the environment should be significantly lower and more precise because of the direct availability of data and the ability to accumulate results.

The issue with the Open Data layer is to organize and index the data to provide easy access and ensure that the data is current. This is a significant but manageable problem that can be addressed. Once the data is accessible, then CEDSS tools can be provided by the private sector.

The open data layer is not meant to be restrictive. For example, other data systems such as those based on OLAP [14] or SOLAP [15] could be incorporated as required.

Science and Engineering Layer. The Science & Engineering layer in Figure 1 shows the CEDSS tools that have been developed or are under development by the authors' organizations. Many of these tools have become mainstream in the last decade and so are relatively new to the environmental management community. Of course the software platform in Figure 1 is an open platform and CEDSS tools from other suppliers can be added as they become available.

Box (1) summarizes the functionality of the <u>CANWET</u>^{TM} (<u>CAN</u>adian <u>W</u>atershed <u>E</u>valuation <u>T</u>ool) [9]. Since 2004, versions of CANWET^{TM} have been used in Ontario, Canada to:

- Develop the Lake Simcoe Protection Plan [7];
- Complete Tier 1 and 2 water budget/water taking and other source water protection related projects;
- Identify new infrastructure solutions and sustainable community planning policies associated with Ontario's Places to Grow Act [6];
- Prepare regional Official Plan directives, such as the County of Simcoe's Infrastructure Visioning Strategy [8]; and
- Complete project related climate change impact assessments.

This part of the diagram, Box (1) in Figure 1, represents CANWET^{TM}5, which is a web-based cloud version.

Box (2) in Figure 1 shows the first version of the ISWMSFF^{TM} (Integrated Stormwater <u>M</u>anagement and <u>F</u>lood <u>F</u>orecasting tool) that is being upgraded to a web-based cloud application with added functionality.

Box (3) in Figure 1 shows the possibility of using Predictive Models for Biological Stream Health through measurements of water course shape, fish and plant populations, and benthic data. The Flowing Water Information System (FWIS) [1], a joint project of the Ontario Ministry of Natural Resources and Forestry, Ontario's Conservation Authorities, the University of Waterloo Computer Systems Group and the Centre for Community Mapping contains this data for Ontario. The <u>THREATS</u> software (<u>T</u>he <u>H</u>ealthy <u>R</u>iver <u>E</u>cosystem <u>A</u>ssessmen<u>T</u> <u>S</u>ystem) [3] also captures information about stream health and is available to be integrated into the layer.

Boxes (4) and (5) in Figure 1 identify significant problems that we have encountered. We do not have accurate floodplain maps considering the extreme weather events that are occurring more frequently. These floodplain maps could be developed using digital elevation mapping acquired by Lidar[1]-equipped aircraft. The system could use available in-stream/river bathymetry data. Alternatively, blue-green spectrum Lidar technology, capable of penetrating the reflective water surface, may be used to develop bathymetric data for river channels, which can then be integrated with terrestrial elevation data. Real-time flood forecasting is another area where significant progress could be made, based on analyzing snowmelt and weather data, thereby anticipating the next Spring floods such as those that occurred in Muskoka and Calgary in Canada in 2013. Once accurate floodplain maps are created they can be used to model opportunities for flood mitigation using low impact development techniques.

Human Interface Layer. The Human Interface layer in Figure 1 allows the user of ISWMS[TM] to view the results of modelling and analysis using tables, graphs and charts. The mapping engine/interface provides the ability to outline the area to be studied and to visualize the results related to a map. The Web/mobile interface incorporates the maps as well as reports and graphical presentation of the results. The mobile version could also be used for data collection in the field.

3.2   Summary

Many of the components needed to create ISWMS[TM] are available and in a form to be deployed as a web-based cloud system. What is needed is a restricted deployment to investigate and answer the questions in Section 2.3. Organizing, aggregating and accessing open data from multiple levels of government and related agencies is a relatively new endeavour. However, the authors are aware of similar thinking about big data associated with the National Oceanic and Atmospheric Administration (NOAA) in the United States [12].

4   The Initial Operation of the Cloud

Where can we deploy the ISWMS[TM] CEDSS Platform in Section 3.1 to refine the related concepts? Since the consortium building the platform is based in Ontario,

---

[1] Lidar is a remote sensing technology that measures distance by illuminating a target with a laser and analyzing the reflected light. Lidar is used to make high-resolution digital elevation maps.

Canada it would be appropriate to choose two pilot sites in that province. Two choices could be:

1. The South Georgian Bay/Lake Simcoe Source Water Protection Region, including the Lake Simcoe Basin, Nottawasaga River Basin (to the west of Lake Simcoe) and Severn-Sound Basin (draining into Georgian Bay); and
2. The Six Nations Community near Brantford, Ontario.

The Lake Simcoe and Nottawasaga River and Severn Sound Basins study would focus on improvements to stormwater management and source water protection plans, while the Six Nations project would provide an opportunity to investigate flooding and related source water protection, which is also a paramount issue for remote First Nations communities.

4.1   South Georgian Bay/Lake Simcoe Source Water Protection

The Lake Simcoe, Nottawasaga River and Severn-Sound watersheds are shown in the map in Figure 2. CANWET^{TM}has now been used extensively in these three watersheds to evaluate in-stream assimilative capacity, set nutrient targets and determine the potential for use of rural and urban management practices to achieve these targets. These existing models will be added to the platform and brought into the CANWET^{TM}web-based environment such that subscribers will be able to view simulation results and evaluate scenario changes.

The Lake Simcoe Region Conservation Authority (LSRCA) and the Nottawasaga Valley Conservation Authority (NVCA) have expressed strong interest in using the new (5th) version of CANWET^{TM}as a means of evaluating and managing nutrient loads, assessing probable climate change impacts and targeting mitigation efforts to key contributing sources. Both watersheds are experiencing significant development and growth pressures from the communities of Barrie, Orillia, Newmarket, Aurora, Bradford-West Gwillimbury, Innisfil, New Tecumseth, Springwater and Oro-Medonte. These stresses on the watersheds need to be balanced with environmental concerns and economic benefits derived from the natural environment. Creating a standardized and science-based approach to quantifying impacts and mitigation approaches is essential to a sustainable, long-term management approach that achieves results and recognizes the cumulative effects of economic development in Ontario.

In the initial version, the ISWMS^{TM}CEDSS platform will offer a common modelling dataset with ready-to-use calibrated base models. Because of the development pressures in the proposed pilot area basins, it is anticipated that multiple users will use the system to test scenarios such as municipal Low Impact Development (LID), floodway infrastructure changes, private land development, insurance and financial institution flood risk analysis, and/or urban and rural agricultural Best Management Practices (BMPs) changes. These users, as part of the agreement to participate, will consent to "publish" proposed and completed watershed changes thus creating a cumulative database for future modelling. Thus, each application of the base model can reflect up-to-date conditions plus

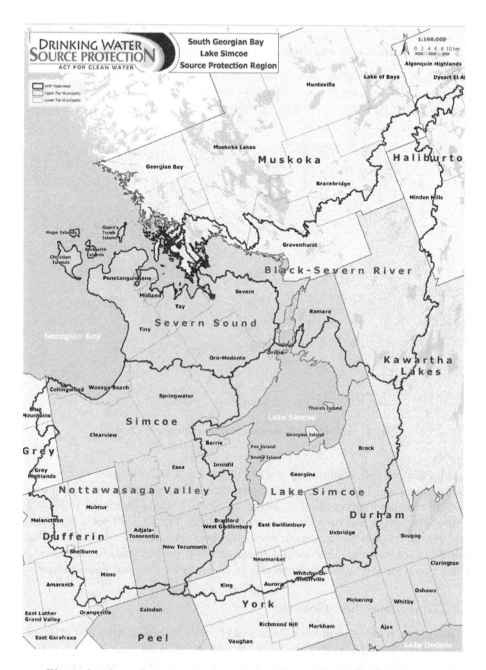

Fig. 2. Southern Georgian Bay Lake Simcoe Source Water Protection Area

pending proposals or master plans. Note that they will not be asked to publish detailed plans, but just the impacts of those plans.

4.2 Six Nations - Watershed Management/Source Water Protection

The second proposed pilot area is the Six Nations community shown in the map in Figure 3. The community encompasses 183.20 sq. km with a network of waterways including between 50-70 drainage catchments and two main watercourses.

Fig. 3. Six Nations Community in Southwestern Ontario

The Six Nations community wants to develop a science-based stormwater management and flood mitigation strategy that includes calibrating computer models with field survey data collected by Sixth Nations members and using state-of-the-art digital mapping and satellite data. The CEDSS tools from this project can be used for long-term watershed oversight, source water protection and flood forecasting. This second demonstration project could establish the groundwork for source water protection in remote First Nations communities.

4.3 Governance and Growth

By setting up these pilot projects, we should be able to implement a first attempt at answering the operational questions in Section 2.3. This project is not exclusive; other organizations and businesses will be encouraged to use and add both data and software tools.

An operational and growth advisory group will be established with representatives from Provincial ministries, Federal agencies and local agencies such as

municipalities and WMAs, First Nations, climate change and adaptation specialists, environmental engineering firms, the development and insurance industries.
The operational and growth advisory group will be responsible for the development of an operational plan. The initial project should take two (2) years to
complete. However, results and partial deployments should start being available
after the first year of operation.

5   Governance and Operations

There are multiple factors at work in setting up the environment in Figure 1.

- Current mapping and environmental data is primarily captured by governments and WMAs and their mandate will always make them a significant
  factor in data collection. However, this is likely to change as more intensive
  land development and land use occur. The private sector will require data
  more quickly than in the past as there will be a pressing need to adapt
  plans to climate change and also to follow the latest environmental laws and
  guidelines. Thus the data gathering cycle will need to be shortened.
- Once mapping and environmental data are collected they need to be assembled and organized into meaningful data sets.
- A large number of the CEDSS tools will be developed and operated by private sector engineering and consulting firms as most of the tasks undertaken
  will not be directly related to government.
- Development of the Open Data store and the CEDSS tools will require
  advanced software technologies.

Based on these observations there needs to be a partnership between the
public and private sectors to construct and oversee the operation of the platform
outlined in Figure 1.

Although the details need to be determined, the partnership should have two
basic components.

1. Governments and NGOs will still continue to collect and supply mapping and
   environmental data according to their mandate. This data will be augmented
   by the private sector where needed.
2. The private sector will:
   (a) Construct and operate the Open Data store; and
   (b) Develop and operate the analysis and simulation software tools.

6   Growth of the Cloud - Provincial ISWMS™CEDSS Platform Operations Road Map

Once proven through testing within the LSRCA and NVCA watersheds and the
Six Nations community the ISWMS™CEDSS platform will be made available
for application across broader jurisdictions.

6.1   ISWMS™CEDSS Platform Deployment Schedule

The ISWMS™CEDSS platform will be developed in stages. The first part of the platform will include the fifth version of the CANWET™(called CANWET 5) a web-based cloud version. Then the ISWMS™CEDSS platform will add the following functions.

- Flood Warning, Flood Forecasting and Flood Damage Assessments tools;
- Hydrologic Analysis for Watershed Planning and Water Resources Design;
- Integrated Water Balance with Nutrient/Sediment/Pathogen/Bacteria Loadings and In-stream Temperature and Dissolved Oxygen Modelling;
- Instream and Lake/Reservoir Routing Capabilities (Quantity and Quality);
- Groundwater (Shallow System) Modelling;
- Predictive Modelling of Urban and Rural BMP Effectiveness;
- Predictive Modelling of Urban Low Impact Development (LID) Options;
- Canadian Climate Change Impact Modelling; and
- Integration of a Rainfall-Snowmelt/Canadian Climate Change Impact Tool.

CANWET 5 and other tools need access to large amounts of Open Data to function. Much of the initial data for CANWET 5 has been developed for earlier studies within the LSRCA and NVCA watersheds. Providing a better automated framework for accessing and maintaining databases of Open Data will be a high priority task within this entire project.

Future additions to the initial ISWMS™CEDSS platform will include:

- Watershed Health Database (Canada) and Cumulative Stress Assessment Tool and includes THREATS [3] and FWIS [1];
- Flood Hazard Identification and Flood Plain Mapping;
- Ecosystem Services Platform;
- Watershed Nutrient Trading and Offsetting Management Tool; and
- Reservoir and Lake Capacity Modelling Tool.

Of course both other private and public sector organizations will be encouraged to participate by providing software tools and data.

7   Conclusions

This paper describes a cloud-based cumulative effects decision support system (CEDSS) for environmental impact analysis and management using both Web-based and mobile systems and based on Open Data. The Open Data will be provided by federal, provincial and municipal governments, WMAs, business and research organizations. The Open Data will be accessible by all participants. The CEDSS tools to manipulate the data for purposes of sustainable management may be openly available, open source or proprietary in nature and organizations and businesses will be encouraged to add to this toolkit.

The entire cloud platform will be operated by the private sector while ensuring that both the public and private sectors' objectives are met. It is proposed that the initial version of the cloud platform will be applied in the South Georgian Bay/Lake Simcoe Source Water Protection Region and Six Nations territory watersheds, as initial case studies.

References

1. Centre for Community Mapping (COMAP): Flowing Waters Information System - FWIS (2014). http://www.comap.ca/fwis/
2. Cowan, D., Alencar, P., McGarry, F.: Perspective on open data: issues and opportunities. In: Software Summit 2014, International Conference on Software Science, Technology and Engineering (SWSTE). IEEE Computer Society Press (2014)
3. Dube, M.: Development of the healthy river ecosystem assessment system (THREATS) for assessing and adaptively managing the cumulative effects of man-made developments on canadian freshwaters (2012). http://www.cwn-rce.ca/project-library/project/development-of-the-healthy-river-ecosystem-assessment-system-threats-for-assessing-and-adaptively-managing-the-cumulative-effects-of-man-made-developments-on-canadian-freshwaters
4. Ross, E.R.: The Cumulative Impacts of Climate Change and Land Use Change on Water Quantity and Quality in the Narragansett Bay Watershed, May 2014. http://scholarworks.umass.edu/masters_theses_2/111/
5. Government of Ontario: The Greenbelt Act, 2005 (2005). http://www.mah.gov.on.ca/Page195.aspx
6. Government of Ontario: Places to Grow Act 2005 (2005). https://www.placestogrow.ca/index.php?option=com_content&task=view&id=4&Itemid=9
7. Government of Ontario: Lake Simcoe Protection Act, 2008 (2008). http://www.e-laws.gov.on.ca/html/statutes/english/elaws_statutes_08l23_e.htm
8. Greenland Consulting Engineers: County of simcoe water and wastewater visioning strategy, February 2012. http://www.simcoe.ca/Planning/Documents/Water%20and%20Wastewater%20Visioning%20Strategy%20Final%20Draft%20-%20February%202012.pdf
9. Greenland International Consulting: CANadian Watershed Evaluation Tool (2014). http://www.grnland.com/index.php?action=display&cat=17
10. Insurance Bureau of Canada: The financial management of flood risk, May 2014. http://assets.ibc.ca/Documents/Natural%20Disasters/The_Financial_Management_of_Flood_Risk.pdf
11. KPMG: Water Damage Risk and Canadian Property Insurance Pricing, February 2014. http://www.cia-ica.ca/docs/default-source/2014/214020e.pdf
12. NOAA: NOAA Big Data Project (2015). https://data-alliance.noaa.gov/
13. The Resource Innovation Group: Toward a Resilient Watershed, January 2012. http://www.theresourceinnovationgroup.org/storage/watershed-guide/Watershed%20Guidebook%20final%20LR.pdf
14. Thomsen, E.: OLAP Solutions: Building Multidimensional Information Systems. Wiley, April 2002
15. Université Laval: Spatial OLAP Components, November 2009. http://spatialolap.scg.ulaval.ca/datastructure.asp

Mining Climate Change Awareness on Twitter: A PageRank Network Analysis Method

Ahmed Abdeen Hamed[(⊠)] and Asim Zia

EPSCoR, University or Vermont, Burlington, VT, USA
{ahamed,azia}@uvm.edu
http://www.uvm.edu/~ahamed

Abstract. Climate change is one of this century's greatest unbalancing forces that affect our planet. Mining the public awareness is an essential step towards the assessment of current climate policies, dedication of sufficient resources, and construction of new policies for business planning. In this paper, we present an exploratory data mining method that compares two types of networks. The first type is constructed from a set of words collected from a Climate Change corpus, which we consider as ground-truth (i.e., base of comparison). The other type of network is constructed from a reasonably large data set of 72 million tweets; it is used to analyze the public awareness of climate change on Twitter.

The results show that the social-language used on Twitter is more complex than just single word expressions. While the term climate and the hashtag (#climate) scored a lower rank, complex terms such as ("Climate Change") and ("Climate Engineering") were more dominant using hashtags. More interestingly, we found the (#ClimateChange) hashtag is the top ranked term, among all other features, used on Twitter to signal climate familiarity expressions. This is indeed striking evidence that demonstrates a great deal of awareness and provides hope for a better future dealing with Climate Change issues.

Keywords: PageRank · Network analysis · K-H Networks · Bigrams Networks · Measurement · Awareness · Climate change · Twitter

1 Introduction

Climate change is a real fact [33]. The governmental efforts spent for climate change gravitate to focus on providing public goods and growing the public awareness [7]. Inevitably, measuring the public awareness on climate change has become an important indicator for governments and policy-makers [43]. In this pursuit, numerous efforts around the globe have strived to capture the awareness using different methods and approaches; Lorenzoni et al. studied how the general public in the UK perceive climate change challenges the barriers to engaging with it [28]. The authors based their work on qualitative data that provided in-depth understanding of how the UK public makes sense of climate change. Curry [9] designed a survey to study people's attitudes toward climate change mitigation.

© Springer International Publishing Switzerland 2015
O. Gervasi et al. (Eds.): ICCSA 2015, Part I, LNCS 9155, pp. 16–31, 2015.
DOI: 10.1007/978-3-319-21404-7_2

His study particularly focused on the public opinions concerning energy and carbon capture by answering 17 survey questions. Semenza et al [40], also designed a survey to study the climate change public awareness and concerns. The survey targeted individuals who live Portland OR and Houston TX in the period from June and September 2007. Sampei et al. analyzed Japanese newspaper coverage of global warming from January 1998 to July 2007, and how the public opinion was impacted by such coverage [38].

With social media exploding in popularity, public opinions have become accessible for virtually any topic. Excitingly, this generated a great deal of interest in using Twitter data to mine and analyze sentiments for various purposes and applications: Mitchell et al. conducted a detailed analysis of correlations between real-time expressions of individuals made across the United States and a wide range of emotional and health characteristics. Conover et al. studied the political polarization on Twitter by analyzing networks of hashtags features [8]. Their analysis exhibits a highly segregated partisan structure for the right-learning versus left-leaning users. Hamed et al. used tweet features (words and hashtags) to design an intelligent recruitment system for smoking cessation purposes [20,21]. Bollen et al. used twitter moods and sentiment analysis to predict stock market prices [5]. Signorini et al. used Twitter to track the levels of outbreak activities and measure public concern in the US during outbreaks of particular diseases (e.g., H1N1) [41]. Undoubtedly, studying the public opinion expressed on Twitter is an exciting computational problem to solve; yet it presents a great human effort to meaure and quantify. Dodds et al. used human evaluators to assign a score of emotional content of the words and sentiments for measuring of happiness studies [14]. This method is expensive, slow, and prone to human subjectivity errors.

In this paper, we show a computational method that is both systematic and objective to measure the climate change awareness, using networks of word features. Words are powerful means. They make up the rich ontology and the literature that we currently possess. At the time of writing, it was not feasible to get access to the Climate Change Ontology vocabulary [15]. This leaves text mining literature as an interesting option. Literature expresses a tremendous range of knowledge in the text within. This fact, though useful, also represents a computational challenge in terms of extracting highly representing word features [22]. Using means of text mining becomes necessary to acquire those features. Constructing feature-based networks of such features establishes a notion of ground-truth to compare against. Mining significant word features, designing data models and algorithms is evident [16,22,24,39]. Text mining is popular in solving many computational problems in biology, medical science, social sciences, and earth sciences [23,29,30,34,46]. Here, we extracted word features from literature gathered from the Nature Climate Change journal. This computational task is a step in measuring the public awareness of climate change on Twitter. The premise of this paper is to compare the ranks of top literature words literature with counterparts from Twitter, with the following goal in mind: *Mining the various aspects of climate change and highlighting which aspect(s) are popular on Twitter.*

2 Data Description and Methods

This section is to deliver a good understanding of what data was used, how it was processed, and how features were selected. Two data sets were collected independently: (1) **Literature Source** – we searched the web portal of Nature Climate Change web portal for the keyword "climate" between year 1990 -2014. The search produced 3,263 articles from which we extracted the titles. (2) **Twitter Source** – We harvested 72 million tweets using the the Twitter streaming API, which provides a 1% free sample from all the public tweets, in the period of July 2014 and December 2014.

As for the preprocessing for the literature titles, we applied a noise removal task to eliminate the stop words. Titles are rich sources of keywords and have been used in various text mining tasks [13,26,35]. Titles are lengthy and they often contain noise words that are not relevant to the study. Such words (e.g., and, this, were, that) do not contribute to the analysis and they have to be eliminated. After applying a preprocessing step for noise elimination, we then analyzed the pure titles to select the features that best describe the data set. These features are what we used to construct the Ground-truth Network (GTN) that is described in sections to follow. The data also contained duplicates, which we ended up removing so the features selected are not biased by such duplications. This in turn reduced the number of titles to 2,084. Table 1 shows a sample of the titles gathered from the Nature's web portal.

Table 1. shows a sample of the titles gathered from Nature's web portal when searching for the keyword "climate". Here we provide raw titles, as they were originally published, and without the noise removal step. These titles show some of the significant features extracted using TF-IDF such as the term ("climate, carbon").

Blanket peat biome endangered by climate change
Breaking the climate change communication deadlock
Largest teach-in ever focuses US on climate change
A blind spot in climate change vulnerability assessments
Atmospheric chemistry: A new player in climate change
A meta-analysis of crop yield under climate change and adaptation
Adapting to climate change through urban green infrastructure
Assessment of the first consensus prediction on climate change
Climate-society feedbacks and the avoidance of dangerous climate change
High Arctic wetting reduces permafrost carbon feedbacks to climate warming
Physically based assessment of hurricane surge threat under climate change

Using the common TF-IDF Information Retrieval scheme, we can selecte the best features that describe the corpus of titles. The TF.IDF measure [37] is a term weighting scheme that is used to describe the product of term frequency and the inverse document frequency. This measure is considered one of the most commonly used term weighting metric in Information Retrieval [2], text mining and summarization [36,47]. The TF-IDF scheme can be described as follows:

$$tf - idf_{t,d} \ = \ tf_{t,d} \times idf_t \ = \ tf_{t,d} \times log(\frac{N}{df_t}) \tag{1}$$

When applying the TF-IDF measure against the data set we gathered 2260 word features; many of these features were duplicates. Filtering out the redundant features reduced the number of unique features to 712 instead. From these features, we collected the top two features of each title. This fosters a meaningful word-pairs mechanism that facilitates the construction of networks. This pairing process was further constrained but adding this following condition: For each title in the corpus, a word-pair must be adjacent (a bigram) in order to be consider in the GTN network. Domain bigram networks have proven informative [45] and have been used to model textual content in various natural languages processing studies. [25,27]. Bigrams are a much richer model than the ("Bag of Words") [4,42] which has been previously used to construct Twitter K-H networks [17–21].

The process of constructing the bigrams generated 364 bigrams with 283 unique features. Such features were stored as a Comma Separated Value file and was imported as a network using the igraph Python library to construct the GTN as follows: For each bigram, we treated its sides left hand side (LHS) and right hand side (RHS) as vertices connected by a directed edge from LHS to RHS preserving the sequence of the words as it appears in the title. As for the edge weight, we assume a weight of 1, giving it full credit since this is the ground-truth/benchmarking knowledge we computed from literature to compare Twitter ranks against. By repeating this process for all bigram, we achieved a networks of a 283 vertices and 726 directed and weighted edges. Table 2 shows a sample list of the bigrams generated from the literature titles retrieved from the Nature Climate Change journal. Figure 1 shows the network that we constructed from the bigrams. The network exhibits a reasonably large portion of connected terms. The remaining terms, though connected as pairs, are isolated from the connected larger part of the network. These pairs such as ("blind spot", and "cloudy picture") do not have an associated contributing rank to them. They are not pruned here, however, to test whether they receive better attention when analyzing the tweet data set. It is also clear to see that the terms ("climate") and ("carbon") appears to be pivotal in the network, which indeed demonstrate an indication of how significant these terms are when measuring their corresponding PageRank.

To obtain a rich understanding of the climate change awareness on Twitter, we curated a 72 million tweet data set. Each tweet was processed in same manner that was applied against the titles, and was also encoded in the ("UTF-8") standard. After the noise removal was completed, we extracted extracted all the bigrams using the same features that were computed for the GTN. This process produced a ≈ 321,000 non-unique bigrams; some are more frequent than others. We used frequencies to consolidate similar bigrams, and assigned as weight to the edges that connect the two sides of the the bigrams. From the unique set of bigrams, we

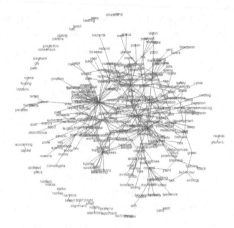

Fig. 1. shows the network of bigrams computed from the literature titles. It is worth-noting that the term "climate" and "carbon" have the most connected term in this network.

Table 2. Sample Literature Bigrams Computed Using TF-IDF

(climate,change), (global,warming), (target,power), (power,plants)
(global,ocean), (ocean,warming), (2008,cooling), (2020,emissions)
(sustainable,future),(fight,climate),(blind,spot),(solar,power)
(challenging,political),(political,climate),(cloudy, picture)
(extreme,heat),(climate,risks),(climate,justice),(conservation,measures)
(global,forest),(forest,carbon),(carbon,balance),(flood,risk)
(challenging,climate),(carbon,budget),(climategate,inquiry),(past,revealed)
(sea,ice),(marine,invaders),(consensus,prediction),(summer,heat)
(dry,southern),(aerosols,impacts),(cool,polar),(summer,extremes)
(biomass,burning),(burning,black),(ocean,acidification),(carbon,feedbacks)
(snow,albedo),(change,communication),(emissions,gap),(cities,resilience)

constructed a Twitter bigram network which we refer to as (TBN). By construct-ing the network this way, we have means of comparing the rank of each term in both networks (i.e., GTN vs TBN). This comparison also guarantees the identi-fication of the top ranked terms in each network and measures the difference in ranking for each term. By knowing the top terms in the literature, we can then easily investigate whether the same terms receive similar attention and score a similar ranking on Twitter. This process indeed constitutes the notion of public awareness that we seek to establish in the experiment section below.

2.1   PageRank Network Analysis

The PageRank algorithm, also known as Google's search algorithms, is the most prominent measure used to compute the global importance of web pages and

rankings performed by search engines [6]. The algorithm was further classified as one of the top 10 data mining algorithms in terms of influence and
importance [44]. The popularity of this algorithm has inspired new uses beyond
searching the web. The PageRank score has become one of the most significant
centrality measures to identify influential and significant players of any network.
Beyond the prime use in searching the World Wide Web, a weighted version of
PageRank has been applied to citation networks to identify the pivotal authors
for a given research field [11,12]. In the world of life sciences and biodiversity,
Google's algorithm was also used as a conservation tool in assessing the species
that are highly endangered. This was achieved by computing the ranks of each
species in a network constructed from existing ecological food webs [3]. It is noteworthy, knowing that PageRank is also a powerful tool for Natural Language
Processing (NLP), word association [10] and semantic networks applications.
Mihalcea et al. [32] explored the applicability of PageRank to semantic networks.
The research shows this family of graph-based ranking algorithms (e.g., HITS,
and PageRank) can contribute significantly to word sense disambiguations. This
contribution is related to the fact that PageRank solidifies the selected meaning
from a set of possible meanings, for a given word in the network, by assigning
the highest rank. This is a powerful idea we use to identify the significant words
in the GTN network.

PageRank is based on the in-degree value (number of incoming edges), and
the out-degree value of each vertex. This is the main reason for constructing the
GTN as a directed bigram network. If the word ("ocean") was followed by the
word ("acidification"), then there will be an outgoing edge $\Rightarrow$ from ("ocean") and
the at same time, there will be an incoming edge $\Leftarrow$ going into ("acidification").
In a larger and more complex network, the number of incoming edges can be
viewed as a vote cast that each term receives [44]. The ranking algorithm does
not stop at counting the incoming edges, but also considers the rank of the
terms casting the vote (i.e, linking to the term). This is how PageRank precisely
determines the most influential climate concepts in the networks. The algorithm
uses a probability value, also called the damping factor, to jump to another vertex
in the graph with a default value set to (0.85). The PageRank formula [44] can
be described as follows:

$$P(i) = (1 - d) + d \sum_{(j,i) \in E} \frac{P(j)}{O_i} \qquad (2)$$

Figure 2 shows a hypothetical network of seven terms (A, B, C, D, E, F, G) and
their PageRanks (.06, .1, .08, .32, .25, .06, .14) respectively. It is clear to see that
the terms that have not received votes from other terms are the lowest in rank
such as (A and F) with a rank equals (.06). On the other hand, the terms that
received votes from prestigious other terms scored the highest. When inspecting
the term (D) we find that it received three incoming edges from (B, C, E). The
rank of E of a value equals (.25) have positively impacted the rank and the score
of the (D), which scored (.32).

The notion of awareness can be derived from comparing how important
a word is (i.e., climate concept) in the literature with its counterpart terms

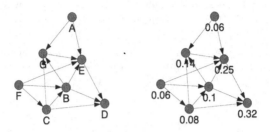

Fig. 2. shows a basic labeled networks to the left and the PageRank scores for each vertex, of the same network, to the right

extracted from social media. It is intuitive to agree if a term is important in the literature, it is expected to be as important on Twitter; this how the awareness can be quantified. Otherwise, it is perceived that the general public are not either aware of the term or perhaps they might believe that the concept is not as important as it appears in the literature. This process of comparison can indeed produce results that can lead to a better understanding of the climate change awareness on Twitter. PageRank enables this comparison by producing comparable ranks for each climate concept that exists in the networks.

3   Experiments

We present two different types of experiments: (1) A network comparison between the keywords in the GTN and TBN: here we compare the rank of the same word in the two different networks and measure the rank difference. (2) A network comparison between the GTN and K-H Network: here we compare the hashtags in the K-H network with the word that are identical to the word features (e.g., "climate" vs #climate).

3.1   GTN vs TBN PageRank Network Analysis

After constructing the ground-truth network (GTN) and the Twitter bigram networks (TBN), we now can compare the PageRank of each term in both networks respectively. As stated above, we used the bigram frequencies as an edge weight of the TBN. The weight value can be assigned to each pair p in the set of all bigrams P as follows:

$$\forall p \in P : l_w = \frac{p_f}{n} \tag{3}$$

where l_w is the weight assigned for each link, p_f is the number of occurrences for each bigram and n is the total number of bigrams derived from the TBN. Table 3 show the top 30 ranked terms in the ground-truth network. The table also shows how these terms were ranked in the TBN network constructed from the tweet data set. Here we observe that the top terms are Climate, Carbon, which also confirms the observation we made on the visualization in Figure 1. These terms, however, did not show to be highly ranked in the TBN (19, 70) respectively, as the table demonstrate . The term ("Weather") scored the exact rank at 30 in both literature and Twitter. Figure 3 shows two WordClouds, the one to the left is a representation of the GTN while the one to the right is a representation of the TBN. The term size reflects the importance in the ranking.

Fig. 3. GTB WordCloud on the left side compared with TBN WordCloud on the right side. The size of the word corresponds to the order so the bigger words represents higher ranks. In the GTN: the top ranked terms are Climate, Carbon, Emission, and Temperature in the GTN. the TBN WordCloud Science, Technology, Health and Change are the top ranked. This highlights that Twitter words are more broad compared with the highly technical terms in the literature.

Figure 4 also shows the distribution of PageRanks of network. The GTN appears to be normally distributed, with the maximum value near the middle of the distribution. The TBN, on the other hand show a closer ranking distribution where the frequencies were established in a frequency range between (5 and 25). Additionally, the TBN logs show linearity which may suggest a power-law distribution as opposed to the distribution of GTN that seems to be converging around a value of zero.

3.2   GTN vs K-H PageRank Network Analysis

Hashtags are special informational devices that people use on social media to express an opinion, describe an entity, or mention a location. There are endless types and forms of hashtags [21]. While a hashtag can be made of a single word (e.g., #Hot), other hashtags can be made of a few words, numbers, or even symbols (e.g., #BusinessIntelligence, #420, and #HealthCareReform).

It is important to consistently compare networks with similar characteristics to achieve the highest level of fair comparison possible. However, when hashtags are involved, it is not possible to construct word-hashtag bigrams since hashtags can be placed anywhere in the tweet and they do not necessarily follow any grammatical rules. In fact, it is more accurate to assume informal hashtag constructs instead. Since we previously computed ground-truth features and

Table 3. shows the order of to the top 30 terms found in the ground-truth network and their corresponding orders in the twitter network. The first row shows the actual words, the second row show corresponding ranks in the GTN (literature) network. The third row shows the ranks of the same words occurring in the TBN (Twitter) network. Due to the page limitation, we only included the top 30 terms. The analysis shows that the term ("Climate") is ranked number 1 while it was found to be ranked number 19. Some terms did not make it to TBN network ("Impact") and ("Response"), which entails that such terms were not paired during the bigram process.

	Climate	Carbon	Temperature	Global	Change	Emissions	Energy	Warming
Literature	1	2	3	4	5	6	7	8
Twitter	19	70	98	37	1	42	16	51

	Forest	Costal	Ocean	Tropical	Science	Ice	Human	Summer
Literature	9	10	11	12	13	14	15	16
Twitter	57	127	75	120	3	34	58	23

	Adaptation	Policy	Reef	Food	future	dioxide	power	glacier
Literature	17	18	19	20	21	22	23	24
Twitter	87	21	124	11	39	104	14	122

	record	response	ecosystem	impacts	sustainable	weather	-	-
Literature	25	26	27	28	29	30	-	-
Twitter	13	N/A	113	N/A	101	30	-	-

bigrams, we handled this issue in two steps: (1) comparing hashtags that are identical to such features constitutes a fair ground of comparison. Accordingly, we searched the tweet data set for hashtags that are identical to the features and bigrams: (Carbon → #Carbon) and ("Global Warming" → #GlobalWarming). (2) The other issue is related to connecting the ground-truth word features to the tweet hashtags, which was handled using Association Analysis experiments. The approach lends itself naturally to a type of network that guarantees directional and weighted edges. This was done by parsing each tweet in the data set, searched for each of the known features we previously gathered (e.g., carbon, warming, emission). Each feature we found was paired up with all the hashtags that existed in the the tweet. A tweet such as this: ("#Florida officials ban the term climate change.") produces the following pairs (climate, #Florida) and ("change, #Florida"). Using Apriori [1], an Association Analysis algorithm, we generated rules from the transaction pairs with their actual confidence values which we naturally used as a weight for the edge that connects the left hand side of each rule with its right hand side.

Association Analysis experiments depend on a minimum confidence configuration parameter that must be set at the beginning of each experiment. Due to the novelty of this research, It is not known which confidence level to set this parameter. Therefore, we performed a series of experiments with different minimum confidence configurations to test which network is a better representation for the comparison task as follows: (minConf: 0.10, 0.25, 0.35, 0.45, 0.55, 0.65, 0.75, 0.85, 0.95, 1.0). The rules that were generated from each experiment then used to make up a K-H network that can be compared with the

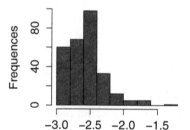

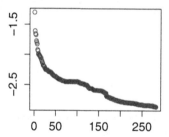

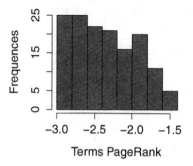

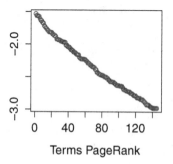

Fig. 4. compares the GTN Term PageRanks logs with TBN counterparts. We here present the histogram to display how the term frequencies are distributed. The top GTN histogram drawn in blue shows the highest values are distributed near the middle while there are some missing terms around the end. The histogram, demonstrated in green shows that the frequencies decreased as the ranks increased. The ranks in the TBN are also much lower than the GTN counterparts. While the max rank frequency was 25 in the TBN, it was observed near 90 in the GTN. Additionally, the PageRank behavior of the TBN appears to be more linear while the PageRanks of GTN appear to be converging towards a specific value. This demonstrates significant rank differences between the literature terms and what their Twitter counterparts have. This explains that fact that the social language on Twitter is not as technical or scientific as the language used in the climate literature, which is expected. This also explains why some of the bigrams in the GTN network did not hold in the tweet data set.

ground-truth network. Table 4 shows experiments, configuration, and a number of rules produced for each configuration.

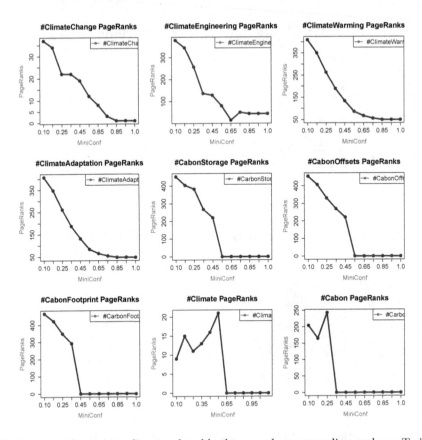

Fig. 5. shows the various climate related hashtags and corresponding ranks on Twitter derived from the dataset. For each hashtag, various association analysis experiments were performed to determine the actual rank of each term using different minimum confidence configurations. The plots show several observations that are noteworthy: (1) Single hashtag terms score low ranks when comparing with their corresponding ranks in the ground-truth network using a very low confidence score. Specifically, (#climate, and #carbon) hashtags has a rank of 9 and 200 respectively. However, the traffic dies out as the minimum confidence score increase, hence the sudden drop in the ranking of their specific plots. (2) Complex hashtags such as #ClimateChange, #ClimateEngineering and #ClimateAdaptation show a rising trend as the minimum confidence reaches %100. This demonstrates how reliable such terms in measuring the awareness of climate change compared with single word hashtags #climate. (3) The most exciting observation that must be highlighted is that the analysis shows that #ClimateChange hashtag scored the top term in the K-H Network when the minimum confidence approached %100. This is indeed in line with the term ranking specified by the ground-truth network and shows striking evidence of a great deal of climate change awareness on twitter. (4) The carbon related hashtags demonstrate that the general public are aware of climate change, however, they might not be as aware of what it entails. All carbon related terms score a much lower rank given that the carbon was the second highest ranked in the ground-truth.

Table 4. shows the Association Analysis experiment configurations with number of rules returned from each experiment using Apriori algorithm. The first column shows the experiment ID used. The second column shows the minimum confidence parameters used in each experiment. The third column shows the number of rules generated by the experiment using such configuration and the last column shows the total number of unique terms found in the the rules generated. The table gives a better understanding how the change of minimum confidence can eventually affect the network construction and the ranking of terms.

Experiments Number	Mini Conf	Number of Rules Generated	Unique Terms/Hashtags
Exp:1	.10	1232	561
Exp:2	.15	785	550
Exp:3	.25	448	456
Exp:4	.35	258	269
Exp:5	.45	161	246
Exp:6	.55	89	150
Exp:7	.65	61	111
Exp:8	.75	45	85
Exp:9	.85	38	74
Exp:10	.95	35	74
Exp:11	1.0	35	74

4 Discussion and Conclusions

In this paper, we explored the climate change awareness on Twitter using a means of Information Retrieval and PageRank analysis of multiple types of networks. The climate change awareness problem has been explored in a large magnitude of studies. However, it has never been explored using social media data such as Twitter. At the time of doing the analysis of this paper, we were unable to secure a formal source of climate change vocabulary (e.g., ontology). This created the need of generating a set of word features from literature. Accordingly, we selected Nature Climate Change journal, as one of the most specialized and highly ranked journals in this field, to extract the much needed features. We gathered the most relevant article titles that were publicly available from the web portal using the search query "Climate" in the period between 1990 and 2014. Using this means of Information Retrieval and Data Mining (PageRank and Association Analysis) we were able to construct various types of networks, some of which were for the purpose of establishing a notion of ground-truth.

From the various experiments we performed, we were able to mine the top influential textual constructs used on Twitter to signal climate change realization. The analysis demonstrates that though there is indeed a great deal of awareness of climate change, the social-language used on Twitter is expressed using more complex terms than single words. Specifically, when comparing a term in the GTN with its counterpart in the TBN, we find a large gap in the ranking. This becomes clear in observing the top terms in the GTN network and contrasting their ranks with the same terms in the TBN. As shown in table 3, while the top term is climate, the term comes as number 19 in the TBN. Similarly, the second and third top terms (i.e., carbon and temperature) scored 70 and 98 in the TBN. This clearly demonstrates a large gap in the level of awareness if only words were considered.

On Twitter, hashtags are commonly used textual constructs in place of grammatically correct contexts. Hashtags are powerful devices than rather textual contents because because of their ability to group the tweets in common and the wide range of knowledge they may represent. Such important devices must be incorporated in the analysis to fairly mine and measure the climate change awareness on Twitter. Using the newly emerging K-H network, we conducted a series of association analysis experiments that indeed lead to demonstrate a surprising amount of awareness of the topic. It turned out that the compound hashtags that are made of more than one word are more powerful than single word hashtags. For example, the (#Climate) hashtag has shown to be less significant in capturing climate change expressions on Twitter. On the other hand, a more complex hashtags such as (#ClimateChange, #ClimateEngineering and #ClimateApatation) are shown to be more dominant and ever lasting given the increase of the minimum confidence parameter.

One of the most interesting observations found is that the (#ClimateChange) was identified as the top ranked hashtag among all hashtags and words that made up the K-H network. Indeed this reflects positively on the amount of awareness that persists on social media in general and Twitter on particular. However, it is important to point out that the awareness traffic was slim and not persistent for more scientific or technical terms. Though the term ("Carbon") is the second highest ranked from the scientific point of view, we find little awareness about the term. This can be explained as the general public lack a deeper understanding of the various dimensions of climate change. It can also be associated with the current climate change events that took place at the time of collecting the data. It is possible to find more general hashtags such as (#ClimateChange or #Climate2014) during the UN Climate Summit than finding more specific terms.

In the future direction of this paper, we will investigate the climate change event identification which will also include the temporal dimension that is lacking in this current paper. It is particularly important to understand which event is more influential by measuring the number of socialization and impressions made against their corresponding hashtags. Another significant aspect of this research, which will necessarily capture a precise measure of the awareness, is the study of the the evolution of the climate change social-language. A question such as this one begs itself: Will the social-language that people use today to refer to climate change to refer to the problem: ("Climate Disruption"), as Jeff McMahn of Forbes [31] suggests? This idea can be more formalized by the development of an ever-evolving climate change social ontology. In addition, understanding the public opinions related to healthcare, food industry would certainly be another necessary direction to explore. Indeed, we have only scratched the surface of the topic of mining and measuring the Climate Change awareness on social media.

However, the analysis of this research suggests strong evidence of awareness and gives much hope for the efforts to follow accordingly.

Acknowledgments. This research is partly funded by Vermont EPSCoR award number EPS-1101713. The authors would like to acknowledge the data collection efforts by Alexa Ayer of UVM Rubenstein. We would like to acknowledge Dr. Ilan Kelman of University College of London for the the valuable discussions. A special thanks to Dr. Ibrahim Mohammed of EPSCoR for his analytical insights.

References

1. Agrawal, R., Srikant, R.: Fast algorithms for mining association rules in large databases. In: Proceedings of the 20th International Conference on Very Large Data Bases, VLDB 1994, pp. 487–499. Morgan Kaufmann Publishers Inc., San Francisco (1994). http://dl.acm.org/citation.cfm?id=645920.672836
2. Aizawa, A.: An information-theoretic perspective of tfidf measures. Information Processing and Management **39**(1), 45–65 (2003). http://www.sciencedirect.com/science/article/pii/S0306457302000213
3. Allesina, S., Pascual, M.: Googling food webs: Can an eigenvector measure species' importance for coextinctions? PLoS Comput Biol **5**(9), e10004942009 (2009). http://dx.doi.org/10.1371%2Fjournal.pcbi.1000494
4. Bekkerman, R., Allan, J.: Using bigrams in text categorization (2003)
5. Bollen, J., Mao, H., Zeng, X.: Twitter mood predicts the stock market. Journal of Computational Science **2**(1), 1–8 (2011)
6. Brin, S., Page, L.: The anatomy of a large-scale hypertextual Web search engine. Computer Networks and ISDN Systems **30**(1–7), 107–117 (1998)
7. Callaway, J.M.: Adaptation benefits and costs: are they important in the global policy picture and how can we estimate them? Global Environmental Change **14**(3), 273–282 (2004). http://www.sciencedirect.com/science/article/pii/S0959378004000366. the Benefits of Climate Policy
8. Conover, M., Ratkiewicz, J., Francisco, M., Gonçalves, B., Menczer, F., Flammini, A.: Political polarization on twitter. In: ICWSM (2011)
9. Curry, T.E.: Public awareness of carbon capture and storage: a survey of attitudes toward climate change mitigation. Ph.D. thesis, Massachusetts Institute of Technology (2004)
10. De Deyne, S., Storms, G.: Word associations: Network and semantic properties. Behavior Research Methods **40**(1), 213–231 (2008). http://dx.doi.org/10.3758/BRM.40.1.213
11. Ding, Y.: Topic-based pagerank on author cocitation networks. J. Am. Soc. Inf. Sci. Technol. **62**(3), 449–466 (2011). http://dx.doi.org/10.1002/asi.21467
12. Ding, Y., Yan, E., Frazho, A., Caverlee, J.: Pagerank for ranking authors in co-citation networks. J. Am. Soc. Inf. Sci. Technol. **60**(11), 2229–2243 (2009). http://dx.doi.org/10.1002/asi.v60:11
13. Do, T.D., Hui, S.C., Fong, A.C.M.: Associative feature selection for text mining. International Journal of Information Technology **12**(4) (2006)
14. Dodds, P.S., Danforth, C.M.: Measuring the happiness of large-scale written expression: Songs, blogs, and presidents. Journal of Happiness Studies **11**(4), 441–456 (2010)

15. Esbjörn-Hargens, S.: An ontology of climate change. Journal of Integral Theory and Practice **5**(1), 143–174 (2010)
16. Forman, G.: An extensive empirical study of feature selection metrics for text classification. The Journal of machine learning research **3**, 1289–1305 (2003)
17. Hamed, A.A.: An Exploratory Analysis of Twitter Keyword-Hashtag Networks and Their Knowledge Discover Applications. Ph.d. dissertation, University of Vermont (2014)
18. Hamed, A.A., Wu, X.: Does social media big data make the world smaller? an exploratory analysis of keyword-hashtag networks. In: IEEE BigData Congress (2014)
19. Hamed, A.A., Wu, X., Fandy, T.: Mining patterns in big data k-h networks. In: ACS/IEEE International Conference on Computer Systems and Applications, AICCSA 2014, Doha, Qatar (2014), November 10–13, 2014
20. Hamed, A.A., Wu, X., Fingar, J.: A twitter-based smoking cessation recruitment system. In: ASONAM (2013)
21. Hamed, A.A., Wu, X., Rubin, A.: A twitter recruitment intelligent system: association rule mining for smoking cessation. Social Netw. Analys. Mining **4**(1) (2014). http://dx.doi.org/10.1007/s13278-014-0212-6
22. Hearst, M.A.: Untangling text data mining. In: Proceedings of the 37th Annual Meeting of the Association for Computational Linguistics on Computational Linguistics, pp. 3–10. Association for Computational Linguistics (1999)
23. Jensen, L.J., Saric, J., Bork, P.: Literature mining for the biologist: from information retrieval to biological discovery. Nature reviews genetics **7**(2), 119–129 (2006)
24. Jing, L.P., Huang, H.K., Shi, H.B.: Improved feature selection approach tfidf in text mining. In: Proceedings of 2002 International Conference on Machine Learning and Cybernetics, vol. 2, pp. 944–946. IEEE (2002)
25. Kam, X.N.C., Stoyneshka, I., Tornyova, L., Fodor, J.D., Sakas, W.G.: Bigrams and the richness of the stimulus. Cognitive Science **32**(4), 771–787 (2008). http://dx.doi.org/10.1080/03640210802067053
26. Kolchinsky, A., Abi-Haidar, A., Kaur, J., Hamed, A.A., Rocha, L.M.: Classification of protein-protein interaction full-text documents using text and citation network features. IEEE/ACM Trans. Comput. Biol. Bioinformatics **7**(3), 400–411 (2010). http://dx.doi.org/10.1109/TCBB.2010.55
27. Levenbach, G.J.: A dutch bigram network. Word Ways **21**(11) (1998). http://digitalcommons.butler.edu/wordways/vol21/iss3/11
28. Lorenzoni, I., Nicholson-Cole, S., Whitmarsh, L.: Barriers perceived to engaging with climate change among the uk public and their policy implications. Global environmental change **17**(3), 445–459 (2007)
29. Macintyre, G., Jimeno Yepes, A., Ong, C.S., Verspoor, K.: Associating disease-related genetic variants in intergenic regions to the genes they impact. PeerJ **2**, e639 (2014). https://dx.doi.org/10.7717/peerj.639
30. Marsi, E., Oztürk, P., Aamot, E., Sizov, G., Ardelan, M.V.: Towards text mining in climate science: extraction of quantitative variables and their relations. In: Proceedings of the Fourth Workshop on Building and Evaluating Resources for Health and Biomedical Text Processing (2014)
31. McMahn, J.: Forget global warming and climate change, call it 'climate disruption', March 2015
32. Mihalcea, R., Tarau, P., Figa, E.: Pagerank on semantic networks, with application to word sense disambiguation. In: Proceedings of the 20th International Conference on Computational Linguistics, COLING 2004. Association for Computational Linguistics, Stroudsburg (2004). http://dx.doi.org/10.3115/1220355.1220517

33. Neil Adger, W., Arnell, N.W., Tompkins, E.L.: Successful adaptation to climate change across scales. Global environmental change **15**(2), 77–86 (2005)
34. Pang, B., Lee, L.: Opinion mining and sentiment analysis. Foundations and trends in information retrieval **2**(1–2), 1–135 (2008)
35. Pardalos, P., Boginski, V.L., Vazacopoulos, A.: Data mining in biomedicine, vol. 7. Springer (2008)
36. Radev, D.R., Jing, H., Sty, M., Tam, D.: Centroid-based summarization of multiple documents. Information Processing and Management **40**(6), 919–938 (2004). http://www.sciencedirect.com/science/article/pii/S0306457303000955
37. Salton, G., Buckley, C.: Term-weighting approaches in automatic text retrieval. Information Processing and Management **24**(5), 513–523 (1988). http://www.sciencedirect.com/science/article/pii/0306457388900210
38. Sampei, Y., Aoyagi-Usui, M.: Mass-media coverage, its influence on public awareness of climate-change issues, and implications for japans national campaign to reduce greenhouse gas emissions. Global Environmental Change **19**(2), 203–212 (2009)
39. Sebastiani, F.: Machine learning in automated text categorization. ACM computing surveys (CSUR) **34**(1), 1–47 (2002)
40. Semenza, J.C., Hall, D.E., Wilson, D.J., Bontempo, B.D., Sailor, D.J., George, L.A.: Public perception of climate change: voluntary mitigation and barriers to behavior change. American journal of preventive medicine **35**(5), 479–487 (2008)
41. Signorini, A., Segre, A.M., Polgreen, P.M.: The use of twitter to track levels of disease activity and public concern in the us during the influenza a h1n1 pandemic. PloS one **6**(5), e19467 (2011)
42. Tan, C.M., Wang, Y.F., Lee, C.D.: The use of bigrams to enhance text categorization. Inf. Process. Manage. **38**(4), 529–546 (2002). http://dx.doi.org/10.1016/S0306-4573(01)00045-0
43. Whitmarsh, L.: Behavioural responses to climate change: Asymmetry of intentions and impacts. Journal of Environmental Psychology **29**(1), 13–23 (2009)
44. Wu, X., Kumar, V., Ross Quinlan, J., Ghosh, J., Yang, Q., Motoda, H., McLachlan, G.J., Ng, A., Liu, B., Yu, P.S., Zhou, Z.H., Steinbach, M., Hand, D.J., Steinberg, D.: Top 10 algorithms in data mining. Knowl. Inf. Syst. **14**(1), 1–37 (2007). http://dx.doi.org/10.1007/s10115-007-0114-2
45. Xie, X., Jin, J., Mao, Y.: Evolutionary versatility of eukaryotic protein domains revealed by their bigram networks. BMC Evolutionary Biology **11**(1), 242 (2011). http://dx.doi.org/10.1186/1471-2148-11-242
46. Ye, N., et al.: The handbook of data mining, vol. 24. Lawrence Erlbaum Associates Mahwah, NJ (2003)
47. Zhang, W., Yoshida, T., Tang, X.: A comparative study of tf*idf, LSI and multiwords for text classification. Expert Systems with Applications **38**(3), 2758–2765 (2011). http://www.sciencedirect.com/science/article/pii/S0957417410008626

Assessing Patterns of Urban Transmutation Through 3D Geographical Modelling and Using Historical Micro-Datasets

Teresa Santos, Antonio Manuel Rodrigues$^{(\boxtimes)}$, and Filipa Ramalhete

CICS.NOVA Interdisciplinary Centre of Social Sciences, Faculdade de Cincias Sociais e Humanas, Universidade Nova de Lisboa, Lisbon, Portugal
{teresasantos,amrodrigues}@fcsh.unl.pt, framalhete@netcabo.pt
http://www.cics.nova.fcsh.unl.pt/

Abstract. The increasing volume of empty houses in historical cities constitute a challenge in times of economic crisis and acute housing needs. In order build coherent guidelines and implement effective policies, it is necessary to understand long-term patterns in city growth. The present work analyses urban dynamics at the micro level and present clues concerning transmutation in Lisbon, Portugal, using 3D geographical modelling to estimate potential housing supply. The recent availability of detailed demographic historical micro-datasets presents an opportunity to understand long-term trends.

Integrating cartographic and altimetric data, vacant houses of the city are mapped and attributes like area, volume and number of floors are estimated. Then, the potential for social housing is evaluated, based on state owned buildings morphology. Exploratory Spatial Data Analysis (ESDA) help to highlight trends at a finer scale, using advanced geovisualization techniques. The challenge of working with distinct data sources was tackled using Free and Open Source (FOSS) Geographical Database Management Systems (GDBMS) PostgreSQL (and spatial extension PostGIS); this facilitated interoperability between datasets.

Keywords: Urban transmutation · Exploratory Spatial Data Analysis (ESDA) · Dasymetric mapping · 3D data · Historical micro-datasets

1 Introduction

Post-industrialism and the shift in the division of labour towards service activities have strong implications in land-use. Urban functions adapt dynamically, which have lasting effects on human footprints. Throughout their life-cycle, although planned to a given extend, parts of the urban landscape suffer morphological as well as functional changes. New residential neighbourhoods emerge while others decline as population moves outwards. Central locations usually transmute as to give rise to dominant service areas. As vital as these dynamic

© Springer International Publishing Switzerland 2015
O. Gervasi et al. (Eds.): ICCSA 2015, Part I, LNCS 9155, pp. 32–44, 2015.
DOI: 10.1007/978-3-319-21404-7_3

processes are in order to understand when, how and why a city evolves in certain directions, a long-term monitoring system is not rarely hampered by the non-availability of adequate data, particularly for small-area analysis.

The objective of the present article is to examine significant patterns of urban population and buildings/dwellings' dynamics in Lisbon, capital city of Portugal in order to aid decision-making at the local level concerning rehabilitation policies. In the new Lisbon Master Plan (2014), urban rehabilitation over new construction is privileged. The municipality classifies the entire urban area as historic district, assigns credits for building rehabilitation and penalizes those who leave their crumbling heritage [10]. Furthermore, creating conditions to attract more people thought programs for affordable housing is part of the strategic vision for the city.

Promoting urban rehabilitation and regeneration, besides having a social impact in vulnerable communities, is also beneficial to local economies. This is achieved through new refurbishment activity, the engagement of local contractors, tradespeople and labourers, as well as generating new council tax revenue.

The regular census of population is the primary micro-data source of information for small-areas, and thus the primary source when there is a need to characterize small urban areas. This is so because: (i) information is collected for the whole of the population at a giver time-stamp, (ii) it is geographically aggregated for very small spatial units, and (iii) its general availability allows cross-validation of scientific results.

As a census of population, attributes are collected which are related to the living conditions of individuals. Figure 1 represents the density of dwellings per resident (or partially resident) building for the study area. It also represents concentric arches, centred in "Praca do Comercio", the historical CBD of the city. Two main waterways cross the heart of the city (white arrows). These coincide with arteries, along which the Lisbon have evolved, particularly since the mid 1800's. The airport area and the Monsanto Park are large areas with a small number of residential buildings and are also highlighted.

When data is available for different census exercises, dynamic analysis of information at the census tracts level allows the cross-counting of buildings. This allows for identifying residential blocks which were no longer counted. This may naturally be due to the simple fact that an individual building may cease to exist (ie. is demolished), which can easily be demonstrated through the analysis of aerial photos. However, most negative growth rates in the number of buildings are the result of two distinct phenomena: tertiarization and derelition.

In order to identify significant patterns of urban transmutation, Exploratory Spatial Data Analysis (ESDA) play an important role when investigating the empirical distributions of variables of interest [3–6, 8]. In the present study, the most relevant information concerned population densities and age-structure, and dwelling densities per building as well as buildings' age. Data availability for the 1991, 2001 and 2011 census exercises allows the longitudinal exploration of explanatory patterns. Given that the shape of census tracts between all three time-stamps does not coincide, an ex-ante exercise is performed in order to

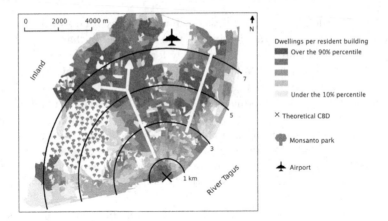

Fig. 1. Representation of Lisbon's main axis emanating from the historical centre, which coincide with two natural waterways

guarantee geometrical symmetry between datasets, following Rodrigues etal. [14–16]. The resulting dataset forms part of the larger databank "Comuns", which is the outcome of in-house research efforts and is a wide source of open containing historical micro-datasets for Continental Portugal [12,13].

Two extra sources of information are: (i) a geographical dataset available at the city hall, containing addresses of unoccupied buildings and (ii) the number of residents on housing waiting lists. Through geographic modelling, we identify the area occupied by empty houses in the city, thus assessing the potential return of empty homes to the useful housing stock. Then, from the initial vacant dataset, the spatial clusters of buildings owned by the local and central government are mapped using ESDA techniques. From this analysis, two outputs are produced: the map of empty houses area in the city (figure 2), and the number of dwellings available for social housing.

The estimation of the number of floors available at each building is only possible due to the 3D dataset. In fact, the altimetric information retrieved from a LiDAR data set, allows infering the mean height of each vacant building, and use it to estimate the number of foors.

The growing availability of micro for large metropolitan areas present an opportunity to understand local structure and dispersion of human phenomena which shape the urban built environment over the years. The techniques developed during the present study, because of the use Free and Open Source (FOSS) tools as well as census open-data and the growing availability of raw 3D data, may be easily replicated in different geographical settings. This may, over time, help build a coherent body of knowledge concerning rehabilitation potential and transmutation patterns.

Using ESDA for pattern and cluster analysis based on longitudinal data, mixed with estimation of potential and effective housing supply present an original and easily replicable methodology with potential long-term benefits.

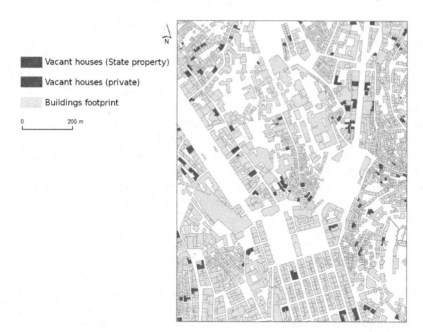

Fig. 2. Vacant buildings available in Lisbon: snapshot from part of the historical downtown

The availability and treatment of geographical 3D data also opened new opportunities as models approach reality. Cenus-based micro time-series conceptually introduced the fourth dimension to the analysis.

The rest of the article is organized as follows: Section 2 describes existing housing needs and effective demand at the municipal level; section 3 describes the methodology developed and applied in the present study; section 4 analysis Lisbon demographic dynamics from macro to micro scales. Section 5 jointly compares demographic and vacant buildings datasets with the objective of quantifying potential housing supply. Section 6 concludes.

2 Housing Needs

According to a diagnosis on housing needs and dynamics (2008/2013), based on family data, there is a need for 200 000 dwellings in Portugal as well as 190 000 others needing urgent rehabilitation interventions. These numbers show that, in spite of the construction boom the country experienced in the last decades, there is not a clear relation between the number of existent buildings and the resolution of housing needs. These needs result from the difficulty that the low income population faces to access to the housing market (either as rentals or owners). As part of the welfare state policy and as a direct result of housing as an institutional right, most municipalities are also social landlords and manage a housing park which tries to solve these needs.

In Lisbon, in spite of the various rehousing programs held in the past by the municipality, there is still a need for social housing. The municipal regulation considers that applying for a social dwelling is open to those Lisbon residents: a) whose family does not own a house/apartment on the Lisbon Metropolitan Area; b) are not already financially supported for housing purposes with public funds; c) who have a family income within the value defined by law for social aid. The number of seniors or dependent individuals (children or handicapped) is also within the criteria to apply to a municipal dwelling.

Within this context, the demand is significant. From April 2012 to December 2014 the municipality received about 7000 appliances for municipal housing. The present study bridges the gap between local demand and the need to have access to detailed information concerning potential rehabilitation and housing supply at the municipal level.

3   Methodology

Several geographic datasets were used for the present study: the buildings footprint (2006), in vector format, retrieved from the 1:1000 scale Municipal Cartography updated for 2006 [17], the census of vacant buildings and dwellings (2009) from the Lisbon Municipality authorities, the Census tracts datasets for 2011, and sets from the database "Comuns", containing historical micro-datasets for continental Portugal. This later source allowed a detailed analysis of demographic dynamics in the study area.

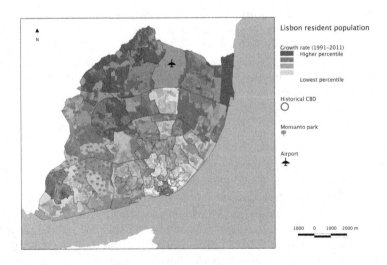

Fig. 3. Resident population (growth rate 1991-2011)

Altimetic data included two raster files: a Digital Terrain Model (obtained from elevation points and contours, derived from 1998 cartography) and, a Digital surface Model (obtained from a Light Detection and Ranging LiDAR data from 2006). The altimetric data was used to generate, a normalized Digital Surface Model (nDSM) with 1m2 of pixel resolution [17]. The nDSM is a raster file with height of all elements above ground.

The historical micro-datasets were built using dasymetric mapping techniques, which, through a filtering algorithm, redistributed demographic data from distinct geometries (different areas) to common minimum areas. The buildings' footprint were used as ancillary data. This were filtered according to function (residential, services and industrial) in order to build a coherent weighting scheme. In practice first a source geometry (ie. 2011 census tracts) was intersected with a target geometry - common minimum geometries from the 1991 and 2001 census [7], and the area occupied by residents was estimated according to the proportion occupied by resident buildings. Information was re-allocated and aggregated according to the set of estimated weights.

In relation to vacant buildings, starting from a point dataset containing vacant dwellings, the footprint of the vacant buildings was constructed as those areas (polygons) within a maximum distance of 2 metres from any point event of interest (totally vacant buildings).

For representing geographical data, other than orthodox choroplet maps, a method is used which highlight clusters (or "hot spots"). This is achieved through the calculation of spatial lags for the variables of interest. For variable X, the spatial lag for observation i is computed using the expression:

$$x_i^{lag} = \sum_{j=1}^{k} w_{ij} x_j \;,\tag{1}$$

where k is the set of i's neighbours and w_{ij} is the ij observation which represents the geographical proximity between areas i and j. Proximity measures in this case are represented in a weights matrix which takes the form:

$$W = \begin{cases} w_{ij} = 0 & \text{if } i = j \\ w_{ij} = f_{ij} & \text{if } f_{ij} \neq 0 \\ , w_{ij} = 0 & \text{if } f_{ij} = 0 \end{cases}\tag{2}$$

where f_{ij} represents the border between Geographical units i and j. In practice, areas are considered to be neighbours if they are in some respect contiguous. This is the most general form of a contiguity spatial weights matrix. In the present case, when $f_{ij} \neq 0$, then $w_{ij} = 0$. The result is a standard binary weights matrix.

4   Lisbon Demographic Dynamics (1991-2011): from Macro to Micro

Between 1991 and 2011, the population of the city of Lisbon have been decreasing (table 1). Yet, after a 15% fall in the last decade of the $20^{t}h$, numbers have almost remained stable in the following decade. Also, from 2001 to 2011, residents aged between zero and four years old have increased 12%, which is a sign that average age may decrease in the future as population. So far, between 2001 and 2011, average age of residents have increased from 44 to 45 years old; hence, a reversal is expected in the near future.

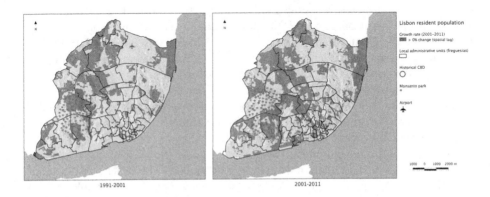

Fig. 4. Resident population growth rates (1991-2001 and 2001-2011) calculated as spatial lags for each census tract

Table 1. Aggregate evolution of resident population, young children and elderly

	Census year		
	1991	2001	2011
Total resident population	663394	564657 (-15%)	551394 (-2%)
Residents age 0-4	-	21287	23766 (+12%)
Elderly population (+65)	-	133304	131576 (-1%)

Figure 3 allows identifying patterns of residential population dynamics over a 20 years span. Datasets are taken from databank Comuns, hence the aggregation level is the minimum common areas, as defined in the 2001 census exercise. The map captures a clear growth pattern in the outer ring, evidence of a long-term trend of outward moving of residents. Nonetheless, there is an interesting trend in a cluster located around the city old Central Business District (CBD).

When we break down the growth rates into two distinct periods (1991-2001 and 2001-2011), and recalculate values according to the mean of each area's neighbours (spatial lags), some other trends emerge. In figure 7 dark gray patches represent blocks where resident population growth was positive. When comparing the first and the second period of analysis, it is possible to infer a strengthening of re-population along a central corridor as well as in the western part of the city, along a strip of the riverbank. To the East, the new residential area of "Oriente", which emerged after the rehabilitation of a large and old industrial site in 1998 with the World Expo clearly emerges as the new residential (and services) growth pole. Over the last decade, growth in this area have in fact been contagious to neighbouring areas, in an interesting textbook development.

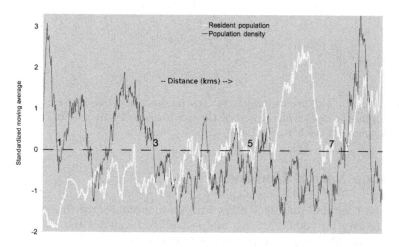

Fig. 5. Distribution of resident population and population density as distance increases from the historical/theoretical Central Business District - CBD (distance in Kms represented in the horizontal axis). Original values were tranformed in two steps: first, they were standardized to a standard normal distribution; second, the series were smoothed taking the average for each block of five observations - moving averages. Such transformations facilitate comparison between series and identification of trends.

As the city grew away from the historical CBD and along two main corridors, distinct housing types and changing planning habits have created singular patterns both in terms of number of residents and population density. Figure 5 shows the distribution of these two variables as one moves linearly away from the CBD. The pattern of population density is positively correlated to its growth rate (see figure 3) and to the density of dwellings per resident building (see figure 1) - numbers are higher near the center and in the fringes of the city. Note that actual values were transformed using the standardization of the moving average, so that trends become more clear.

In relation to the age of construction, figure 6 shows that as we move away from the center, the number of older buildings clearly decreases. Buildings constructed prior to 1945 are mainly located near the CBD and in the 3 km arch. An opposite, yet not so clear trend, may be observed in subsequent periods.

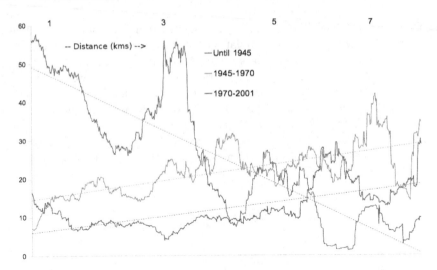

Fig. 6. Number of buildings according to construction age per census tract as distance increases from the historical/theoretical Central Business District - CBD (distance in Kms represented in the horizontal axis)

5   Lisbon Vacant Housing

The Lisbon City Hall published a list with the unoccupied dwellings and buildings in the city for 2009. For each address, is also available the owner (public or private sector) and the status of the building (totally or partially empty). From this list only totally vacant buildings were selected. Based on the address, a geolocation process was applied and a map of the vacant buildings localization (x, y) was produced and used to identify the respective building footprints. Figure 7 represents, through an empirical distribution function, the distribution of vacant dwellings in Lisbon. There is a strong concentration in and around the historical core. This indicates a strong potential for re-population of this area which, as mentioned previously is a strong political priority.

Using the buildings footprint and the altimetric information, a geographic database with information about the owner, area and height was produced for each vacant building (Figure 8). The use of altimetric 3D data obtained in the nDSM allows extracting the mean height for every building. This information, generally not available in cartographic data sets, is then used as a proxy for the number of available floors at each vacant building. For this study, a height value of 3 m for each storey was used to estimate the number of floors. From this analysis, the Potential Housing Supply Area in the city is mapped. Then, for evaluating social housing supply, only those buildings that are public properties were investigated and the number of floors is estimated. These represent available accommodations that can be used to fulfil social housing demand. All information was aggregated into a single databank, supported by PostgreSQL geographical data types (using PostGIS extension)

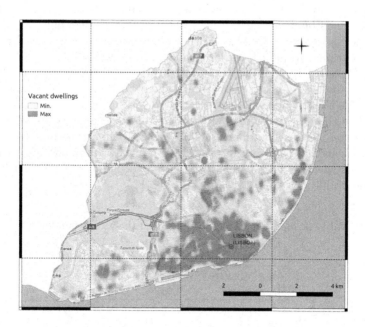

Fig. 7. Vacant dwellings kernel density estimation (contextual geographical information from MapQuest - http://www.mapquest.com/)

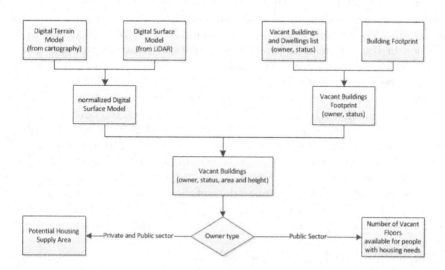

Fig. 8. Methodology to assess the potential housing offer based on 3D geographical modelling

The total number of vacant buildings in the city is 2114, with 434 948 m^2 . The corresponds built volume is 4 371 744 m^3. The average building occupies 206 m^2 and has 3 floors (Table 2). Such buildings constitute a wasted resource and can contribute to the process of neighbourhood decline.

It is interesting to compare the location of vacant houses with the spatial variation in the rate of change for average age of resident population. Comparing those neighbourhoods where population is getting younger at a higher rate (classified as those areas whose value of the rate of change is above the 80% percentile) and the location of clusters of vacant dwellings, there is little correlation, which indicate a strong potential for rejuvenation of the historical city as two complementary forces are clearly in play: demand from younger families and political will.

Table 2. Morphologic characteristics of the vacant buildings

	Area (m^2)	Volume (m^3)	Floors
Total	434948	4371744	6503
Mean	206	2068	3

6   Concluding Remarks

In times of economic crisis, acute housing needs are a concern. The present work describes a methodology to estimate potential supply from the housing market, by examining the distribution of vacant dwellings/buildings in Lisbon city. Furthermore, a detailed analysis of the transmutation patterns and demographic evolution in the city helps to understand long-term trends as to adapt policy guidelines. These findings constitute relevant information for policy-makers and other pertinent stakeholders.

Evaluation housing supply, using volume data instead of the typical 2D information on building area, revealed to be very useful. In fact, the number of floors is a more researched estimation of available dwellings than using just areal information. In this context, the use of altimetric data obtained from LiDAR sensor proved to be an efficient alternative to the traditional photogrammetric methodologies, very demanding in human resources, expensive and time consuming. Also, the existent of historical micro-datasets is of paramount importance as source of detailed investigation of the phenomena of interest. In this respect, Exploratory Spatial Data Analysis proved efficient in pattern recognition of geographical trends.

The findings represent an exploratory analysis of data concerning empty houses in Lisbon city. No attempted regarding structural conditions (i.e., the housing quality was not evaluated) was made, and we assumed that all vacant houses present in the municipality's list were conformed. Historical cities, like Lisbon, can gain from a sustainable development, though regeneration and rehabilitation of the existing empty houses. Such actions constitute valuable contributions for public spaces qualification, by limiting new constructions thus promoting the presence of green and permeable areas. Ultimately, there are trends which point to future demographic regeneration and rejuvenation. It is up to local authorities to decide future spatial trends through effective urban planning.

Acknowledgments. This article was partly funded by Portuguese national funds through FCT Portuguese Foundation for Science and Technology in the framework of projects PEsT-UID/SOC/04647/2013, SFRH/BPD/76893/2011 and SFRH/BPD/66012/2009.

References

1. Regulamento do Regime de Acesso à Habitação Municipal (2009)
2. O Parque Habitacional e a sua Reabilitação: Análise e Evolução 2001–2011 (2013)
3. Anselin, L.: Spatial Econometrics: Methods and Models. Kluwer Academic Publishers (1988)
4. Anselin, L.: Spatial Regression Analysis in R A Workbook (2005)
5. Bivand, R.: Spatial econometrics functions in R: Classes and methods. Journal of Geographical Systems **4**(4), 405–421 (2002). http://www.springerlink.com/Index/10.1007/s101090300096
6. Fingleton, B.: Spatial Autoregression. Geographical Analysis **41**(2009), 385–391 (2009)
7. Geirinhas, J.A.: Conceitos e Metodologias: BGRI - Base Geográfica de Referenciação de Informação. Revista de Estudos Regionais (2001)
8. Goodchild, M.F.: What Problem? Spatial Autocorrelation and Geographic Information Science. Geographical Analysis **41**(4), 411–417 (2009). http://doi.wiley.com/10.1111/j.1538-4632.2009.00769.x
9. Guerra, I., Pereira, S.M., Fernandes, M., Botelho, P., Marques, P., Mateus, A., Primitivo, S., Caetano, A., Cabral, C., Pereira, M., Portas, N., Marques, T.S., Matos, F., Ferreira, E.: Relatório 1: Diagnóstico de Dinâmicas e Carências Habitacionais. Tech. rep., CET-ISCTE / IRIC / A.Mateus e Associados
10. de Lisboa, C.M.: Lisbon Master Plan. Tech. rep. (2014). http://www.cm-lisboa.pt/viver/urbanismo/planeamento-urbano/plano-diretor-municipal/pdm-em-vigor
11. Madeira, C.A.C.L.: A Reabilitação Habitacional em Portugal - Avaliação dos Programas RECRIA, REHABITA, RECRIPH E SOLARH. Master thesis, Universidade Técnica de Lisboa (2009)
12. Neves, B., Rodrigues, A.M., Santos, T., Freire, S.: Unlocking geographical information from academia: an open source WebGIS solution. In: Proceedings of the VI Jornadas de SIG Libre. SIGTE - Servei de Sistemes d'informció Geogràfica i Teledetecció, Girona (2012). http://www.fcsh.unl.pt/e-geo/sites/default/files/dl/
13. Rodrigues, A.M., Neves, B., Rebelo, C.: Terra communis (tComm): a free data provider for historical census micro-data. In: Proceedings of the VII Jornadas de SIG Libre. SIGTE - Servei de Sistemes d'informció Geogràfica i Teledetecció, Girona (2013). http://www.sigte.udg.edu/jornadassiglibre2013/uploads/articulos_13/a9.pdf

14. Rodrigues, A.M., Santos, T., de Deus, R.F., Pimentel, D.: Land-use dynamics at the micro level: constructing and analyzing historical datasets for the portuguese census tracts. In: Murgante, B., Gervasi, O., Misra, S., Nedjah, N., Rocha, A.M.A.C., Taniar, D., Apduhan, B.O. (eds.) ICCSA 2012, Part II. LNCS, vol. 7334, pp. 565–577. Springer, Heidelberg (2012). http://link.springer.com/chapter/10.1007/978-3-642-31075-1_42#

15. Rodrigues, A.M., Santos, T., Pimentel, D.: Asymmetrical-mapping based methodology: constructing historical datasets for Portuguese census tracts. In: Gensel, J., Josselin, D., Vandenbroucke, D. (eds.) Proceedings of the AGILE 2012 International Conference on Geographic Information Science, Avignon. Multidisciplinary Research on Geographical Information in Europe and Beyond, pp. 24–27 (2012)

16. Rodrigues, A.M., Santos, T., Pimentel, D.: Socio-economic dynamics within a middle-size municipality: testing the strengths of historical micro-datasets. In: Gestão integrada de territórios intermunicipais: o papel dos Sistemas de Informaçãao Geográfica. 7as Jornadas de Gestão do Território - Livro de resumos, Tomar (2012). http://www.fcsh.unl.pt/e-geo/sites/default/files/dl/

17. Santos, T.: Producing Geographical Information for Land Planning Using VHR Data: Local Scale Applications. LAP LAMBERT Academic Publishing (2011)

Potential Nitrogen Load from Crop-Livestock Systems: An Agri-environmental Spatial Database for a Multi-scale Assessment

Marco Vizzari[1(✉)], Alessandra Santucci[2], Luca Casagrande[3], Mariano Pauselli[1], Paolo Benincasa[1], Michela Farneselli[1], Sara Antognelli[1], Luciano Morbidini[1], Piero Borghi[1], and Giacomo Bodo[2]

[1] Department of Agricultural, Food, and Environmental Sciences, University of Perugia, Perugia, Italy
{marco.vizzari,mariano.pauselli,paolo.benincasa, michela.farneselli,sara.antognelli,luciano.morbidini, piero.borghi}@unipg.it
[2] ARPA (Agenzia Regionale Protezione Ambientale), Umbria, Italy
{a.santucci,g.bodo}@arpa.umbria.it
[3] T4E S.r.l., Perugia, Italy
l.casagrande@t4e.it

Abstract. The EU "Water" Directive establishes a common European framework for the environmental protection of inland, coastal and marine waters. Environmental pressures related to agri-livestock systems are still a major concern among the general public and policy makers. In this study, carried out in Umbria region, Italy, a novel spatial database for a multi-scale analysis was designed and implemented integrating different agricultural and livestock farming datasets. Beyond descriptive indicators about agricultural and livestock farming systems, this database allows to assess, at different geographic levels of investigation (cadastral sheets, municipalities, provinces, entire region, Nitrogen Vulnerable Zones, bodies of groundwater, sub-basins), the potential nitrogen crop supply, the potential nitrogen availability from livestock manure, and, by means of a scenario analysis, the total potential nitrogen load. These indicators appear to be very relevant to support decision making and to pursue the environmental objectives established by EU and national regulations.

Keywords: Water framework directive · Non-point source pollution · Environmental planning · Agri-environmental indicators · Crop-livestock data integration

1 Introduction

The environmental impact of high crop-livestock concentrations appears particularly significant where it coincides with weaker policy standards and poor manure management strategies [1]. Livestock sludge, in accordance with best agricultural practices, can be used successfully as an organic fertilizer for crops, ensuring its optimal

© Springer International Publishing Switzerland 2015
O. Gervasi et al. (Eds.): ICCSA 2015, Part I, LNCS 9155, pp. 45–59, 2015.
DOI: 10.1007/978-3-319-21404-7_4

disposal [2]. In many cases, however, agronomic and environmental damage can result from the improper use of sludge, such as damage to the soil, degradation of the soil structure due to the levels of certain cations (K^+, Na^+), salinization, alterations in soil pH, alteration of the soil microbial population, and accumulation of heavy metals [3,4,5]. These processes tend to lead to degradation of the agronomic potential of agricultural lands, and to the pollution of ground and surface water [6,7,8,9,10].

The element causing the greatest number of management problems with regard to impact on near-surface and deep aquifers is the excess of nitrogen (N) [4], [11,12,13]. This form of pollution is caused by N fertilisation, but its importance may change with the fertilizer-N rate and source, and by fertilization technique [14,15,16,17,18,19] including the technique used for spreading livestock manure and sewage sludge [5], [18,19,20,21]. It continues to represent a significant problem, especially in regions of intensive livestock farming, in all European Union Member States [4], [22,23].

The EU Water Framework Directive (WFD) - integrated river basin management for Europe - establishes a common framework at European level for the protection of inland (surface and groundwater), coastal and marine waters. The main objective is to improve water quality (also preventing further deteriorations), through a gradual reduction of pollution, promoting a sustainable water use [22]. The Directive obliges Member States to protect, enhance and restore the environmental quality of water bodies and to achieve ecological chemical and quantitative objectives of those water bodies not significantly disturbed by human activities. In Italian law, the Legislative Decree (LD) n. 152/2006 transposes the EU directive and requires all the Italian Regions to implement a local Water Protection Plan (WPP) which constitutes a sectorial plan within the wider River Basin Management Plan defined by same WFD. WPP represents the planning tool that contains measures and programs aimed to water quality protection and environmental objectives achievement as required by the WFD. To this aim, knowledge of human pressures on water bodies is critical in almost all key stages of the WPP implementation and application process [24,25,26,27].

Thus, the characterization of an updated picture about the pressures due to human activities (including crop-livestock activities) on water seems essential for supporting water sector planning both at regional and district levels. The analysis of pressures must be obviously consistent with the criteria of the WFD and its Italian transposition, which require the assessment of pressures at body of surface and groundwater water scale. Article 2.10 of WFD defines "body of surface water" as a discrete and significant element of surface water such as a lake, a reservoir, a stream, river or canal, part of a stream, river or canal, a transitional water or a stretch of coastal water. The application of this definition requires the sub-division of river basins into 'discrete and significant elements' by the regional environmental agencies. In addition, the Nitrogen Vulnerable Zones (NVZ), identified by the European Directive 91/616/EEC should be included in the analysis considering the need to evaluate the causes of the trophic status of some Umbrian surface water-bodies, and the incidence of the nitrogen coming from agriculture and livestock in determining some critical issues.

To these aims, the estimation of potential nitrogen loads should be based on a spatial approach and coherent data which meet some key criteria: 1. high spatial and temporal accuracy to analyse the nitrogen pressure sources at the different scales (including the more detailed ones) and to compare the estimations with the results from water monitoring; 2. high updating frequency for ensuring the reproducibility of the analysis with a timing consistent with the obligations arising from the implementation of the major European Directives (WFD, but also i.e. 91/676, 91/271). In this regard spatial databases are useful tools for integrating different data, making them spatially informative [28]. If structured using proper user-interfaces and simulation models, they become key components of wider Decision Support Systems [29] able to help decision-makers to answer different specific questions by facilitating the use of data in an interactive way (see e.g. [30,31,32]).

In this framework the main objectives of the present study can be identified in: a) design and implementation of an agri-environmental spatial database for the Umbria region territory, containing detailed multi-temporal information about the crops and livestock systems; b) assessment and analysis of environmental pressures due to Nitrogen potential load from mineral and manure fertilization at different geographic levels and scales of interest.

2   Materials and Methods

The present study was developed for the Umbria region a 8.456 km² wide area located in the central Italy (Fig. 1). Its morphology is featured mainly by the valley of the Tiber stream (flowing in north-south direction) and its tributaries, by the Trasimeno Lake, and by the Apennine mountains along its Eastern part.

Fig. 1. Localization and hydromorphology of the Umbrian study area

2.1   Spatial DB Design and Implementation

The database design answered to the user-requirement specifications as well as the said criteria linked to WPP implementation. The main user's need was to integrate data coming from different and available data sources to analyse the Umbria agri-livestock system and quantify the Nitrogen potential load (N_{PL}) generated by this system, at the different geographic levels and scales of interest (administrative units: cadastral sheets, municipalities, provinces, entire region; other geographical areas of interest: NVZ, bodies of groundwater, sub-basins). Two other general users' requirements were considered: to build a solid instrument to retrieve, in a fast and efficient way, the agri-environmental data and all related parameters and to create a tool which is not static and where all the output are dynamic in relation to the input data. Before the implementation, the following points were considered in developing the database design: a) select effective and consistent data; b) discover conceptual relationships in the data; c) identify future data needs for the spatial DB update; d) determine how the data is used and generated to determine the DB reliability.

The DB implemented for this purpose was called "AGUA" (Agri-environmental Geo-database of UmbriA). Different datasets, grouped in three main blocks, were collected and processed (Table 1):

AGUA was implemented using PostgreSQL, a high performing and reliable open-source platform. In the implementation phase, three general criteria were followed: a) develop an efficient and versatile scheme, with low redundancy; b) achieve a solid relational-structure with primary and foreign keys and dictionary tables; c) keep the maximum quantity of information as possible during data integration, ensuring maximum efficiency of all procedures of calculation; d) define automatic procedures for conversion of source datasets in order to allow an easier DB update even after its distribution. The final general model of AGUA is represented in Fig. 2.

In the implementation phase, a matching process between SIAN, PSR, PUA, SEP, COM, and cadastral sheets geometries was based on cadastral identifiers, while CCT was linked to SIAN and PSR datasets using the crops typologies. A semi-automatic process for linking BDN, COM, CBP datasets, based on three subsequent steps, was developed: a) match between VAT and/or tax codes (54 farms); b) match by comparisons between farm names based on fuzzy algorithms and subsequent manual verification (95 farms); c) manual link by comparison of table data (e.g. farm name, farm type, localization) and/or verification of the farm position on digital ortophoto of the year 2011 (93 farms).

CAS, the smallest geographical unit included in AGUA, were the key spatial elements (average area of 110 ha) used to localize all data with cadastral identifiers. To spatially relate these data to the various AOI a simple SQL script to overlay cadastral sheets and the other AOI was implemented in order to calculate the percentage of each sheet falling within the different AOI. Such percentages were subsequently used to compute the various indicators for the AOI multiplying and aggregating the values calculated at CAS level.

Table 1. Datasets included into AGUA

Block	Code	Name	Description
Crop data	SIAN	National Agricultural Information System	Farms data, crop types, and related areas at cadastral parcel level (years 2011, 2012, 2013).
	PSR	Rural Development Plan: Measure 2.1.4, Action A – integrated farming and action B: organic farming	Farms data, crop types, and related areas regarding the Agri-environmental schemes application at cadastral parcel level (N limitations due to integrated and organic crop management techniques).
	PUA	Agronomic use plans	Farms data, crop types, and related areas of N application plans at cadastral parcel level
	SEP	Sewage sludge spreading permissions	Quantities and cadastral parcels plans of sewage sludge spread at a farm level
	CCT	Crop coefficients table	Coefficients of N supply to crops defined by experts agronomists, differentiated according to crop type and farming technique (conventional, integrated and organic)
Livestock data	BDN	National Livestock Database data	Farm data, coordinates, numbers and types of reared animals (year 2013).
	COM	Farm Communications	Date and quantity of livestock manure or sewage sludge spread at farm level
	CBP	Compost and biogas producers data	Sewage sludge production at farm level
	LCT	Livestock coefficient tables	Potential N supply for unit of live weight, differentiated according to the type and age of reared species retrieved from official regulations of bibliography
Geographical data	CAS	Cadastral sheets	Parcels groupings of Umbria Region (7727 geometries, average area equal to 109 hectares)
	AOI	Geographical areas of interest	NVZ, sub-basins, bodies of groundwater

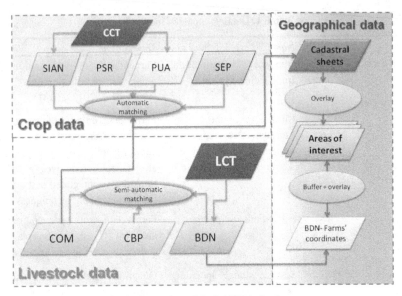

Fig. 2. General model of AGUA database

The indicators associated to the various hierarchical administrative units (in ascending order: municipalities, provinces, region) were computed by simple aggregations of the values calculated at CAS. On the other hand, to allow the localization of N_{AL} of the livestock farms for which the COM were not available, an advanced buffering procedure was developed in AGUA implementing the (1):

$$r = \sqrt[2]{\frac{N_{AL}}{N_{MAX}} \frac{TA_{CS}}{UA_{CS}} \frac{10000}{\pi}} \tag{1}$$

Where r is the buffer radius in meters, N_{AL} is the N availability from livestock manure and sewage sludge at a farm level, N_{MAX} is the N maximum allowed load (170 kg/ha in NVZ, 340 kg/ha in the other zones), TA_{CS} is the total area of the cadastral sheet containing the farm, UA_{CS} is the Utilized Agricultural area within the cadastral sheet. Thus the circular geometries, representing the area where livestock manure and sewage sludge can potentially be spread, become larger with the increase of N_{AL} at farm level and of the TA_{CS}/UA_{CS} ratio that reflects the average availability of areas for sludge spreading. The location of the livestock farm in a NVZ, considering the minor N_{MAX}, as expected, increases the radius of the circle. Through this approach, in the absence of COM, each surface potentially used by farmers for livestock wastes spreading was calculated assuming an optimal management of available agricultural areas around each individual farm. Also in this case a SQL script perform a final overlay of the buffers and the other AOI in order to calculate the percentage of each buffer falling within the different AOI and CAS.

In order to develop the spatial analysis requested by the study several dynamic views based on SQL scripts were developed. The dynamic character of the views allowed to obtain indicators calculated in real time, updatable by changing the input tables or geometries.

2.2   Potential Nitrogen Load Assessment

N_{PL} was assessed considering the two main factors of the N balance: the N supply to crops (N_{SC}) and the N availability from livestock manure and sewage sludge (N_{AL}). Thus the estimation of potential crop-livestock loads and their geographic analysis were developed by three subsequent steps, corresponding to three different levels of analysis: a) N_{SC} estimation; b) N_{AL} estimation; c) Scenario analysis for N_{PL} assessment by comparisons and integration of N_{SC} and N_{AL} values.

N_{SC} Estimation

N_{SC} was computed at cadastral parcel level using the SIAN, PSR, PUA data, and coefficients from CCT, according to three different levels of progressive data integration and analysis:

I.   SIAN level: $N_{SC\,I}$ was calculated at CAS level by the following equation:

$$N_{SC\,I} = \sum_{1}^{n} CA_{SIAN\,n} CCT_{C\,n} \tag{2}$$

where n is the crop type, CA_{SIAN} is the n-th crop area retrieved from SIAN and CCT_C is cultural coefficients associated with conventional farming techniques;

II.   SIAN-PSR level: $N_{SC\,II}$ was calculated by equation (3), which sums $N_{SC\,OI}$ (4), only available for those parcels classified as integrated/organic by PSR, with N_{SC} (2), calculated for the other areas not included in PSR.

$$N_{SC\,II} = N_{SC\,OI} + N_{SC\,I} \tag{3}$$

$$N_{SC\,OI} = \sum_{1}^{n} CA_{SIAN\,n} CCT_{OI\,n} \tag{4}$$

where n is the crop type, CA_{SIAN} is the n-th crop area retrieved from SIAN and CCT_{OI} is cultural coefficients associated with organic or integrated farming techniques.

III. SIAN-PSR-PUA level: $N_{SC\,III}$ was calculated by (5), which considers the N supply declared by farmers for crop areas included in PUA where available. To this quantities was added the $N_{SC\,OI}$ where available, and $N_{SC\,I}$ for other areas.

$$N_{SC\,III} = \sum N_{PUA} + N_{SC\,OI} + N_{SC\,I} \tag{5}$$

This level was implemented only partially due to incompleteness of available PUA data.

The three levels of N_{SC} were aggregated at CAS level for the years 2011, 2012, 2013. Considering the different total areas associated to the three years under investigation, the final N_{SC} indicators at CAS level was computed by means of a weighted average sum of the three annual values.

N_{AL} Estimation

Similarly to the N_{SC} calculation, three different levels of data integration were implemented to estimate the different quotas of total N_{AL}:

I.   COM/BDN/CBP level: $N_{AL\,I}$ was calculated at cadastral sheet level by (6) for those farms included in BDN where COM and CBP were available. This estimation was very accurate since considered both the declared N_{AL} at farm level and the detailed spread plans:

$$N_{AL\,I} = \sum C_{COM\,CBP} \times LIV_{BDN} \tag{6}$$

where $C_{COM\,CBP}$ are COM-CBP based coefficients at a farm level, while LIV_{BDN} is the total live weight calculated for each livestock typology. The information provided in COM and CBP, about the treatments performed on sludge and its final destination, were used to define additional coefficients at farm level for a more accurate N_{AL} calculation.

II.   BDN level: for each livestock farm not associated with COM and CBP $N_{AL\,II}$ was determined by (7). The spatial localization of this quota of N_{AL} was based on the buffering procedure described previously.

$$N_{AL\,II} = \sum C_{LCT} \times LIV_{BDN} \tag{7}$$

where C_{LCT} is an official nitrogen conversion coefficients from LCT, while LIV_{BDN} is the farm consistency

III. SEP level: $N_{AL\,III}$ was estimated at a cadastral sheet level by (8),

$$N_{AL\,III} = \sum C_{SEP} \times N_{SEP} \tag{8}$$

where N_{SEP} is N from SEP data and C_{SEP} is a coefficient retrieved from bibliography [33].

The three levels of N_{AL} were summed to obtain the total N_{AL} from livestock farming and sewage sludge spreading (9). To quantify the reliability and the update level of the final results, the calculation procedures were implemented in order to allow the classification of the final N_{AL} outputs according to the type of reared species, the year of the data, the source dataset (BDN-COM-CBP, BDN, SEP) as well as the localization method (cadastral sheets indicated in spreading plans or buffers).

$$N_{AL} = N_{AL\,I} + N_{AL\,II} + N_{AL\,III} \tag{9}$$

Scenario Analysis for N_{PL} Calculation

Since reliable and detailed information about the integration of livestock manure with chemical fertilizers have been missing, three scenarios characterized by different management strategies of N_{AL} (given $N_{SC\,II}$ levels) were hypothesized in order to assess, at the various geographic levels of interest, the total N_{PL} arising from the entire crop-livestock system:

- Sc1: $N_{PL\,I}$ is calculated as the higher value between N_{SC} and N_{AL};
- Sc2: $N_{PL\,II}$ is calculated by (10) if $N_{PL\,II} > N_{PL\,I}$, or by (11) if $N_{PL\,II} < N_{PL\,I}$;

$$N_{PL\,II} = N_{SC\,II} + \frac{1}{2}N_{AL} \qquad (10)$$

$$N_{PL\,II} = N_{PL\,I} \qquad (11)$$

- Sc3: $N_{PL\,III}$ is equal to the (12)

$$N_{PL\,III} = N_{SC\,II} + N_{AL} \qquad (12)$$

Sc1 is based on the hypothesis that farmers tend to prevent any risk of N deficiency to the crop, even exceeding with the fertilization rate. As a consequence, in case of high availability of livestock manures, even skilled farmers may be inclined to use them in overabundance as compared to the nutrient rates they would provide by using mineral fertilizers. However, this is an "optimal" scenario since in the different AOI, N_{AL} exceeds rarely, and only with very limited quantities, N_{SC}. Sc2 considers that sometimes farmers provide all the N demanded by crops through chemical fertilizers, taking into account only partially the nutrient already supplied by manure. This can be considered an "intermediate" scenario. Sc3 is the most pessimistic hypothesis and probably causes an over-estimation of total potential load at AOI level, since it hypothesizes that additional N amounts, provided with chemical fertilizers are calculated without considering the supply from N_{AL}. The scenario Sc3 represents the maximum potential impact associated with agricultural and livestock production and, as such, it is actually the worst case of our simulations.

3   Results

According to the objectives, AGUA is able to calculate all the indicators either for the different administrative units (cadastral sheets, municipalities, provinces, entire region) or the other geographical areas of interest (NVZ, bodies of groundwater, subbasins). To show such capability, the potential N_{SC} and N_{AL} per hectare of utilized agricultural area (UAA) are here mapped for three different levels of interest (Fig. 3). The scenarios for the analysis of total crop-livestock N_{PL} are reported only at subbasin level (Fig. 4). The results are always represented geographically using classes corresponding to the quartiles of the distribution and upper outliers. The extreme values of distributions are identified by robust statistical methods that can be considered reliable even in case of skewed distributions [34,35].

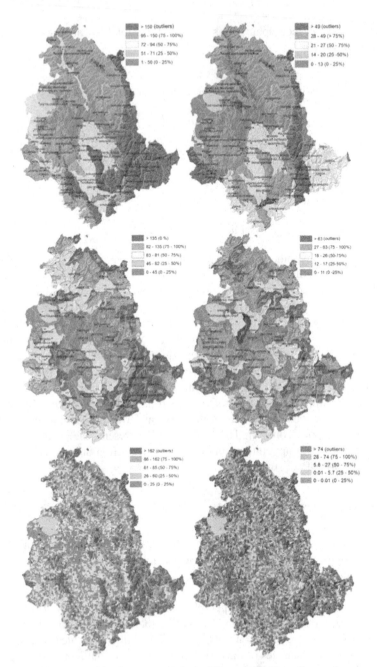

Fig. 3. N crop supply (SIAN-PSR level) (*left*) and N available from livestock manure (*right*) assessed at body of groundwater *(top)*, sub-basin *(center)*, and cadastral sheet *(bottom)* level in Umbria

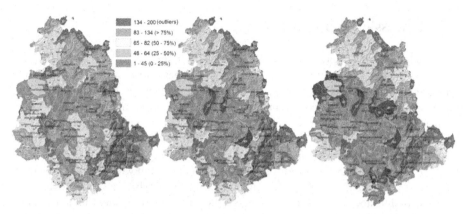

Fig. 4. Potential N loads (Kg/ha) in scenario 1 (*left*), 2 (*middle*), and 3 (*right*) in the Umbrian body of surface water. *The class division is that statistically defined for scenario 1.*

The largest N amounts used per hectare of UAA are localized, as expected, in the plain and low hills areas of the region. At the cadastral sheet level, the greatest reductions in the nutrient quantities due to the application of agri-environmental schemes shows a good correlation with the agricultural areas where they were applied (Pearson's r = 0.65, p <0.001), but the entity of such reduction is obviously linked to crop types. The analysis of the N availability from livestock and sewage sludge spreading allows highlighting only one upper outlier sub-basin where the higher concentration of swine and cattle breeding farms can be observed. However also the sub-basins included in 4th quartile include areas characterized by high livestock farming intensity.

According to the first scenario, the potential N loads are generally equal to the N crop supplies (estimated for the SIAN-PSR level), except in few cases where a surplus of supply from livestock N availability was found. As a consequence, in overall and relative terms, the classification of sub-basins belonging to the 75th percentile follows, with few exceptions, the N crop supply. Sc2 and Sc3 allow the effective identification of contexts where an inappropriate assessment of the fertilising power of livestock manure, and the consequent surplus of chemical fertilizers provided to crops, may generate relevant environmental pressures.

In the light of the criteria used for the definition of these three scenarios, the highest increase of the total potential loads, going from the "best" to "worst" scenario, is observed clearly in the sub-basins where the livestock N availability is higher. Sc2 and Sc3, despite based on a simplified approach, show situations that results not far away from being impossible, as very often livestock wastes, are disposed in high quantities in lands adjacent to the farms, where, in many cases, mineral fertilizers were spread too, without considering the real amount of N supplied by livestock manure. These practices, besides being economically inefficient, generate evident environmental issues related to excessive nutrients loadings.

4   Discussion and Conclusions

As already pointed out previously [17], [36,37,38,39], the role of agri-environmental indicators, characterized by suitable spatial and temporal resolution, is increasingly important to support decision making. In this regard, the methodology implemented in this study allowed to effectively integrate, for the first time and at regional level, detailed crop-livestock data, coming from very different sources by means of an innovative multidisciplinary approach that brought together agronomic, livestock, environmental and geo-informatics knowledge. The results allow to effectively identifying, at different scales and levels of analysis, the contexts in which the N loads from crop-livestock systems are potentially high, and the areas where a non-rational management of livestock wastes could generate the most relevant environmental pressures. The calculation procedures applied in the study allow producing additional information useful to assess the reliability and the update level of the crop-livestock indicators, even if a further integration of reliable methodologies for assessing input and output data quality should be developed.

The value of this spatial DB is also demonstrated by the fact that it is currently being used by the Regional Agency for Environmental Protection in Umbria, as a standard tool not only to support the WPP development, but also to monitor water resources in a more effective way, both in the short and medium terms. In this regard the system could be further improved to implement a more reliable DSS starting from the development of a more user-friendly interface. It will be essential in the future, to integrate an agro-environmental simulation model (e.g. AGNPS, CROPSYST, SWAT) paired with the spatial database developed in the study, in order to better estimate the quantities of nutrients coming from crop-livestock non-point sources, also considering reliable pedo-climatic spatial data. This would allow to improve significantly the variety and accuracy of indicators calculated by the system. With the aim to increase computational performance of AGUA and to allow a more interactive and intuitive exploration of data, it would be interesting to integrate the new SOLAP (Spatial On-Line Analytical Processing) technologies. As shown in different previous applications (e.g. [40,41]), SOLAP supports a faster and more straightforward spatio-temporal exploration and analysis of data by means of a multidimensional approach and different aggregation strategies available in cartographic, tabular, and diagram displays [42].

As highlighted by different authors (see e.g. [43,44]), the regulatory apparatus, information systems, institutions and organizations, and control procedures play a fundamental role in the analysis and monitoring of the crop-livestock system, and in the overall verification of its agri-environmental sustainability. However these components, despite being effective and relying on technologically advanced systems, will never be able to completely prevent or highlight improper behaviour of individual farm operators which play an extremely important role from the landscape and environmental point of view, often without a true consciousness. In the near future a crucial match will be definitely played on the side of the environmental protection in managing crop-livestock activities, environment and landscape, not only in the early stages of implementation of the tools available to institutions, organizations and professionals, but also in terms of communication, training, participation for improving people's agro-environmental awareness.

Acknowledgements. This research was developed within the framework of a project entitled *"Analisi delle pressioni esercitate sulla matrice acqua dalle attività agro-zootecniche nel territorio della regione Umbria"* funded by ARPA Umbria (Agreement between ARPA Umbria and DUT/Department of Agricultural, Food, and Environmental Sciences, University of Perugia, 2 may 2013).

References

1. EEA (European Environment Agency): Europe's environment. The fourth assessment. State of the Environment Report no. 1 (2006). Office for Official Publications of the European Communities, Luxembourg (2007)
2. Martinez, J., Dabert, P., Barrington, S., Burton, C.: Livestock waste treatment systems for environmental quality, food safety, and sustainability. Bioresource technology **100**, 5527–5536 (2009)
3. Bonazzi, G., Fabbri, C., Valli, L.: Allevamenti a basso impatto ambientale. CRPA, Centro Ricerche Produzioni Animali, Regione Emilia Romagna. Informatore Agrario Edizioni, Bologna (2003)
4. EEA (European Environment Agency): EEA Signals 2009, key environmental issues facing Europe. Office for Official Publications of the European Communities, Luxembourg (2009)
5. Halberg, N., Van der Werf, H.M.G., Basset-Mens, C., Dalgaard, R., de Boer, L.J.M.: Environmental assessment tools for the evaluation and improvement of European livestock production systems. Livestock Production Science **96**, 33–50 (2005)
6. Steinfeld, H., Mooney, H.A., Schneider, F., Neville, L.: Livestock in a changing landscape, vol. 1: Drivers, consequences, and responses. Island Press, Washington DC (2010)
7. Burton, C., Martinez, J.: Contrasting the management of livestock manures in Europe with the practice in Asia: What lessons can be learnt? Outlook on Agriculture **37**, 195–201 (2008)
8. Mantovi, P., Piccinini, S., Baldoni, G.: Fanghi di depurazione, gli effetti a lungo termine su colture e terreni. Agricoltura, Regione Emilia-Romagna (2005)
9. Steinfeld, H., Gerber, P., Wassenaar, T., Castel, V., Rosales, M., de Haan, C.: Livestock's long shadow: environmental issues and options. FAO (Food and Agriculture Organization), Rome (2006)
10. Sutton, M., Howard, C., Erisman, J.: The European nitrogen assessment: sources, effects and policy perspectives. Cambridge University Press, Cambridge (2011)
11. Teira-Esmatges, M.R., Flotats, X.: A method for livestock waste management planning in NE Spain. Waste management **23**, 917–932 (2003)
12. EEA (European Environment Agency): Agriculture and environment in EU-15. The IRENA indicator Report. EEA (European Environment Agency) Report no. 6/2005, Office for Official Publications of the European Communities, Luxembourg (2005)
13. FAO (Food and Agriculture Organization): The state of food and agriculture. Livestock in the balance. FAO, Food and Agriculture Organization, Rome (2009)

14. Benincasa, P., Guiducci, M., Tei, F.: The nitrogen use efficiency: meaning and sources of variation - Case studies on three vegetable crops in central Italy. Horttechnology **21**, 266–273 (2011)

15. Tosti, G., Benincasa, P., Farneselli, M., Tei, F., Guiducci, M.: Barley–hairy vetch mixture as cover crop for green manuring and the mitigation of N leaching risk. European Journal of Agronomy **54**, 34–39 (2014)

16. Farneselli, M., Benincasa, P., Tosti, G., Pace, R., Tei, F., Guiducci, M.: Nine-year results on maize and processing tomato in an organic and in a conventional low input cropping system. Italian Journal of Agronomy **8**, 9–13 (2013)

17. Brouwer, F.: Nitrogen balances at farm level as a tool to monitor effects of agri-environmental policy. Nutrient cycling in Agroecosystems **52**, 303–308 (1998)

18. Oenema, O., Kros, H., De Vries, W.: Approaches and uncertainties in nutrient budgets: implications for nutrient management and environmental policies. European Journal of Agronomy **20**, 3–16 (2003)

19. Parris, K.: Agricultural nutrient balances as agri-environmental indicators: an OECD perspective. Environmental Pollution **102**, 219–225 (1998)

20. Oenema, O.: Nitrogen budgets and losses in livestock systems. International Congress Series **1293**, 262–271 (2006)

21. Vizzari, M., Modica, G.: Environmental Effectiveness of Swine Sewage Management: A Multicriteria AHP-Based Model for a Reliable Quick Assessment. Environmental management **52**, 1023–1039 (2013)

22. APAT (Italian Agency for Environmental Protection and Technical Services): L'inquinamento da nitrati di origine agricola nelle acque interne in Italia. APAT (Italian Agency for Environmental Protection and Technical Services) Report no. 50/2005 (2005)

23. Ju, S., DeAngelis, D.L.: Nutrient fluxes at the landscape level and the R* rule. Ecological Modelling **221**, 141–146 (2010)

24. Cingolani, L., Charavgis, F.: Monitoraggio qualitativo dei corsi d'acqua superficiali individuati nel Piano Stralcio per il Lago Trasimeno. Regione Umbria (2004)

25. Kesner, B.T., Meentemeyer, V.: A regional analysis of total nitrogen in an agricultural landscape. Landscape Ecology **2**, 151–163 (1989)

26. Öborn, I., Edwards, A., Witter, E., Oenema, O., Ivarsson, K., Withers, P.J.A., Nilsson, S.I., Richert Stinzing, A.: Element balances as a tool for sustainable nutrient management: a critical appraisal of their merits and limitations within an agronomic and environmental context. European Journal of Agronomy **20**, 211–225 (2003)

27. Provolo, G., Riva, E., Serù, S.: Gestione e riduzione dell'azoto di origine zootecnica. Soluzioni tecnologiche e impiantistiche. Quaderni della ricerca. ERSAF (Ente Rregionale per I Servizi all'Agricoltura e alle Foreste). Regione Lombardia (2007)

28. Goodchild, M.F., Parks, B.O., Steyaert, L.T. (eds.): Environmental Modeling with GIS. Oxford University Press, Oxford (1993)

29. Andreu, J., Capilla, J., Sanchís, E.: AQUATOOL, a generalized decision-support system for water-resources planning and operational management. Journal of Hydrology **177**, 269–291 (1996)

30. Mysiak, J., Giupponi, C., Rosato, P.: Towards the development of a decision support system for water resource management. Environ. Model. Softw. **20**, 203–214 (2005)

31. McLain, R., Poe, M., Biedenweg, K., Cerveny, L., Besser, D., Blahna, D.: Making Sense of Human Ecology Mapping: An Overview of Approaches to Integrating Socio-Spatial Data into Environmental Planning. Human Ecology **41**, 651–665 (2013)

32. Coutinho-Rodrigues, J., Simão, A.: A GIS-based multicriteria spatial decision support system for planning urban infrastructures. Decision Support Systems **51**(3), 720–726 (2011)

33. CRPA (Centro Ricerche Produzioni Animali) L'uso dei fanghi di depurazione, Agricoltura 2/2009, 53–66, Regione Emilia-Romagna (2009)
34. Tukey, W.: Exploratory Data Analysis. Addison-Wesley (1977)
35. Vandervieren, E., Hubert, M.: An adjusted boxplot for skewed distributions. In: Antoch, J. (ed.) Proceedings in Computational Statistics 1933–1940. Springer-Verlag, Heidelberg (2004)
36. Yli-Viikari, A., Hietala-Koivu, R., Huusela-Veistola, E., Hyvönen, T., Perälä, P., Turtola, E.: Evaluating agri-environmental indicators (AEIs) - Use and limitations of international indicators at national level. Ecological Indicators 7, 150–163 (2007)
37. Bodo, G., Tamburi, L.: Analisi delle modalità di utilizzo agronomico dei reflui zootecnici. Piano di Tutela delle Acque della Regione Umbria, Regione Umbria (2005)
38. Jain, D.K., Tim, U.S., Jolly, R.: Spatial decision support system for planning sustainable livestock production. Computers, Environment and Urban Systems 19, 57–75 (1995)
39. Mendes, A., Soares da Silva, E., Azevedo Santos, J. (eds.): Efficiency Measures in the Agricultural Sector. Springer Science+Business Media, Dordrecht (2013)
40. Bimonte, S., Bertolotto, M., Gensel, J., Boussaid, O.: Spatial OLAP and Map Generalization. International Journal of Data Warehousing and Mining 8, 24–51 (2012)
41. Boulil, K., Le Ber, F., Bimonte, S., Grac, C., Cernesson, F.: Multidimensional modeling and analysis of large and complex watercourse data: an OLAP-based solution. Ecological Informatics 24, 90–106 (2014)
42. Rivest, S., Bédard, Y., Marchand, P.: Towards better support for spatial decision-making: defining the characteristics of Spatial On-Line Analytical Processing (SOLAP). Geomatica, the Journal of the Canadian Institute of Geomatics 55, 539–555 (2001)
43. Pennacchi, F., Cortina, C., Massei, G., Vizzari, M.: Valutazione del programma agro-ambientale della Regione Umbria – Studio di una procedura di valutazione. Department of Economic and Estimative Sciences, University of Perugia, Perugia (2001)
44. Vizzari, M., Mennella, V., Maraziti, F.: Rischio ambientale nel bacino del lago Trasimeno. Vulnerabilità del territorio e impatti legati alla gestione dei liquami suinicoli. Faculty of Agriculture, University of Perugia, Perugia (2008)

An Empirical Multi-classifier
for Coffee Rust Detection in Colombian Crops

David Camilo Corrales[1,3(✉)], Apolinar Figueroa[2],
Agapito Ledezma[3], and Juan Carlos Corrales[1]

[1] Grupo de Ingeniería Telemática, Universidad del Cauca,
Campus Tulcán, Popayán, Colombia
{dcorrales,jcorral}@unicauca.edu.co
[2] Grupo de Estudios Ambientales, Universidad del Cauca,
Carrera 2 no. 1A-25 - Urbanización Caldas, Popayán, Colombia
apolinar@unicauca.edu.co
[3] Departamento de Ciencias de la Computación e Ingeniería,
Universidad Carlos III de Madrid, Avenida de la Universidad 30, 28911, Leganés, Spain
ledezma@inf.uc3m.es

Abstract. Rust is a disease that leads to considerable losses in the worldwide
coffee industry. In Colombia, the disease was first reported in 1983 in the
department of Caldas. Since then, it spread rapidly through all other coffee departments in the country. Recent research efforts focus on detection of disease
incidence using computer science techniques such as supervised learning algorithms. However, a number of different authors demonstrate that results are not
sufficiently accurate using a single classifier. Authors in the computer field
propose alternatives for this problem, making use of techniques that combine
classifier results. Nevertheless, the traditional approaches have a limited performance due to dataset absence. Therefore we proposed an empirical multi-
classifier for coffee rust detection in Colombian crops.

Keywords: Coffee rust · Classifier · Multi-classifier · Dataset

1   Introduction

Rust is the main disease that attacks the coffee crop and it causes losses up to 30%
in susceptible varieties of Arabica Coffee species in Colombia. This disease is
found in most of the world's coffee-producing countries, and it was first reported in
Colombia in 1983 in the department of Caldas. Since then, it spread rapidly through
other coffee departments in the country [1]. Thenceforth The National Centre for
Coffee Research (Cenicafé) supplied Colombian coffee farmers with the Castillo
variety. This variety incorporates genetic attributes of rust resistance, which
improves grain size, quality and productivity compared to the Caturra variety.
However, despite of disease resistant plants availability, three-quarters of the area
in Colombia planted with coffee varieties is still susceptible. Meaning that plants

O. Gervasi et al. (Eds.): ICCSA 2015, Part I, LNCS 9155, pp. 60–74, 2015.
DOI: 10.1007/978-3-319-21404-7_5

are vulnerable to rust attack, depending on environmental conditions and crop agronomy [2].

Since coffee rust has led to considerable losses in the industry worldwide, recent Brazilian supervised learning researchers have focused on detection of the incidence of the disease using simple classifiers as decision trees, support vector machines and bayesian networks [3, 4, 5, 6, 7, 8]. They made use of numerical values of the infection rates which were mapped into two categories (or classes). The first option of the binary infection rate was with value 1 for infection rates equal or greater than 5 percentage points (pp) and 0 otherwise. The second option was created, with value 1 for infection rates equal or greater than 10 pp, and 0 otherwise.

Meanwhile experts in computer science demonstrated that using a simple classifier is not accurate enough [9]. This indicates that approaches such as those mentioned above [3, 4, 5, 6, 7, 8], which address the rust incidence rate detection using simple classifiers, lack of accuracy needed for predictions. Authors in the computer field suggested as an alternative solution to make use of techniques that combine classifier results [10, 11]. Nevertheless the traditional approaches have a limited performance (Bagging, Random subspace, Rotation forest, Stacking, inter alia) due to dataset absence to construct accurate classifiers [5, 6, 12, 13, 14].

Therefore, we proposed an empirical multi-classifier for coffee rust detection in Colombian crops. The remainder of this paper is organized as follows: Section 2 describes the data collection and the algorithms used; Section 3 the algorithms used in the multi-classifier proposed; Section 4 presents results and discussion and Section 5 conclusions and future work.

2   Background

This section describes the data collection process and the generation of the dataset used in experiments, introduces algorithms which comprise the multi-classifier.

2.1   Data Collection

The data used in this work were collected [12] trimonthly for 18 plots, closest to weather station at the Technical Farm (Naranjos) of the Supracafé, in Cajibio, Cauca, Colombia (21°35'08"N, 76°32'53"W), during the last 3 years (2011-2013). The dataset includes 147 examples from the total of 162 available ones. The remaining 15 samples were discarded due to problems in the collection process.

The dataset is composed of 13 attributes that are divided in 3 categories: Weather conditions (6 attributes), Physic crop properties (3 attributes), and crop management (4 attributes). Below are describe the 13 attributes (Table 1):

Table 1. Dataset for Incidence Rate of Rust Detection

Attributes for Incidence Rate of Rust Detection	
Weather Conditions	Relative humidity average in the last 2 months (RHA2M), Hours of relative humidity > 90% in the last month (HRH1M), Temperature variation average in the last month (TVA1M), Rainy days in the last month (RD1M), Accumulated precipitation in the last 2 months (AP2M), Nightly accumulated precipitation in the last month (NAP1M).
Physic Crop Properties	Coffee Variety (CV), Crop age (CA), Percentage of shade (PS).
Crop Management	Coffee rust control in the last month (CRC1M), Coffee rust control in the last 3 months (CRC3M), Fertilization in the last 4 months (F4M), Accumulated coffee production in the last 2 months (ACP2M).

In this sense, the class was defined as, the Incidence Rate of Rust (IRR). IRR is calculated by following a unique methodology in Colombian coffee crops collection developed by Cenicafé [2] for a plot with area lower or equal of one hectare. The steps of the methodology are presented below:

1. The farmer must be standing in the middle of the first furrow and he has to choose one coffee tree and pick out the branch with greater foliage for each level (high, medium, low); the leaves of the selected branches are counted as well as the infected ones for rust.

2. The farmer must repeat the step 1 for every tree in the plot until 60 trees are selected. Take in consideration that the same number of trees must be selected in every furrow (e.g. if plot has 30 furrows, the farmer selects two coffee trees for each furrow).

3. Finished the step 1 and 2, the leaves of the coffee trees selected (LCT) are added as well as the infected leaves of rust (ILR). Later it must be computed the Incidence Rate of Rust (IRR) using the following formula:

$$IRR = \frac{ILR}{LCT} \, 100 \qquad (1)$$

The collection process and IRR computation spend large amount of money and time, for this reason the IRR samples are limited (trimonthly for 18 plots). This process and its samples are considered very important, since it provides coffee crops rust approximation.

2.2   Supervised Learning Techniques

To evaluate the empirical multi-classifier for coffee rust detection in Colombian crops were determined the three following base classifiers: Backpropagation neural network (BPNN), Regression Tree (M5) and Support Vector Regression (SVR). This section provides a short description of base classifiers mentioned above and briefly reviews the four main ensemble methods, including: Bagging, Random subspace, Rotation forest and Stacking.

Base Classifiers
These classifiers learn by examples that map input vectors into one of several desired output classes. That is, a pattern classifier can be created through the training or learning process. The learning process of creating a classifier is to calculate the approximate distance between input–output examples and make correct output labels of the training set. This process is called the model generation phase. When the model is generated, it can classify an unknown instance into one of the learned classes in the training set [15]. Below are presented the base classifiers of the empirical multi-classifier for coffee rust detection

Backpropagation neural network
Backpropagation neural network (BPNN) is a feed forward neural network used to capture the relationship between the inputs and outputs [16]. The neural network is trained using backpropagation algorithm [17], where the error in the output layer is propagated backwards to adjust the weights in the hidden layers. The error in neuron p in the hidden layer is obtained using

$$\delta_p = O_p(1 - O_p) \sum_q \delta_q W_{pq}(n + 1) \tag{2}$$

The error δ_p is used to adjust the weights connecting to neuron p in the hidden layer. This process is repeated for all the hidden layers. Application of all inputs once to the network and adjusting the weights is called an epoch [18]. In this work, a three layer feedforward neural network is used with a learning rate $\alpha = 0.3$ and momentum applied to the weights during updating of $\beta = 0.2$.

Regression Tree
M5 algorithm constructs a regression tree by recursively splitting the instance space. The splitting condition is used to minimize the intra-subset variability in the values down from the root through the branch to the node. The variability is measured by the standard deviation of the values that reach that node from the root through the branch [19]. The standard deviation reduction (SDR) is calculated as follows

$$SDR = sd(T) - \sum_i \frac{|T_i|}{|T|} \; x \; sd(T_i) \tag{3}$$

Where T, is the set of examples that reach the node, T_i are the sets that are resulted from splitting the node according to the chosen attribute and sd is the standard deviation. We defined the minimum proportion of the variance on all the data that needs to be present at a node of 0.001.

Support Vector Regression
Support Vector Regression (SVR) is a supervised learning algorithm based on statistical learning theory and structural risk minimization principle [20, 21]. It can be expressed as the following equation:

$$f(x) = w^t \varphi(x) + b \tag{4}$$

Where $\varphi(.)$ is a non-linear mapping which takes the input data points into a higher dimensional feature space, w is a vector in the feature space and b is a scalar threshold [22]. For our multi-classifier was used a Gaussian radial basis function, soft margin parameter $C = 5.0$ and insensitive cost function parameter $\epsilon = 0.01$.

Ensemble Methods
An ensemble of classifiers is a collection of several classifiers whose individual decisions are combined in some way to classify the test examples [23]. In the literature, there are a number of ensemble methods, e.g. Bagging, Random subspace, Rotation forest, Stacking, Cascading, Boosting, etc. Next are presented the four main ensemble methods.

Bagging
In its standard form, the bagging (Bootstrap Aggregating) algorithm [24] generates M bootstrap samples $T_1, T_2, \ldots, T_M$ randomly drawn (with replacement) from the original training set T of size n. From each bootstrap sample T_i (also of size n), a base classifier C_i is induced by the same learning algorithm. Predictions on new observations are made by taking the majority vote of the ensemble C^* built from $C_1, C_2, \ldots, C_M$. As bagging resamples the training set with replacement, some instances may be represented multiple times while others may be left out. Since each ensemble member is not exposed to the same set of instances, they are different from each other. By voting the predictions of each of these classifiers, bagging seeks to reduce the error due to variance of the base classifier [25].

Random subspace
The random subspace method (RSM) is an ensemble construction technique, in which the base classifiers $C_1, C_2, \ldots, C_M$ are trained on data sets $T_1, T_2, \ldots, T_M$ constructed with a given proportion of attributes picked randomly from the original set of features F. The outputs of the models are then combined, usually by a simple majority voting scheme. The author of this method suggested to select about 50% of the original features. This method may benefit from using random subspaces for both constructing

and aggregating the classifiers. When the data set has many redundant attributes, one may obtain better classifiers in random subspaces than in the original feature space. The combined decision of such classifiers may be superior to a single classifier constructed on the original training data set in the complete feature space. On the other hand, when the number of training cases is relatively small compared with the data dimensionality, by constructing classifiers in random subspaces one may solve the small sample size problem [25, 26].

Rotation forest

Rotation forest [27] refers to a technique to generate an ensemble of classifiers, in which each base classifier is trained with a different set of extracted attributes. The main heuristic is to apply feature extraction and to subsequently reconstruct a full attribute set for each classifier in the ensemble. To this end, the feature set F is randomly split into L subsets, principal component analysis (PCA) is run separately on each subset, and a new set of linear extracted attributes is constructed by pooling all principal components. Then the data are transformed linearly into the new feature space. Classifier C_i is trained with this data set. Different splits of the feature set will lead to different extracted features, thereby contributing to the diversity introduced by the bootstrap sampling [25].

Stacking

Stacking is an ensemble technique that uses a meta-learner for determining which classifiers are reliable and which are not. Stacking is usually employed to combine models built by different inducers. The idea is to create a meta-dataset containing a tuple for each tuple in the original dataset. However, instead of using the original input attributes, Stacking uses the classifications predicted by the base-classifiers as the input attributes. The target-attribute remains as in the original training-set. A test instance is first classified by each of the base-classifiers. These classifications are fed into a meta-level training-set to produce a meta-classifier. The meta-classifier that has been produced combines the different predictions into a final prediction. In order to avoid over-fitting of the meta-classifier, the instances used for training the base-classifiers should not be used to train the meta-classifier. Thus the original dataset should be partitioned into two subsets. The first subset is reserved to form the meta-dataset while the second subset is used to build the base-level classifiers. Consequently, the meta-classifier predictions reflect the true performance of base-level learning algorithms [28].

3   Empirical Multi-classifier for Coffee Rust Detection

The empirical multi-classifier for coffee rust detection is based on Cascade Generalization method, where are used sequentially a set of classifiers, at each step performing a modification of the original dataset [29]. In this manner, the main idea is focused on the use of multiple classifiers in such a way that each of the classifiers (BPNN, M5 and SVR) covers a different part of the dataset. All of this with the objective to integrate the classification results and produce the final classification (Fig. 1). In addition, we used the interquartile range and k-mean algorithms to improve the performance in the dataset.

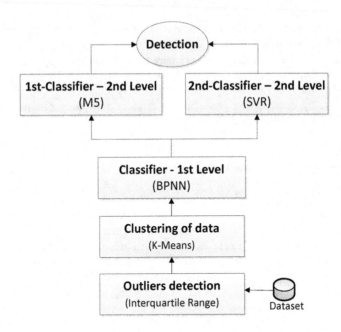

Fig. 1. Workflow for coffee rust detection

The *outliers detection module*, detects and removes the values that have an abnormal behavior of an attribute into the dataset through Interquartile Range Method (IR). This method compute the quartile Q_1 , Q_2 , Q_3 , which split a sort dataset in four parts [30], after, computing the interquartile range (IQR), which is the difference among the Third quartile (Q_3) and First quartile (Q_1). The IQR is a measure of noise for the data set. Points that are beyond the quartiles by half IQR's will be deemed potential outliers [31]. Below is presented the mathematic representation:

$$x < q_1 - 1.5 \, x \, IQR \ \lor \ x > q_3 + 1.5 \, x \, IQR; \ \forall \, x \ \in \ \mathbb{R} \qquad (5)$$

Where, x is the observation to evaluate and $q_1 - 1.5$ x IQR denote the lower inner fences and $q_3 + 1.5$ x IQR the upper inner fences. Hence points beyond these fences are potential outliers. In this case the Interquartile Range Method was applied to the class - Incidence Rate of Rust (IRR). The outcomes obtained to apply IR, presented as lower inner fence -2.0156 and upper inner fence 11.84, thus, the last 9 observations (139 - 147) are removed.

On the other hand, the *clustering of data module*, builds clusters from data set (leaving out the IRR class) with K-means algorithm (k=3). The k-means algorithm partitions a set of data into a number k of disjoint clusters by looking for inherent patterns in the set. Let us suppose that X represents the available set of samples. Each sample can be represented by an m-dimensional vector in the Euclidean space $\mathbb{R}^m$. Thus, in the following, $X = \{x_1, x_2, ..., x_3\}$ will represent a set of n samples, where the generic sample xi is a m-dimensional vector [32]. Each cluster is a subset of X and contains samples with some similarity. The distance between two samples provides a measure of similarity: it shows how similar or how different two samples are.

In the k-means approach, the representative of a cluster is defined as the mean of all the samples contained in the cluster [32].

Once the k-means is used each cluster is transformed in a data training set. The basic idea is to know the meaning of the clusters and define an expert classifier for each cluster. Next is explained the interpretation of the clusters through Bayesian Network and Decision Tree.

We used a Bayesian network [33] to build a conditional probability distribution for clusters generated by k-means (k=3) and four main attributes: coffee rust control in the last month (CRC1M), coffee rust control in the last 3 months (CRC3M), fertilization in the last 4 months (F4M) and incidence rate of rust (IRR), as can be seen in Table 2.

Table 2. Percentage of conditional probability distribution for clusters generated by k-means

Cluster	Coffee rust control in the last month		Coffee rust control in the last 3 months		Fertilization in the last 4 months		Incidence Rate of Rust	
	Yes	Not	Yes	Not	Yes	Not	<7.18 %	≥7.18%
C_1	45.3%	54,7%	98.3%	1.7%	32.8%	67.2%	75%	25%
C_2	1%	99%	27.9%	72.1%	20.2%	79.8%	66.66 %	33.33 %
C_3	88.6%	11.4%	58.8%	41.2%	99.1%	0.9%	98.70%	1.30%

For cluster C_1, CRC1M and CRC3M have opposite probability distribution (54.7% indicates that it was not done a coffee rust control in the last month while 98.3 % shows that it was done a coffee rust control in the last 3 months), which means that C_1 contains contradictory instances. For another part, for C_2, the attributes of CRC1M, CRC3M and F4M present a similar behavior, indicating that it was not done rust controls and fertilizations on coffee crops (probability distribution of "Not" values: 99%, 72.1% and 79.8% respectively), whereas probability distribution of CRC1M, CRC3M and F4M attributes of C_3 indicates the use of coffee rust controls and fertilizations (probability distribution of "Yes" values: 88.6%, 58.8% and 99.1% respectively); for this reason the Incidence Rate of Rust is less than 7.18% (probability distribution of IRR < 7.18 % = 98.70%).

To test out the outcomes obtained by Bayesian Network, we used a C4.5 Decision Tree (pruning the irrelevant attributes) [34], as can be seen in Fig. 2.

In Fig. 2, the C4.5 decision tree accounted 3 attributes: Coffee rust control in the last month (CRC1M), Hours of relative humidity > 90% in the last month (HRH1M), and Crop Age (CA) in months. The distribution of instances is founded in the leaves. In this sense the rule obtained for cluster C_1 (Fig. 2) does not contain the necessary attributes to know the meaning of C_1 (CRC1M and HRH1M); unlike of C_2 where conditions promote the appearance of rust, because there was not done a coffee rust control in the last month, high hours of relative humidity > 90% in the last month (> 341 hours), and older crops (age > 50 months). Finally C_3 can be interpreted as the youngest crops (age < 48 months) that are resistant to rust without regard for conditions of relative humidity and coffee rust controls performed.

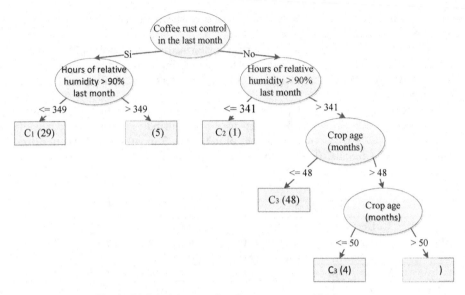

Fig. 2. C4.5 decision tree for clusters generated by k-means

Based on the foregoing, we interpret the C_2 cluster as the cases where conditions induce high losses in crops caused by rust (IRR $\geq 7.18\%$), while C_3 cluster presents the cases with low risk of losses in crops (IRR $< 7.18\%$).

To choose the k suitable, the k-means algorithm was tested with k $=$ 2, 3, 4. The clusters are displayed with CLUSPLOT graphical library, which creates a bivariate plot visualizing a partition (clustering) of the data [24]. All observation (instances) are represented by points in the plot, using Principal Components Analysis (relative to the first two principal components) [27]. The clusters $\{C_1, C_2, ..., C_n\}$; k $=$ n, are again represented as ellipses, which are based on the average and the covariance matrix of each cluster; and their size is such that they contain all the points of their cluster (Fig. 3).

Fig. 3. k-means algorithm with k = 2, 3, 4 displayed with CLUSPLOT

The Fig. 3a presents the k-means algorithm with k $=$ 2, where the points of C_1 (red color) has misclassified in correspondence to C_2 (blue color), in this manner we cannot differentiate the C_1 from the C_2. On the other hand, k $=$ 3 (Fig. 3b), C_2 (pink color) and C_3 (blue color) are completely distinct. However, most of the points of C_1 (red color) was misclassified as C_3. Lastly, k = 4 (Fig. 3c), C_1 (red color), C_2 (blue color) and C_4 (pink color) are overlapped, while the C_3 (green color) differ of the other clusters.

In this sense, we define the k-means algorithm with $k = 3$ (Fig. 3b), because its clusters C_2 (pink color) and C_3 (blue color) are completely distinct. C_1 (red color) has misclassified points due to its belonging to C_2 and C_3, besides the contradictory instances found by Bayesian Network and C.4 Decision Tree. For this reasons, the observations of C_1 were deleted (29 instances) to avoid the noise. The new dataset has 109 instances, 52 of C_2 and 57 of C_3.

When the clusters are defined we create the three classifiers (In Fig. 1 *classifier - 1st level,* and *classifiers - 2nd level*):

The *classifier - 1st level* use a Backpropagation neural network (BPNN) responsible for deciding which classifier of second level will detect the incidence rate of rust (IRR). The *1st-classifier-2nd level* trains a regression tree (M5) with instances of C_2 to detect the incidence rate of rust greater than 7.18%; whereas the *2nd-classifier-2nd level* trains a support vector regression (SVR) with instances of C_3 with aim to detect the incidence rate of rust less than 7.18%.

4   Experimental Results

This section reports a number of experiments carried out to select the base classifiers used to detect the coffee rust incidence rate. Here we compare the results obtained by the empirical multi-classifier against classical approaches as: simple classifiers and ensemble methods.

4.1   Selection of Base Classifiers

The families of supervised learning algorithms assessed were: Support Vector Machines, Neural Networks, Bayesian Networks, Decision Trees, and K nearest neighbors. The selection criteria of these classifiers are based on previous surveys which show that are the most suitable for classification and predictions tasks [35, 36], especially in the detection of crops diseases and pest [37]. With the dataset introduced in Section 2.1, we used a 10-fold cross validation to estimate the scores reported in the following figure and tables.

Classifier - 1st Level
We tested the most relevant algorithms of supervised learning for classification tasks as Support Vector Machine (SVM), Backpropagation Neural Network (BPNN), Naive Bayes (NB), C4.5 Decision Tree, and K nearest neighbors (K-NN) [37] to choose the *classifier - 1st level,* computing precision, recall and F- measure as seen in Table 3.

Table 3. Precision, Recall and F-measure for SVM, ANN, BN, DT and K-NN

Measures	Supervised learning algorithms				
	SVM	**BPNN**	**NB**	**C4.5**	**K-NN**
Precision	99.3%	99.3%	88.6%	96.3%	97.9%
Recall	99.3%	99.3%	87.4%	96.3%	97.8%
F- measure	99.3%	99.3%	87.5%	96.3%	97.8%

The supervised learning algorithms in Table 3 present values greater than 88.6% for Precision, 87.4% for Recall and 87.5% for F-measure, which indicates low false positives and good outcomes to classify new instances. However, these measures are insufficient to choose the *classifier - 1st level* inasmuch as SVM and BPNN have the same values (Precision, Recall and F-measure 99.3%). For this reason we used the ROC Curve for evaluating the SVM and BPNN such as shows in Fig. 4.

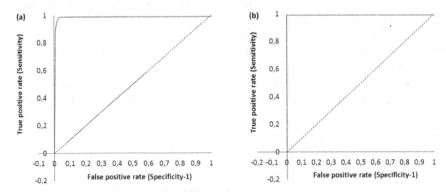

Fig. 4. Roc curves for (a) SVM (b) BPNN

In this regard, the curves showed a good performance of SVM (Fig. 4a) and BPNN (Fig. 4b) to classify correctly new instances, with an area under a curve of 99.1% for SVM and 100% of BPNN. Based on the above, the Backpropagation neural network was selected as *classifier - 1st level.*

Classifiers-2nd Level

We tested the main learning algorithms for prediction tasks as Support Vector Regression (SVR), Multilayer Perceptron (MP), Radial Basis Function Network (RBFN), K Nearest Neighbors Regression (K-NN R) and Regression Tree (M5) [37] to choose the *Classifiers - 2nd level* through Pearson's Correlation Coefficient (PCC), Mean Absolute Error (MAE) and Root Mean Squared Error (RMSE) as seen in Table 4 and Table 5.

Table 4. Comparison: SVR, MP, RBF, K-NN R, M5 for selection of *1st-classifier-2nd* level

Measures	Supervised learning algorithms				
	SVR	MP	RBF	K-NN R	M5
PCC	0.43	0.39	-0.19	0.32	0.51
MAE	3.17%	3.62%	3.50%	3.22%	1.83%
RMSE	3.81%	4.83%	4.36%	4.04%	2.16%

In Table 4, the M5 regression tree is approaching a directly proportional relation (PCC = 0.51) among Incidence Rate of Rust Detected (IRRP) and Incidence Rate of Rust Real (IRRR). In that manner M5 presents the least difference among IRRP and IRRR with MAE = 1.83% and RMSE = 2.16%. In accordance with the above, the M5 regression tree was selected as *1st-classifier-2nd level.*

Table 5. Comparison: SVR, MP, RBF, K-NN R, M5 for selection of 2nd-classifier-2nd level

Measures	Supervised learning algorithms				
	SVR	MP	RBF	K-NN R	M5
PCC	0.36	0.32	0.27	0.37	0.08
MAE	1.20%	1.22%	1.24%	1.31%	1.49%
RMSE	1.73%	1.91%	1.74%	1.86%	2.08%

In Table 5, the K-NN R and SVR algorithms have the highest value for positive correlations (PCC= 0.37 and PCC = 0.36 respectively), however, SVR present the least difference among IRRP and IRRR (MAE = 1.20% and RMSE = 1.73%) respect to K- NN R (MAE = 1.31% and RMSE = 1.86%). Based on the above, SVR was selected as *2nd-classifier-2nd level*.

4.2   Evaluation of the Empirical Multi-classifier for Coffee Rust Detection

This section presents the results obtained by the empirical multi-classifier against simple classifiers and classical ensemble methods.

Empirical Multi-classifier vs. Simple Classifiers
Table 6 compares the outcomes obtained by the empirical multi-classifier against simple classifiers as Support Vector Regression (SVR), Back Propagation Neural Network (BPNN), and Regression Tree (M5) which were tested in [12] with the same dataset.

Table 6. Comparison of empirical multi-classifier and simple classifiers

Measures	Supervised learning algorithms				
	Empirical multi-classifier		Simple classifiers		
	1st-classifier-2nd level	*2nd-classifier-2nd level*	SVR	BPNN	M5
PCC	0.51	0.36	0.29	0.35	0.22
MAE	1.83%	1.20%	2.28%	2.34%	2.55%
RMSE	2.16%	1.73%	3.38%	3.31%	3.50%

The outcomes obtained by empirical multi-classifier are better than simple classifiers; especially for *2nd-classifier-2nd level* where the instances are closer to each other respect to instances of *1st-classifier-2nd level* (Fig. 3 for k = 3).

Empirical Multi-classifier vs. Classical Ensemble Methods
Table 7 compares the outcomes obtained by the empirical multi-classifier against classical ensemble methods as Bagging, Random subspaces, Rotation forest and Stacking. The classical ensemble methods as Bagging used M5 as base classifier, Random subspaces: K- NN R, Rotation forest: M5, Stacking three base classifiers: BPNN, K- NN R, M5 and SVR as meta-learner. We choose the four ensemble methods as the best outcomes to use the dataset explained in section 2.1.

Table 7. Comparison of empirical multi-classifier and classical ensemble methods

| Measures | Supervised learning algorithms | | | | | |
| | Empirical multi-classifier | | Classical ensemble methods | | | |
	1st-classifier-2nd level	2nd-classifier-2nd level	Bag-ging	Ran. Sub-spaces	Rot. Forest	Stacking
PCC	0.51	0.36	0.27	0.25	0.24	0.14
MAE	1.83%	1.20%	2.38%	2.38%	2.43%	2.41%
RMSE	2.16%	1.73%	3.34%	3.52%	3.37%	3.43%

The outcomes obtained by empirical multi-classifier are better than classical ensemble methods. Bagging is the ensemble method with better results; nevertheless, simple classifiers as BPNN (PCC = 0.35; RMSE = 3.31%) and SVR (MAE = 2.28%) outperformed the results obtained by Bagging (PCC = 0.27; MAE = 2.38%; RMSE = 3.34%).

5   Conclusions and Future Work

This paper presented an empirical multi-classifier for coffee rust detection in Colombian crops. Our multi-classifier proposal outperformed the classical approaches as simple classifiers and ensemble methods in terms of PCC (0.51 of *1st-classifier-2nd level* and 0.36 of *2nd-classifier-2nd level* respect to 0.35 of BPNN and 0.27 of Bagging), MAE (1.83% of *1st-classifier-2nd level* and 1.20% of *2nd-classifier-2nd level* respect to 2.28% of SVR and 2.38% of Bagging) and RMSE (2.16% of *1st-classifier-2nd level* and 1.73% of *2nd-classifier-2nd level* respect to 3.31% of BPNN and 3.34% of Bagging) which use the same dataset of coffee rust. The limitation encountered during this study was the absence of data from actual coffee crop. Especially in rust incidence rate samples due to the expensive collection process that requires big efforts in money and time. Accordingly, the results obtained on this study are not very precise.

In future studies we intend to tackle the insufficient data using different approaches such as synthetic data and incremental learning. This will allow an existing classifier be updated using only new individual data instances, without having to re-process past instances [38]. Also we will propose the use of weather time series data which are automatically capture each five minutes by weather station. We will analyze its behavior with the rust infection rate.

Acknowledgments. The authors are grateful for the technical support of the Telematics Engineering Group (GIT), Environmental Study Group (GEA) of the University of Cauca, Control Learning Systems Optimization Group (CAOS) of the Carlos III University of Madrid, Supracafé, AgroCloud project of the RICCLISA Program and Colciencias for PhD scholarship granted to MsC. David Camilo Corrales.

References

1. Cristancho, M., Rozo, Y., Escobar, C., Rivillas, C., Gaitán, Á.: Razas de roya (2012)
2. Rivillas-Osorio, C., Serna-Giraldo, C., Cristancho-Ardila, M., Gaitán-Bustamante, A.: La roya del cafeto en Colombia, impacto, manejo y costos de control. Cenicafé (2011)
3. Luaces, O., Rodrigues, L.H.A., Alves Meira, C.A., Quevedo, J.R., Bahamonde, A.: Viability of an alarm predictor for coffee rust disease using interval regression. In: García-Pedrajas, N., Herrera, F., Fyfe, C., Benítez, J.M., Ali, M. (eds.) IEA/AIE 2010, Part II. LNCS, vol. 6097, pp. 337–346. Springer, Heidelberg (2010)
4. Luaces, O., Rodrigues, L.H.A., Alves Meira, C.A., Bahamonde, A.: Using nondeterministic learners to alert on coffee rust disease. Expert Systems with Applications **38**, 14276–14283 (2011)
5. Cintra, M.E., Meira, C.A.A., Monard, M.C., Camargo, H.A., Rodrigues, L.H.A.: The use of fuzzy decision trees for coffee rust warning in Brazilian crops. In: 2011 11th International Conference on Intelligent Systems Design and Applications (ISDA), pp. 1347–1352 (2011)
6. Pérez-Ariza, C.B., Nicholson, A.E., Flores, M.J.: Prediction of Coffee Rust Disease Using Bayesian Networks. In: Andrés Cano, MG.-O., Nielsen, T.D. (eds.) The Sixth European Workshop on Probabilistic Graphical Models. DECSAI, University of Granada, Granada (Spain) (2012)
7. Meira, C.A.A., Rodrigues, L.H.A., Moraes, S.: Modelos de alerta para o controle da ferrugem-do-cafeeiro em lavouras com alta carga pendente. Pesquisa Agropecuária Brasileira **44**, 233–242 (2009)
8. Meira, C., Rodrigues, L., Moraes, S.: Análise da epidemia da ferrugem do cafeeiro com árvore de decisão. Tropical Plant Pathology **33**, 114–124 (2008)
9. Li, L., Zou, B., Hu, Q., Wu, X., Yu, D.: Dynamic classifier ensemble using classification confidence. Neurocomputing **99**, 581–591 (2013)
10. Ranawana, R., Palade, V.: Multi-Classifier Systems: Review and a roadmap for developers. Int. J. Hybrid Intell. Syst. **3**, 35–61 (2006)
11. Ghosh, J.: Multiclassifier systems: back to the future. In: Roli, F., Kittler, J. (eds.) MCS 2002. LNCS, vol. 2364, pp. 1–15. Springer, Heidelberg (2002)
12. Corrales, D.C., Ledezma, A., Peña, A., Hoyos, J., Figueroa, A., Corrales, J.C.: A new dataset for coffee rust detection in Colombian crops base on classifiers. Sistemas y Telemática **12**, 9–22 (2014)
13. Jain, R., Minz, S.: Ramasubramanian: Machine Learning for Forewarning Crop Diseases. Journal of the Indian Society of Agricultural Statistics **63**, 97–107 (2009)
14. Korada, N.K., Kumar, N.S.P., Deekshitulu, Y.: Implementation of Naive Bayesian Classifier and Ada-Boost Algorithm Using Maize Expert System. International Journal of Information Sciences and Techniques (IJIST) **2** (2012)
15. Mitchell, T.: Machine learning. McGraw-Hill (1997)
16. Poh, H.L.: A neural network approach for marketing strategies research and decision support, vol. Ph.D. Thesis. Stanford University (1991)
17. Haykin, S.S.: Neural networks: a comprehensive foundation. Prentice Hall (2003)
18. Suhasini, A., Palanivel, S., Ramalingam, V.: Multimodel decision support system for psychiatry problem. Expert Systems with Applications **38**, 4990–4997 (2011)
19. Bonakdar, L., Etemad-Shahidi, A.: Predicting wave run-up on rubble-mound structures using M5 model tree. Ocean Engineering **38**, 111–118 (2011)
20. Vapnik, V.: The Nature of Statistical Learning Theory. Springer, New York (2000)

21. Vapnik, V.N.: An overview of statistical learning theory. IEEE Transactions on Neural Networks **10**, 988–999 (1999)
22. Balasundaram, S., Gupta, D.: Training Lagrangian twin support vector regression via unconstrained convex minimization. Knowledge-Based Systems **59**, 85–96 (2014)
23. Skurichina, M., Duin, R.P.: Bagging, boosting and the random subspace method for linear classifiers. Pattern Analysis & Applications **5**, 121–135 (2002)
24. Breiman, L.: Bagging Predictors. Machine Learning **24**, 123–140 (1996)
25. Marqués, A.I., García, V., Sánchez, J.S.: Two-level classifier ensembles for credit risk assessment. Expert Systems with Applications **39**, 10916–10922 (2012)
26. Tin Kam, H.: The random subspace method for constructing decision forests. IEEE Transactions on Pattern Analysis and Machine Intelligence **20**, 832–844 (1998)
27. Rodriguez, J.J., Kuncheva, L.I., Alonso, C.J.: Rotation Forest: A New Classifier Ensemble Method. IEEE Transactions on Pattern Analysis and Machine Intelligence **28**, 1619–1630 (2006)
28. Menahem, E., Rokach, L., Elovici, Y.: Troika – An improved stacking schema for classification tasks. Information Sciences **179**, 4097–4122 (2009)
29. Gama, J., Brazdil, P.: Cascade Generalization. Machine Learning **41**, 315–343 (2000)
30. McAlister, D.: The Law of the Geometric Mean. Proceedings of the Royal Society of London **29**, 367–376 (1879)
31. Grubbs, F.: Procedures for Detecting Outlying Observations in Samples. Technometrics **11**, 1–21 (1969)
32. Mucherino, A., Papajorgji, P., Pardalos, P.: Clustering by k-means. In: Du, D.-Z. (ed.) Data Mining in Agriculture, vol. 30, pp. 47–56. Springer, New York (2009)
33. Araujo, B.S.: Aprendizaje automático: conceptos básicos y avanzados: aspectos prácticos utilizando el software Weka. Pearson Prentice Hall, España (2006)
34. Quinlan, J.R.: C4.5: programs for machine learning. Morgan Kaufmann Publishers Inc (1993)
35. Kotsiantis, S.B.: Supervised machine learning: a review of classification techniques. In: Proceedings of the 2007 Conference on Emerging Artificial Intelligence Applications in Computer Engineering: Real Word AI Systems with Applications in eHealth, HCI, Information Retrieval and Pervasive Technologies, pp. 3–24. IOS Press (2007)
36. Bhavsar, H., Ganatra, A.: A Comparative Study of Training Algorithms for Supervised Machine Learning. International Journal of Soft Computing and Engineering (IJSCE) **2**, 74–81 (2012)
37. Corrales, D.C., Figueroa, A., Corrales, J.C.: Toward detecting crop diseases and pest by supervised learning. Revista Ingeniería y Universidad **19** (2015)
38. Schlimmer, J., Granger Jr, R.: Incremental learning from noisy data. Machine Learning **1**, 317–354 (1986)

Workshop on Approaches or Methods of Security Engineering (AMSE 2015)

A Fast Approach Towards Android Malware Detection

Hongmei Chi[(✉)] and Xavier Simms

Department of Computer and Information Sciences,
Florida A&M University, Tallahassee, FL 32307, USA
hchi@cis.famu.edu

Abstract. The proposed research compares the feasibility of three well known machine learning algorithms on the detection of malware on the Android platform. Once accuracy is at an acceptable level, these algorithms performance are further enhanced to decrease analysis time, which can lead to faster detection rates. The framework makes use of powerful GPU's (Graphics Processing Unit) in order to reduce the time spent on computation for malware detection. Utilizing MATLAB's parallel computing kit, we can execute analysis at a much higher speed due to the increased cores in the GPU. A reduced computation time allows for quick updates to the user about zero day malware, resulting in a decreased impact. With the increase in mobile devices unending, quick detection will become necessary to combat mobile malware, and with Android alone reaching its 50 billionth app downloads will be no small task.

Keywords: Malware · Vulnerability detection · Andriod · GPU · Data mining

1   Introduction

The popularity of smart phones has increased dramatically in the past decade; from 405 million devices sold in 2000 [2], to a projected 2.5 billion by 2016 [1]. Even with clusters of nodes, malware analysis can take more than thirty hours for a few thousand applications [4]. Global mobile traffic grew by 70 percent in 2012, with Android exceeding the bandwidth consumption of iOS. The smartphones of today are able to carry out almost any activity that was historically done using a personal computer. Mobile devices are being used for online shopping, paying bills, navigation, video conferencing, social networking, etc. In order for our smartphones to provide this vast array of functionality they use software designed to fulfill a specific purpose, called applications and commonly referred to as an "app". Apps can be created by software development companies, like traditional personal computer software or by individuals and independent developers.

The smartphone is arguably the most utilized device on the planet. It is used for a variety of task including some that require secure transmission, such as bank transactions and transfers. It is predicted that the phone market this year will grow to over 1.8 billion devices [3]. With the Android operating system predicted to hold the lion's share of the mobile market at 79% by the end of the year [4], mobile security on this platform is a high priority. Utilizing well refined machine learning algorithms,

© Springer International Publishing Switzerland 2015
O. Gervasi et al. (Eds.): ICCSA 2015, Part I, LNCS 9155, pp. 77–89, 2015.
DOI: 10.1007/978-3-319-21404-7_6

malware patterns can be identified across large data sets on application specific data-points. The algorithms are highly computation intensive and can take numerous hours to complete analysis on large datasets. Utilizing parallel computing, the computational workload can be moved to a GPU to increase performance and detection times. This results in faster user notification and a decreased infection footprint. By the end of this chapter you will understand the motivation and necessity for this research, and the challenges associated with securing nearly 2 billion mobile devices filled with personal data.

Traditional malware detection techniques such as, signature and behavior based detection require analysis of huge datasets that take too long for the results to be distributed to the user. The user will receive notification of new malware, after the application has been installed on the device, rendering such techniques useless. This paper will give a brief introduction to Android applications and the common malware that threatens everyday users. Current detection and mitigation methods of mobile malware will also be discussed, along with the proposed framework.

2   Related Work

Research in mobile security deterrence is a very active topic and security experts are working diligently to combat this rise of malware. So far there have been two mean approaches for detection of malware: static analysis and dynamic analysis. Static analysis refers to testing for vulnerabilities by examining the code without executing the application, while dynamic analysis focuses testing the application during run time. This section will focus on some of the related work on the detection of malware in Android applications using both methods, static and dynamic analysis.

Crowdroid[7] took an approach that involved going beyond the application layer of the Android platform. Instead they went down to the operating system level and extracted systems calls from the Linux kernel. The framework creates as many datasets as application identifiers, so the more Crowddroid users the better the database will be able to provide information on benign and malicious apps. Behavior patterns are then analyzed using clustering algorithms. Zhou et al[4] propose a less resource intensive approach in order to decrease the size of the parameters that need to be scanned and decrease the overall time needed to detect malware on large Android markets. The proposed DroidRanger leverages a crawler to collect Android apps from existing Android markets and save them to a local repository. Unlike the other solutions that we have seen, where key indicators include API Calls and service request, the authors of this paper rely on unique information that is embodied in each app.

While the above research does provides accurate detection of some malware variants, this alone is not enough to combat the consistent threat that mobile malware poses. The most challenging part of deterring malware is detecting new threats early and notifying the user base in a timely manner. In order to notify users before new malware can quickly propagate the amount of time complete analysis must be reduced. The proposed research provides a practical approach to reducing the time needed to complete analysis, using a algorithm that provides fast detection of apps that pose a threat to user data and ignoring those that do not. Each app that has the ability to access and transmit that receives a threat value of 'true' while those that do not have said abilities receive a threat value of 'false'. The apps that receive the positive negative result can be removed from the app pool that has to receive more thorough analysis, thus reducing the total time needed for analysis and updating users of potential threats while ignoring those that pose no threat.

3   Motivational Example

The increasing use of mobile devices as the main for all aspects of our daily communication and access to web services has created a huge vector for cyber-criminal activity. Numerous articles, research papers and news blogs [2,3,7] have all highlighted the fact that the trends of increasing mobile malware are prevalent and here to stay. As the popularity and capabilities of mobile device increased in the past years, so has the sophistication of malware for said devices. Traditionally PC based malware have made its way to the mobile arena. Malware such a Zues has morphed into the mobile version of Zitmo [3]. Zitmo main premise is the detection of mobile transaction numbers used to authorize money transfers through SMS messages. With more companies now issuing mobile devices such as; tablets and smartphones the risk to confidential business information has increased dramatically. The network administrator now needs to monitor company resources that have persistent connections to the internet and, is not usually on the premise. The proposed solution allows for network administrators to collect statistics on all of their mobile devices and receive updates on newly detected malware signatures. By allowing remote analysis of company mobile devices, the network administrator can be notified when a malicious app is installed and mitigation can be handled quickly, before exfiltration of sensitive data takes place. Take the corporate scenario below as an example;

- Each department issues mobile devices to select employees.
- Each device has the Android client app, that collects application data for analysis
- The device uploads a log file for desired applications to the remote server once per day.
- Analysis of the log file is carried out once per day and updated daily.
- Each device will receive warning of a perceived threat, if it is downloaded or installed.

This allows for the quick detection of malware on Android mobile devices. The network administrator can be notified to the existence of malware on company devices, and decide to disconnect such devices from the network before propagation occurs. Employees can be notified on apps that are not to be installed, creating a more secured digital environment. With mobile malware now being monetized, sensitive corporate data can be sold to the highest bidder, who may also be a competitor.

Although malware is typical on any mobile device, reports suggest that attackers predominately design malware aimed at the Android platform. In 2011 mobile malware grew by 155%, while from March 2012 to 2013 there was a 614% increase, with 92% of all malware targeting the Android platform [5]. With the number of devices shipped expected to surpass 2.6 billion units in 2016 [1], this presents a huge issue that must be addressed for the longevity of the Android platform and mobile computing in general.

4   Malware Detection

There are five main common malware detection techniques;

- Signature-based malware detection, where sequence of bytes within an application source code is scanned, in order to identify malicious code.
- Specification-based malware detection, makes use of a predefined set of rules that identifies normal behavior, any deviations outside the rule set can trigger an alarm.
- Behavioral-based detection: This approach generates a data-store of malicious behaviors by studying a distinct set of malware families and training a classifier to distinguish between malicious and innocent programs.
- Data mining malware detection: This method detects patterns in large amounts of data, such as user patterns in order to detect future instances. This approach must have a large training data set of malicious and legitimate applications in order to run data mining algorithms.
- Cloud Based Malware Detection: Cloud based detection allow for applications to be scanned securely in a cloud environment. Google makes used of cloud detection in its Play store. Google uses the Bouncer services to scan all applications for malware, prior to making it available on the Play market.

The proposed approach makes use of machine learning identify patterns in datasets produced by legitimate and benign application. Three machine learning algorithms are tested using a large data set of legitimate and spam messages as input. The method with the highest accuracy rate will then be used to group applications into clusters of malware and legitimate apps. This process will then be improved by using the GPU to carryout calculations, instead of the CPU. Results will be evaluated and presented in Section 8.

5   Approach

The proposed approach takes into consideration the time required to compute clustering analysis on large datasets of malware [4]. By the year 2016, sales forecast predict that there will be over 2.5 billion devices shipped [1]. With the increased number of devices and variations in malware families, security firms will have to carry out much high volumes of malware analysis. This approach proposes the use of GPU's to accelerate the clustering process and reduce the time it takes to detect new variants of malware through behavioral and signature statistics. A client side application will run on the user's phone, logging application specific data that will be stored in a database for analysis. APK (Android Application Packages) files will also be downloaded to a remote server, in order to retrieve detailed data about the installation file so that any changes made can be realized. The data variables collected will be stored on the remote server in a database. The database will be queried in order to create input files for analysis purposes; three types of analysis algorithms will be used to verify the accuracy and performance of the different algorithms. This approach is continuous in the sense that datasets must be updated frequently for effective identification of new and different malware variants.

Fig. 1. Displays the dataflow diagram of the proposed framework

This research aims to present data on detection of malware infested applications on the Android platform using clustering analysis. Clustering analysis of datasets are a computation intensive task and usually takes up large CPU and memory resources, this is can be countered using the multiple cores of GPU parallel processing. The main contributions of this research are:

- Offer data on the results of clustering analysis for malware detection on the Android platform

- Provide data on the performance of CPU-only, as compared to parallel computing with the aid of the GPU

- Provide data on the integrity of the clustering model, with comparisons of other known implementations

Clustering analysis is adopted in many industries for detection of similar groups of objects in large unstructured datasets. E-commerce sites use clustering to identify different types of customers, search engines cluster results into similar groups, while biologist use clustering to identify different gene pools. Large datasets used for clustering consume vast amounts of computing resources. In order to reduce the needed resources and decrease computation time, the GPU can be deployed for quick calculations. The faster a piece of malware is detected, the smaller the impact to the mobile community.

6   Proposed Framework

The proposed framework has a basic two-tier client-server setup. The Android application which will be installed on different test devices will collect specific application data related to resource usage. This data will include application permissions, bandwidth consumption, memory, and battery usage. User will be allowed to add and remove applications to be monitored. Once an application is selected, a service then runs in the background of the mobile device which periodically updates usage statistics of apps. The second tier of the proposed framework is the remote server, which is responsible for the storage and distribution of data, as well as execution of malware analysis utilizing the on board GPU.

The AndroidMalDec (Android Malware Machine Learning) application is a data collection application that requires users to register using an email address before you can allow the app to collect usage data. The application allows user who are not signed up to search specific apps and retrieve analysis reports before choosing to install. This allows the user to an idea of the safety of a particular app before they decide to install it. All applications have a unique package name that allows for identification. Storing resource usage data of running applications, the device and its OS allows for a wide comparative analysis across many devices, OS and usage patterns. This data can yield important information about performance benchmarks, usage anomalies and malware detection of applications running across various manufactures devices and different versions of the Android OS (Operating System).

The Remote server could be considered the most critical part of the proposed framework. The remote server is responsible for; storage of data, analysis of data and the distribution of results. The remote server will run Linux Ubuntu and will host the MySQL database, and PHP scripts used to connect the application and to its data source. The dashboard portal website that can be used to get an overall view of application performance and usage patterns across all of Android devices under the assigned email address or username is also hosted on the server. The remote server also runs MATLAB with the additional parallel computing kit in order to make use of the onboard NVidia GTX470 GPU (Graphics Processing Unit). The server is a HPZ800 upgraded with an additional from 500GB of storage to 1TB, and memory increased from 8GB to 16GB. Due to the small scale of users in the test environment this set-up will exceed the requirements needed to execute machine learning analysis on small amounts of data. For testing purposes a Linksys WRT54G router is used to reproduce provider access to internet and remote server to execute SQL statements and analysis. The Android devices connect to the wireless router to upload usage statistics and retrieve results from the remote server.

The server is responsible for the majority of the functionality, from analysis to storage. The components of the remote server include;

- The MySQL database was designed to store application variable in an efficient manner to reduce duplicate data. This database consists of eight tables with various relationships to map devices to apps and users to devices. The growth of the database depends solely on the share numbers of people utilizing the AndroidMalDec app.

- PHP scripts are created to handle the Android application request, to insert data into the data base for app statistics and to retrieve data when searching. This required the creation of PHP functions for database connections, new user data entry and app data entry. The Netbeans IDE proved to be a very good editor for PHP programming.

- MATLAB is required for its parallel computing capabilities. The parallel computing kit allows for the use of GPU calculation. MATLAB functions were created in order to load app data in the GPUArray format. This allows for data to be moved from system memory to GPU memory, which has a much higher bandwidth. The machine learning algorithms (K-means, Fuzzy C and Naïve Bayes) could then be executed on the GPU instead of the CPU for accelerated analysis. This allowed for a performance comparison and evaluation.

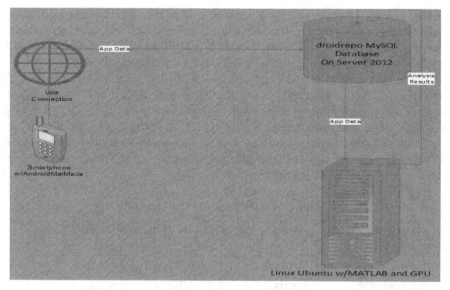

Fig. 2. The system architecture of the proposed framework

- The permission list from each application is a very important variable for determining the possibility of malware. However the string format in which it was retrieved could not be used for the K-means algorithm because strings were not supported. This required a PHP parser to be created in order to produce a readable format that was compatible with the machine learning algorithms. A function to produce Boolean values for permissions was created.

- The dash board portal is a website that provides registered users with a way to get a more detailed overview of the performance and usage patterns of apps across all of their devices. This could be useful for IT administrators, developers who want bench marks on different devices and even families who are trying to manage data usage on a shared plan. PHP was used to create the dashboard portal and it is also hosted on the web server.

7   Layout

The most common method of mobile malware detection involves executing static or dynamic analysis on every application installed on the mobile device. While these methods are effective at detecting anomalies in application behavior, they utilize resources inefficiently by scanning apps that are known to not be malicious. Grading each application before malware analysis takes place in order to reduce the app pool size that is to be analyzed can decrease the time needed to produce results. Assigning a malware probability value to each application allows the analysis to be completed in a shorter time frame due to the decrease size of the app pool. In order to accomplish this device and application data must first be recorded and stored, then analyzed.

Setup: Due to the complexity of simulating mobile infrastructure, the hardware requirements for this experiment were numerous in terms of host machines and other devices. There were a total of two host machines, three mobile devices and one WiFi router used for the study; excluding the devices used for development.

The mobile devices play an instrumental role in this simulation. To provide the most realistic mobile environment, actual devices were chosen to be the installation point of various apps instead of using emulators which would not be able to provide real data usage and other statistics. The variety of Android powered devices also played a vital role, where it was important to have different manufacturers, models and operating system versions. The table below list the different Android mobile devices used for this simulation and their specifications;

Table 1. Mobile devices used for experiment

Device Manufacturer	Device Model	Device Memory	Android Version
Asus	Nexus 7	1 GB RAM	4.4.2
Sony	Xperia Z	2 GB RAM	4.2.2
HTC	Sensation 4G	768 MB	4.0.3

Analysis Host Machine: In order to utilize the data retrieved from different devices and applications it must be analyzed to make useful information and results. The data is stored on a separate host machine to reduce the workload of the database server and to create a more decentralized solution. For this approach a separate machine was chosen to execute the analysis of the data stored on the database server. The analysis host machine was also a HP Z800, with an upgraded GPU (Graphics Processing Unit). For the analysis machine the GPU was upgraded to a NVIDIA GeForce GTX 480 instead of the factory GTX 470. The storage was the standard 500 Gigabytes and the memory was also the factory 8 Gigabytes, with the processor being the same as the database server's six-core Intel Xeon. Windows server 2012R makes hosting the database and accessing the PHP (Hypertext PreProcessor) files much easier and efficient due to its built-in web hosting features. Once IIS (Internet Information Services) is installed, you can then utilize Microsoft's web platform installer to install all necessary components, including PHP and MySQL.

In order to simulate a wireless network and allow the mobile devices the ability to connect to the database server to upload data, a wireless router was necessary. The router also was necessary for the purpose of downloading legitimate apps from the

Google Play store and allowing applications to utilize the internet in order to generate traffic for app analysis purposes. The router used was a simple Linksys WRT54G.

Development Tools: Android Studio is a new Android development environment based on IntelliJ. It provides an integrated development environment that offers all the necessary tools to build Android applications. It was chosen because of its cohesiveness and ease of building activities. It is similar to Eclipse with the ADT plug-in but it manages your application activities java code and XML files simultaneously.

Netbeans was used to develop the PHP scripts that were necessary to allow the Android devices to query the database. It provides capabilities to develop for several languages including Java, PHP and C++. Similar to Android Studio, Netbeans can be downloaded in a bundled package including all the necessary plug-ins for different languages, with the only prerequisite software necessary being JRE (Java Runtime Environment) 7 or later.

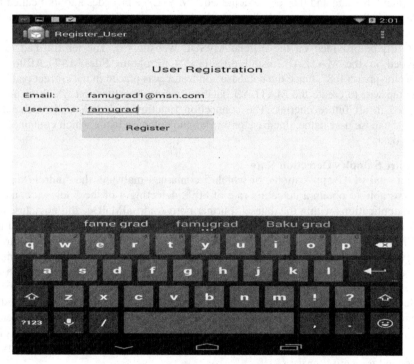

Fig. 3. AndroidMalDec application home screen

AndroidMal Android App Design: The AndroidMal Android application contains five activities that assist with the collection of user, device and application data. The application also contains two separate classes for handling responses from the PHP scripts and populating the list of installed applications on any given device. In total there are seven Java classes that make up the functionality of the application, with a total of 1271 lines of code not including XML files.

The home screen (see in Fig. 3) allows a user to login to the main application or register and create an account. This activity contains two buttons and text boxes, with intents to either go to the user home activity or the register user activity. The unique identifier for users is their email address, which is required for both logging in and registering. An asynchronous task is used in order to retrieve the email from the text field and then insert into the database using an HTTP request, and a PHP post method which is accepted by the corresponding PHP script. The user email is used as the variable in the select query to retrieve user name, user ID and user date.

8 Results

Analysis of the collected data was executed using MATLAB. This section covers the setup of the database connection, the construction of datasets, functions created to manipulate the data and the results achieved. MATLAB provides a toolkit called the database toolbox. In order to take advantage of these capabilities you must first install a database connector. The database connector used was the popular Jconnector and JDBC driver provided on the official MySQL website [5]. The jar file had to be unzipped in the MATLAB install directory "C:\Program Files\MATLAB\mysql-connector-java-5.0.8". Once the database connector was placed in the correct path, the next step was to create the MATLAB database connection function which would be included in all future queries. The connection function included the host machine name, database user name, database password and database name which contained the stored data.

Malware Samples Detection Rate
With a total of 41 applications, of which 5 contained malware, the AndroidMalDec tool was able to obtain a detection rate of 80%, detecting 4 of the 5 known compromised applications within the dataset. Furthermore AndroidMalDec did not cause any false negatives, and detected all applications that contained the permissions and necessary to transmit and access user's private data. For a more rigorous analysis to be achieved AndroidMalDec could have easily achieved a 100% detection rate by simply decreasing the malware probability threshold to 40%, where by any application receiving a malware probability greater than 40% would require further analysis. Table 3 shows the results from the AndroidMalDec analysis, and those highlighted in green are successful malware detections, while those in red represent failed detections. The applications in blue are all legitimate applications from the google play store. A 'True' under the scan column indicates that the application has the ability to access and transmit user data.

Table 2. Sample output data from All Apps function

App ID	App Package Name	App Name	Bandwidth
54	'com.yahoo.mobile.client.android.mail'	'Yahoo Mail'	13412669
79	'com.weather.weather'	'The Weather Channel'	205462
80	'com.gamelio.drawslasher'	'Blood vs Zombie'	0
81	'com.king.candycrushsaga'	'Candy Crush Saga'	3321949
87	'com.crazyapps.angry.birds.rio.unlocker'	'Angry Birds Rio Unlocker'	6986822
92	'com.wapit.carbuzz'	'CarBuzz'	8813122
93	'com.snapchat.android'	'Snapchat'	6292659
94	'com.ebay.mobile'	'eBay'	466089
95	'kagegames.apps.dwbeta'	'Dog Wars - Beta'	3204
97	'duchm.sa'	'Plants vs. Zombies? 2'	3803

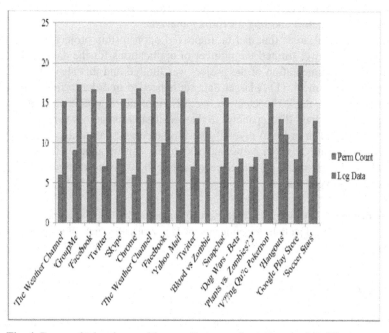

Fig. 4. Bar graph showing positive results comparing perm count and data usage

9   Conclusions

The main contributions of the proposed research can be broken down into two main down into sub categories of the AndroidMalDec application and the remote server. The AndroidMalDec application has two main components: (1) the app usage collection component allows you to register your account and requires you to enter information such as; email user name and password. Once the account is created the user can then select which apps to upload usage statistics from, once the apps are selected no further user input is needed. Background services run periodically to collect data for the selected apps. (2) The app search component allows non-registered users to search existing apps that have been profiled to get a general overview of the possibility of malware, or performance bench marks and other data.

Modern mobile operating systems continue to be targeted for their abundance of personal data stored and used daily for various reasons. While ongoing research has proven to be successful in detecting most threats, criminals are continuously adapting approaches in order to circumvent new detection methods. With mobile operating systems now becoming the norm and merging their usability for more stationary devices such as desktops and laptops, it is imperative that detection time for new malware variants be decreased. The AndroidMalDec experiment was able to decrease the time for malware analysis by simply removing those that applications that do not pose a significant threat to user data. While AndroidMalDec does not protect against all forms of malware; such as premium rate abusers and rootkit exploit, it was able to attain an 80% detection rate for data stealers malware, which was the kind of malware it was designed to detect and flag for further analysis.

There are several areas that can be improved upon in this project. Limitations in devices and manpower limited the number of applications for the data set. The time frame for the implementation of the project was limited and this also had an adverse effect on the experiment. (1) A larger data set is needed in order to fine tune analysis which may increase the 80% detection rate. (2) Application versions can be included in the dataset to address discrepancies in the permission list across versions. (3) More devices are needed in order to diversify performance and data usage patterns of different makes and models. (4) New analysis engine can be added with existing data sets solely to bench mark performance of applications using different devices, OS versions and app versions.

Acknowlededgments. This work has been partially supported by U.S. Department of Education grant P120A090122 and partially supported by the U.S. Army Research Office grant W911NF-13-1-0382.

References

1. Canalys, Mobile device market to reach 2.6 billion units by 2016, Press release 2013/056, February 22, 2013. http://www.canalys.com/newsroom/mobile-device-market-reach-26-billion-units-2016
2. Aaron, S.: Pew Internet Project. Pew Research center (June 2012)
3. Carolina, M., Lillian, T., Roberta, C., Ranjit, A., Hut, N.T., Tracy, T., Annette, Z.: Forecast: Devices by Operating System and User Type, Worldwide, 2010–2017, 2Q13 Update, Garter Research

4. Zhou, Y., Wang, Z., Zhou, W., Jiang, X.: Hey, you, get off of my market: detecting malicious apps in official and alternative android markets. In: Proceedings of the 19th Annual Network and Distributed System Security Symposium (February 2012)
5. Juniper Networks Mobile Threat Center: Third annual mobile threats report: March 2012 through March 2013 (March 2013). http://www.juniper.net/us/en/local/pdf/additional-resources/3rd-jnpr-mobile-threats-report-exec-summary.pdf
6. Vanja, S.: SophosLabs, When Malware Goes Mobile: Causes, Outcomes and Cures. http://www.sophos.com/enus/medialibrary/Gated%20Assets/white%20papers/Sophos_Malware_Goes_Mobile.pdf
7. Burguera, I., Zurutuza, U., Nadjm-Tehrani, S.: Crowdroid: behavior-based malware detection system for android. In: Proceedings of the 1st ACM Workshop on Security and Privacy in Smartphones and Mobile Devices. ACM (2011)

A Novel Prototype Decision Tree Method
Using Sampling Strategy

Bhanu Prakash Battula[1], Debnath Bhattacharyya[2],
C.V.P.R. Prasad[3], and Tai-hoon Kim[4(✉)]

[1] Department of CSE, Vignan College, Guntur, AP, India
battulaphd@gmail.com
[2] Department of Computer Science and Engineering,
Vignan's Institute of Information Technology, Visakhapatnam, AP, India
debnathb@gmail.com
[3] Research Scholar, Acharya Nagarjuna University, Guntur, Andhra Pradesh, India
prasadcvpr@gmail.com
[4] Department of Convergence Security, Sungshin Women's University, 249-1,
Dongseon-dong 3-ga, Seoul 136-742, Korea
taihoonn@daum.net

Abstract. Data Mining is a popular knowledge discovery technique. In data mining decision trees are of the simple and powerful decision making models. One of the limitations in decision trees is towards the data source which they tackle. If data sources which are given as input to decision tree are of imbalance nature then the efficiency of decision tree drops drastically, we propose a decision tree structure which mimics human learning by performing balance of data source to some extent. In this paper, we propose a novel method based on sampling strategy. Extensive experiments, using C4.5 decision tree as base classifier, show that the performance measures of our method is comparable to state-of-the-art methods.

Keywords: Knowledge discovery · Data mining · Classification · Decision trees · Sampling strategy

1 Introduction

In Machine Learning community, and in data mining works, classification has its own importance. Classification is an important part and the research application field in the data mining [1].

A decision tree gets its name because it is shaped like a tree and can be used to make decisions. —Technically, a tree is a set of nodes and branches and each branch descends from a node to another node. The nodes represent the attributes considered in the decision process and the branches represent the different attribute values. To reach a decision using the tree for a given case, we take the attribute values of the case and traverse the tree from the root node down to the leaf node that contains the decision [2]. "A critical issue in artificial intelligence (AI) research is to overcome the

© Springer International Publishing Switzerland 2015
O. Gervasi et al. (Eds.): ICCSA 2015, Part I, LNCS 9155, pp. 90–97, 2015.
DOI: 10.1007/978-3-319-21404-7_7

so-called-knowledge-acquisition bottleneck" in the construction of knowledge-based systems. Decision tree can be used to solve this problem. Decision trees can acquire knowledge from concrete examples rather than from experts [3]. In addition, for knowledge-based systems, decision trees have the advantage of being comprehensible by human experts and of being directly convertible into production rules [4].

A decision tree not only provides the solution for a given case, but also provides the reasons behind its decision. So the real benefit of decision tree technology is that it avoids the need for human expert. Because of the above advantages, there are many successes in applying decision tree learning to solve real-world problems.

To summarize, the contributions of this paper are as follows:

1. A sampling strategy is extended in the decision learning model.
2. Empirical evaluation on a wide variety of real world datasets, and establishing the superiority of the new framework.
3. Analyzing the performance of the methods using the measures of diversity.

The paper is organized as follows. In Sect. 2 we present the recent advances in decision tree learning. This will directly motivate the main contribution of this work presented in Sect. 3, where we propose a new framework for sampling strategic learning. Evaluation criteria's for decision tree learning is presented in section 4. Experimental results are reported in Sect. 5. Finally, we conclude with Sect. 6 where we discuss major open issues and future work.

2   Recent Advances in Decision Trees

In Data mining, the problem of decision trees has also become an active area of research. In the literature survey of decision trees we may have many proposals on algorithmic, data-level and hybrid approaches. The recent advances in decision tree learning have been summarized as follows:

A parallel decision tree learning algorithm expressed in MapReduce programming model that runs on Apache Hadoop platform is proposed by [5]. A new adaptive network intrusion detection learning algorithm using naive Bayesian classifier is proposed by [6]. A new hybrid classification model which is established based on a combination of clustering, feature selection, decision trees, and genetic algorithm techniques is proposed by [7]. A novel roughest based multivariate decision trees (RSMDT) method in which, the positive region degree of condition attributes with respect to decision attributes in rough set theory is used for selecting attributes in multivariate tests is proposed by [8].

A novel splitting criteria which chooses the split with maximum similarity and the decision tree is called mstree is proposed by [9]. An improved ID3 algorithm and a novel class attribute selection method based on Maclaurin-Priority Value First method is proposed by [10]. A modified decision tree algorithm for mobile user classification, which introduced genetic algorithm to optimize the results of the decision tree algorithm, is proposed by [11]. A new parallelized decision tree algorithm on a CUDA (compute unified device architecture), which is a GPGPU solution provided by NVIDIA is proposed by [12]. A Stochastic Gradient Boosted Decision Trees based method is proposed by [13]. A modified Fuzzy Decision Tree for the fuzzy rules extraction is proposed by [14].

Obviously, there are many other algorithms which are not included in this literature. A profound comparison of the above algorithms and many others can be gathered from the references list.

3   The Proposed Method

In this section, the proposed approach is presented.

The proposed approach follows a sampling strategic approach for continuous improvement. The decision tree performs classification in two stages. In the first stage it builds model from the training instances available and in the second stage it validates the testing instances using the build model. The efficiency of the decision tree is evaluated on the testing instances. If a normal or balance data source is provided as input to the decision tree then the model build is efficient enough to classify the testing instances with considerable efficiency.

If the data source provided to the decision tree is of imbalance nature i.e; Let us consider the dataset is of binary class. One class has predominantly more number of instances than the other class; then we may say that type of dataset as an imbalance dataset. The instances in one class can be 95% and in other class it can be 5%. If the decision tree uses the dataset for both training and testing and it follows training-testing strategy of 66-33% or 10 Fold cross validation (CV) there is a great chance that the training set will contain instances of only one class(class of 95% instances). The model build by decision tree using training instances of only one class may not be an efficient model. In the validation phase when the above build model is used for testing instances then definitely the model will encounter some of the instances which it has not seen, then the question comes, "IF IT HAS SEEN NO INSTANCES, HOW CAN IT KNOW?". We proposed A Novel Prototype Decision Tree Method using Sampling Strategy as our problem for investigation.

We designed a sampling strategy which can solve the above limitation of decision trees. One of the solutions is to allow decision trees to build an efficient model by using the instances of all the classes in the dataset. If a binary imbalance datasets encountered in the decision tree learning process the selective sampling can be performed to the class which has very less percentage of instances.

The above said strategy is implemented in the proposed system. In the initial stage the decision tree learning process will initiate with the identification of data source as normal or imbalance dataset. A threshold (Imbalance ratio) value is provided for classification of the data source as a normal or imbalance dataset.

In the next stage, if the data source is identified as an imbalance dataset then the class with less percentage of instances is identified and the proposed sampling strategy is implemented. The resampling is done by replication and hybridized instances. The percentage of synthetic instances generated will range from 0 – 100 % depending upon the percentage of difference of majority and minority classes in the original dataset. The synthetic minority instances generated can have a percentage of instances which can be a replica of the pure instances and reaming percentage of instances are of the hybrid type of synthetic instances generated by combing two or more instances from the pure minority subset. In the next and final phase a base algorithm is used to evaluate the improved dataset.

4   Experimental Design and Evaluation Criteria's

We used the open source tool Weka [16] and implemented our proposed model. In order to test the robustness of our method it is compared to existing methods C4.5 [17], Classification and Regression Trees (CART) [18], Functional Trees [FT], Reduced Error Pruning Tree (REP), and SMOTE[19] in our experiments.

In order to compare the classifiers, we use 10-fold cross validation. In 10-fold cross validation, each dataset is broken into 10 disjoint sets such that each set has (roughly) the same distribution. The classifier is learned 10 times such that in each iteration a different set is withheld from the training phase, and used instead to test the classifier. We then compute the measures as the average of each of these runs.

To assess the classification results we count the number of true positive (TP), true negative (TN), false positive (FP) (actually negative, but classified as positive) and false negative (FN) (actually positive, but classified as negative) examples. It is now well known that error rate is not an appropriate evaluation criterion when there is class imbalance or unequal costs. In this paper, we use AUC, Precision, F-measure, TP Rate and TN Rate as performance evaluation measures.

Let us define a few well known and widely used measures:

The Accuracy (ACC) measure is computed by equation (1) ,

$$AUC = \frac{1 + TP_{RATE} - FP_{RATE}}{2} \tag{1}$$

The Area under Curve (AUC) measure is computed by equation (2),

$$ACC = \frac{TP + TN}{TP + FN + FP + FN} \tag{2}$$

The Precision measure is computed by equation (3),

$$\Pr ecision = \frac{TP}{(TP) + (FP)} \tag{3}$$

The F-measure Value is computed by equation (4),

$$F - measure = \frac{2 \times \Pr ecision \times \mathrm{Re}\,call}{\Pr ecision + \mathrm{Re}\,call} \tag{4}$$

The True Positive Rate measure is computed by equation (5),

$$TruePositi\ veRate = \frac{TP}{(TP)+(FN)} \tag{5}$$

The True Negative Rate measure is computed by equation (6),

$$TrueNegati\ veRate = \frac{TN}{(TN)+(FP)} \tag{6}$$

DATASETS USED IN DECISION TREE LEARNING

Table 1 summarizes the datasets used in the proposed study from UCI [15].

Table 1. Summary of imbalanced datasets

S.no Datasets	# Ex.	# Atts.	Class (_,+)	IR
1. Abalone19	4174	9	(32; 1412)	1:130
2. Abalone19-18	731	9	(42; 689)	1:17
3. Shuttle-c0-vs-c4	1829	10	(123:1706)	1: 14
4. Vowel0	988	14	(90:898)	1: 10
5. Yeast-0-5	528	9	(51:477)	1: 9.4

The details of the datasets are given in table 1. For each data set, S.no., name of the dataset, number of instances, Classes, imbalance ratio (IR) are descried in the table for all the datasets.

5   Results

In this section, we carry out the empirical comparison of our proposed algorithm with the benchmarks. Our aim is to answer several questions about the proposed learning algorithms in the scenario of two-class imbalanced problems.

1. In first place, we want to analyze which one of the approaches is able to better handle a large amount of imbalanced data-sets with different IR, i.e., to show which one is the most robust method.

2. We also want to investigate their improvement with respect to classic decision tree methods and to look into the appropriateness of their use instead of applying a unique preprocessing step and training a single method. That is, whether the trade-off between complexity increment and performance enhancement is justified or not. Given the amount of methods in the comparison, we cannot

afford it directly. On this account, we compared the proposed algorithm with each and every algorithm independently. This methodology allows us to obtain a better insight on the results by identifying the strengths and limitations of our proposed method on every compared algorithm.

Table 2 shows the detailed experimental results of the mean classification accuracy, AUC, Precision, Recall, F-measure of C4.5, CART, FT, REP, SMOTE and Proposed Algor. on all the data sets. From Table 2 we can see that the performance of accuracy of our proposed model achieved substantial improvement over C4.5, CART, FT, REP and SMOTE on most data set which suggests that the proposed model is potentially a good technique for decision trees.

Table 2. Summary of tenfold cross validation performance for proposed algorithm on all the datasets

Datasets	C4.5	CART	FT	REP	SMOTE	Proposed
			Accuracy			
Abalone19	99.23±0.096	99.23 ±0.096	99.23±0.096	99.21±0.114	91.21±2.639●	99.55±0.203
Abalone19-18	93.982±2.053●	94.51±1.338●	95.40±1.559●	94.31±1.581●	98.46±0.173○	97.07±1.679
Shuttle-c0-vs-c4	99.94±0.16	100.0±0.000	99.94±0.165	100.00 ±0.000	100.0±0.000	99.94±0.181
Vowel0	98.92±1.064●	98.23±1.394●	98.28±1.224●	98.29±1.453●	99.12±0.882	99.25±0.991
Yeast-0-5vs4	90.21±3.22●	91.06±2.927●	92.46±2.810●	91.08±3.060●	87.89±3.762●	95.05±1.725
			AUC			
Abalone19	0.500±0.000●	0.500±0.000●	0.500±0.000●	0.510±0.053●	0.745±0.098○	0.685±0.144
Abalone19-18	0.623±0.143●	0.605±0.123●	0.818±0.118●	0.631±0.134●	0.511±0.047●	0.805±0.149
Shuttle-c0-vs-c4	1.000±0.001	1.000±0.000	1.000±0.000	1.00±0.000	1.000±0.000	1.000±0.001
Vowel0	0.966±0.050●	0.949±0.065●	0.960±0.061●	0.957±0.052●	0.984±0.019○	0.968±0.054
Yeast-0-5_vs_4	0.720±0.172○	0.749±0.150○	0.769±0.110○	0.744±0.159○	0.851±0.075○	0.698±0.143
			Precision			
Abalone19	0.000±0.000●	0.000±0.000●	0.000±0.00●	0.00±0.000●	0.705±0.222○	0.297±0.236
Abalone19-18	0.384±0.034●	0.343±0.418●	0.669±0.353○	0.288±0.405●	0.010±0.100●	0.624 ±0.323
Shuttle-c0-vs-c4	0.993±0.023○	1.000±0.00○	1.000±0.000○	1.000±0.00○	1.00±0.00○	0.989±0.032
Vowel0	0.952±0.068○	0.915±0.090●	0.924±0.077●	0.923±0.102●	0.977±0.036○	0.946±0.099
Yeast-0-vs_4	0.510±0.241○	0.529±0.032○	0.683±0.244○	0.510±0.332○	0.672±0.125○	0.255±0.118
			Recall			
Abalone19	0.000±0.000●	0.000±0.000●	0.000 ±0.000●	0.000±0.000●	0.412±0.163○	0.221±0.183
Abalone19-18	0.194±0.214●	0.155±0.198●	0.360±0.225●	0.138±0.204●	0.002±0.017●	0.550±0.304
Shuttle-c0-vs-c4	1.000±0.000	1.000±0.000	0.992±0.025	1.000±0.000	1.00±0.000	1.000±0.000
Vowel0	0.933±0.082○	0.898±0.112●	0.892±0.111●	0.902±0.111●	0.972±0.036○	0.923±0.117
Yeast-0-vs_4	0.413±0.226○	0.352±0.225○	0.475±0.204○	0.351±0.239○	0.657±0.155○	0.275±0.131
			F-measure			
Abalone19	0.000±0.000●	0.000±0.000●	0.000±0.000●	0.000±0.000●	0.494±0.159○	0.240±0.182
Abalone19-18	0.242±0.250●	0.201±0.224●	0.441±0.238●	0.175±0.245●	0.003±0.029●	0.559±0.277
Shuttle-c0-vs-c4	0.996±0.012	1.000±0.000○	0.996±0.013	1.000±0.000○	1.00±0.000○	0.994±0.018
Vowel0	0.940±0.061○	0.901±0.081●	0.902±0.072●	0.906±0.081●	0.974±0.026○	0.932±0.092
Yeast-0-vs_4	0.431±0.198○	0.402±0.236○	0.534±0.190○	0.396±0.248○	0.652±0.112○	0.254±0.106

● Bold dot indicates the win of proposed method; ○ Empty dot indicates the loss of proposed method.

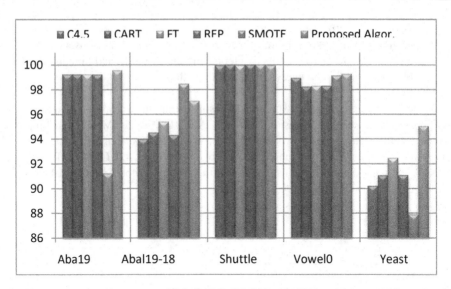

Fig. 1. Test results on accuracy on C4.5, CART, FT, REP, SMOTE and Proposed Algor. for all datasets.

The proposed method had also gained significantly improvement in terms of AUC over C4.5, CART, FT, REP and SMOTE and is comparable to two state-of-the-art technique for decision trees. The performance of the proposed method is improved on almost all the datasets for the measures of precision, recall and f-measure.

Figure 1 shows the detailed pictorial representation of the accuracy results for all the compared algorithms C4.5, CART, FT, REP and SMOTE on all the data sets. From Table 2 and Figure 1 we can see that our proposed approach had given a proper solution for the investigated question.

Finally, we can say that the proposed model is one of the best alternatives to handle class imbalance problems effectively in decision trees. This experimental study supports the conclusion that the a proper sampling strategy can improve the performance of decision when dealing with imbalanced data-sets, as it has helped the proposed method to be the best performing algorithm when compared with five classical and well-known algorithms.

6   Conclusion

In this paper, we proposed a sampling strategy for decision trees. The proposed algorithm mimics human learning approach. We posited that without building proper model the decision trees cannot perform better. Applying human learning in machine spaces will lead to an improved performance due to dynamic plaining. To test this hypothesis we ran experiments on 5 widely available datasets from UCI. We then compared this method with traditional benchmark algorithms. From these results it is apparent that our proposed approach is a competitive one amongst the benchmarks.

References

1. Juanli, H., Deng, J., Sui, M.: A new approach for decision tree based on principal component analysis. In: Proceedings of Conference on Computational Intelligence and Software Engineering, pp. 1–4 (2009)
2. Bergsma, S.: Large-scale semi-supervised learning for natural language processing. PhD Thesis, University of Alberta (2010)
3. Durkin, J.: Expert systems: design and development. Prentice Hall, Englewood Clis (1994)
4. Quinlan, J.: Programs for Machine Learning. Morgan Kaufmann, San Mateo, CA (1993)
5. Purdila, V., Pentiuc, S.-G.: MR-Tree - A Scalable MapReduce Algorithm for Building Decision Trees. Journal of Applied Computer Science & Mathematics, 16(8) (2014). Suceava
6. Farid, D.M., Harbi, N., Mohammad Zahidur, R.: Combining naive bayes and decision tree for adaptive intrusion detect. International Journal of Network Security & Its Applications (IJNSA), 2(2) (April 2010)
7. Mohammad, K., Mahmood, A.: The Use of Genetic Algorithm, Clustering and Feature Selection Techniques in Constrcution of Decision Tree Models for Credit Scoring. International Journal of Managing Information Technology (IJMIT) 5(4) (November 2013). doi:10.5121/ijmit.2013.5402
8. Dianhong, W., Xingwen, L., Liangxiao, J., Xiaoting, Z., Yongguang, Z.: Rough Set Approach to Multivariate Decision Trees Inducing? Journal of Computers, 7(4) (April 2012)
9. Xinmeng, Z., Shengyi, J.: A Splitting Criteria Based on Similarity in Decision Tree Learning. Journal of Software, 7(8) (August 2012)
10. Ying, W., Xinguang, P., Jing, B.: Computer Crime Forensics Based on Improved Decision Tree Algorithm. Journal of Networks, 9(4) (April 2014)
11. Dong-sheng, L., Shujiang, F.: A Modified Decision Tree Algorithm Based on Genetic Algorithm for Mobile User Classification Problem. Scientific World Journal, Article ID 468324, 11 (2014). Hindawi Publishing Corporation. http://dx.doi.org/10.1155/2014/468324
12. Win-Tsung, L., Yue-Shan, C., Ruey-Kai, S., Chun-Chieh, C., Shyan-Ming, Y.: CUDT: A CUDA Based Decision Tree Algorithm. Scientific World Journal, Article ID 745640, 12 (2014). Hindawi Publishing Corporation. http://dx.doi.org/10.1155/2014/745640
13. Tarun, C., Jayashri, V.: Fault Diagnosis in Benchmark Process Control System Using Stochastic Gradient Boosted Decision Trees. International Journal of Soft Computing and Engineering (IJSCE), 1(3) (July 2011). ISSN: 2231-2307
14. Ganga Devi, S.V.S.: Fuzzy Rule Extraction for Fruit Data Classification. Compusoft, An international journal of advanced computer technology, 2(12) (December 2013)
15. Hamilton, A., Asuncion, D., Newman.: UCI Repository of Machine Learning Database (School of Information and Computer Science). Univ. of California, Irvine (2007). http://www.ics.uci.edu/~mlearn/MLRepository.html
16. Witten, I.H., Frank, E.: Data Mining: Practical machine learning tools and techniques, 2nd edn. Morgan Kaufmann, San Francisco (2005)
17. Quinlan, J.: Induction of decision trees. Machine Learning 1, 81–106 (1986)
18. Breiman, L., Friedman, J., Olshen, R., Stone, C.: Classification and Regression Trees. Wadsworth, Belmont (1984)
19. Chawla, N.V., et al.: Synthetic Minority Over-sampling Technique. Journal of Artificial Intelligence Research. 16, 321–357 (2002)

A Tight Security Reduction Designated Verifier Proxy Signature Scheme Without Random Oracle

Xiaoming Hu[✉], Hong Lu, Yan Liu, Jian Wang, and Wenan Tan

College of Computer and Information Engineering,
Shanghai Second Polytechnic University, Shanghai 201209, China
xmhu@sspu.edu.cn

Abstract. Most of existing designated verifier proxy signature (DVPSt) schemes which are proved to be secure in the standard model is constructed based on Water's identity-based encryption. Therefore, security reduction efficiency of these schemes is very low and it may decrease these schemes' security. In this study, we propose a new DVPSt scheme and present a detailed security proof of the proposed DVPSt in the standard model. We also address a tight security reduction of the proposed scheme based on the gap bilinear Diffie-Hellman assumption. Compared with other DVPSt schemes, our proposed DVPSt has two advantages, i.e., the computational cost is lower and security reduction is tighter. Therefore, our proposed DVPSt scheme is very suitable for application in some communication network situations in where the resources are limited.

Keywords: Information security · Communication network · Designated verifier proxy signature · Standard model · Gap bilinear Diffie-Hellman

1 Introduction

In a proxy signature (ProSig) scheme which first was addressed by Mambo et al. in 1996 [1], there is two participates, i.e., the original signer (call Oliver) and the proxy signer (call Peter). Oliver can transfer his signing right to Peter in any time when Oliver wants to do so. After Oliver delegates his signing capability to Peter, Peter can generate valid ProSig on behalf of Oliver. Anyone with the public keys of Oliver and Peter can check the validity of the ProSig. ProSig can be applied in many environments, such as e-commerce, distributed systems and so on. Up to now, much research work on ProSig has been done and a lot of ProSig schemes have been presented, such as [2–10].

However, in some application scenarios, the public-verifiable property of the ProSig may be not desired. For instance, the signed message may be privacy and sensitive, for example a bill of health or a contact, etc. The ProSig receiver

© Springer International Publishing Switzerland 2015
O. Gervasi et al. (Eds.): ICCSA 2015, Part I, LNCS 9155, pp. 98–109, 2015.
DOI: 10.1007/978-3-319-21404-7_8

expects that the ProSig can only be checked himself. In order to solve this problem, Dai et al.[11] in 2003 proposed a DVPSt scheme by combining ProSig and designated verifier signature (DeVeSig) which was first addressed by Jakobsson et al.[12]. In a DVPSt scheme, only the designated verifier (call Dick) can check the validity of the received ProSig. And Dick can not transfer the proof of the signature which produced during verification to any third party. Since Dick can produce valid ProSig that are indistinguishable from the ProSig produced by Peter, which is called non-transferability in general. However, Wang et al.[13] showed in their literature that Dai et al.'s DVPSt scheme was not secure and there existed a forgery attack in their scheme.

Followed by Dai et al.'s work, many DVPSt schemes have been presented [14–21]. However, most of these DVPSt schemes only provided security proof in the random oracle model which has suffered much criticism. In 2009, based on Water's identity-based encryption [22], Yu et al.[15] addressed the first DVPSt scheme which is secure in the standard model. However, in their DVPSt scheme, the computational cost is very high and also the efficiency of security reduction is very low. In 2013, Ming et al.[19] proposed a DVPSt scheme which had multi-warrant and they also provided a security proof in the standard model. Similarly, their scheme also suffered the same attck problems as Yu et al.'s scheme since Ming et al.'s DVPSt scheme also was constructed based on Water's identity-based encryption. In order to fix this problem, in this study, based on [23]'s scheme, we address a new and efficient DVPSt scheme. By the comparison with other schemes, our major contributions are as follows.

(1) We present a new DVPSt scheme. What's more, based on Gap Bilinear Diffie-Hellman problem, we provide a detailed security proof in the standard model. The proof shows that our new scheme not only is secure in the standard model but also the security reduction is very tight.
(2) We make a performance comparison in terms of computational load with other similar schemes. The comparison shows that our proposed DVPSt scheme has better computational performance and save almost half of other similar schemes' computational load.

2   Preliminaries

In this section, we review some concepts which will be used in our following paper.

2.1   Bilinear Map

Let G_1 generated by g be a cyclic additive group and G_2 be a cyclic multiplicative group. G_1 and G_2 both have the same prime order q. $e: G_1 \times G_1 \to G_2$ is called a bilinear map if it possesses:

(1) Bilinearity: for any $x, y \in Z_q$, $h_1, h_2 \in G_1$, $e(h_1^x, h_2^y) = e(h_1, h_2)^{xy}$.
(2) Non-denegracy: there exists $h \in G_1$, such that $e(h, h) \neq 1$.
(3) Computability: for $h_1, h_2 \in G_1$, $e(h_1, h_2)$ can be efficiently computational.

2.2   Bilinear Diffie-Hellman (BDH) Problem

(h, h^a, h^b, h^c) is a random four tuple where $h, h^a, h^b, h^c \in G_1$ and $a, b, c \in Z_q^*$ is unknown, the BDH problem is to compute $e(h, h)^{abc}$.

2.3   Decisional Bilinear Diffie-Hellman (DBDH) Problem

(h, h^a, h^b, h^c, U) is a random five tuple where $h, h^a, h^b, h^c \in G_1$ and $a, b, c \in Z_q^*$ is unknown and $U \in G_2$, the DBDH problem is to decide whether $U = e(h, h)^{abc}$ holds.

For the input (h, h^a, h^b, h^c, U), the DBDH oracle outputs *true* if $U = e(h, h)^{abc}$ holds. Otherwise, output *false*.

2.4   Gap Bilinear Diffie-Hellman(GBDH) Problem and GBDH Assumption

(h, h^a, h^b, h^c) is a random four tuple where $h, h^a, h^b, h^c \in G_1$ and $a, b, c \in Z_q^*$ is unknown, the GBDH problem is to compute $e(h, h)^{abc}$ with the help of the DBDH oracle.

The GBDH assumption is said to (ζ, ϖ) hold, if no algorithm can solve the GBDH problem in G_1 and G_2 with running time at most ϖ and probability at least ζ.

2.5   DVPSt Scheme

A DVPSt scheme includes six algorithms DVPS = {SysSet, KeyGen, Delegate, ProxySig, SigVerify, SignSimul}.

SysSet: For a given system security parameter, SysSet outputs *params* as the system public parameters.

KeyGen: Input *params*, KeyGen produces the public-private key pair for any user.

Delegate: Input *params*, the private key of Oliver and a warrant w, Delegate generates a delegation on the warrant w.

ProxySig: input *params*, the private key of Peter, the delegation on w and a message m that is signed, ProxySig outputs a ProSig σ.

SigVerify: Input *params*, the public key of Oliver, the public key of Peter and the ProSig σ on m with w, SigVerify outputs *true* if σ is valid, or it outputs *false*.

SignSimul: Input *params*, the private key of Dick, a message m which is signed with the warrant w, Dick outputs a simulated ProSig σ which is valid.

2.6   Security Model for DVPSt Scheme

In a DVPSt scheme, there exist three types of adversaries. We assume that the adversary is $\mho$. The original signer is S_{or}, the proxy signer is S_{pr} and the designated verifier is S_{de}.

Type 1 of adversary: $\mho$ only knows the public key of S_{or} and the public key of S_{pr}.

Type 2 of adversary: $\mho$ knows the public keys of S_{or} and S_{pr}. And $\mho$ knows the private key of S_{or}.

Type 3 of adversary: $\mho$ knows the public keys of S_{or} and S_{pr}. And $\mho$ knows the private key of the S_{pr}.

According to the information obtained by $\mho$ it can be saw that if a DVPSt scheme is unforgeable against Type 2 and 3 of adversaries, it also is unforgeable against Type 1 of aversary. Therefore, we only need to address the the security model of type 2 and 3 of adversaries, respectively.

(1) The unforgeability against type 2 adversary is described by a game performed between a challenger Φ_2 and a type 2 adversary $\mho_2$.

Setup: Φ_2 runs the SysSet algorithm to construct the system parameter *params* and sets the public-private key pairs (pk_o, sk_o), (pk_p, sk_p) and (pk_d, sk_d) of S_{or}, S_{pr} and S_{de} respectively. Φ_2 sends *params* and (sk_o, pk_o, pk_p, pkd) to $\mho_2$.

Query: $\mho_2$ makes the following query in polynomial times respectively. $\mho_2$ submits signing queries to Φ_2, Φ_2 produces a ProSig σ and returns σ to $\mho_2$; $\mho_2$ makes verifying queries to Φ_2, Φ_2 checks the signature and returns *true* if it is valid or returns *false*.

Forgery: $\mho_2$ outputs a forged DVPSt σ^* on a message m^* with a warrant w^*, which meets: σ^* is a correct ProSig on m^* with w^*; m^* and w^* has not made a signing query.

Define Suc_{t2} to be the successful probability of $\mho_2$ in the above game. We say that $\mho_2$ can $(\varpi, \zeta, \eta_s, \eta_v)$ break a DVPSt if Φ_2 performs time at most ϖ and the successful probability at least ζ after making at most η_s times signing queries, η_v times verifying queries.

(2) The unforgeability against type 3 adversary is described by a game performed between a challenger Φ_3 and a type 3 adversary $\mho_3$.

Setup: Φ_3 runs the SysSet algorithm to produce the system parameter *params* and sets the public-private key pairs (pk_o, sk_o), (pk_p, sk_p) and (pk_d, sk_d) of S_{or}, S_{pr} and S_{de} respectively. Φ_3 sends *params* and (sk_p, pk_o, pk_p, pk_d) to $\mho_3$.

Query: In this phase, $\mho_3$ can make three types query, i.e., delegating query, signing query and verifying query. Φ_3's answers for signing query and verifying query are similar to these of type 2. For delegating query, Φ_3 uses the Delegate algorithm to generate a delegation and returns it to $\mho_3$.

Forgery: $\mho_3$ outputs a forged DVPSt σ^* on m^* with the warrant w^*, which holds: w^* has not made a delegating query; σ^* is a valid ProSig on m^* and w^*; m^* and w^* has not made a signing query.

Define Suc_{t3} to be the successful probability of $\mho_3$ in the above game. We say that a type 3 adversary $\mho_3$ can $(\varpi, \zeta, \eta_w, \eta_s, \eta_v)$ break a DVPSt if $\mho_3$ performs time at most t and the successful probability at least ϵ after making at most q_w times delegating queries, at most q_s times signing queries, q_v times verifying queries.

3   Our DVPSt Scheme

In this section, we address the proposed DVPSt scheme. Our DVPSt scheme consists of six components: SysSet, KeyGen, Delegate, ProxySig, SigVerify, Sign-Simul and involves three participants: the original signer Oliver, the proxy signer Peter and the designated verifier Dick. Each component works as follows.

SysSet: Define a hash function $F: \{0,1\}^* \rightarrow \{0,1\}^n$ and let $h \in G_1$ to be a generator. $e: G_1 \times G_1 \rightarrow G_2$ is a bilinear map. $X = (x_i)$ is a vector of n-length where $x_i \in Z_q^*$, $1 \le i \le n$. The public parameters $params=\{G_1, G_2, e, F, X, h\}$.

KeyGen: Oliver selects a random element $sk_o \in Z_q^*$ and computes $pk_o = h^{sk_o}$. Then the public-private key pair of Oliver is (sk_o, pk_o). Using the same method, Peter and Dick get (sk_p, pk_p) and (sk_d, pk_d) respectively.

Delegate: Given warrant $w \in \{0,1\}^n$, Oliver picks randomly $k_1 \in Z_q^*$ and computes the delegation as follows:

$$I_w = F(w||k_1), \sigma_w = (\prod_{i\in(I_w)} x_i)^{sk_o},$$

where $I_w[i]$ denotes the i-th bit of I_w and (I_w) denotes the set of all i with $I_w[i]=1$. Oliver sends the delegation (σ_w, k_1) to Peter.

ProxySig: Assume $m \in \{0,1\}^n$ is a message which is signed. Peter picks randomly $k_2 \in Z_q^*$ and computes $I_m = F(w||m||k_2)$. Then Peter generates the ProSig $(\sigma_1, \sigma_2, \sigma_3)$ where

$$\sigma_1 = e(\sigma_w(\prod_{i\in(I_m)} x_i)^{sk_p}, pk_d), \sigma_2 = k_1, \sigma_3 = k_2.$$

Peter sends $(\sigma_1, \sigma_2, \sigma_3)$ to Dick.

SigVerify: Given a ProSig $(\sigma_1, \sigma_2, \sigma_3)$ on m with w, Dick first computes

$$I_w = F(w||\sigma_2), I_m = F(w||m||\sigma_3).$$

Then Dick checks the following equation:

$$\sigma_1 = e((\prod_{i\in(I_w)} x_i)^{sk_d}, pk_o)e((\prod_{i\in(I_m)} x_i)^{sk_d}, pk_p).$$

If the above equation holds, Dick accepts the ProSig or he rejects it.

SignSimul: Dick can use his private key to produce a valid ProSig of a message m and the warrant w as the following method.

Dick picks two random numbers $k_1', k_2' \in Z_q^*$ and computes

$$I_w' = F(w||k_1'), I_m' = F(w||m||k_2').$$
$$\sigma_1' = e((\prod_{i\in(I_w)} x_i)^{sk_d}, pk_o)e((\prod_{i\in(I_m)} x_i)^{sk_d}, pk_p).$$

Then $(\sigma_1', \sigma_2' = k_1', \sigma_3' = k_2')$ is the simulating ProSig. The ProSig produced in the above is correct because

$$\sigma_1 = e(\sigma_w(\prod_{i \in (I_m)} x_i)^{sk_p}, pk_d)$$

$$= e((\prod_{i \in (I_w)} x_i)^{sk_o}(\prod_{i \in (I_m)} x_i)^{sk_p}, pk_d)$$

$$= e((\prod_{i \in (I_w)} x_i)^{sk_o}, pk_d)e((\prod_{i \in (I_m)} x_i)^{sk_p}, pk_d)$$

$$= e((\prod_{i \in (I_w)} x_i)^{sk_d}, h^{sk_o})e((\prod_{i \in (I_m)} x_i)^{sk_d}, h^{pk_p})$$

$$= e((\prod_{i \in (I_w)} x_i)^{sk_d}, pk_o)e((\prod_{i \in (I_m)} x_i)^{sk_d}, pk_p)$$

4 Security Analysis

Theorem 1. *If there has a type 2 of adversary $\mho_2$ who can $(\varpi, \zeta, \eta_s, \eta_v)$ break our DVPSt scheme, then we can construct another algorithm Φ_2 who can (ϖ', ζ') solve the given GBDH problem in (G_1, G_2), where*

$$\zeta' \geq \frac{1}{2}\zeta(1 - \zeta_{hc}),$$

$$\varpi' = \varpi + (n + 5 + 2\eta_s + 6\eta_v)l_{ex} + (n + 3 + \eta_s + 4\eta_v)l_{mul} + (3 + \eta_s + 4\eta_v)\tau_p,$$

l_{ex}, l_{mul} and l_{pair} denote one exponentiation, multiplication and pairing operation in G_1 respectively. ζ_{hc} denotes the hash collision probability. η_s and η_v denotes the maximal number of signing queries and verifying queries respectively.

Proof. Φ_2 is given a random instance of the GBDH problem (h, h^a, h^b, h^c) where a, b and c is not known by Φ_2. Φ_2's goal is to obtain $e(h, h)^{abc}$. Our proof is based on the idea used in [23] and [24].

Setup: Φ_2 selects a random number $\gamma \in Z_q^*$ and picks $2n$ random elements $\alpha_i, \beta_i \in Z_q^*$, where $1 \leq i \leq n$. Compute the n-length vector $X = (x_i)$ where $x_i = h^{a(-1)^{\alpha_i}}h^{\beta_i}$. Set the public-private key pair of Oliver is $(pk_o = h^\gamma, sk_o = \gamma)$. Peter's $(pk_p = h^b, sk_p = b)$ and Dick's $(pk_d = h^c, sk_d = c)$. Send $\{G_1, G_2, e, F, X, h\}$ and (pk_o, pk_p, pk_d, sk_o) to the adversary $\mho_2$. For convenience of analysis, we define two functions: $G(I_x) = \sum_{i \in (I_x)}(-1)^{\alpha_i} \mod q$ and $L(I_x) = \sum_{i \in (I_x)} \beta_i \mod q$.

Query: In this stage, $\mho_2$ makes signing query and verifying query in polynomial times. Φ_2 responds as follows.

Signing Query: When $\mho_2$ requests a signing query on a message m_i with the warrant w_i, Φ_2 does as follows.

— Φ_2 picks randomly $k_{i,1}, k_{i,2} \in Z_q^*$ and computes

$$I_{w_i} = F(w_i || k_{i,1}), I_{m_i} = F(w_i || m_i || k_{i,2}).$$

If $G(I_{m_i}) \neq 0$, Φ_2 fails and aborts. If $G(I_{m_i})=0$, Φ_2 computes

$$\sigma_{i,1} = e((\prod_{i\in(I_{w_i})} x_i)^\gamma (h^b)^{L(I_{m_i})}, pk_d),$$

and returns $(\sigma_{i,1}, \sigma_{i,2} = k_{i,1}, \sigma_{i,3} = k_{i,2})$ to $\mho_2$. $(\sigma_{i,1}, \sigma_{i,2}, \sigma_{i,3})$ is a valid ProSig on m_i and w_i. Since

$$\sigma_{i,1} = e((\prod_{i\in(I_{w_i})} x_i)^\gamma (h^b)^{L(I_{m_i})}, pk_d)$$

$$= e((\prod_{i\in(I_{w_i})} x_i)^\gamma h^{abG(I_{m_i})} h^{bL(I_{m_i})}, pk_d)$$

$$= e((\prod_{i\in(I_{w_i})} x_i)^\gamma (h^{aG(I_{m_i})} h^{L(I_{m_i})})^b, pk_d)$$

$$= e((\prod_{i\in(I_{w_i})} x_i)^\gamma (\prod_{i\in(I_{m_i})} x_i)^b, pk_d)$$

Verifying Query: When $\mho_2$ makes a verifying query on a ProSig $(\sigma_{i,1}, \sigma_{i,2}, \sigma_{i,3})$ with m_i and w_i, Φ_2 does as follows.

— Φ_2 computes

$$I_{w_i} = F(w_i||\sigma_{i,2}), I_{m_i} = F(w_i||m_i||\sigma_{i,3}).$$

If $G(I_{m_i}) = 0$, Φ_2 checks if the following equation meets

$$\sigma_{i,1} = e((\prod_{i\in(I_{w_i})} x_i)^\gamma (h^\beta)^{L(I_{m_i})}, pk_d),$$

If the above equation holds, Φ_2 returns *true* or Φ_2 returns *false*. If $G(I_{m_i}) \neq 0$, Φ_2 computes

$$\sigma_{i,0} = (\frac{\sigma_{i,1}}{e(h^a, h^c)^{\gamma G(I_{w_i})} e(h, h^c)^{\gamma L(I_{w_i})} e(h^b, h^c)^{L(I_{m_i})}})^{G(I_{m_i})^{-1}}.$$

Φ_2 sends $(\sigma_{i,0}, h, h^a, h^b, h^c)$ to BDBH oracle. If BDBH oracle returns 1, then Φ_2 returns true to $\mho_2$. If BDBH oracle returns 0, then Φ_2 returns false to $\mho_2$.

Remark 1. If the ProSig $(\sigma_{i,1}, \sigma_{i,2}, \sigma_{i,3})$ on m_i with the warrant w_i submitted by $\mho_2$ is valid, then $\sigma_{i,0} = e(h, h)^{abc}$ must be satisfied. This is because if $(\sigma_{i,1}, \sigma_{i,2}, \sigma_{i,3})$ is valid ProSig, then it satisfies

$$\sigma_{i,1} = e((\prod_{i\in(I_{w_i})} x_i)^c, pk_o)e((\prod_{i\in(I_{m_i})} x_i)^c, pk_p),$$

where $I_{w_i} = F(w_i||\sigma_{i,2})$, $I_{m_i} = F(w_i||m_i||\sigma_{i,3})$. Thus, it can get

$$\sigma_{i,1} = e((\prod_{i\in(I_{w_i})} x_i)^c, pk_o)e((\prod_{i\in(I_{m_i})} x_i)^c, pk_p)$$

$$= e((\prod_{i\in(I_{w_i})} x_i)^\gamma(\prod_{i\in(I_{m_i})} x_i)^b, h^c)$$

$$= e((h^{aG(I_{w_i})}h^{L(I_{w_i})})^\gamma(h^{aG(I_{m_i})}h^{L(I_{m_i})})^b, h^c)$$

So, finally we obtain

$$\sigma_{i,0} = e(h,h)^{abc} = (\frac{\sigma_{i,1}}{e(h^a,h^c)^{\gamma G(I_{w_i})}e(h,h^c)^{\gamma L(I_{w_i})}e(h^b,h^c)^{L(I_{m_i})}})^{G(I_{m_i})^{-1}}.$$

Forgery: If $\mathfrak{U}_2$ does not fail in the above game, then finally $\mathfrak{U}_2$ outputs a forged ProSig which is valid. We assume that the forged ProSig is $(\sigma_1^*, \sigma_2^*, \sigma_3^*)$ on m^* and w^*. Since $(\sigma_1^*, \sigma_2^*$ is a valid ProSig, it meets the following equation

$$\sigma_1^* = e((\prod_{i\in(I_{w^*})} x_i)^c, pk_o)e((\prod_{i\in(I_{m^*})} x_i)^c, pk_p),$$

where $I_{w^*} = F(w^*||\sigma_2^*)$, $I_{m^*} = F(w^*||m^*||\sigma_3^*)$. From the above equation, it can get

$$\sigma_1^* = e((\prod_{i\in(I_{w^*})} x_i)^c, g^\gamma)e((\prod_{i\in(I_{m^*})} x_i)^c, h^b),$$

$$= e((\prod_{i\in(I_{w^*})} x_i)^\gamma, h^c)e((\prod_{i\in(I_{m^*})} x_i)^b, h^c),$$

$$= e((h^{aG(I_{w^*})}h^{L(I_{w^*})})^\gamma, h^c)e((h^{aG(I_{m^*})}h^{L(I_{m^*})})^b, h^c)$$

If $G(I_{m^*})=0$, then Φ_2 aborts. If $G(I_{m^*}) \neq 0$, then Φ_2 computes

$$e(h,h)^{abc} = (\frac{\sigma_1^*}{e(h^a,h^c)^{\gamma G(I_{w^*})}e(h,h^c)^{\gamma L(I_{w^*})}e(h^b,h^c)^{L(I_{m^*})}})^{G(I_{m^*})^{-1}}.$$

So, finally Φ_2 solves the given GBDH problem.

The successful probability analysis of Φ_2 is that Φ_2 does not fail in the above whole simulation. Let L1 denote that Φ_2 does not abort in the signing query, L2 denote that Φ_2 does not abort in verifying query and L3 denote that Φ_2 does not abort in the forgery phase. Thus, we can get $\epsilon' = \Pr[\overline{abort}] = \Pr[L1\wedge L2\wedge L3]$. Since L1, L2 and L3 are three independent events, $\Pr[L1\wedge L2\wedge L3]=\Pr[L1]\Pr[L2]\Pr[L3]$. According to the result from [23] and [24], for 128-bit security level, it needs averagely 21 time selections of $k_{i,2} \in Z_q^*$ to make $G(I_{m^*})=0$ where $I_{m^*} =F(w_i||m_i||k_{i,2})$. On the other hand, if we exclude the hash collision, then the probability of $G(I_{m^*}) \neq 0$ where $I_{m^*} = F(w_i||m_i||\sigma_3^*)$ is at most 1/2. Therefore, $\zeta' = \Pr[\overline{abort}] \geq \frac{1}{2}\zeta(1-\zeta_{hc})$.

The computation time is mainly consumed by pairing operation, exponentiation operation and multiplication operation in G_1. Therefore, the total time cost is $\varpi' = \varpi + (n+5+2\eta_s+6\eta_v)l_{ex} + (n+3+\eta_s+4\eta_v)l_{mul} + (3+q_s+4q_v)l_{pair}$.

Using the similar method to the above, it can obtain the following result.

Theorem 2. *If there has a type 3 adversary $\mho_3$ who can $(\varpi, \zeta, \eta_w, \eta_s, \eta_v)$ break our DVPSt scheme, then we can construct another algorithm Φ_3 who can (ϖ', ζ') solve the given GBDH problem in (G_1, G_2), where*

$$\zeta' \geq \frac{1}{2}\zeta(1 - \zeta_{hc}),$$

$$\varpi' = \varpi + (n + 5 + \eta_w + 2\eta_s + 6\eta_v)l_{ex} + (n + 3 + \eta_s + 4\eta_v)l_{mul}$$
$$+ (3 + \eta_s + 4\eta_v)l_{pair},$$

l_{ex}, l_{mul} and l_{pair} denote one exponentiation, multiplication and pairing operation in G_1 respectively. ϵ_{hc} denotes the hash collision probability. η_w denotes the maximal number of delegating queries, η_s denotes the maximal number of signing queries and η_v denotes the maximal number of verifying queries.

Proof. The proof is very similar to that of Theorem 1. The main difference is that the adversary $\mho_3$ knows the private key of Peter in advance instead of the private key of Oliver. Since $\mho_3$ does not know the private key of Oliver, $\mho_3$ can not produce the delegation himself. So, in this case, $\mho_3$ is allowed to make delegating query to Φ_3 in query stage. The detailed description of main difference is as follows.

Setup: Φ_3 picks randomly $\gamma \in Z_q^*$ and sets the public-private key pair of Peter is $(pk_p = h^\gamma, sk_p = \gamma)$. The public-private key pair of Oliver is $(pk_o = h^b, sk_o = b)$ and The public-private key pair of Dick is $(pk_d = h^c, sk_d = c)$. The other system parameters are set as theorem 1. Send $\{G_1, G_2, e, F, X, h\}$ and (pk_o, pk_p, pk_d, sk_p) to the adversary $\mho_3$.

Query: $\mho_3$ makes delegating query, signing query and verifying query in polynomial times. Φ_3 answers as following method.

Delegating Query: When $\mho_3$ submits a delegating query on the warrant w_i, Φ_3 works as follows.

—— Φ_3 picks a random number $k_{i,1} \in Z_q^*$ and computes $I_{w_i} = F(w_i||k_{i,1})$.
—— If $G(I_{w_i}) \neq 0$, Φ_3 aborts. If $G(I_{w_i}) = 0$, Φ_3 computes

$$\sigma_{w_i} = (pk_o)^{L(I_{w_i})} = h^{abG(I_{w_i})}h^{bL(I_{w_i})} = (h^{aG(I_{w_i})}h^{L(I_{w_i})})^b = (\prod_{i \in (I_{w_i})v_i})^{sk_o}.$$

Φ_3 returns $(\sigma_{w_i}, k_{i,1})$ to $\mho_3$.

Signing Query: For a message m_i with the warrant w_i from $\mho_3$, Φ_3 works as follows.

—— Φ_3 generates a delegation $(\sigma_{w_i}, k_{i,1})$ on w_i using the above *Delegating query*.
—— Then Φ_3 picks randomly $k_{i,2} \in Z_q^*$ and computes $I_{m_i} = F(w_i||m_i||k_{i,2})$.
—— Φ_3 computes $\sigma_{i,1} = e(\sigma_{w_i}(\prod_{i \in (I_{m_i})})^\gamma, pk_d)$, and sets $\sigma_{i,2} = k_{i,1}$ and $\sigma_{i,3} = k_{i,2}$. Return $(\sigma_{i,1}, \sigma_{i,2}, \sigma_{i,3})$ to $\mho_3$.

Verifying Query: For a ProSig $(\sigma_{i,1}, \sigma_{i,2}, \sigma_{i,3})$ on m_i and w_i, Φ_3 works as that of theorem 1.

Forgery: Finally, Φ_3 outputs a forged ProSig $(\sigma_1^*, \sigma_2^*, \sigma_3^*)$ on m^* with w^*. Similarly to that of theorem 1, Φ_3 first computes $I_{w^*} = F(w^*||\sigma_2^*)$. If $G(I_{w^*}) = 0$, then Φ_3 aborts. If $G(I_{w^*}) \neq 0$, then Φ_3 computes

$$e(h,h)^{abc} = (\frac{\sigma_1^*}{e(h^a,h^c)^{\gamma G(I_{m^*})}e(h,h^c)^{\gamma L(I_{m^*})}e(h^b,h^c)^{L(I_{w^*})}})^{G(I_{w^*})^{-1}}.$$

Using the same method to that of theorem 1, we can get $\zeta' \geq \frac{1}{2}\zeta(1-\zeta_{hc})$, $\varpi' = \varpi + (n+5+\eta_w+2\eta_s+6\eta_v)l_{ex} + (n+3+\eta_s+4\eta_v)l_{mul} + (3+\eta_s+4\eta_v)l_{pair}$.

Theorem 3. *The proposed DVPSt scheme is non-transferable.*

Proof. The ProSig generated by Peter is

$$\sigma_1 = e((\prod_{i\in(I_w)} x_i)^{sk_o}(\prod_{i\in(I_m)} x_i)^{sk_p}, pk_d), \sigma_2 = k_1, \sigma_3 = k_2.$$

and the ProSig simulated by Dick is

$$\sigma_1 = e((\prod_{i\in(I_w)} x_i)^{sk_d}, pk_o)e((\prod_{i\in(I_m)} x_i)^{sk_d}, pk_p), \sigma_2 = k_1, \sigma_3 = k_2.$$

So the ProSig produced by Peter and the ProSig constructed by Dick have identical distribution. Therefore, our ProSig scheme is non-transferable.

5   Efficiency Analysis

From the above construction of our ProSig scheme it can find that our DVPSt scheme spends four exponentiation computations and $4n$ multiplication computations and three pairing computations in G_1. Next, we compare our DVPSt scheme with other related schemes which present the security proof in the standard model as our DVPS scheme. Compared with other DVPSt schemes [15,19], our scheme has better efficiency in computational cost. We provide a comparison result in Table 1. On the other hand, from the Table 1, we can see that our scheme has an efficient security reduction with $\zeta' \geq \frac{1}{2}\zeta(1-\zeta_{hc})$ based on GBDH problem, which is more efficient than that of other DVPSt schemes [15,19].

Table 1. The comparison with other DVPSt schemes

scheme	computational cost	security reduction
[15]	$10l_{ex} + 5nl_{mul} + 5l_{pair}$	$\zeta' \geq \frac{\zeta}{3(n+1)^3(3(\eta_s+\eta_w))^{\eta_v}+2}$
[19]	$14l_{ex} + 4nl_{mul} + 7l_{pair}$	$\zeta' \geq \frac{\zeta}{(n+1)^3(3(\eta_s+\eta_w))^{\eta_v}+2}$
Our	$4l_{ex} + 4nl_{mul} + 3l_{pair}$	$\zeta \geq \frac{1}{2}\zeta(1-\zeta_{hc})$

6   Conclusions

In this study, we addressed a DVPSt scheme and also provided a detailed security analysis with a tight reduction based on the gap bilinear Diffie-Hellman problem. Further, we make an computational comparison between our DVPSt scheme and other DVPSt schemes which are secure in the standard model. The analysis presents that our DVPSt scheme is more efficient than other similar schemes.

Acknowledgments. This work was supported by the Innovation Program of Shanghai Municipal Education Com-mission under Grant No.14ZZ167, the National Natural Science Foundation of China under Grant No.61103213 and the Construct Program of the Key Discipline in SSPU (fourth): Soft-ware Engineering under Grant No. XXKZD1301.

References

1. Mambo, M., Usuda, K., Okmamoto, E.: Proxy signatures for delegation signing operation. In: Proc. 1996 ACM CCS conf., pp. 48–57. ACM Press, New York (1996)
2. Shao, Z.: Proxy Signature Schemes based on Factoring. Information Processing Letter **85**(3), 137–143 (2003)
3. Huang, X., Susilo, W., Mu, Y., Wu, W.: Proxy signature without random oracles. In: Cao, J., Stojmenovic, I., Jia, X., Das, S.K. (eds.) MSN 2006. LNCS, vol. 4325, pp. 473–484. Springer, Heidelberg (2006)
4. Zhou, Y., Cao, Z.-F., Chai, Z.: An efficient proxy-protected signature scheme based on factoring. In: Chen, G., Pan, Y., Guo, M., Lu, J. (eds.) ISPA-WS 2005. LNCS, vol. 3759, pp. 332–341. Springer, Heidelberg (2005)
5. Schuldt, J.C.N., Matsuura, K., Paterson, K.G.: Proxy signatures secure against proxy key exposure. In: Cramer, R. (ed.) PKC 2008. LNCS, vol. 4939, pp. 141–161. Springer, Heidelberg (2008)
6. Sun, Y., Xu, C., Yu, Y., Mu, Y.: Strongly Unforgeable Proxy Signature Scheme Secure in The Standard Model. Journal of Systems and Software **84**(9), 1471–1479 (2011)
7. Zhang, J., Mao, J.: Another Efficient Proxy Signature Scheme in The Standard Model. Journal of Information Science and Engineering **27**, 1249–1264 (2011)
8. Zhang, J., Liu, C., Yang, Y.: An Efficient Secure Proxy Verifiably Encrypted Signature Scheme. Journal Network and Computer Applications **33**, 29–34 (2010)
9. Boldyreva, A., Palacio, A., Warinschi, B.: Secure Proxy Signature Schemes for Delegation of Signing Rights. Journal of Cryptology **25**, 57–115 (2012)
10. Bao, H., Cao, Z.: Group-proxy Signature Scheme: a Novel Solution to Electronic Cash. Journal of Intelligent Systems **22**(2), 95–110 (2013)
11. Dai, J., Yang, X., Dong, J.: Designated-receiver proxy signature scheme for electronic commerce. In: Proc. 2003 SMC conf., pp. 384–389. IEEE Press, New York (2003)
12. Jakobsson, M., Sako, K., Impagliazzo, R.: Designated verifier proofs and their applications. In: Maurer, U.M. (ed.) EUROCRYPT 1996. LNCS, vol. 1070, pp. 143–154. Springer, Heidelberg (1996)
13. Wang, G.: Designated-verifier proxy signatures for e-commerce. In: Proc. 2004 ICME conf., pp. 1731–1734. IEEE Press, New York (2004)

14. Lu, R., Cao, Z., Dong, X.: Designated verifier proxy signature scheme from bilinear pairings. In: Proc. 2006 ACMCCS conf., 2006, pp. 40–47. IEEE Press, New York (2006)
15. Yu, Y., Xu, C., Zhang, X., Liao, Y.: Designated Verifier Proxy Signature Scheme without Random Oracles. Computers and Mathematics with Applications **57**(8), 1352–1364 (2009)
16. Shim, K.A.: Short Designated Verifier Proxy Signatures. Computers and Electrical Engineering **37**(2), 180–186 (2011)
17. Zhang, X., Huang, S., Xin, X.: Improved proxy signature for designated verifier. In: Proc. 2011 CIS conf., pp. 231–233. IEEE Press, New York (2011)
18. Lin, H., Wu, T., Huang, S.: An Efficient Strong Designated Verifier Proxy Signature Scheme for Electronic Commerce. Journal of information science and engineering **28**(4), 771–785 (2012)
19. Ming, Y., Jin, Q., Zhao, X.: Designated Verifier Proxy Signature Scheme with Multi-warrant in The Standard Model. Journal of Information & Computational Science **10**(7), 2097–2107 (2013)
20. Hsu, C., Lin, H.: Pairing-based Strong Designated Verifier Proxy Signature Scheme with Low Cost. Security and Communication Networks **5**(5), 517–522 (2012)
21. Zhou, C.: Attacks on a Strong Designated Verifier Proxy Signature Scheme. Advanced Science and Technology Letters **48**, 7–12 (2014)
22. Waters, B.: Efficient identity-based encryption without random oracles. In: Cramer, R. (ed.) EUROCRYPT 2005. LNCS, vol. 3494, pp. 114–127. Springer, Heidelberg (2005)
23. Kang, L., Tang, X., Lu, X., Fan, J.: A Short Signature Scheme in The Standard Model. Cryptology ePrint Archive: Report 2007/398 (2007)
24. Tian, H., Jiang, Z., Liu, Y., Wei, B.: A systematic method to design strong designated verifier signature without random oracles. Cluster Comput **16**, 817–827 (2013)

Workshop on Advances in Information Systems and Technologies for Emergency Preparedness and Risk Assessment (ASTER 2015)

CBR Method for Risk Assessment on Power Grid Protection Under Natural Disasters: Case Representation and Retrieval

Feng Yu[✉], Xiangyang Li, and Shiying Wang

School of Management, Harbin Institute of Technology, Harbin, China
{fengyu,xiangyangli}@hit.edu.cn, shiying_W@126.com

Abstract. Risk analysis is always the pivotal part of emergency preparedness for critical infrastructure protection such as power grid and traffic network. The main contribution of this paper is to employ case-based reasoning (CBR) method (combines case representation and retrieval) to illustrate a risk assessment framework for protecting power grid. It focuses on two core key parts: (1) Using ontology model to express precursors of risk, the described semantic network contains sub-concepts to outline selected precursors from hazards, environment, responders and physical system and (2) by analysis of emergency scenario precursors with sub-concept similarity and fuzzy value similarity calculation, the potential risks could be recognized to assist to retrieve the past knowledge, and effective and feasible actions would be taken to decrease the threats or cut disaster loss for power grid. Via a case study, the result shows that the proposed method extends the applicability of conventional CBR technique to numbers of real-world settings.

Keywords: Case-based reasoning (CBR) · Emergency management · Power grid protection · Risk assessment · Case representation and retrieval

1 Introduction

Learning from experience is a fundamental way that helps individuals or organizations to improve and avoid previous mistakes [1]. However, in the meanwhile, as the key mainstay of energy for national product, poor safety risks under natural disasters always pose threat to power grid. Once electric power facilities are damaged, its harm will cause the chain reaction on other critical infrastructure networks such as water, traffic, communication network. According to disaster preparedness workbook [2], critical infrastructure (CI) protection involves activities taken to prevent or minimize damage to critical infrastructure. It requires, first, that a repository assesses vulnerability or risk of

This work is supported by the National Science Foundation of China (NSFC) under the Grant Nos. 91024028, 91024031, 91324018, 71172156 and 71473058.

© Springer International Publishing Switzerland 2015
O. Gervasi et al. (Eds.): ICCSA 2015, Part I, LNCS 9155, pp. 113–125, 2015.
DOI: 10.1007/978-3-319-21404-7_9

power grid to floods, earthquakes, hurricanes and other natural disasters. Second, it contains actions or tasks to prevent or reduce the impact of disasters. Preventive or protective work takes a variety of forms: installing sensors for detecting, preparing the sufficient emergency resources, recognize the vulnerability and so on as a contingency plan. Furthermore, the protection strategy is a part of contingency plan which includes discrete lists of facts, resources, procedures, priorities and options. These items are brought together to form a coherent working strategy that will guide policies and actions in a pre-disaster situation and also on day-to-day basis.

In the context of uncertainty and complexity, historical cases could be treated as the most critical resource for emergency decision-making. Past risk reduction knowledge should be incorporated into the new risk assessment with help of artificial intelligence (AI) tools like case-based reasoning (CBR) technique [3]. The CBR system provides a method simulating the human way of cognition and learning which generates the proper solution for the target problem by adapting previously successful similar cases. Furthermore, these related cases illustrated by domain knowledge are stored in case base which a memory which contains a collection of emergency cases that is used in the context of the CBR methodology for the purpose of performing a reasoning task.

A classical CBR technique called 4R model is composed of four sequential steps which refers to: (1) retrieval: obtain the most relevant case by using suitable similarity measure, (2) reuse: reuse information and knowledge of previous cases, (3) revise: use soft computing method to adapt the suggested solution as necessary, and (4)

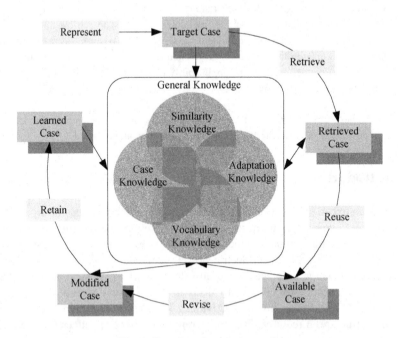

Fig. 1. Case-based reasoning cycle

retain: store the valuable cases and update the case base[4], and also include representation and the classification of case base [5] (see Figure 1). Adapting CBR technique is able to increase the amount of available knowledge and give emergency decision-makers effective psychological hints and inspiration under time pressure [6]. Especially, when dealing with the complex emergencies, time window is compressed extremely and it is difficult to grasp the situation of disaster. Therefore, similar cases become a powerful and effective experience and knowledge for inferring the unfolding disaster scenarios and its risk management plan.

In this paper, CBR tools are introduced to identify possible risks under natural disaster that would be able to discover available clues and information as feedback used to strengthen power grid. The rest of the paper is organized as follows: The next section discusses the related work. In section 3, for expressing case, ontology model is proper to describe the sub-concepts and relationship amongst a vast number of precursors. Section 4 presents the case retrieval method including two components, local similarity with sub-concept similarity and fuzzy value similarity and global similarity. Section 5 shows a case study as explanation in Shenzhen City, China. The paper closes with ideas on future steps and conclusions that summarize the main view of us. We aim to test the feasibility of the CBR method on improving and sustaining the capability of preparing for and protecting against all natural hazards for emergency operation center. It is hoped that the question will be resolved with our proposed approach.

2   Related Works

Given the irregular, complicated and uncertain evolution of risk scenario caused, it could not be denied that source cases still have value on getting the solutions by the instruction of the historical cases. Therefore, much of the CBR researchers have put effort on risk assessment. In this section, we review the directly relative studies for risk analysis based on CBR, in particular, in area of emergency management.

In safety management of construction, Mendes et al. [7] aimed to generate proper coastal well design planning by referencing source cases which can reduce risk. Zhao et al. [8] proposed a learning HAZOP expert system (hazard & operability) using CBR technique which the case was expressed by ontology. With respect to construction hazard identification, Goh and Chua [9] focused on two of the core parts of CBR (representation & retrieval) to automatically search the most similar case for facilitating systematic feedback of past knowledge. As can we see from the researches, successful cases indicate that the workable measures can avoid the risk event. And effective experience would assist to dealing with new similar scenario and obtain the same consequence.

Especially, in the area of emergency management, CBR approach always integrated with compatible knowledge representation models and soft computing methods to analyze risks. Balducelli and D'Esposito [10] developed related AI tools and applied to fire risk management and contingency planning, and the CBR method combined genetic algorithm and analog simulation to optimize emergency resources management and evacuation. Remm [11] proposed a CBR approach to estimate the risk of enterobiasis in Estonia's nursery school and prove the validity of previous events in keep disease away. Lu, et al. [12] referenced Goh's work and put forward the method on subway risk management that comparing the precursors between source cases and target case. Consider emergency scenario caused by natural disasters, emergency risk is involved in the whole process of incident development that is the case showing the relevance among each phase.

To sum up, Goh and Chua's and Lu, et al.'s frameworks are the inspiration for our work, but we improve retrieval calculation to make the most similar case more effective. However, few studies have yet been done on using historical cases for emergency preparedness and risk management. For this reason, the purpose of this paper is to discuss the CBR representation and retrieval method for power grid protection.

3   Case Representation Model

The intelligibility and degree of structure of case representation directly have impact on the validity and efficiency of reasoning results. The original and simple way to represent a case is by using feature-value pairs [13]. In the field of emergency, most of the case frames are organized based on case three-tuple which is represented as: case::=<problem, solution, evaluation>. In this research, Ontology model is used to express case. The term "ontology" is defined as the conceptualization of terms and relations between the concepts explicitly and formally in a domain [14]. Ontologies enable us to construct case structure with domain knowledge, and it involves a set of key concepts and semantic relationships. There are a variety of standard ontology languages such as Web Ontology Language (OWL), Resource Description Framework (RDF) and Extensive Markup Language (XML).

3.1   Concepts of Disaster Scenario for Power Grid

Before describing an emergency case totally, the concept sets of disaster scenario for power grid should be established. A disaster scenario involves in the case that is a logical and sequential entity of relative features. Gilboa's model [15] presented that the case structure contains problems, solutions and the results description formed as $C \equiv P \times S \times R$, so in this study, with respect to risk management, case three concept sets, "Risk Precursors", "Safety Risk", and "Risk Measures", are proposed. Suppose that a case C consists of risk precursor space $RP = rp_1 \times rp_2 \times ... \times rp_i \times ...rp_{n1}$, $RP \in \mathbb{R}^{n1}$; safety risk space $SR = sr_1 \times sr_2 \times ... \times sr_j \times ...sr_{n2}$, $SR \in \mathbb{R}^{n2}$; and risk solution space $RM = rm_1 \times rm_2 \times ... \times rm_k \times ...rm_{n3}$, $RM \in \mathbb{R}^{n3}$. It is a three-tuple structure. So, let us assume that a case is a u-tuple $(\{rp_i\}, sr_j, \{rm_k\}) \in \mathbb{R}^{n1} \times \mathbb{R}^{n2} \times \mathbb{R}^{n3}$, where $\{rp_i\} \in \mathbb{R}^{n1}$ is the set of value of RP observed in this case for one specific risk $sr_j \in \mathbb{R}^{n2}$ and corresponding several solutions $\{rm_k\} \in \mathbb{R}^{n3}$.

Precursor as suitable description for reasons of risk is the most important element in the concept system of disaster scenario. A broad definition of a precursor was the conditions, incidents, clues, and signals that result in risk occurs [16]. For power grid management, our previous studies had put effort on finding and decomposing the failure mode which can give rise to failed emergency management [17-18]. Now, it shows that power grid system with similar precursors tends to generate similar risks due to similar factor combination under natural disaster scenario. To put it another way, historical cases for risk management in phase of emergency preparedness could provide enough precursors and corresponding actions to mitigate or even avoid. The precursors come from the hazard, environment, physical system, and responders.

3.2   Ontology Expression of Potential Risk Case

Using a semantic network to describe the connection between precursors, power grid (PG) risk and PG safety measures can depict the relationship among them and make sure that the specific precursors and risk can point to the right solutions. In the ontology model, we divide two parts called conceptual level and instance level. Then the relationship between precursors and risk is shown as "hasRisk", and the connection of the two with safety measure is "hasMeasure". The PGS Ontology (Power Grid Safety Ontology) as the domain ontology outlines the static and physical descriptive knowledge or objects, and the instance of power line trip under typhoon inherit the property of conceptual model with real content. Therefore, the ontology model of a case can be constructed for risk assessment as shown in Figure 2.

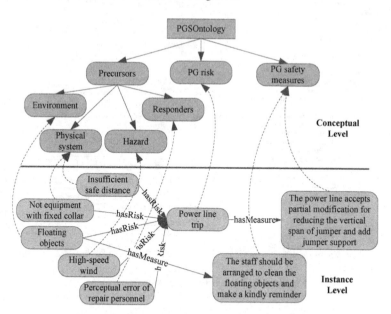

Fig. 2. Ontology model of power line trip instance

4   Case Retrieval Method

As we known, the goal of retrieval is to determine the case that is most similar to the new problem. Retrieval begins when the new problem is readily available and completes when a case is retrieved. It obtains the most relevant case by using hybrid similarity measure [19-20]. In this research, two precursors are compared by using semantic network similarity measure and fuzzy value similarity measure.

4.1   Precursors Semantic Network

Before calculating the similarity of the case, precursors would be treated as the feature sets to construct semantic network. On the basis of investigations in Shenzhen city, the four major components (environment, hazards, physical system and responders) contribute to the power grid risk, which can be decomposed into lots of sub-concepts related to causes as precursor base. Here, we use Protege 4.3 to outline the semantic network of power grid risk precursors under natural disasters, and take an example of "repair personnel" in Figure 3 [12].

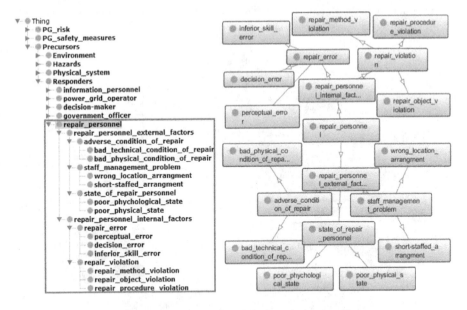

Fig. 3. A semantic network for "repair personnel"

4.2   Similarity Scoring Approach

The similarity scoring approach in retrieval process has three main steps. First, the right case group should be located in case base based on relevance (let assume that the case base for power grid has been well-designed). Relevance can be defined by emergency decision-makers depending on the disaster scenario. After that, the sub-concept similarity should be computed by comparing with the depth in semantic network and the degree of path overlapped. Then, the fuzzy value similarity due to a match on the precursor's contribution on power grid risk can be calculated. At last, each precursor's similarity as local similarity can be integrated as global similarity.

Relevance

Relevance marked as $R_{t,s}$ between the source case and target case refers to the similarity of domain of the two cases. In this research, we define the type and indexes of natural disaster determine the relevance.

$$R_{t,s} = \begin{cases} 1, & if \text{ the same hazard with same indexes} \\ \mu, & if \text{ the same type of hazard with similar indexes} \\ 0, & if \text{ different type of hazard} \end{cases} \tag{1}$$

Local Similarity

Concept Similarity

Before comparing the concept similarity, the number of precursor involves the number of sub-concepts of environment, hazards, physical system, and responders. Here, we use h_1, h_2, h_3, h_4 and g_1, g_2, g_3, g_4 to represent the number of the four components of target case and source case respectively [9,12].

$$Sim(rp_x^{tf}, rp_y^{sf}) = \sum_{su=1}^{m1} \omega_{su} \Big/ \Big(\sum_{su=1}^{m1} \omega_{su} + \sum_{dv=1}^{m2} \omega_{dw} \Big) \tag{2}$$

$$F_{t,s} = \sum_{y=1}^{g1} \varphi_{1i} \max\{Sim(rp_{1x}^{tf}, rp_{1y}^{sf})\} + \sum_{y=1}^{g2} \varphi_{2i} \max\{Sim(rp_{2x}^{tf}, rp_{2y}^{sf})\}$$
$$+ \sum_{y=1}^{g3} \varphi_{3i} \max\{Sim(rp_{3x}^{tf}, rp_{3y}^{sf})\} + \sum_{y=1}^{g4} \varphi_{4i} \max\{Sim(rp_{4x}^{tf}, rp_{4y}^{sf})\} \tag{3}^{[1]}$$

Where, $Sim(rp_x^{tf}, rp_y^{sf})$ refers to the sub-concept similarity of each precursor, ω_{su} is weight of the common sub-concept u, $u = 1, 2, ..., m_1$, and ω_{dw} is weight of sub-concept w, $w = 1, 2, ..., m_2$, that is part of only rp_x^{tf} or rp_y^{sf}. $F_{t,s}$ is sub-concept similarity of target case and source case, φ_i is the weight of precursor.

---

[1] The subscript of the variables added into 1,2,3,4 like φ_{1i} and rp_{3x}^{tf} represents the corresponding component (1.hazard, 2.environment, 3.physical system, and 4.responders).

Furthermore, in the semantic network, the sub-concept in higher level has higher impact on determining the type of precursor. By calculating the sub-concept similarity, the weight would be distributed by the Eq. (4) following:

$$\omega_z = 1 - \sum_{l=1, w \neq z}^{m3} LN_l \Bigg/ \sum_{l=1}^{m3} LN_l \tag{4}$$

Where ω_z is the sub-concept z, and LN_l means the number of level of sub-concept l, $l = 1, 2, ..., m_3$.

Moreover, the weight of φ_i can be calculated by the Eq. (5):

$$\varphi_{1i}, \varphi_{2i}, \varphi_{3i}, \varphi_{4i} = \frac{1}{\max(h_1, g_1) + \max(h_2, g_2) + \max(h_3, g_3) + \max(h_4, g_4)} \tag{5}$$

Fuzzy Value Similarity

The value of each precursor refers to the degree of contribution to power grid risk. However, it is difficult to express by using crisp number at pre-disaster stage. In this research, triangular fuzzy number is employed to measure the capacity of precursors to cause risk event.

In Eq. (6), the merits of the inverse exponential function used as the similarity function are: it can better match human notions of similarity and it can also better satisfy the properties and requirements of similarity measure, such as symmetry, reflexivity and multiplicative transitivity, etc. [20].

$$\begin{cases} Sim(rp_x^{tv}, rp_y^{sv}) = \exp\left(-\dfrac{\sqrt{(m_t - m_s)^2 + (n_t - n_s)^2 + (l_t - l_s)^2}}{\max \sqrt{(m_t - m_s)^2 + (n_t - n_s)^2 + (l_t - l_s)^2}}\right) \\ Sim(rp_x^{tf}, rp_y^{sf}) = 1 \end{cases} \tag{6}$$

$$\begin{cases} V_{t,s} = \displaystyle\sum_{i=1}^{m4} \delta_i Sim(rp_x^{tv}, rp_y^{sv}) \\ \delta_i = 1/m_4, \ Sim(rp_x^{tf}, rp_y^{sf}) = 1 \end{cases} \tag{7}$$

Where, $Sim(rp_x^{tv}, rp_y^{sv})$ refers to the fuzzy value similarity of each precursor, $\max \sqrt{(m_t - m_s)^2 + (n_t - n_s)^2 + (l_t - l_s)^2}$ is the maximum of risk value of source case and target case, $m_t, n_t, l_t \in [0,1]$ and $m_s, n_s, l_s \in [0,1]$ is the value of triangular fuzzy number of source case and target case respectively. $V_{t,s}$ is the fuzzy value similarity of target case and source case, δ_i is the weight of precursors with the same sub-concept, $i = 1, 2, ... m_4$.

Global Similarity

A similarity measure for a problem space P is a function: $Sim : P \times P \rightarrow [0,1]$. The global similarity can be calculated by Eq. (8) below.

$$Sim(C_t, C_s) = f(R_{t,s}, F_{t,s}, V_{t,s}), \quad R_{t,s}, F_{t,s}, V_{t,s} \in [0,1] \tag{8}$$

$$f(R_{t,s}, F_{t,s}, V_{t,s}) = \begin{cases} F_{t,s}\varepsilon_F + V_{t,s}\varepsilon_V, & \text{if } R_{t,s} = 1 \\ R_{t,s}(F_{t,s}\varepsilon_F + V_{t,s}\varepsilon_V), & \text{if } R_{t,s} \in (0,1) \\ 0, & \text{if } R_{t,s} = 0 \end{cases} \tag{9}$$

Where $Sim_{t,s}$ represents the global similarity of target case and source case, also illustrate the extent of availability of the similar case. Furthermore, ε_F and ε_V are weight of local similarity and fuzzy value similarity severally.

5   Case Study

5.1   Background

Shenzhen power gird is oversize urban power network with the largest power load density, advanced reliability of power supply and provincial power grid scale in China. Shenzhen always struck by typhoon for around five times per year. However, until to now, there is still no effective way to make full use of the historical cases for power grid protection. In year of 2012, the attack from No.1208 typhoon "Vincent" was a serious incident to let emergency management realize that the emergency preparedness for power grid is not enough. So, this case study shows the application on power gird risk assessment which aims to illustrate how the most similar case can be found out based on proposed ontology model and similarity method.

By collecting the cases in Shenzhen City, in order to guarantee $R_{t,s} = 1$ and reduce the complexity of retrieval, we select the code "T120801" case base (T1208 means No. 1208 typhoon "Vincent", and 01 means the number of power line trip accident), which includes 9 cases that belong to power line trip category (see in Table 1). Given a target case typhoon "Usagi" of 110kv Yangli II Line, the similarity between the source cases and target case can be calculated.

Table 1. The cases in "T120801" base

Code	Power line name	Frequency	Fault location
SC$_1$	110kv Wusha II Line	2	N26
SC$_2$	110kv Baide Line	1	/
SC$_3$	110kv Liliyi I Line	2	/
SC$_4$	110kv Longli Zhizhan II Line	1	N32-N33
SC$_5$	110kv Zhengji Line	1	N4
SC$_6$	110kv Danfen II Line	1	N10
SC$_7$	220kv Lingwan Line	1	N22
SC$_8$	500kv Lingkun II Line	1	N71
SC$_9$	500kv Lingshen II Line	5	N108

5.2   Results and Discussions

The target case (TC) mentioned above has four precursors, "wrong location arrangement of repair personnel", "insufficient safe distance", "high speed wind", and "automatic system failure for 110kv line", and also the precursors of the first source case (SC$_1$) as comparison shown in Table 2 (SC$_1$ is an example for illustrating the processes).

As can we see from the Figure 4, it shows the two precursors, "short-staffed arrangement of repair personnel" and "wrong location arrangement of repair personnel". The round frame represents the same sub-concept, and the box is the different concept. According to Eq. (2) and (4), the weights of the five-level structure are 0.33, 0.27, 0.20, 0.13, 0.07, and the local sub-concept similarity of the two precursors is 0.87.

Table 2. The precursors of SC$_1$ and TC

Component	SC$_1$	Contrib.	TC	Contrib.
Environment	floating objects	(0.6,0.7,0.9)	/	/
Hazards	high-speed wind	(0.5,0.6,0.7)	high-speed wind	(0.5,0.6,0.7)
Physical system	insufficient safe distance	(0.5,0.7,0.8)	insufficient safe distance	(0.5,0.6,0.7)
	automatic system failure for 110kv line	(0.3,0.4,0.5)	automatic system failure for 110kv line	(0.3,0.5,0.6)
Responders	short-staffed arrangement of repair personnel	(0.3,0.5,0.6)	wrong location arrangement of repair personnel	(0.6,0.8,0.9)
	perceptual error of repair personnel	(0.2,0.3,0.4)		

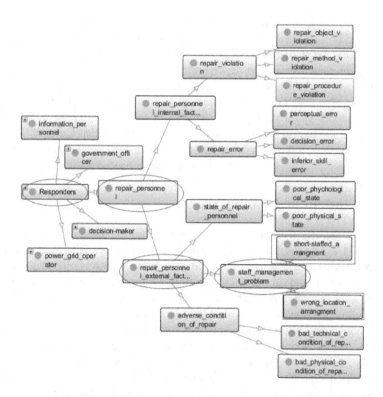

Fig. 4. Comparison between "short-staffed arrangement of repair personnel" and "wrong location arrangement of repair personnel"

After that, the local sub-concept similarity of precursor of "automatic system failure for 110kv line" is 1.00, so the second equation in Eq. (6) is satisfied with local fuzzy value similarity 0.61 ($\max \sqrt{(m_t - m_s)^2 + (n_t - n_s)^2 + (l_t - l_s)^2} = 0.28$). Therefore, using Eq. (2)-(9) circularly, we can obtain the final result shown in Table 3. The global similarity between SC_1 and TC is 0.62, but based on the method of [12] the result is 0.65. As it shown, the more real similarity can be achieved.

Table 3. The results of similarity

$Sim(rp_2^{tf}, rp_2^{sf})$	$Sim(rp_2^{tf}, rp_2^{sf})$	$Sim(rp_2^{tf}, rp_2^{sf})$	$Sim(rp_2^{tv}, rp_2^{sv})$	$Sim(rp_2^{tv}, rp_2^{sv})$
1	1,1,1,1	0.87	0.92	0.78,0.61
φ_{1i}	φ_{2i}	φ_{3i}	φ_{4i}	δ_i
1/6	1/6	1/6	1/6	1/3
LSS of Com.2	LSS of Com.3	LSS of Com.4	LFS of Com.2	LFS of Com.3
0.17	0.17,0.17	0.14	0.32	0.26
ε_F	ε_V	LSS of cases	LFS of cases	GS of cases
0.5	0.5	0.65	0.58	0.62

Note: LSS=local sub-concept similarity, LFS=local fuzzy value similarity, GS=global similarity, and Com. is the abbreviation of component.

Furthermore, the global similarities of other eight source cases (0.58, 0.42, 0.55, 0.39, 0.60, 0.35, 0.21 and 0.28 respectively) are less than the SC_1, so the most similar case is SC_1. Given a similarity baseline, the safety risk and safety measures for power grid can be automated evaluated depending on the first source case in Table 1. According to the survey and research we carried out in Shenzhen City for target case, the target line exactly existed risk with frequency of 2 and fault location in N89, and the solutions assuredly had a similarity with those of SC_1, which prove this method is valid to a certain extent. However, the difficulty is obvious that the new problem is not exactly like that in the case base, and even if the most similar and suitable case can be found out, it cannot be used exactly in the same way as it was used in the past.

6   Conclusions and Future Work

CBR tools are the effective and feasible ways of acquiring experience with great reference value. Case representation and retrieval are the core parts to guide emergency decision-makers toward available knowledge to protect power grid. We extend the similarity measure based on ontology model by adding fuzzy number to describe the precursors' contribution to risk. Therefore, the result is more exact than previous studies, which embodies more practical value. Also, we just introduce this method for protecting power grid and it can be employed to other critical infrastructures like traffic network. The most considerable limitation is how to deal with the time window, that is to say, if the precursors cause the risk in a short time or case base is too large to retrieve rapidly, the proposed method may be inefficient.

The scope of the previous methods is presently limited to the risk assessment for power grid. So, this is the next step to discuss the method of case reuse and revise after obtain the most similar source case. Furthermore, in the context of more complex scenario, interdependency between different critical infrastructures should be considered in risk analysis in future research work. It is also worthy to explore and expand the proposed method with the consideration of resilience for enhancing the availability of retrieval.

References

1. Goh, Y.M., Chua, D.K.H.: Case-based Reasoning Approach to Construction Hazard Identification: Case Representation and Retrieval. Journal of Construction Engineering and Management **136**(2), 170–178 (2009)
2. Patkus, B., Schnare, R.E.: Disaster Preparedness Workbook for Cultural Institutions Within the Military. Department of Defense Legacy Resource Management Program, U.S. Naval War College Library (2009)
3. Schank, R., Abelson, R.: Scripts, Plans, Goals and Understanding. Erlbaum, Hillsdale (1977)
4. Aamodt, A., Plaza, E.: Case-based Reasoning: Foundational Issue. Methodological Variation, and System Approaches, AI Com-Artificial Intelligence Communication **7**(1), 39–59 (1994)

5. Gavin, F., Sun, Z.H.: R5 Model for Case-Based Reasoning. Knowledge-Based Systems **16**, 59–65 (2003)
6. Kleindorfer, P., Saad, G.: Managing Disruption Risks in Supply Chains. Production and Operations Management **14**(1), 53–68 (2005)
7. Mendes, J.R.P., Morooka, C.K., Guilherme, I.R.: Case-based Reasoning in Offshore Well Design. Journal of Petroleum Science and Engineering **40**(1–2), 47–60 (2003)
8. Zhao, J.S., Cui, L., Zhao, L.H., Qiu, T., Chen, B.Z.: Learning HAZOP Expert System by Case-based reasoning and Ontology. Computer Chemistry Engineering **33**(1), 371–378 (2009)
9. Goh, Y.M., Chua, D.K.H.: Case-based Reasoning for Construction Hazard Identification: Case Representation and Retrieval. Journal of Construction Engineering and Management **135**(11), 1181–1189 (2009)
10. Balducelli, C., D'Esposito, C.: Genetic Agents in an EDSS System to Optimize Resources Management and Risk Object Evacuation. Safety Science **35**(1–3), 59–73 (2000)
11. Remm, M., Remm, K.: Case-based Estimation of the Risk of Enterobiasis. Artificial Intelligence Medical **43**(3), 167–177 (2008)
12. Lu, Y., Li, Q.M., Xiao, W.J.: Case-based Reasoning for Automated Safety Risk Analysis on Subway Operation: Case Representation and Retrieval. Safety Science **57**, 75–81 (2013)
13. Vong, C.M., Wong, P.K., Ip, W.F.: Case-based Expert System Using Wavelet Packet Transform and Kernel-based Feature Manipulation for Engine Ignition System Diagnosis. Engineering Application Artificial Intelligence **24**(7), 1281–1294 (2011)
14. Gruber, T.R.: A Translation Approach to Portable Ontology Specifications. Knowledge Acquisition **5**(2), 199–220 (1993)
15. Gilboai, I., Schmeidler, D.: Case-based Decision Theory. Quarterly Journal of Economics **110**, 605–639 (1995)
16. Phimister, J.R., Bier, V.M., Kunreuther, H.C.: Accident Precursor Analysis and Management: Reducing Technological Risk through Diligence. National Academy Press, Washington, D.C. (2004)
17. Feng, Y., Xiangyang, L., Yingxiong, L.: Discovery method on failure factors of collaborative allocation under power grid emergency. In: International Conference on Collaboration Technologies and Systems, U.S. San Diego, pp. 281–286 (2013)
18. Feng, Yu., Li, X., Yue, G.: Fault-tree Analysis for Power Gird Emergency Logistics System under Large-scale Natural Disaster. Applied Mechanics and Materials **986–987**, 311–314 (2014)
19. Tadrat, J., Boonjing, V., Pattaraintakorn, P.: A New Similarity Measure in Formal Concept Analysis for Case-based Reasoning. Expert Systems with Applications **39**(1), 967–972 (2012)
20. Fan, Z., Li, Y., Wang, X., Liu, Y.: Hybrid Similarity Measure for Case Retrieval in CBR and Its Application to Emergency Response towards Gas Explosion. Expert Systems with Applications **41**(5), 2526–2534 (2014)

Urban Power Network Vulnerability Assessment Based on Multi-attribute Analysis

Jun Li[✉], Xiang-yang Li, and Rui Yang

School of Management, Harbin Institute of Technology, Harbin 150001, China
lijun_85@live.cn, xiangyangli@hit.edu.cn, yangrui49@163.com

Abstract. Power network is usually the vital part of critical infrastructure protection. Confirming the level of vulnerability of the power network forms the basis of making a sound protection mechanism. This paper attempts to build a power network vulnerability analysis framework by introducing the multi-attribute analysis logic into power network vulnerability assessment and studying the power network vulnerability under the multi-attribute, dynamic and static combined conditions. Compared to one-attribute vulnerability analysis, multi-attribute analysis proves that a comprehensive physical vulnerability could reveal the power network vulnerability in the case of multi-attribute integration as a whole, and then we can use the result to confirm the hazard level for different parts of the power network. Finally, this paper illustrated the feasibility of the method through assessing the power network vulnerability of the city of Q.

Keywords: Power network · Vulnerability · Multi-attribute analysis · Complex systems

1 Introduction

Conducting the vulnerability assessments on the power network is an important method to enhance capacity of disaster handling. Considering the definition of vulnerability, there are some subtle differences, for example, vulnerability refers to the possibility that individuals or groups are exposed to hazards [1], vulnerability is the degree to which systems and subsystems are exposed to dangers [2], and so on. But, there is one thing in common; it is that vulnerability is the nature of susceptibility to damage and loss when the hazard-affected bodies are in a specific environment.

Power network is typically a complex network. Generally speaking the study of the critical infrastructure could be conducted by two different methods: the first one is abstracting the critical infrastructure as a complex network, where all the vulnerability assessment research is based on the complex network theory [3-5]; the second one is assessing the vulnerability of the critical infrastructure subclass, such as power

This work is supported by the National Science Foundation of China (NSFC) under the Grant Nos. 91024028, 91024031, 91324018, 71172156 and 71473058.

O. Gervasi et al. (Eds.): ICCSA 2015, Part I, LNCS 9155, pp. 126–140, 2015.
DOI: 10.1007/978-3-319-21404-7_10

network and road network [6-8]. In the assessment of power network vulnerability, network nodes and edges are usually identified by complex network theory [9], or sometimes combined with other theory to overcome the problem of biasness of complex network theory [10]. In the assessment of power operation, transient energy function approach is a way to analyze transient stability of a power system from the view of system energy [11], furthermore, artificial intelligence approach is a way to analyze whether the system is stable [12]. More researches have been conducted for other aspects of vulnerability assessment, such as studying the vulnerability under the disaster [13] and simple physical vulnerability [14]. Vulnerability is an important indicator to measure the health of power network, and any single vulnerability assessment could not reflect the comprehensive problem of power equipment. This text starts from the basic concept, assessed by three different perspectives, and then provides some models to assess the different parts of vulnerability. This paper argues that vulnerability can be seen from two perspectives, one is the degree of susceptibility to damage, and the other one is that if parts of the network are broken down, then the impact on the whole network comes from the parts.

2   Multi-attribute Composition of Physical Vulnerability

Vulnerability contains three aspects: running state vulnerability, component vulnerability and structural vulnerability. Indeed these three aspects are different attributes of vulnerability and all of them are parts of assessment system. We can only measure the vulnerability after one disaster has happened, but we couldn't perform a destructive experiment to measure the vulnerability when there is no problem with the equipment before any disaster happened, so vulnerability research is a forecast evaluation.

The study focuses on the two latter aspects, they are both direct observation of the power network equipment, rather than the running state and mechanism. So we named these two directly observable vulnerabilities as physical vulnerability. There is no uniform definition for the composition of vulnerability. The structural vulnerability and state vulnerability are generally illustrated separately [15]. Physical vulnerability is defined as the possibility of change of the physical state [16]. We divided the physical vulnerability into three areas: inherent vulnerability, which means the vulnerability with no hazard and includes natural aging; structural vulnerability which may occur due to networks and induced vulnerability which occurs due to the external influence such as deliberate destructions, natural disasters and technical errors. Inherent vulnerability evaluates the object itself, structural vulnerability evaluates the correlation and induced vulnerability evaluates the interaction between the object and external influences. Since the three vulnerabilities are not in the same dimensionality, they could be seen as the components of a vulnerability vector which could be assessed in a three-dimensional coordinate system in order to avoid the shortcoming of one-attribute assessment. We rename inherent vulnerability as inherent component, structural vulnerability as network structure component and induced vulnerability as induced component. Fig.1 shows the composition of physical vulnerability. V_c Represents inherent component, V_n represents network structure component, V_d represents induced component. The three components constitute a curved surface. Every point in the curved surface represents the vulnerability in one condition.

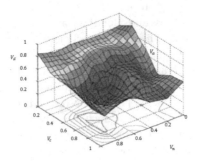

Fig. 1. Composition of physical vulnerability

3   Vulnerability Assessment Model

3.1   The Basic Model

The results will be more accurate if we calculate completely in accordance with the actual situation. But it will also bring some adaptability problems to the model, and paying too much attention to details may lead to a deviation from the actual situation. Therefore, we proposed the following three basic assumptions to solve the problem:

- Hypothesis 1. Each equipment or groups in the grid are abstracted into a node, each overhead line between two equipment or groups are abstracted into an edge.
- Hypothesis 2. In this paper, the health status of the worst equipment represents the health status of the groups; generally the health of one group depends on the shortest board of the bucket.
- Hypothesis 3. Power grid construction and maintenance are governed by different areas. In the same area ,the overhead lines and equipment are the same as each other, the anti-disaster index and depreciation rate of every overhead line and equipment are the same, and they are all different form each other in different areas.

We build a basic relation model:

$$V_a = f(V_c, V_n, V_d) \tag{1}$$

Where V_c represents inherent component, V_n represents network structure component, V_d represents induced component and V_a represents the vulnerability vector.

3.2   Inherent Component of Vulnerability

Inherent component describes the damage of nodes or lines which are caused by the equipment and line aging and changes in the mechanical properties. Inherent component is composed of inner vulnerability and environmental pollution vulnerability, inner vulnerability means the damage which due to design, material selection, unreasonable structure, the defect of processing and installation and so on. Environmental pollution vulnerability means the damage which due to natural environment and pollution. So node inherent component V_c can be divided into inner vulnerability V_i and environmental pollution vulnerability V_e.

Inner Vulnerability

Inner vulnerability could be measured through two ways. One is to predict the health of the equipment by design index, but this method will not be accurate after a few years because it is much dependent on the design index which could be changed when the equipment has been repaired. The other one is to monitor the health of the equipment every year or month to find the main vulnerable node, but this method only concerns the instantaneous state and doesn't pay attention to the trend of health. To combine the two methods could avoid the shortcomings of each one.

We could use CBRM (Condition Based Risk Management by EA Company) to predict the health or risk:

$$HI_t = HI_0 \times e^{B \times (T-T_0)} \tag{2}$$

$$B = \frac{\ln HI_b / \ln HI_0}{n} \times k \tag{3}$$

HI represents health index, range from 0-10, 0 is the best and 10 is the worst. When $0 < HI < 3.5$, it means the equipment is in the early stage of aging and the equipment performance is good; when $3.5 < HI < 5.5$, it means the equipment's failure rate begun to rise; when $5.5 < HI < 7$, it means the equipment aging is serious; when $7 < HI < 10$, it means the equipment is in worst state and it needs to be replaced. HI_t represents the health index in t years. HI_0 presents the initial health state, generally $HI_0 = 0.5$. HI_b represents the health index when the failure rate begins to rise quickly, generally $HI_b = 5.5$ or 6.7. B is aging constant which is estimated, T is date for assessment, T_0 is operational date, and k is environmental correction factor and n is life expectancy.

For monitoring the health, we could refer to "Guide for Condition Evaluation of Overhead Transmission Line (China)", SI is health index, range from 0-100. When $SI_0 = 100$, it means the equipment is in best health state, SI_t is the health state to assess:

$$SI_t = SI_0 - SI_T \tag{4}$$

We combined SI_t with HI_t, with SI_t as an adjusted value,

$$V_i = \beta HI_t + \gamma \overline{D} = \beta HI_0 \times e^{B \times (T-T_0)} + \gamma \left(SI_0 - SI_T \right) \tag{5}$$

Environmental Pollution Vulnerability

Environmental pollution has a harmful effect on the grid. The influence mainly comes from two types of pollution, one is air pollution, the other is acid rain, and so environmental pollution vulnerability could be shown as:

$$V_e = (\lambda_{aqi} E_{aqi} + \lambda_{ac} E_{ac}) \times SI_0 \tag{6}$$

V_e represent the environmental pollution vulnerability index, range from 0-100, the conversion rate which is leaded by air pollution is λ_{aqi}, the conversion rate which is leaded by acid rain is λ_{ac}. We can choose the representative index to assess the vulnerability, such as (7), (8). AQI is air quality index, PH is PH value, and p_c is the probability of acid rain.

$$E_{aqi} = AQI \tag{7}$$

$$E_{ac} = p_c(7 - PH) \tag{8}$$

Combined (5) and (6), inherent component could be represented as (9):

$$V_c = f(V_i, V_e) \tag{9}$$

3.3   Network Structure Component of Vulnerability

Power network is a complex network, structural vulnerability could reflect the importance of node and line in the network. Connectivity factor, path length, clustering coefficient and centrality could be used to measure structural vulnerability, but these factors are too simple to measure accurately between to similar nodes or lines. This paper argues that network structure component could be assessed by network efficiency index:

$$E = \frac{1}{N(N-1)} \sum_{i \neq j} \frac{1}{d_{ij}} \tag{10}$$

E represents network efficiency index, range from 0-1; N is the number of nodes; d_{ij} is the shortest path between two nodes, when there is no connection between two node, $d_{ij}=0$. So the loss of network efficiency is:

$$E_l = E_t - E_{t+n} = \frac{1}{N(N-1)} \sum_{i \neq j} \left(\frac{1}{d_{ij}^t} - \frac{1}{d_{ij}^{t+n}} \right) \tag{11}$$

3.4   Induced Component of Vulnerability

Induced component is a main part of physical vulnerability. When the different parts of power grid are exposed to the same disaster, the vulnerability of them is different, and when the same part of the power grid are exposed to the same disaster with different level, the vulnerability will also be different. So we could show V_d as:

$$V_d = f(I_d, D_d, G_d) \tag{12}$$

I_d is anti-disaster ability, D_d is disaster grade, G_d is geological condition. The greater I_d, G_d the better, the greater D_d the worse, the range of them is 0-10:

$$V_d = \frac{D_d}{I_d \times G_d} \times SI_0 \qquad (13)$$

So, the range of V_d is 0-10.

3.5   Assessment Method of Power Network Physical Vulnerability

Inherent component is composed of inner vulnerability and environmental pollution vulnerability (9), so we could rewrite (1) as:

$$V_a = f(V_i, V_e, V_n, V_d) \qquad (14)$$

The dimensions of variables in the model are different, so we need to normalize the variables to 0-1. V_a is made up of multi-attribute, so vulnerability assessment is a multiple attribute decision making problem. Weighted sum method, weighted product method and TOPSIS method could be used to solve the multiple attribute decision making problems. Because weighted sum method concerns maximum value, weighted product method concerns minimum value, both of them couldn't reflect the overall characteristics individually. Whereas TOPSIS method can reflect the trend to optimal values, so this paper chooses TOPSIS (Technique for Order Preference by Similarity to an Ideal Solution) method [17] to evaluate V_a. The process of TOPSIS method is: build normalized matrix, multiply by weight matrix $w = (w_i, w_e, w_n, w_d)^T$ to get positive ideal solution and negative ideal solution, compute   the Euclidean distance, and in the end compute proximity C. The greater is the worse for C.

Through (2)-(14), we could build a decision matrix to compute the weight of every variable by information entropy method. First of all we build a decision matrix which is made up of V_i, V_e, V_n, V_d, and then normalize it to get matrix R, column vector of R is $(r_{1i}, r_{2i}, \ldots, r_{mi})^T, (j=1,2,\ldots,n)$, which could be seen as information distribution, so the entropy of V is

$$E_j = -k \sum_{i=1}^{m} r_{ij} \ln r_{ij}, k = 1/\ln m, j = 1, 2, \cdots, n \qquad (15)$$

Degree of distinction is

$$L_j = 1 - E_j, j = 1, 2, \cdots, n \qquad (16)$$

Weight is

$$w_j = \frac{L_j}{\sum_{j=1}^{n} L_j}, j = 1, 2, \cdots, n \qquad (17)$$

And then we get the optimization ranking through TOPSIS.

4   The Application

Take an example of Q city in China, the power network topological graph is depicted in Fig.2, there are 32 nodes and 38 lines. This example studied the vulnerability of nodes in the network; put aside the lines, because the vulnerability of lines is the same with nodes.

Fig. 2. 110KV and above power network topological graph of Q city

4.1   Node Inherent Component of Vulnerability Analysis

Inner Vulnerability Analysis

Eight companies are responsible for the maintenance and construction of power grid in Q city. Related information of the entire node is in Table 1.

Table 1. Nodes Information of power network in Q city in 2013

Node	Area	Life	HI_b	$T-T_0$	Node	Area	Life	HI_b	$T-T_0$	Node	Area	Life	HI_b	$T-T_0$
1	A	25	5.5	15	12	B	25	5.5	14	23	F	25	5.5	15
2	A	25	5.5	10	13	C	25	5.5	8	24	G	35	5.5	15
3	A	25	5.5	16	14	C	25	5.5	20	25	G	35	5.5	26
4	A	25	5.5	18	15	C	25	5.5	8	26	H	35	5.5	8
5	A	25	5.5	13	16	C	25	5.5	16	27	H	35	5.5	16
6	A	25	5.5	9	17	D	25	5.5	18	28	H	35	5.5	11
7	A	25	5.5	10	18	E	35	5.5	19	29	H	35	5.5	12
8	A	25	5.5	13	19	E	35	5.5	8	30	H	35	5.5	5
9	B	25	5.5	19	20	E	35	5.5	20	31	H	35	5.5	7
10	B	25	5.5	23	21	F	25	5.5	15	32	H	35	5.5	6
11	B	25	5.5	17	22	F	25	5.5	4					

According to Table 1, (2) and (3) we could get HI of the power network in Q city, as in Table 2.

Table 2. HI of nodes(CBRM) in 2013

Node	B	HI_t	Node	B	HI_t	Node	B	HI_t	Node	B	HI_t
1	0.0984	2.1870	9	0.0984	3.2415	17	0.0984	2.9377	25	0.0703	3.1076
2	0.0984	1.3373	10	0.0984	4.8044	18	0.0703	1.9002	26	0.0703	0.8772
3	0.0984	2.4130	11	0.0984	2.6625	19	0.0703	0.8772	27	0.0703	1.5390
4	0.0984	2.9377	12	0.0984	1.9821	20	0.0703	2.0386	28	0.0703	1.0831
5	0.0984	1.7964	13	0.0984	1.0984	21	0.0984	2.1870	29	0.0703	1.1619
6	0.0984	1.2120	14	0.0984	3.5765	22	0.0984	0.7411	30	0.0703	0.7105
7	0.0984	1.3373	15	0.0984	1.0984	23	0.0984	2.1870	31	0.0703	0.8177
8	0.0984	1.7964	16	0.0984	2.4130	24	0.0703	1.4346	32	0.0703	0.7622

The aging of node 10 and node 14 are obvious, and failure rate has begun to rise. The workers should pay more attention to maintenance. The HI of node 4, node9, node17, node25 are over or close to the threshold of 3, the worker should pay attention to them. Refer to "Guide for Condition Evaluation of Overhead Transmission Line (China)" to assess the health of power network. According to the impact on the security operation, we divided state into four grades, their weights are weight 1, weight 2, weight 3, weight 4, the coefficients of them are 1,2,3,4. According to the deterioration degree of the state, we divided it into four grades, I, II, III and IV, the basic deduction are 2,4,8,10, the actual deduction is equal to the basic deduction multiply by weight, as Table 3.

Table 3. State rating

deterioration degree of state	Basic deduction	Weight 1	Weight 2	Weight 3	Weight 4
I	2	2	4	6	8
II	4	4	8	12	16
III	8	8	16	24	32
IV	10	10	20	30	40

Refer to Guide for Condition Evaluation; every node could be measured by 5 states: insulating property(a), contact interface(b), accuracy(c), temperature rise(d) and exterior(e). According to (4), we could assess 32 nodes vulnerability of the power network in Q city, the result is in Table 4.

Table 4. SI of nodes in 2013

Node	a(W4)	b(W4)	c(W4)	d(W3)	e(W2)	SI$_t$	Node	a(W4)	b(W4)	c(W4)	d(W3)	e(W2)	SI$_t$
1	8					1.6	17			8	6		2.8
2			8	6	4	3.6	18			16			3.2
3				12		2.4	19	16			6	8	6
4	16		16			6.4	20				6	4	2
5		8				1.6	21		8				1.6
6	16			6		4.4	22	32		16			9.6
7				6		1.2	23		8		12		4
8			16		4	4	24				6	4	2
9		8		6		2.8	25		16				3.2
10			8			1.6	26		16	8		4	5.6
11	16			6	4	5.2	27				12		2.4
12			8	12		4	28			8	6	4	3.6
13		16				3.2	29		8			16	4.8
14				24	4	5.6	30	16	*		6		4.4
15					16	3.2	31			8		16	4.8
16	8	16				4.8	32			16		6	4	5.2

Environmental Pollution Vulnerability Analysis

The main damages of environmental pollution are pollution flashover and corrosion. The reason of pollution flashover is air pollution, the particulates and water in the air. So when we analyzed pollution flashover state, AQI could be a good indicator. The main reason of corrosion is acid rain, so the frequency and PH are appropriate indicators to represent the corrosion degree. We set the proportionality coefficient $\lambda_{aqi}=0.00015$, $\lambda_{ac}=0.0001$, according to (6) (7) (8) to assess the environmental pollution vulnerability. The result is in Table 5.

Table 5. Environmental pollution vulnerability and the influence factors

Area	AQI	PH	Frequency	E$_{aqi}$	E$_{ac}$	V$_e$
A	69	5.91	24.0	0.0104	0.002616	1.2966
B	75	5.52	29.0	0.0113	0.004292	1.5542
C	88	5.27	30.0	0.0132	0.00519	1.839
D	85	4.98	33.5	0.0128	0.006767	1.9517
E	72	5.04	29.5	0.0108	0.005782	1.6582
F	70	5.62	30.5	0.0105	0.004209	1.4709
G	65	5.98	30.5	0.0098	0.003111	1.2861
H	86	5.76	31.5	0.0129	0.003906	1.6806

4.2   Network Structure Component of Vulnerability Analysis

Complex network theory is main method of network structure research. With the development of power grid topology model, many evaluation indicators were presented, such as centrality with weight, percentage of load loss and economical index of load loss. In this paper, we assessed the vulnerability by network efficiency index. Degree of node means the number of the nodes which is connected with it. The greater the degree, the more important the node is in the network, and the more vulnerable it is. Table 6 shows the degree of every node.

Table 6. The degree of every nodes in power network

Node	1	2	3	4	5	6	7	8	9	10	11	12	13	14	15	16
Degree	1	6	1	1	3	3	2	2	4	1	2	4	2	4	2	1
Node	17	18	19	20	21	22	23	24	25	26	27	28	29	30	31	32
Degree	1	2	1	1	2	2	4	6	2	5	2	1	3	2	4	1

The degree could reflect the importance of the node and the loss if the node breaks down. But in fact, one node with greater degree may be not more vulnerable, because little flow may go through it and it may have a high anti-disaster index. We can see that in Table 6, many nodes have the same degree, but factually their vulnerability is not the same. So we use "network efficiency index" to reflect that the vulnerability is much better. In (10), we use Pajek and R software to solve the problem. When the network is complete, the network efficiency index is 0.337694. Whereas when we delete the nodes one at a time, then the network efficiency index will change as in Table 7.

Table 7. Network efficiency when delete the nodes(%)

Node	V_n	Node	V_n	Node	V_n	Node	V_n
1	31.875	9	29.036	17	32.183	25	30.447
2	26.151	10	32.12	18	30.287	26	27.605
3	31.875	11	30.277	19	32.13	27	31.048
4	31.875	12	26.249	20	31.768	28	32.193
5	31.606	13	30.456	21	31.666	29	30.669
6	31.101	14	27.928	22	31.686	30	31.732
7	31.937	15	30.456	23	29.455	31	29.939
8	31.272	16	31.769	24	24.322	32	32.032

We can see that some nodes have the same degree, but all the nodes have different network efficiency in Table 6 and 7. By comparing the two Tables we can find that even though the nodes have the same degree, when they are deleted the influences are different, it means they have different vulnerability. So network efficiency is more suitable for assessing vulnerability.

4.3   Induced Component of Vulnerability Analysis

Induced component of vulnerability is closely related to anti-disaster index, disaster grade and geological condition. We can assess the vulnerability from the three factors. Q city is in the east of China, the main natural disasters in the city are rainstorm, icing and landslide. By taking rainstorm as example, according to (13) we can get:

Table 8. Induced component of vulnerability

Node	I_d	D_d	G_d	V_d	Node	I_d	D_d	G_d	V_d	Node	I_d	D_d	G_d	V_d
1	7.5	6.82	9	10.1	12	7.5	7.8	9	11.56	23	8	7.01	7	12.52
2	7.5	6.82	9	10.1	13	7.5	8.32	9	12.33	24	8	7.01	4	21.91
3	7.5	6.82	6	15.16	14	7.5	8.32	9	12.33	25	8	7.01	6	14.6
4	7.5	6.82	5	18.19	15	7.5	7.8	9	11.56	26	8	7.17	6	14.94
5	7.5	6.82	9	10.1	16	7.5	7.8	9	11.56	27	8	7.17	6	14.94
6	7.5	6.82	5	18.19	17	7.5	7.43	9	11.01	28	8	7.17	9	9.96
7	7.5	6.82	9	10.1	18	7.5	7.01	9	10.39	29	8	7.17	9	9.96
8	7.5	7.8	6	17.33	19	7.5	6.1	9	9.04	30	8	7.17	9	9.96
9	7.5	7.8	9	11.56	20	7.5	7.01	9	10.39	31	8	7.17	9	9.96
10	7.5	7.8	9	11.56	21	8	7.01	7	12.52	32	8	7.17	9	9.96
11	7.5	7.8	9	11.56	22	8	7.01	7	12.52					

4.4   Summary of Power Network Vulnerability in Q City

According the components of vulnerability analysis in the first three sections of this chapter, we found that the greater V_i, V_e, V_n, V_d are, the more vulnerable the equipment is. Because the dimensions of these variables are different, we should normalize them first and the normalization results are in Table 9. Where V_c is composed of V_i and V_e.

Table 9. Vulnerability of Q city

Node	Area	V_i	V_e	V_n	V_d	Node	Area	V_i	V_e	V_n	V_d
1	A	0.211	0.016	0.040	0.083	17	D	0.375	1.000	0.001	0.153
2	A	0.217	0.016	0.768	0.083	18	E	0.265	0.559	0.242	0.105
3	A	0.285	0.016	0.040	0.475	19	E	0.295	0.559	0.008	0.000
4	A	0.580	0.016	0.040	0.711	20	E	0.215	0.559	0.054	0.105
5	A	0.161	0.016	0.075	0.083	21	F	0.211	0.278	0.067	0.270
6	A	0.246	0.016	0.139	0.711	22	F	0.483	0.278	0.064	0.270
7	A	0.080	0.016	0.033	0.083	23	F	0.348	0.278	0.348	0.270
8	A	0.298	0.016	0.117	0.645	24	G	0.138	0.000	1.000	1.000
9	B	0.413	0.403	0.401	0.196	25	G	0.419	0.000	0.222	0.433
10	B	0.544	0.403	0.009	0.196	26	H	0.272	0.593	0.583	0.458
11	B	0.477	0.403	0.243	0.196	27	H	0.174	0.593	0.145	0.458
12	B	0.321	0.403	0.755	0.196	28	H	0.184	0.593	0.000	0.072
13	C	0.163	0.831	0.221	0.256	29	H	0.263	0.593	0.194	0.072
14	C	0.616	0.831	0.542	0.256	30	H	0.182	0.593	0.059	0.072
15	C	0.163	0.831	0.221	0.196	31	H	0.219	0.593	0.286	0.072
16	C	0.422	0.831	0.054	0.196	32	H	0.235	0.593	0.020	0.072

We could get a 32×4 matrix from Table 9 and normalize it to get matrix R:

$$R^T = \begin{bmatrix} 0.022 & 0.023 & 0.030 & 0.061 & 0.017 & 0.026 & 0.008 & 0.031 & 0.044 & 0.057 & 0.050 & 0.034 & 0.017 & 0.065 & 0.017 & 0.045 \\ 0.001 & 0.001 & 0.001 & 0.001 & 0.001 & 0.001 & 0.001 & 0.001 & 0.032 & 0.032 & 0.032 & 0.032 & 0.065 & 0.065 & 0.065 & 0.065 \\ 0.006 & 0.110 & 0.006 & 0.006 & 0.011 & 0.020 & 0.005 & 0.017 & 0.057 & 0.001 & 0.035 & 0.108 & 0.032 & 0.078 & 0.032 & 0.008 \\ 0.010 & 0.010 & 0.056 & 0.084 & 0.010 & 0.084 & 0.010 & 0.076 & 0.023 & 0.023 & 0.023 & 0.023 & 0.030 & 0.030 & 0.023 & 0.023 \end{bmatrix}$$

$$\begin{bmatrix} 0.040 & 0.028 & 0.031 & 0.023 & 0.022 & 0.051 & 0.037 & 0.015 & 0.044 & 0.029 & 0.018 & 0.019 & 0.028 & 0.019 & 0.023 & 0.025 \\ 0.079 & 0.044 & 0.044 & 0.044 & 0.022 & 0.022 & 0.022 & 0.000 & 0.000 & 0.047 & 0.047 & 0.047 & 0.047 & 0.047 & 0.047 & 0.047 \\ 0.000 & 0.035 & 0.001 & 0.008 & 0.010 & 0.009 & 0.050 & 0.143 & 0.032 & 0.083 & 0.021 & 0.000 & 0.028 & 0.008 & 0.041 & 0.003 \\ 0.018 & 0.012 & 0.000 & 0.012 & 0.032 & 0.032 & 0.032 & 0.118 & 0.051 & 0.054 & 0.054 & 0.008 & 0.008 & 0.008 & 0.008 & 0.008 \end{bmatrix}$$

According to (15) we can get $E_i=0.971723296$, $E_e=0.88845254$, $E_n=0.836229805$, $E_d=0.90269825$, and then we can get $W_i=0.0705338$, $W_e=0.2782453$, $W_n=0.4085103$, $W_d=0.2427106$ from (17). We can use the data to get C through TOPSIS method, so the results of the comprehensive physical vulnerability are in Table 10 and Fig 3.

Table 10. The comprehensive physical vulnerability of power network in Q city

P	V_a	P	V_a	P	V_a	P	V_a
1	0.113125	9	0.446939	17	0.263494	25	0.327289
2	0.598571	10	0.318055	18	0.279972	26	0.533667
3	0.172201	11	0.374838	19	0.190261	27	0.18933
4	0.339106	12	0.656886	20	0.156011	28	0.133399
5	0.0943	13	0.2505	21	0.136003	29	0.249855
6	0.195556	14	0.624965	22	0.299126	30	0.145346
7	0.029472	15	0.249881	23	0.378293	31	0.293824
8	0.211921	16	0.287423	24	0.658097	32	0.161661

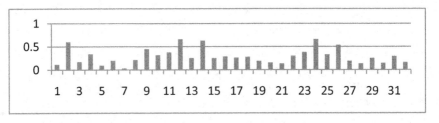

Fig. 3. The comprehensive physical vulnerability of power network in Q city

According to Table 10 and Fig 3 we can find the vulnerability of every node of power network in Q city, and then get their rankings. Compared to the four single attribute vulnerability analyses we can get Fig 4. The dotted line shows the single attribute vulnerability of the nodes, and solid line shows the comprehensive physical vulnerability. Each one can range from 0-1, the greater value the more vulnerable. Their values cannot be compared with each other as it only shows the rank in every kind of vulnerability assessment. We can see that there is a big difference among them, and comprehensive physical vulnerability is much better for measuring.

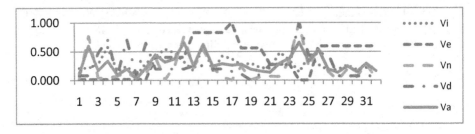

Fig. 4. Comparison of all kinds of vulnerability assessments

According the comparison of all attribute vulnerability, we can find that comprehensive physical vulnerability is more suitable for assessment. The main reasons are: the comprehensive vulnerability could reflect the vulnerability of the equipment itself and the vulnerability of individual in total. It is a multi-angle analysis. The comprehensive vulnerability assessment is also a dynamic and static combined analysis. The comprehensive vulnerability curve is more flat than the single attribute vulnerability curves and reflects the actual difference.

5   Conclusions

This paper proposes a multi-attribute vulnerability assessment method. This method contains inherent component, network structure component and induced component. It includes real time assessment which is based on observation and dynamic assessment which is based on historical data. It is a multi-attribute, dynamic and static combined assessment method, and it can avoid the one-sidedness problem of previous methods. We found that comprehensive physical vulnerability can give an objective and reasonable assessment result in the example of Q city, and the result of vulnerability assessment is useful for power network protection, through which we could ascertain the vulnerability of every node in the network and have a protection prioritization list, so that we can draw up the protection plan more scientifically. We should not determine which node should be paid more attention only by vulnerability, at the same time, risk and importance should also be taken into account, in the future research, risk, importance, reliability and robustness of the power network will all be considered to construct a comprehensive power network protection plan and assessment frame.

References

1. Cutter, S.L.: Living with risk: The Geography of Technological Hazards. Edward Arnold, London (1993)
2. Tunner II, B.L., Kasperson, R.E., Matson, P.A., et al.: A framework for vulnerability analysis in sustainability science. PNAS 100(14), 8074–8079 (2003)
3. Mishkovski, I., Biey, M., Kocarev, L.: Vulnerability of complex networks. Communications in Nonlinear Science and Numerical Simulation 16, 341–349 (2011)
4. Grubesic, T.H., Matisziw, T.C.: A typological framework for categorizing infrastructure vulnerability. Geojurnal 78(4), 287–301 (2013)
5. Bompard, E., Masera, M., Napoli, R., Xue, F.: Assessment of Structural Vulnerability for Power Grids by Network Performance Based on Complex Networks. In: Setola, R., Geretshuber, S. (eds.) CRITIS 2008. LNCS, vol. 5508, pp. 144–154. Springer, Heidelberg (2009)
6. Bompard, E., Cuccia, P., Masera, M., Fovino, I.N.: Cyber vulnerability in power systems operation and control. In: Lopez, J., Setola, R., Wolthusen, S.D. (eds.) Critical Infrastructure Protection. LNCS, vol. 7130, pp. 197–234. Springer, Heidelberg (2012)

7. Taylor, M.A.P., Sekhar, S.V.C., D'Este, G.M.: Application of Accessibility Based Methods for Vulnerability Analysis of Strategic Road Networks. Networks and Spatial Economics 6(9), 267–291 (2006)
8. Chuanfeng, H., Chao, Z., Liang, L.: Critical Inrastructure Network Mdel of Chain Reaction. System Simulation Technology 6(2), 121–125 (2010)
9. Ming, D., Pingping, H.: Vulnerability Assessment to Small-world Power Grid Based on Weighted Topological Model. Proceedings of the CSEE 28(10), 20–25 (2008)
10. Jun, L., Xiangyang, L., Baishang, Z.: Study on Critical Infrastructure Network Risk with Natural Disaster. Chinese Journal of Management Scinece 22(9), 66–73 (2014)
11. Weihua, C., Quanyuan, J., Yijia, C.: Risk-based Vulnerability Assessment in Complex Power Systems. Power System Technology 29(4), 12–17 (2005)
12. Moulin, L.S., Silva, A.: Support vector machines for transient stability analysis of large-scale power system. IEEE Transactions on Power Systems 19(2), 818–825 (2004)
13. Yongxiu, H., Jiang, Z., Tao, L., Haiying, H.: Risk Assessment of Natural Disaster in Urban Electric Power Network Planning. Transactions of China Electro technical Society 26(12), 205–210 (2011)
14. SGCC: Guide for Condition Evaluation of Overhead Transmission Line (January 21, 2008)
15. Tao, L., Xingyuan, F., Xialing, X.: Summary of Power System Vulnerability Assessment Methods. Journal of Electric Power Science and Technology 25(4), 20–24 (2010)
16. Zhigang, L.: Analysis on Physical Vulnerability and Research on Geological Disaster in Yunan Power Grid, pp. 22–26. Wuhan University of Techonogy, Wuhan (2013)
17. Qiyuan, J.: The Comparative Study of Several Main Methods in the Applications for Multiple Attribute Decision Making. Mathematical Modeling and Its Applications 1(3), 16–28 (2012)

Workshop on Advances in Web Based Learning (AWBL 2015)

Teaching-Learning Environment Tool to Promote Individualized Student Assistance

Rafael Santos[1], Bruno Nogueira Luz[1,2], Valéria Farinazzo Martins[3],
Diego Colombo Dias[1], and Marcelo de Paiva Guimarães[1,4(✉)]

[1] Programa de Mestrado em Ciência da Computação, FACCAMP,
Campo Limpo Paulista, Brasil
rafael@renovaci.com,
{bnogueira.luz,diegocolombo.dias,marcelodepaiva}@gmail.com
[2] Instituto Federal de São Paulo, São Paulo, Brasil
[3] Faculdade de Computação e Informática,
Universidade Presbiteriana Mackenzie, São Paulo, Brasil
valfarinazzo@gmail.com
[4] Universidade Aberta do Brasil, UNIFESP, São Paulo, Brasil

Abstract. To develop an effective teaching-learning process for a group of students respecting their individual learning pace is a challenging task for teachers. To assist students individually, it is necessary to identify each student's difficulty and take appropriate teaching action. This paper presents an assisted learning tool based on the web that monitors and reports the student's learning behavior for the teacher. This tool, called eTutor, also performs preconfigured actions (i.e., displays a video or text) according to the current state of student learning. We tested this tool in two different topics for two groups of students. The evaluation showed that this tool promotes student assistance, helping the teachers to be closer to their students.

1   Introduction

Learning is a fundamental and challenging activity for human beings, providing satisfaction for all, especially when students take ownership of knowledge and apply it in different contexts [1]. Each student has a different learning pace [2, 3], so persistence is necessary to create situations that allow the students to achieve mastery of a practice or understanding of a subject [4]. Generally, students who actively participate in classes (i.e., exposing doubts, trying to complete exercises, not being absent, and not giving up) reach the learning goal at some point. Teachers can maximize the students' ability to learn by assisting them directly, regardless of their physical proximity (either through distance learning or classroom teaching). This is one of the assumptions emphasized by authors who discuss the student's zone of proximal development [5, 6].

Due to the disparity of the learning pace of each student, it is essential that teachers be aware of students' learning behavior (i.e., when students are using a learning web tool, it is important to know whether they are interacting with the environment). Teachers can carry out several actions in order to enhance learning when they know the students' learning behavior. For example, they can resolve questions, adapting their strategies to the context, taking individual questions and sharing with everyone; most student questions are common to all. However, monitoring the students' learning behavior is a challenging task for teachers, because it is influenced by several variables, such as the number of students per class and the differences in the learning pace of each one [2, 7].

This paper presents the assisted learning tool, eTutor, which aims to promote individual assistance to students, whether reporting their learning behavior to the teacher in real time or not (either in distance or classroom learning). This web-based tool also helps the development of learning situations, tailoring content to the students (i.e., offering a video or text). The eTutor tool was designed to allow the teacher to provide private assistance to each student during a course with small groups of students (10–30). We maintain that students who work with private assistance can learn more quickly than those in a typical classroom. Reiser, Anderson, and Farrell [8] show that students with private human tutoring could learn approximately four times more than students attending traditional classroom lectures.

Technological tools can monitor students' learning and offer some help to them, for instance, providing texts, multimedia, and simulations. Several intelligent tutoring systems have been proposed [9, 10, 11, 12, 13, 14, 15, 16], which are computer applications to assist human teaching. Commonly, these tools do not aim to provide details about students' learning behavior to teachers; they try to adapt the material content to each student. In our system, information about the students' learning behavior is collected, interpreted, and delivered to the teacher. This can help not only to highlight problems during the teaching-learning process but can also hint at complementary content. Information analyzed by eTutor allows teachers to be closer to their students. This approach allows the teacher to act at the moment when a student experiences difficulty in working on an exercise. It also helps the student to become confident.

The eTutor tool has a Student Modeling Module that uses fuzzy logic [17, 18, 19] to trigger actions according to the actual student's learning behavior and the status of the Interactive Learning Objects (ILOs). ILOs are a set of items created, having a clear educational purpose and including the following features: reusability (can be adapted), interoperability (can be supported by any system), accessibility (can easily be stored and retrieved), manageability (can be updated over time) and interactivity (can provide the items' status and automatically generate actions). These objects are an extension of the Sharable Content Object Reference Model (SCORM) [20] of a traditional learning object. The eTutor tool provides a web interface to create, manage, and distribute ILOs among the students.

The paper is organized as follows: the introduction explains the motivation and idea behind eTutor; section 2 presents eTutor's architecture and the details of its implementation; section 3 shows the methodology used to test our tool; Section 4 shows tests performed with this tool, as well as the results obtained; and finally, the implications of this work are discussed in section 5.

2   The eTutor Architecture

The purpose of eTutor is to bring teachers closer to their students, allowing each student to follow his or her own learning pace. Figure 1 depicts the architecture overview of eTutor:

- Teacher Web Interface, which allows teachers to create and/or modify ILOs and teach distance learning or present courses;
- Student Web Interface, which allows students to attend the classes;
- Student Modeling Module, which uses the following, to monitor the student's learning behavior:
  - Interactive Learning Objects (ILOs), which consist of educational content that reports its current status and engages in preconfigured actions;
  - Fuzzy logic, which analyzes the ILO status to trigger preconfigured actions (for example, by sending an alert to the teacher and/or suggesting a video to the student);
- Web technologies, which are the back-end solution implemented.

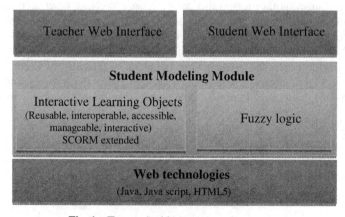

Fig. 1. eTutor - Architecture overview

2.1   Web Technologies

Recent advances in web technologies have allowed sophisticated applications to be written. Major browsers currently support HTML 5 [21], which makes it the de facto markup language for developing complex web applications. This made it the chosen language in which to develop the eTutor. Currently, it is possible to run complex applications on the web and even on mobile devices, high quality image and video and which are highly responsive, with immediate feedback. With HTML 5 and hardware improvement (portable and desktop devices), a whole range of new applications can be created.

All ILOs and eTutor information (i.e., information about students, teachers, and courses) is stored in an internet server and is accessed through an HTTP REST protocol. Both the website and the applications (student and teacher interface) use this same protocol, keeping the back end simple. The response is sent in the JavaScript Object Notation (JSON) standard, which is a way to store information in an organized way, completely language independent, and in a lightweight data-interchange format.

2.2   Student Modeling Module

This module reflects the actions of the teacher and student interface, generated from the analysis of the students' learning behavior. The two independent components work together to monitor the environment.

2.2.1 Interactive Learning Objects

The advancement of the Internet has driven several mechanisms capable of providing learning and knowledge through sets of didactic educational materials that are reusable and shareable, such as the Learning Objects [22]. According to L'Allier [23], a Learning Object is defined as "the smallest independent instructional structure that contains a target, i.e. a learning activity," which consists of a digital component, the basis of a course, unit, or lesson that can be reused to create other unique instructional structures. A Learning Object's main features are as follows: 1) flexibility; it is designed to be reused; 2) personalization; customization of content allows Learning Objects' recombination; 3) interoperability; it allows the definition of design specifications' development and presentation, based on organizational needs; and 4) it increases the significance of knowledge; through reuse many times and in different situations, the contents of Learning Objects are consolidated naturally as timed steps.

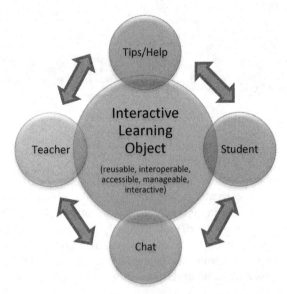

Fig. 2. Operation of an Interactive Learning Object

A Learning Object in this paper is also considered as a unit able to report its current status, which allows a learning tool to trigger preconfigured actions. For our purpose, we expanded the SCORM standard, adding 26 new elements (e.g., period with no interaction, level of difficulty, and help: text, video, and audio). Thus, we propose the Interactive Learning Objects (ILOs) that are focused on interaction. The compatibility between SCORM and ILOs was kept to allow the export/import of objects along with the tools that support them. Kemezinski et al. [24] have already described a methodology for building interactive learning objects; however, they did not provide details about the construction of these objects or whether they follow a standard such as the SCORM [20]. Figure 2 depicts the concept of an ILO. In eTutor, this object is preconfigured by the teacher with some parameters, for example, when the teacher will be alerted that the student is not interacting with the object or when tips such as texts and videos will be offered to the student. The teacher can inspect the student ILO and contact the student at any time.

The created ILOs and their current status are stored on a server. When a class starts, the teacher can select an ILO and send a copy of it to the student's computer, or the students can access the previous ILO.

2.2.2 Fuzzy Logic

The Student Modeling Module uses fuzzy logic to trigger actions, according to the ILO status, that were preconfigured by the teacher. For example, if the student does not add any information within five minutes in a descriptive exercise, then the eTutor offers some tips and how-to videos; if the situation persists, the teacher is alerted. This module is similar to the adaptive model described by Seters, Ossevoort, Tramper, and Goedhart [25], which integrates previous information (such as downtime and tips already used) in order to select appropriate learning content to be presented to the student or which even carries out actions such as alerting the teacher.

This module is based on input and output variables, which represent the final outcome of the student's performance (and the intervention, or not, in the learning process). The following input variable values were used to trigger actions:

- Period of time with no interaction: period during which an ILO goes without receiving interaction from the student;
- Number of characters entered for descriptive answers: percentage of characters that the user typed in terms of the average number of characters per exercise. The number of characters may be low, medium, or high;
- Requests for assistance: number of requests already made by the student, which can be low, medium, or high;
- Exercise difficulty: can be easy, intermediate, or difficult.

Considering these input variables (Table 1), eTutor determines the level of assistance to be offered to the students (Table 2):

- Level 0: no assistance is required;
- Level 1: audio, text, or video are offered;
- Level 2: a similar exercise commented on by the teacher is offered;

- Level 3: the tutor (teacher-designated person who provides help in learning) is alerted about the current status of the student's learning behavior;
- Level 4: the teacher is required to contact the student directly.

There are also penalty factors associated with the levels of assistance used. These penalties are optionally configured by the teacher. Each penalty is associated with a value to be deducted from the grade. The pertinence function used was the trapezoidal, depicted in Equation 1. Tables 1 and 2 present the values for the degree of pertinence, where *a*, *b*, *c*, and *d* are the edges of the trapezoid. The degree of pertinence is a real value in the interval [0,1]. This pertinence function is fundamental to enable the use of fuzzy logic [17, 18, 19].

Table 1. Fuzzy input variables

Input variables		
Variable	**Linguistic terms**	**Relevance degree (range)**
Period of time with no interaction	small	[0,60]
	medium	[58,180]
	high	[170,300]
Number of characters entered	low	[0,10%]
	medium	[9%,30%]
	high	[28%]
Amount of assistance	low	[0,1]
	medium	[1,3]
	high	[2,5]
Exercise difficulty	low	[0,5]
	medium	[5,7.5]
	high	[7.10]

Table 2. Fuzzy output variables

Output variables		
Variable	**Linguistic terms**	**Degree of relevance (range)**
Level of aid	Level 0 – no assistance	[0,1]
	Level 1 – audio, video, and text	[1,2]
	Level 2 – review exercise	[2,3]
	Level 3 – tutor assistance	[3,4]
	Level 4 – teacher assistance	[4,5]

$$trapmf\ (x;a,b,c,d) = \max\left(\min\left(\frac{x-a}{b-a}, 1, \frac{d-x}{d-c} \right), 0 \right)$$

Equation 1. Trapezoidal pertinence function

2.3 Teacher and Student Web Interface

Each ILO consists of some attributes such as name, internet links, and multimedia objects (audio, video, and text). When an ILO does not receive interactions, eTutor alerts the teacher and offers assistance to the student. Figure 3 depicts the main student interface that allows the students to answer a descriptive question.

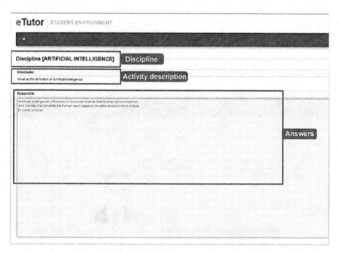

Fig. 3. eTutor: Main student interface

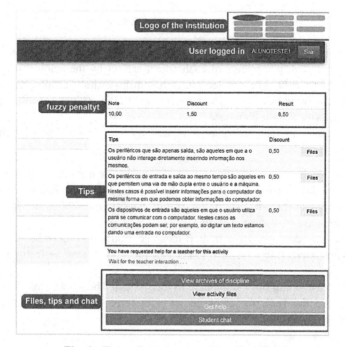

Fig. 4. eTutor: Current status of a student ILO

In order to discourage students from asking for tips when they don't need them, eTutor keeps score and deducts points (optional) for each help request. Figure 4 depicts the current status of a student ILO with the penalties applied, tips requested, and other information, for instance, attached files and chat history.

The teacher uses a specific interface (Figure 5) to monitor the students. In this interface, the teacher can request a copy of a student ILO, which allows verification of the current status. If an intervention is needed, the teacher can contact the student. The teacher interface also has a color system that is associated with the state of student learning.

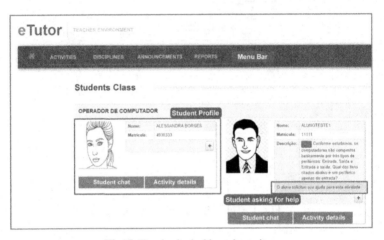

Fig. 5. Teacher's dashboard monitor

One eTutor challenge is to detect the ILO state and trigger the action associated with it. The behavior parameters of each object are set by the teacher at the time of the creation of an ILO. The teacher can also configure the ILO to determine which task the student should do next, linking to the next ILO. This allows the students to start a new activity at their own pace.

While a student attempts to complete an activity that is part of an ILO, the following interactions can occur:

- The teacher can copy the student ILO to his or her interface, which allows inspection of the status. If the teacher decides to intercede, he/she can contact the student (face-to-face in the case of classroom teaching, or via chat/voice, in the case of distance learning);
- The teacher can receive an alert message, notifying him or her that a student is not interacting with an ILO because the student may be facing some difficulty;
- The eTutor can offer some help to the student. The teacher may have preconfigured the ILO, for example, to offer three tips (text content messages) while the activity is being worked on;

- The teacher can share an ILO instance with all students, which allows the start of a discussion about an activity. This can be a new ILO instance or a previous one from a student;
- The students can share their own instances among the other students, promoting collaborative activity, although the teacher should approve;
- Students can request a video or chat session with the teacher/tutor, which may or may not be private. The system manages the requests using a FIFO.

3   Methodology

The eTutor tool was validated in two topics: "Introduction to Information Security" and "Introduction to Hardware and Computation." Both classes were given in different courses and universities. The first course (Class A) had five students enrolled, while the second had twenty (Class B). Before the tests started, the teachers and students were registered on eTutor. Each teacher received training in using eTutor and creating ILOs, including how to follow up the activities. Additionally, we showed the interaction options available to the students.

The teachers were free to create their ILOs as needed to meet their educational goals. In class A, the teacher created one ILO with five questions. In group B, the teacher created one ILO with six questions. Each ILO had at least three tips that were available to students, which were preconfigured to be presented automatically according to input data defined by the teacher and shown by the Student Modeling Module. At the end of the course, both groups answered a questionnaire to analyze the eTutor as an assisted learning tool and with respect to its usability.

4   Results and Discussion

The analysis of the results was performed per class. However, the discussion involves both.

4.1   Introduction to Information Security Class

One teacher and five students participated in the "Introduction to Information Security" class. Answers to the questions given to them after the class were tabulated and are presented in Tables 3 and 4. In all cases, the students received automatic tips from the eTutor environment according to their learning behavior and the preconfiguration done in the ILOs. Only one student felt that the tips had little influence on the outcome; others regarded as great/excellent the interaction provided by the ILOs. All of them had already used other virtual learning environments such as Moodle [26].

Table 3. Student and teacher behavior

Question	Yes	No	I did not request
1- Have you ever used any virtual learning environment?	100%	0%	Does not apply
2- Did you feel accompanied by the professor during the activity?	80%	20%	Does not apply
3- Did you request tips during the activities?	100%	0%	Does not apply
4- When needed, did the teacher offer some assistance?	60%	20%	20%

All students requested tips, and 80% of them judged these tips as excellent. Only 20% of them did not request direct assistance form the teacher during the activities; the other 80% judged the assistance as satisfactory. In total, as shown in Table 4, 100% of the students considered the eTutor tool and the ILOs as Excellent (60%) and Great (40%).

Table 4. eTutor evaluation

Question	Regular	Great	Excellent
5- How do you classify the interaction offered in the activities?	0%	80%	20%
6- How do you classify the eTutor?	0%	40%	60%

4.2 Introduction to Hardware and Computation Class

Three tips were configured for each question of the ILO created. They were to be offered after a period of one minute without student interaction and could also be requested at any time. When the activity was completed, an evaluation questionnaire was given. Tables 5 and 6 present the results.

Table 5. Student and teacher behavior

Question	Yes	No	I did not request
1- Have you ever used any virtual learning environment?	15%	85%	Does not apply
2- Did you feel accompanied by professor during the activity?	85%	15%	Does not apply
3- Did you request tips during the activities?	55%	45%	Does not apply
4- When needed, did the teacher offer some assistance?	65%	5%	30%

Table 6. eTutor evaluation

Question	Regular	Great	Excellent
5- How do you classify the interaction offered in the activities?	25%	30%	45%
6- How do you classify the eTutor?	15%	30%	55%

Considering the results from both classes, the proposal presented in this work had a positive rating of 85% (Great and Excellent). Similarly, the interaction achieved with ILO received a positive approval of 75%. It is noteworthy that only 15% of students had used other Learning Objects in other virtual learning environments. Students who already knew other learning environments reported that using the ILO in eTutor made them feel accompanied by the teacher, and they evaluated the possibilities of interaction as excellent. Of the students, 67% were assisted by the teacher when they requested it, while 33% did not request any assistance. Students who felt accompanied (85%) considered the evaluation of the interaction model offered as great or excellent (83%).

4.3   Discussion

Combining both tests applied to the classes, it is possible to trace a single scenario with respect to eTutor. The results were positive. Although there were significant differences when comparing questions 1 and 3 in both classes, the questions that really aimed to measure the student assistance were 2 and 4, which showed positive results. The direct teaching assistance occurred, on average, in 80% of the activities. Figure 6 depicts the final results.

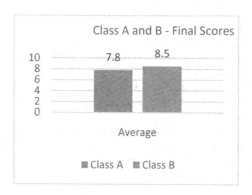

Fig. 6. Final scores, Class A and B

5   Conclusions

This paper presented a tool to assist students based on their learning behavior. This web environment aims to alert the teacher when the students are facing difficulties. It also offers preconfigured help (video, text, and audio). As result, we expected to bring the teachers closer to their students, helping the students to become confident and allowing each one to follow his or her own learning pace, regardless of physical proximity (i.e., whether in distance learning or classroom teaching).

This environment was based on ILOs, which are learning units able to store content as well as their status. We projected the ILOs to be reusable, interoperable, accessible, manageable, and interactive. These objects are an expansion of the SCORM standard to which we added 26 new elements. These new elements allowed the eTutor to report the ILO status to the teacher. The ILOs were created and managed by the teacher, who distributed and monitored them using the teacher interface.

We also presented a case study that used the eTutor in the teaching of two topics, "Introduction to Information Security" and "Introduction to Hardware and Computation." From the observation results, we can conclude that the students felt assisted by the teachers.

As future work, we intend to improve the teacher interface to support other options for multimedia content (i.e., PowerPoint, Flash, augmented reality) for the ILOs. We also intend to investigate how to deal with eTutor usability issues and test this tool with more teachers and groups of students.

References

1. Svinicki, M., McKeachie, W.J.: McKeachie's Teaching Tips: Strategies, Research, and Theory for College and University Teachers. Cengage Learning; 13 edn, 416 pages (2010)
2. Freire, P.: Pedagogia Da Autonomia: Saberes Necessários à Prática Educativa. Paz e Terra, São Paulo (1996)
3. Zabala, A.: A Prática Educativa: Como Ensinar. Artmed, Porto Alegre (1998)
4. Wells, G.: Dialogic Inquiry: Towards a Sociocultural Practice and Theory of Education. Cambridge University Press, Cambridge (1999)
5. Vygotsky, L.S.:. Thinking and speech (n. Minick, Trans.), In: Rieber, R.W., Carton, A.S. (eds.) The collected works of L. S. Vygostky: vol. 1. Problems of general psycology, pp. 239–285. Plenum Press, New York (1987)
6. Vygotsky, L.S.: The problem of age (M. Hall, Trans.). In: Rieber, R.W. (ed.) The collected works of L.S. Vygotsky: (Vol. 5 Child psycology), pp. 187–205. Plenum Press, New York (1998)
7. Guimarães, M.P., Martins, V.F., Dias, D. C.: Uso de Lógica Fuzzy no Auxílio ao Acompanhamento Automático de Alunos utilizando um Ambiente de Aprendizagem. XXIV Simpósio Brasileiro de Informática na Educação (SBIE), doi:10.5753/CBIE.SBIE.2013.707, (2013)
8. Reiser, B.J., Anderson, J.R., Farrel, R.G.: Dynamic student modeling in an intelligent tutoring for lisp programming. In: Proceedings of Ninth International Joint Conference on Artificial Intelligence, pp. 8–14. Morgan Kaufman, Los Altos, CA (1985)
9. VanLehn's, K.: The Behavior of. Tutoring Systems. International Journal of Artificial Intelligence in Education 16(3), 267–270 (2006)
10. Yang, F.: The ideology of intelligent tutoring systems. ACM Inroads. 1(4), 63–65 (2010)
11. Chakraborty, S., Roy, D., Basu, A.: Development of Knowledge Based Intelligent Tutoring System, Chapter 5. TMRF e-Book Advanced Knowledge Based Systems: Model, Applications & Research(Eds. Sajja & Akerkar) 1, 74–100 (2010)
12. Chen, H., Yu, C., Chang, C.: E-Homebook System: A web-based interactive education interface. Computers & Education. 49(2), 160–175 (2007)
13. Huanga, Y., Lina, Y., Chengb, S.: An adaptive testing system for supporting versatile educational assessment. Computers & Education 52(1), 53–67 (2009)
14. Chen, C., Chen, M.: Mobile formative assessment tool based on data mining techniques for supporting web-based learning. Computers & Education 52(1), 256–273 (2009)

15. Yang, Z., Liu, W.: Research and development of web-based virtual online classroom. Computers & Education **48**(2), 171–184 (2007)
16. Chen, C.: Intelligent web-based learning system with personalized learning path guidance. Computers & Education **51**(2), 787–814 (2008)
17. Zadeh, L.A.: Fuzzy sets. Information and Control **8**(3), 338–353 (1965). doi:10.1016/s0019-9958(65)90241-x
18. Bellman, R.E., Zadeh, L.A.: Decision-making in a fuzzy environment. Management science **17**(4), B-141 (1970)
19. Ross, T.J.: Fuzzy Logic with Engineering Applications, 3rd edn. Wiley; 3 edition (March 1, 2010). 606 pages (2010)
20. SCORM. Content Aggregation Model – CAM. 4th Edition SCORM 200 (2009)
21. Sarris, S.: HTML5 Unleashed. Sams Publishing, 1st edn (July 26, 2013).432 pages (2013)
22. Downes, S.: Learning Objects: Resources for Distance Education Worldwide. International Review of Research in Open and Distance Learning **2**(1) (2001)
23. L'Allier, J.J.: A Frame of Reference: NETg's Map to Its Products. Their Structures and Core Beliefs (1997)
24. Kemezinski, A, Costa, I.A., Wehrmeister, M.A., Hounsell, M.S., Vahldck, A.: Metodologia para Construção de Objetos de Aprendizagem Interativos", Anais do 23° Simpósio Brasileiro de Informática na Educação (SBIE) (2012). ISSN 2316-6533
25. Seters, J.R.V., Ossevoort, M.A., Tramper, J., Goedhart, M.J.: The influence of student characteristics on the use of adaptive e-learning material. Computers & Education. **58**, 942–952 (2012)
26. Moodle (2015). https://moodle.org/. Accessed on 20 February 2015

General Tracks

Autonomous Tuning for Constraint Programming via Artificial Bee Colony Optimization

Ricardo Soto[1,2,3], Broderick Crawford[1,4,5], Felipe Mella[1], Javier Flores[1], Cristian Galleguillos[1(✉)], Sanjay Misra[6], Franklin Johnson[7], and Fernando Paredes[8]

[1] Pontificia Universidad Católica de Valparaíso, Valparaíso, Chile
{ricardo.soto,broderick.crawford}@ucv.cl,
{felipe.mella.101,javier.flores.v,cristian.galleguillos.m}@mail.pucv.cl
[2] Universidad Autónoma de Chile, Santiago, Chile
[3] Universidad Científica del Sur, Lima, Perú
[4] Universidad San Sebastián, Santiago, Chile
[5] Universidad Central de Chile, Santiago, Chile
[6] Atilim University, Ankara, Turkey
ssopam@gmail.com
[7] Universidad de Playa Ancha, Valparaíso, Chile
franklin.johnson@upla.cl
[8] Escuela de Ingeniería Industrial, Universidad Diego Portales, Santiago, Chile
fernando.paredes@udp.cl

Abstract. Constraint Programming allows the resolution of complex problems, mainly combinatorial ones. These problems are defined by a set of variables that are subject to a domain of possible values and a set of constraints. The resolution of these problems is carried out by a constraint satisfaction solver which explores a search tree of potential solutions. This exploration is controlled by the enumeration strategy, which is responsible for choosing the order in which variables and values are selected to generate the potential solution. Autonomous Search provides the ability to the solver to self-tune its enumeration strategy in order to select the most appropriate one for each part of the search tree. This self-tuning process is commonly supported by an optimizer which attempts to maximize the quality of the search process, that is, to accelerate the resolution. In this work, we present a new optimizer for self-tuning in constraint programming based on artificial bee colonies. We report encouraging results where our autonomous tuning approach clearly improves the performance of the resolution process.

Keywords: Artificial intelligence · Optimization · Adaptive systems · Metaheuristics

1 Introduction

Constraint Programming (CP) [2] is a programming paradigm used to solve constraint satisfaction and optimization problems. In this context, problems are

© Springer International Publishing Switzerland 2015
O. Gervasi et al. (Eds.): ICCSA 2015, Part I, LNCS 9155, pp. 159–171, 2015.
DOI: 10.1007/978-3-319-21404-7_12

represented by a sequence of variables owning a non-empty domain of possible values and a set of constraints. The solving process is carried out by employing a solver, which creates and explores a search tree of potential solutions. A solution to the problem is the complete assignment of a value to each variable such that all the constraints are satisfied.

In the resolution process, enumeration and propagation strategies are used. The first ones are responsible for choosing the order in which variables and values are going to be selected to instantiate variables and thus creating the tree branches. While propagation techniques are responsible for pruning branches that do not lead to any solution. The selection of enumeration strategies is critical to the performance of the resolution process and the correct choice can greatly reduce the computation cost to find out a solution. However, deciding which strategy would be the right one for a problem is not simple, since each enumeration strategy may have particular behaviors depending on the problem and on the status of the search process. Autonomous Search (AS) is a framework [11] that targets this problem. The idea is to autonomously replace underperforming strategies for more promising ones based on a set of performance indicators. This self-tuning process is commonly supported by an optimizer which attempts to maximize the quality of the search process, that is, to accelerate the resolution.

In this work, we present a new optimizer for self-tuning in constraint programming based on artificial bee colonies. The artificial bee colony algorithm (ABC) is a modern metaheuristic recently proposed by Karaboga [12, 15], based on the intelligent behavior exhibited by bee colonies when they seek food sources. We report encouraging results where our autonomous tuning approach clearly improves the performance of the resolution process. The rest of this work is organized as follows: Section 2 presents the related work. Section 3 and 4 describe the problem and the proposed solution, respectively. Finally, the experimental evaluation is illustrated followed by conclusions and future work.

2   Related Work

A pioneer work in AS for CP is the one presented in [4]. This framework introduced a four-component architecture, allowing the dynamic replacement of enumeration strategies. The strategies are evaluated via performance indicators of the search process, and better evaluated strategies replace worse ones during solving time. Such a pioneer framework was used as basis of different related works. For instance, a more modern approach based on this idea is reported in [7]. This approach employs a two-layered framework where an hyper-heuristic placed on the top-layer controls the dynamic selection of enumeration strategies of the solver placed on the lower-layer. An hyper-heuristic can be regarded as a method to choose heuristics [11]. In this approach, two different top-layers have been proposed, one using a genetic algorithm [6, 19] and another using a particle swarm optimizer [8]. Similar approaches have also been implemented for solving optimization problems instead of pure CSPs [17]. In Section 5 we provide a comparison of our approach with the best AS optimizers reported in the literature.

3   Autonomous Search

As previously presented, AS aims at providing self-tuning capabilities to the solver. In this context, the idea is to autonomously control which enumeration strategy is applied to each part of the search tree during resolution. The replacement of strategies is carried out according to a quality ranking which is provided by a choice function (CF). A CF is mainly composed of performance indicators (see indicators employed in Table 1) of the search process and weights that controls its relevance within the equation. Formally, a CF for a strategy S_j at time t is computed as follows.

$$CF_t(S_j) = \sum_{i=1}^{IN} w_i a_{it}(S_j) \tag{1}$$

where IN corresponds to the indicator set, w_i is a weight that controls the relevance of the ith-indicator and $a_{it}(S_j)$ is the score of the ith-indicator for the strategy S_j at time t. A main component of this model are the weights, which must be finely tuned by an optimizer. This is done by carrying out a sampling phase where the problem is partially solved to a given cutoff. The performance information gathered in this phase via the indicators is used as input data of the optimizer, which attempt to determine the most successful weight set for the CF. The optimizer employed in this work is ABC (see Section 4).

Let us remark that this tuning process is dramatically important, as the correct configuration of the CF may have essential effects on the ability of the solver to properly solve specific problems. Parameter (weights) tuning is hard to achieve as parameters are problem-dependent and their best configuration is not stable along the search [16]. Additional and detailed information about this framework can be seen in [10, 18].

Table 1. Indicators used during the search process

Name	Description
SB	Number of Shallow Backtracks [3]
In1	Represents a Variation of the Maximum Depth. It is calculated as: $CurrentMaximumDepth - PreviousMaximumDepth$
In2	Calculated as: $CurrentDepth - PreviousDepth$. A positive value means that the current node is deeper than the one explored at the previous step

4   Artificial Bee Colony Algorithm

The ABC algorithm [5, 20] is a metaheuristic based on the intelligent behavior of honey bees when they seek new food sources in the environment. The position of the food source in space represents a solution to the problem while the quality of it is associated with the amount of nectar that it has. Within the colony there are three types of bees: employed, onlookers and scouts bees. Employed bees fly over the food source that they are exploiting and return to the hive to share the information collected on the amount of nectar in the dance area. The onlookers bees wait in the hive and choose a food source to exploit based on the dance performed by employed bees. Employed bees, whose food source has been exploited, become scout bees and go out in search of a new source, abandoning the former. This is controlled by a parameter called the *limit* [14], which is the only parameter other than those commonly used in populations based algorithms, such as the colony size (CS) and the maximum number of iterations (MCN). The ABC algorithm was developed under the following assumptions [1]:

1. Half of the colony consists in employed bees, while the other half corresponds to onlookers bees.
2. Each food source is explored by only one employed bee.

The ABC algorithms exposed by D. Karaboga [13] is explained down below. In the initial phase of the algorithm, a set of x_i solution is randomly generated through Eq. 2, where FS is the number of food sources, j corresponds to the dimension of the x_i solution. Dim is the number of variables to optimize, and finally, x_j^{max} and x_j^{min} are the upper and lower limits, respectively, of the j_{th} parameter for the i_{th} solution. Once initialized, food sources are evaluated and the search process of the employed bees, onlookers and scouts is repeated.

$$x_{ij} = x_{ij}^{min} + rand\,(0.1) * \left(x_{ij}^{max} - x_{ij}^{min}\right), i \in [1, FS], j \in [1, Dim] \qquad (2)$$

Each employed bee tries to improve its solution by modifying its dimension by using Eq. 3, where ϕ is random number uniformly distributed between $[-1, 1]$, k is a randomly selected food source different from i, and j is an integer selected randomly from $[1, Dim]$. Once the new food source v_i is generated, it is evaluated via Eq. 4 and compared with x_i for choosing the one with better fitness. In other words, if v_i is better than x_i, the employed bee replaces its food source by the new one and reset to zero its counter. Otherwise, it keeps in memory the position of x_i and its counter is incremented by one.

$$v_{ij} = x_{ij} + \phi\,(x_{ij} - x_{kj}) \qquad (3)$$

$$fit_i = \begin{cases} \frac{1}{1+f_i}, if\, f_i \geq 0 \\ 1 + abs\,(f_i), if\, f_i < 0 \end{cases} \qquad (4)$$

$$p_i = 0.9 * \frac{fit_i}{fit_{best}} + 0.1 \qquad (5)$$

After the employed bee completes their phase, the onlookers bees select one source by Eq. 5. The selection probability p_i of a food source depends on the amount of nectar that it holds, where fit_i is the fitness associated with the food source i, and fit_{best} is the fitness associated with the best food source found so far. Once each onlooker bee has chosen a food source, it finds a new food source in the neighborhood by using Eq. 3, and apply the same selection criteria that employed bees do. When the nectar of a food source is depleted, the employed bee of this source becomes a scout bee in order to search a new one. To determine if a food source has been abandoned at the end of each iteration, the value of the attempt counter is evaluated. If the counter of attempts to improve the food source is greater than the parameter limit, then that food source is replaced by a new one by Eq. 2.

5   Experimental Evaluation

In this section, we illustrate a performance of our proposed approach when it is subjected to evaluation on various classical CSPs, which are named below with their respective instances:

- N-Queens problem with $n = \{8, 10, 12, 20, 50, 75\}$
- Magic Square problem size $n = \{3, 4, 5, 6\}$
- Sudoku puzzle size $n = \{2, 5, 7\}$
- Knights Tournament with $n = \{5, 6\}$
- Quasi Group with $n = \{3, 5, 6\}$
- Langford with size $n = \{2\}$ and $k = \{12, 16, 20, 23\}$

We implement a solver using the $Ecl^i ps^e$ Constraint Programming System version 6.0 and Java. Tests have been performed on a 3.30 GHz Intel Core i3 with 4 GB RAM running Windows 7. The instances are solved to a maximum number of 65535 steps. If no solution is found at this point the problem is set to t.o. (timeout). To determine the quality of a solution, we use the indicators described in table 1. In the experiments, 8 variable selection heuristics were used and 3 value selection heuristics, which when combined, provide a portfolio of 24 enumeration strategies (see Table 2). The results are evaluated in terms of runtime and backtracks, both are widely employed indicators to evaluate performance in constraint programming.

Tables 3, 4, and 5 illustrate the performance in terms of runtime required to find a solution for each enumeration strategy individually (S1 to S24) and the proposed approach (ABC). The results clearly validate our approach, which is the only one in solving all instances of all problems, taking the best average runtime. Tables 6, 7, and 8 depict the results in terms of backtracks, which are analogous to the previous ones. This demonstrates the ability of the proposed approach to correctly select the strategy to each part of the search tree. Finally, we compare the proposed approach with the two previously reported optimized online control systems, one based on genetic algorithm (GA) [8] and the other one based on

Table 2. Portfolio used

Id	Variable ordering	Value ordering
S_1	First variable of the list	min. value in domain
S_2	The variable with the smallest domain	min. value in domain
S_3	The variable with the largest domain	min. value in domain
S_4	The variable with the smallest value of the domain	min. value in domain
S_5	The variable with the largest value of the domain	min. value in domain
S_6	The variable with the largest number of attached constraints	min. value in domain
S_7	The variable with the smallest domain. If are more than one, choose the variable with the bigger number of attached constraints.	min. value in domain
S_8	The variable with the biggest difference between the smallest value and the second more smallest of the domain	min. value in domain
S_9	First variable of the list	mid. value in domain
S_{10}	The variable with the smallest domain	mid. value in domain
S_{11}	The variable with the largest domain	mid. value in domain
S_{12}	The variable with the smallest value of the domain	mid. value in domain
S_{13}	The variable with the largest value of the domain	mid. value in domain
S_{14}	The variable with the largest number of attached constraints	mid. value in domain
S_{15}	The variable with the smallest domain. If are more than one, choose the variable with the bigger number of attached constraints.	mid. value in domain
S_{16}	The variable with the biggest difference between the smallest value and the second more smallest of the domain	mid. value in domain
S_{17}	First variable of the list	max. value in domain
S_{18}	The variable with the smallest domain	max. value in domain
S_{19}	The variable with the largest domain	max. value in domain
S_{20}	The variable with the smallest value of the domain	max. value in domain
S_{21}	The variable with the largest value of the domain	max. value in domain
S_{22}	The variable with the largest number of attached constraints	max. value in domain
S_{23}	The variable with the smallest domain. If are more than one, choose the variable with the bigger number of attached constraints	max. value in domain
S_{24}	The variable with the biggest difference between the smallest value and the second more smallest of the domain.	max. value in domain

particle swarm optimization (PSO) [9]. Table 9 illustrates solving time and number of backtracks required by GA and PSO in contrast with our approach. This comparison shows that ABC and PSO stand out in terms of number of problems solved with a low number of backtracks with respect to GA. Moreover, considering runtime, the ABC algorithm based optimizer is far superior compared to the other ones by requiring half the time or less to find a solution for every problem. A graphical comparison can be seen in Figures 1 and 2.

Table 3. Runtime in ms for strategies S1 to S8

Problem	Strategies							
	S_1	S_2	S_3	S_4	S_5	S_6	S_7	S_8
Q-8	5	5	5	4	2	4	4	2
Q-10	5	8	3	4	4	5	3	4
Q-12	12	11	11	11	13	14	11	10
Q-20	20405	4867	20529	20529	1294	26972	15	93
Q-50	t.o.	t.o.	532	t.o.	t.o.	t.o.	524	t.o.
Q-75	t.o.	t.o.	4280	t.o.	t.o.	t.o.	4217	t.o.
MS-3	1	5	1	1	1	4	1	1
MS-4	14	2340	6	21	21	1500	6	11
MS-5	1544	t.o.	296	6490	t.o.	t.o.	203	1669
MS-6	t.o.	t.o.	t.o.	t.o.	t.o.	t.o.	t.o.	t.o.
S-2	35	30515	10	50	225	1607	10	10
S-5	7453	t.o.	2181	8274	t.o.	t.o.	2247	897
S-7	26882	t.o.	2135	25486	t.o.	t.o.	2187	31732
K-5	1825	t.o	2499	t.o	t.o	t.o	t.o	t.o
K-6	90755	t.o	111200	89854	t.o	t.o	39728	t.o
QG-5	t.o.	t.o.	7510	t.o.	t.o.	t.o.	9465	t.o.
QG-6	45	t.o.	15	45	t.o.	3605	15	t.o.
QG-7	256	8020	10	307	943	16896	10	16
LF 2-12	20	242	4	29	43	32	4	22
LF 2-16	70	70526	231	115	1217	489	237	7
LF 2-20	191	t.o.	546	318	61944	11	553	240
LF 2-23	79	t.o.	286	140	68254	19	285	19
$\overline{x}$	8311	11653.9	7252	8922.3	11163.5	3935.3	2986.3	2315.6

Table 4. Runtime in ms for strategies S9 to S16

Problem	Strategies							
	S_9	S_{10}	S_{11}	S_{12}	S_{13}	S_{14}	S_{15}	S_{16}
Q-8	5	5	4	5	2	4	4	2
Q-10	5	8	7	5	5	5	3	4
Q-12	11	11	11	11	13	14	11	10
Q-20	20349	4780	18	23860	1250	36034	17	87
Q-50	t.o.	t.o.	532	t.o.	t.o.	t.o.	533	t.o.
Q-75	t.o.	t.o.	4336	t.o.	t.o.	t.o.	4195	t.o.
MS-3	1	4	1	1	1	4	1	1
MS-4	13	2366	6	21	21	1495	6	11
MS-5	1498	t.o.	297	6053	t.o.	t.o.	216	1690
MS-6	t.o.	t.o.	t.o.	t.o.	t.o.	t.o.	t.o.	t.o.
S-2	35	29797	10	50	225	1732	10	10
S-5	7521	t.o.	2394	9015	t.o.	t.o.	2310	972
S-7	26621	t.o.	2069	26573	t.o.	t.o.	2094	30767
K-5	1908	t.o	2625	t.o	t.o	t.o	t.o	t.o
K-6	93762	t.o	102387	109157	t.o	t.o	46673	t.o
QG-5	t.o.	t.o.	9219	t.o.	t.o.	t.o.	10010	t.o.
QG-6	40	t.o.	15	45	t.o.	3565	15	t.o.
QG-7	240	13481	10	348	1097	18205	11	15
LF 2-12	20	270	4	29	44	32	5	21
LF 2-16	69	55291	250	118	1273	530	235	8
LF 2-20	185	t.o.	538	312	61345	11	541	237
LF 2-23	79	t.o.	285	140	71209	19	278	19
$\overline{x}$	8464.6	10601.3	5953.3	10337.9	11373.8	4742.4	3358.4	2257

Table 5. Runtime in ms for strategies S17 to S24 and ABC

Problem	Strategies								ABC
	S_{17}	S_{18}	S_{19}	S_{20}	S_{21}	S_{22}	S_{23}	S_{24}	
Q-8	5	5	4	4	4	**2**	4	**2**	230
Q-10	4	7	**2**	4	5	4	3	5	265
Q-12	11	10	11	11	14	13	11	**8**	315
Q-20	22286	4547	16	13135	26515	1249	16	1528	575
Q-50	t.o.	t.o.	**520**	t.o.	t.o.	t.o.	521	t.o.	2770
Q-75	t.o.	t.o.	4334	t.o.	t.o.	t.o.	**4187**	t.o.	10775
MS-3	**1**	**1**	**1**	**1**	**1**	**1**	**1**	**1**	290
MS-4	88	37	99	42	147	37	102	79	400
MS-5	t.o.	t.o.	t.o.	165878	t.o.	153679	t.o.	t.o.	340
MS-6	t.o.	t.o.	t.o.	t.o.	t.o.	t.o.	t.o.	t.o.	**868**
S-2	**5**	18836	30	**5**	100	1710	30	40	287
S-5	t.o.	t.o.	2590	t.o.	t.o.	t.o.	2670	t.o.	335
S-7	3725	t.o.	**338**	5350	t.o.	t.o.	378	9168	465
K-5	1827	t.o	2620	t.o	t.o	t.o	t.o	t.o	**1132**
K-6	96666	t.o	97388	90938	t.o	t.o	40997	t.o	**1961**
QG-5	9743	t.o.	**20**	10507	t.o.	t.o.	21	t.o.	312
QG-6	7075	t.o.	125	6945	t.o.	t.o.	130	t.o.	310
QG-7	**9**	1878	12	**9**	1705	**9**	12	14	262
LF 2-12	18	242	**4**	29	33	43	5	13	265
LF 2-16	66	55687	245	107	510	1297	240	584	279
LF 2-20	170	t.o.	562	294	**11**	58732	569	15437	337
LF 2-23	75	t.o.	272	126	20	73168	276	**10**	360
$\bar{x}$	8339.7	8125	5459.7	17258	2422.1	22303.4	2640.7	2068.4	**1051.5**

Table 6. Backtracks requires for strategies S1 to S8

Problem	Strategies							
	S_1	S_2	S_3	S_4	S_5	S_6	S_7	S_8
Q-8	10	11	10	10	3	9	10	3
Q-10	6	12	4	6	6	6	4	5
Q-12	15	11	16	15	17	16	16	12
Q-20	10026	2539	11	10026	862	15808	11	63
Q-50	>121277	>160845	177	>121277	>173869	>143472	177	>117616
Q-75	>118127	>152812	818	>118127	>186617	>137450	818	>133184
MS-3	0	4	0	0	0	4	0	0
MS-4	12	1191	3	10	22	992	3	13
MS-5	910	>191240	185	5231	>153410	>204361	193	854
MS-6	>177021	>247013	>173930	>187630	>178895	>250986	>202927	>190877
S-2	18	10439	4	18	155	764	4	2
S-5	4229	>89125	871	4229	>112170	>83735	871	308
S-7	10786	>59828	773	10786	>81994	>80786	773	10379
K-5	767	>179097	767	>97176	>228316	>178970	>73253	>190116
K-6	37695	>177103	37695	35059	>239427	>176668	14988	>194116
QG-5	>145662	>103603	8343	>145656	>92253	>114550	8343	>93315
QG-6	30	>176613	0	30	>83087	965	0	>96367
QG-7	349	3475	1	349	4417	4417	1	4
LF 2-12	16	223	1	16	29	22	1	12
LF 2-16	39	24310	97	39	599	210	97	0
LF 2-20	77	>158157	172	77	26314	1	172	64
LF 2-23	26	>157621	64	26	29805	3	64	7
$\bar{x}$	3611.8	4221.5	2381.6	3878.1	5185.8	1786	1327.3	781.8

Table 7. Backtracks requires for strategies S9 to S16

Problem	Strategies							
	S_9	S_{10}	S_{11}	S_{12}	S_{13}	S_{14}	S_{15}	S_{16}
Q-8	10	11	10	10	3	9	10	3
Q-10	6	12	12	6	6	6	4	5
Q-12	15	11	16	15	17	16	16	12
Q-20	10026	2539	11	10026	862	15808	11	63
Q-50	>121277	>160845	177	>121277	>173869	>143472	177	>117616
Q-75	>118127	>152812	818	>118127	>186617	>137450	818	>133184
MS-3	0	4	0	0	0	4	0	0
MS-4	12	1191	3	10	22	992	3	13
MS-5	910	>191240	185	5231	>153410	>204361	193	854
MS-6	>177174	>247013	>174068	>187777	>179026	>251193	>203089	>191042
S-2	18	10439	4	18	155	764	4	2
S-5	4229	>89125	871	4229	>112174	>83735	871	308
S-7	10786	>59828	773	10786	>81994	>80786	773	10379
K-5	767	>179126	767	>97176	>228316	>178970	>73253	>190116
K-6	37695	>177129	37695	35059	>239427	>176668	14998	>194116
QG-5	>145835	>103663	8343	>145830	>92355	>114550	8343	>93315
QG-6	30	>176613	0	30	>83087	965	0	>93820
QG-7	349	3475	1	349	583	4417	1	4
LF 2-12	16	223	1	16	29	22	1	12
LF 2-16	39	24310	97	39	599	210	97	0
LF 2-20	77	>158157	172	77	26314	1	172	64
LF 2-23	26	>157621	64	26	29805	3	64	7
$\bar{x}$	3611.8	4221.5	2382	3878.1	4866.3	1786	1327.8	781.8

Table 8. Backtracks requires for strategies S17 to S24 and ABC

Problem	Strategies								ABC
	S_{17}	S_{18}	S_{19}	S_{20}	S_{21}	S_{22}	S_{23}	S_{24}	
Q-8	10	11	10	10	9	3	10	2	1
Q-10	6	12	4	6	6	6	4	37	1
Q-12	15	11	16	15	16	17	16	13	3
Q-20	10026	2539	11	10026	15808	862	11	1129	11
Q-50	>121277	>160845	177	>121277	>173869	>143472	177	>117616	0
Q-75	>118127	>152812	818	>118127	>186617	>137450	818	>133184	797
MS-3	1	0	1	1	1	0	1	1	0
MS-4	51	42	3	29	95	46	96	47	2
MS-5	>204089	>176414	>197512	74063	>201698	74711	>190692	>183580	16
MS-6	>237428	>176535	>231600	>190822	>239305	>204425	>204119	>214287	303
S-2	2	6541	9	2	89	887	9	12	1
S-5	>104148	>80203	963	>104148	>78774	>101058	963	>92557	17
S-7	1865	>80295	187	1865	>93675	>91514	187	2626	105
K-5	767	>179126	767	>97178	>178970	>228316	>73253	>190116	1
K-6	37695	>177129	37695	35059	>176668	>239427	14998	>160789	90
QG-5	7743	>130635	0	7763	>96083	>94426	0	>95406	1
QG-6	2009	>75475	89	2009	>108987	>124523	89	>89888	0
QG-7	3	845	1	3	773	1	1	1	1
LF 2-12	16	223	1	16	22	29	1	6	1
LF 2-16	39	24592	98	39	210	599	98	239	0
LF 2-20	77	>158028	172	77	1	26314	172	4521	1
LF 2-23	26	>157649	64	26	3	29805	64	0	0
$\bar{x}$	3550.1	3481.6	2054.3	7706.5	1419.5	10252.4	932.4	664.2	61.5

Table 9. Runtime in ms and Backtracks requires for optimizers ABC, PSO and GA

	ABC		PSO		GA	
Problem	Runtime	Backtracks	Runtime	Backtracks	Runtime	Backtracks
Q-8	230	1	4982	3	645	1
Q-10	265	1	7735	1	735	4
Q-12	315	3	24369	1	875	40
Q-20	575	11	52827	11	7520	3879
Q-50	2770	0	1480195	0	6530	15
Q-75	10775	797	t.o.	818	16069	17
MS-3	290	0	2745	0	735	0
MS-4	400	2	15986	0	1162	42
MS-5	340	16	565155	14	1087	198
MS-6	868	303	t.o.	>47209	t.o.	>176518
S-2	287	1	10967	2	15638	6541
S-5	335	17	2679975	13	8202	4229
S-7	465	105	967014	256	25748	10786
K-5	1132	1	4563751	106	21089	50571
K-6	1961	90	t.o.	12952	170325	21651
QG-5	312	1	59158	0	11862	7763
QG-6	310	0	44565	0	947	0
QG-7	262	1	28612	0	795	4
LF 2-12	265	1	10430	1	1212	223
LF 2-16	279	0	20548	0	1502	97
LF 2-20	337	1	28466	1	1409	64
LF 2-23	360	0	30468	3	1287	0
$\bar{x}$	1051.5	61.5	557786.8	675.4	14065.5	5053.6

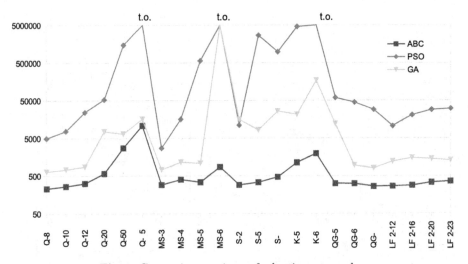

Fig. 1. Comparing runtimes of adaptive approaches

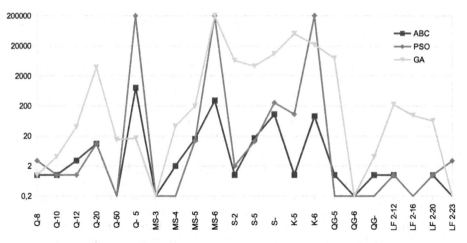

Fig. 2. Comparing backtracks of adaptive approaches

6   Conclusions

Autonomous search is an interesting approach to provide more capabilities to the solver in order to improve the search process based on some performance indicators and self-tuning. In this paper we have focused on the automated self-tuning of constraint programming solvers. To this end, we have presented an artificial bee colony algorithm able to find good CF parameter settings when solving constraint satisfaction problems. The experimental results have demonstrated the efficiency of the proposed approach validating the ability to adapt to dynamic environments and improving as a consequence the performance of the solver during the resolution process. As future work, we plan to test new modern metaheuristics for supporting CFs for AS. The incorporation of additional strategies to the portfolio would be an interesting research direction to follow as well.

Acknowledgements. Cristian Galleguillos is supported by Postgraduate Grant Pontificia Universidad Católica de Valparaíso 2015, Ricardo Soto is supported by Grant CONICYT/FONDECYT/INICIACION/ 11130459, Broderick Crawford is supported by Grant CONICYT/ FONDECYT/ 1140897, and Fernando Paredes is supported by Grant CONICYT/ FONDECYT/ 1130455.

References

1. Akay, B., Karaboga, D.: A modified artificial bee colony algorithm for real-parameter optimization. Information Sciences **192**, 120–142 (2012)
2. Apt, K.R.: Principles of Constraint Programming. Cambridge University Press (2003)

3. Barták, R., Rudová, H.: Limited assignments: a new cutoff strategy for incomplete depth-first search. In: Proceedings of the 20th ACM Symposium on Applied Computing (SAC), pp. 388–392 (2005)
4. Castro, C., Monfroy, E., Figueroa, C., Meneses, R.: An approach for dynamic split strategies in constraint solving. In: Gelbukh, A., de Albornoz, A., Terashima-Marín, H. (eds.) MICAI 2005. LNCS (LNAI), vol. 3789, pp. 162–174. Springer, Heidelberg (2005)
5. Chandrasekaran, K., Hemamalini, S., Simon, S.P., Padhy, N.P.: Thermal unit commitment using binary/real coded artificial bee colony algorithm. Electric Power Systems Research **84**, 109–119 (2012)
6. Crawford, B., Soto, R., Castro, C., Monfroy, E.: A hyperheuristic approach for dynamic enumeration strategy selection in constraint satisfaction. In: Ferrández, J.M., Álvarez Sánchez, J.R., de la Paz, F., Toledo, F.J. (eds.) IWINAC 2011, Part II. LNCS, vol. 6687, pp. 295–304. Springer, Heidelberg (2011)
7. Crawford, B., Soto, R., Castro, C., Monfroy, E., Paredes, F.: An Extensible Autonomous Search Framework for Constraint Programming. Int. J. Phys. Sci. **6**(14), 3369–3376 (2011)
8. Crawford, B., Soto, R., Monfroy, E., Palma, W., Castro, C., Paredes, F.: Parameter tuning of a choice-function based hyperheuristic using Particle Swarm Optimization. Expert Systems with Applications **40**(5), 1690–1695 (2013)
9. Crawford, B., Soto, R., Monfroy, E., Palma, W., Castro, C., Paredes, F.: Parameter tuning of a choice-function based hyperheuristic using particle swarm optimization. Expert Syst. Appl. **40**(5), 1690–1695 (2013)
10. Crawford, B., Soto, R., Montecinos, M., Castro, C., Monfroy, E.: A framework for autonomous search in the eclipse solver. In: Mehrotra, K.G., Mohan, C.K., Oh, J.C., Varshney, P.K., Ali, M. (eds.) IEA/AIE 2011, Part I. LNCS, vol. 6703, pp. 79–84. Springer, Heidelberg (2011)
11. Hamadi, Y., Monfroy, E., Saubion, F.: Autonomous Search. Springer (2012)
12. Karaboga, D., Basturk, B.: Artificial bee colony (abc) optimization algorithm for solving constrained optimization problems. In: Proceedings of the 12th International Fuzzy Systems Association World Congress on Foundations of Fuzzy Logic and Soft Computing, pp. 789–798 (2007)
13. Karaboga, D., Basturk, B.: On the performance of artificial bee colony (abc) algorithm. Soft Computing **8**, 687–697 (2008)
14. Karaboga, D., Ozturk, C., Karaboga, N., Gorkemli, B.: Artificial bee colony programming for symbolic regression. Information Sciences **209**, 1–15 (2012)
15. Karaboga, D., Ozturk, C., Karaboga, N., Gorkemli, B.: A comprehensive survey: artificial bee colony (abc) algorithm and applications. Artificial Intelligence Review **42**, 21–57 (2014)
16. Maturana, J., Saubion, F.: A compass to guide genetic algorithms. In: Rudolph, G., Jansen, T., Lucas, S., Poloni, C., Beume, N. (eds.) PPSN 2008. LNCS, vol. 5199, pp. 256–265. Springer, Heidelberg (2008)
17. Monfroy, E., Castro, C., Crawford, B., Soto, R., Paredes, F., Figueroa, C.: A reactive and hybrid constraint solver. Journal of Experimental and Theoretical Artificial Intelligence **25**(1), 1–22 (2013)

18. Crawford, B., Soto, R., Castro, C., Monfroy, E.: A hyperheuristic approach for dynamic enumeration strategy selection in constraint satisfaction. In: Ferrández, J.M., Álvarez Sánchez, J.R., de la Paz, F., Toledo, F.J. (eds.) IWINAC 2011, Part II. LNCS, vol. 6687, pp. 295–304. Springer, Heidelberg (2011)
19. Soto, R., Crawford, B., Monfroy, E., Bustos, V.: Using autonomous search for generating good enumeration strategy blends in constraint programming. In: Murgante, B., Gervasi, O., Misra, S., Nedjah, N., Rocha, A.M.A.C., Taniar, D., Apduhan, B.O. (eds.) ICCSA 2012, Part III. LNCS, vol. 7335, pp. 607–617. Springer, Heidelberg (2012)
20. Yan, X., Zhu, Y., Zou, W., Wang, L.: A new approach for data clustering using hybrid artificial bee colony algorithm. Neurocomputing **97**, 241–250 (2012)

Monitoring of Service-Oriented Applications for the Reconstruction of Interactions Models

Mariam Chaabane[1], Fatma Krichen[1(✉)], Ismael Bouassida Rodriguez[1,2,3], and Mohamed Jmaiel[1,4]

[1] ReDCAD Laboratory, National School of Engineers of Sfax, University of Sfax, B.P. 1173, 3038 Sfax, Tunisia
{mariam.chaabane,fatma.krichen,bouassida,mohamed.jmaiel}@redcad.org
[2] CNRS, LAAS, 7 avenue du colonel Roche 31400 Toulouse, France
[3] Université de Toulouse, LAAS, 31400 Toulouse, France
[4] Research Center for Computer Science, Multimedia and Digital Data Processing of Sfax, B.P. 275, Sakiet Ezzit, 3021 Sfax, Tunisia

Abstract. This work focuses on software applications designed using service-oriented architectural model. These applications consist of several Web Services from different sources working together, to obtain complex Web Services. Several events occur during the interaction between Web Services, such as degradation of QoS, breakdown of deployed services, etc. These events disrupt the communications. In our work we have developed a service-oriented application for a Smart City, we have deployed a monitoring mechanisms to examine a service oriented application architecture, monitor interactions and build the graphical representation of the architecture using graphs to view the interactions events.

Keywords: Service-oriented applications · Monitoring · Architectural model · Graphical representation

1 Introduction

Software applications designed according to the service-oriented architectural model is constructed by composing several Web Services from different sources and working together, regardless of their location.

At run time, several events occur, namely the degradation of quality of service, failures deployed services, etc. This generates disruption of communications. To overcome this problem, we propose in this paper an approach for the implementation of monitoring mechanisms to examine service-oriented architectures and monitor interactions between their applications. Our challenge is to monitor the flow of messages between Web Services for these applications.

We rely on messages interception principle to extract service quality metrics values. Monitoring results that perceive communications disturbances are stored in a database. They are used to construct the graphical representation of the application architecture and to study the evolution of service quality metric.

© Springer International Publishing Switzerland 2015
O. Gervasi et al. (Eds.): ICCSA 2015, Part I, LNCS 9155, pp. 172–186, 2015.
DOI: 10.1007/978-3-319-21404-7_13

To describe these architectures, we use graphs that represent a natural, intuitive and expressive way, to specify static and dynamic architectures aspects. Graph nodes represent services and edges represent links between them. We generate several types of graphs using the frame work Java Universal Network/Graph Framwork (JUNG). In fact, this facilitates the architecture monitoring to observe interactions from several points of view.

Several research activities have addressed the problem of monitoring to detect service quality degradation. Moreover, a multitude of monitoring approaches have been implemented. However, in the context of orchestrated Web Services, monitoring based on deployment of monitors (called handlers for service-oriented applications) hasn't been proposed for a Web Services orchestrator. However, in our approach, We have used on the one hand a support provided by Axis2 [2] to deploy monitors in atomic Web Services and clients. On the other hand, we have extended the monitors deployed in the orchestration engine Business Process Execution Language (BPEL) [7], Apache-Ode, to monitor Composite Web Services (BPEL process). Several research activities have treated agraphical representation of application architecture construction. However, the combination of service-oriented applications monitoring and architecture dynamic representation using graphs at the run time, has not been proposed in these research activities.

To implement our approach, we used a case study dealing with a smart energy distribution called Smart City. The increase in energy consumption during the last years, creates considerable attention to the implementation of energy distribution smart network. These networks use heterogeneous devices to identify precisely the inhabitants energy consumption of a building, a city or a region.This type of network can encourage energy consumers to rationalize their electricity consumption and enable global economies in a sustainable development perspective. To realize such a network, we have implemented a service-oriented application that implements our approach of monitoring and building architecture graphical representation.

The rest of the paper is organized as follows: Section 2 presents related work. Section 3 defines our approach which is based on two steps: the creation and deployment of monitors in our service-oriented application, then the architecture representation as a graph using recovered measures. Sections 4 and 5 present our case study called Smart City which aims to reduce energy consumption in a city or a region. In Section 6 we have applied our approach to the Smart City case study. Section 7 describes the implementation steps and perform experiments through a set of scenarios.

2   Related Work

Our work is based on monitoring on the one hand, the reconstruction and presentation of architecture on the other. Thus, we present in this section a set of related works which address these two issues.

2.1 Existing Approaches for Monitoring Software Architectures

Erradi et al. [4] propose an approach of monitoring composite Web Services. For this, they present a project called MASC (Manageable and Adaptive Service Compositions) performing monitoring during execution which is based on the WS-Policy4MASC language that extends WS-Policy. In fact, this solution allows monitoring the SOAP messages and orchestration process synchronously and asynchronously. The main advantage of MASC is the separation of basic workflow and monitoring. But it is not integrated into existing standards.

The researchs of Moser et al. [9] present an approach for monitoring of BPEL applications using the Aspect Oriented Programming (AOP). They propose VieDAME system that allows intercepting SOAP messages. In addition, it offers a substitution strategy Web Services during runtime.

In Sun et al. approach [12], the AOP extends the BPEL engine in order to recover interactions data between Web Services. Then these data are analyzed and compared by the properties of Web Services described in EMSC (Extended Message Sequence Charts) and METG (Message Event Transferring Graph).

The approach of Yoo and Lee [5] is based on AOP but it provides a monitoring solution able to consider the system as a single entity. Thus, it can detect the failure points for each function as well as the system as a whole.

The approach of An et al. [1] present a monitoring technology based on the paradigm publish/subscribe. The author offers a scalable solution called SQRT-C operating the OMG Data Distribution Service to be used with a multitude of platforms Cloud.

2.2 Existing Approaches for Representing Software Architectures

Heijstek et al. [6] performed an experiment to compare the effectiveness of visual or textual artifacts in making design decisions of software architecture. The results of this experiment show that neither the diagrams nor the textual descriptions are significantly more effective in this term. Hence, we need more research to understand how the text and diagrams can complement each other.

Krichen et al. [8] present a formal approach called P/S-COM+ to present the Styles of Architecture Publish/Subscribe (PSAS). For this, they have created an Eclipse plugin P/S-CoM'SD which enables the design of this architectural style graphically. In fact, the approach P/S-COM+ extends its previous P/S-COM by specifying the behavioral interaction rules coded on Z notation. However, it does not support the temporal aspect.

The approach of Weinreich and Buchgeher [13] integrates the requirements, scenarios and design decisions in one coherent architectural representation, formally defined, called the LISA model. This approach allows the tracing of connections between requirements, design decisions, other architectural elements and artifacts implemented on the one hand. On the other hand, between the architecture analysis and impact assessment. Nevertheless, the LISA model does not present more complex connections.

The Bouassida et al. work [3], [10], [11] propose an approach to multi-level reconfiguration architectures. This approach presents these architectures as graphs. In addition, it uses the grammar graph to obtain a management model based on rules. These rules manage both transformations architecture within the same level and between different architectural levels.

3 Approach of Monitoring and Construction of Architecture Graphical Representation of Service-Oriented Applications

Our approach is divided into two phases: Monitoring: to monitor the entities that constitute the service-oriented application and their interactions in order to observe the communication disruption that can happen. The construction of architecture graphical representation: to reconstruct the architecture of the application previously described, at the run time, and study data evolution of quality service.

3.1   Approach Architecture and Description

In this section, we present the architecture of our approach. It should be noted here that a Web Service belonging to a service-oriented application can be atomic or composed. In our work, we have used composition of Web Services by orchestration. This type of composition is centralized. An orchestrator of Web Services is considered as the only process holding the central coordination and controlling all the Web Services participating in its composition. In fact, the type of service-oriented application that we develop is composed of three types of entities: the atomic Web Services (Web Service), Web Services orchestrators and Web Services clients (Web Service Client).

In our approach, specifically in the monitoring phase we have designed an architecture that deploys a monitor in each entity. These monitors intercept messages exchanges between the entities. Then they capture data allowing to perceive communications disruptions, to study data evolution of service quality. To accomplish the monitoring phase, monitors save recovered data in a database. The construction phase of the architecture graphical representation, at the run time, requires monitoring data stored in database (Fig. 1). These data are used to present different types of graphs. In fact, a graph node is an entity of an architecture that interacts with another. Arcs illustrate the flow of messages between two entities.

In our approach, a Web Service client (Fig. 1) invokes a Web Services orchestrator in order to obtain a complex service. It invokes a number of atomic Web Services (Fig. 1) in a specific order. In fact, each of them offers a specific service to achieve the desired result. Moreover, monitors attached to each entity intercept SOAP messages. This interception aims to enrich SOAP message and extract pertinents informations to study the data evolution of service quality, such as IP addresses, Web service execution time, message transfer time, etc. Finally, monitors save these informations in a database. In the construction phase of the architecture graphical representation at run time, monitoring data are recovered from database. Then, we perform calculations to obtain necessary data to generate graphs. In fact, this step allows generation of three types of graphs.

3.2   Monitoring Phase

For monitoring phase, one monitor is deployed to each entity from service-oriented application. Monitoring is based on intercepting SOAP messages exchanged between these entities, in order to extract data allowing to study the evolution of service quality and detect disturbances communications. We present in a first, monitoring architecture. Then, monitoring running.

Monitoring Phase Architecture. A monitor is a software entity used to intercept SOAP messages. It enriches SOAP messages with personalized data according to the need of the developer such as the time of sending or receiving message. In our monitoring architecture, we implement a Web service client side monitor (Fig. 1), an orchestrator side monitor (Fig. 1) and a Web Service side monitor (Fig. 1).

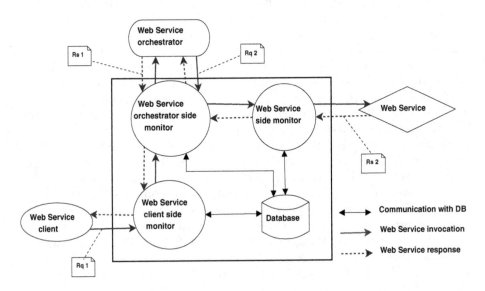

Fig. 1. Monitoring architecture

As indicated in Fig. 1: The Web service client side monitor intercepts messages exchanged between Web service client and Web Services orchestrator. The Web Service side monitor intercepts messages exchanged between Web Services orchestrator and Web Service. And the Orchestrator side monitor intercepts all messages entering and leaving the Web Services orchestrator. All monitors communicate with the database to save the data resulted by the interception of messages.

Monitoring Phase Description. First, a Web Service client sends SOAP request (Rq 1) to Web Service orchestrator and remains blocked because our work is based on synchronous communication. As this request requires a complex service, the orchestrator invokes the necessary atomic Web Services. In fact each invoked atomic Web Services offers a service to achieve the result required by the request (Rq 1). So, orchestrator launches second SOAP message (Rq 2) to invoke atomic Web Service. Then, the invoked Web Service generates a new SOAP response (Rs 2) message containing the result of invocation. After having received all responses from invoked Web Services, orchestrator generates a new SOAP response message (Rs 1) containing the result of invocation, to the client.

We deploy a monitor next to the Web Service, a monitor next to the Web Service client and another next to the Web Service orchestrator to intercept various SOAP messages. All monitors communicate with database to save the intercepted data.

In fact, for a request sent from the client to the orchestrator, the client side monitor extract the message identifier, the IP address of Web Service client, its name and the request sending time from Web Service client. Then, the orchestrator sides monitor extract the IP address of Web Services orchestrator, its name, the name of the called operation and the request receiving time by the Web Services orchestrator. For a response sent from the orchestrator to the client, the orchestrator sides monitor extract response sending time from the Web Services orchestrator. Then, the client side monitor extracts response receiving time by the Web Service client.

The extracted data consist at: the message identifier, the IP address of the source entity, the name of the source entity, The IP address of the target entity, the name of the source entity, the name of the target entity, the name of the target operation, the request sending time from the source entity, the request receiving time by the target entity, the response sending time from the target entity and the response receiving time by the source entity.

3.3   Construction of Architecture Graphical Representation

We choose to use the graphs to construct graphical representation for a service-oriented architecture. For this, we have used monitoring data saved in the database. Then, we have calculated additional variables namely, the Web Service execution time average, request and response transfer time average, Web Service and operation invocation average. In our work, we generate three types of graphs which are: the entity graph, the machines graph and the interaction Bi-Graph that combines the two previous types. These three types of graphs will be used to visualize the messages flow between different entities or machines from several points of view: The first point of view is more focused on Web Services, the second points of view is more focused on machines in which these services are deployed and the third points of view summarizes the two previous points of view. These three points of views are used to study the evolution of service quality data. This facilitates the creation of analysis tools for collected data during the monitoring.

Web Service Graph. Graph node corresponds to Web Service. Links define the relationships between these Web Services. Web Services graph nodes represent the Web Services (Web Service Name, IP, invocations number of Web Service, Average invocations number of Web Service, Web Service average run time). The links represent the messages flow transmitted between two nodes. Links illustrates the messages between Web Services enriched with information concerning the duration of messages and operations invoked and (Target operation name, Invocations number of target operation, Invocations number average of target operation, Request transfer time average, Response transfer time average).

Machines Graph. Graph machines nodes have machines IP addresses. The links describe the messages transmitted between machines, enriched by data concerning Web Services (Source Web Service name, target Web Service name, invocations number of target Web Service, invocations number average of target Web Service). These informations provide IP addresses of used machines, Web Services deployed on each of them and their importance. To explain, source Web Service name and target Web Service name for example, illustrate the Web Services deployed on each machine. Moreover, invocations number of target Web Service and invocations number average of target Web Service illustrates if a machine receives an important number of requests. This reflects the importance of machine and its Web Services based on received messages flow.

Interactions Bi-Graph. Regarding the interactions Bi-Graph, there are two types of nodes: Web Services nodes and machines nodes. However, it has only one type of link representing the messages flow between Web Service nodes.

The data resulting by both previous graphs nodes (IP, Web Service Name, invocations number of Web Service, Average invocations number of Web Service, Web Service average run time) as well as graphs links (Target operation name, Invocations number of target operation, Invocations number average of target operation, Request transfer time average, Response transfer time average). This Bi-Graph provides a better global view of our application architecture.

4 Case Study Smart City

Due to the necessity of lower power consumption, considerable attention has been paid in last years to the construction of intelligent energy distribution networks. Whether newly designed, these buildings are equipped with heterogeneous devices (sensors, actuators, etc.), which are able to monitor and respond to the context change, and provide smarter services. These networks use heterogeneous devices (sensors, actuators, etc.) to identify and rationalize precisely and in real time the energy consumption of the inhabitants of a building, a city or a region.

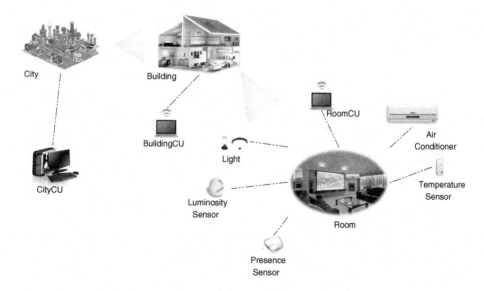

Fig. 2. Case study Smart City

The Fig. 2 shows our case study called Smart City which consists of a number of buildings. Each one includes a number of rooms. Building Control Unit (BuildingCU) is used to manage this structure. As shown in Fig. 2, each room is managed by a Room Control Unit (RoomCU). Sensors (Temperature Sensor, Presence Sensor, etc.) that monitor context parameters and actuators that operate on devices (Light, Air Conditioner, etc.) are adjusted if necessary by the RoomCU. In fact, BuildingCU manages RoomCU managing the various devices in order to reduce energy consumption and satisfy the growing demand for electricity, particularly during peak periods. Moreover, RoomCU allow reducing consumption or even turning off some equipments to avoid overloading network. It is important to indicate that the network administrator using City Control Unit (CityCU) can establish higher prices of electricity consumption at times of peak predictable as hours of the day and season, using data collected by the RoomCU and BuildingCU. This allows encouraging energy consumers to rationalize their electricity consumption and enable global economies in a perspective of sustainable development.

Indeed, during execution, several events occur and disrupt communication. Thus, it is essential for the Smart City manager to monitor this structure during execution to observe for example, affected devices by service quality degradation, exchanged messages between devices and transfer time average, response time of each device, etc.

This justifies the deployment of monitor for each device/Control Unit (CU) to control the messages flow.

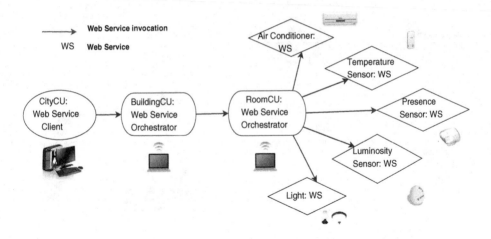

Fig. 3. Case study modeling

In our case study, each device or CU is managed by a software entity from our approach (Web Service, Web services orchestrator, Web Service client).

Modeling case study Smart City is described in Fig. 3. The devices are managed by atomic Web Services (Web Service). The RoomCUs are managed by composed Web Services (Web Services orchestrator). The BuildingCUs are managed by Web Services orchestrators. The CityCUs are managed by Web Services clients.

In fact, in our work, we use a software layer building on top of the infrastructure. The infrastructure is composed by electrical equipment that we have called devices, RoomCUs, BuildingCUs and CityCUs. We have implemented software entities such as Web Services, Web Services orchestrators and Web Services clients that manage this infrastructure. We use machines other than devices for the implementation of the software part, if it is not possible to deploy the Web Service on the device itself. This allows the deployment of more than one entity on each machine. Thus, this layer allows the Smart City manager to enable and disable the CUs or devices as needed.

5   Application of Approach on the Case Study

Our approach is divided into two phases: monitoring and building an architecture graphical representation. This leads to apply this approach to our case study in two steps.

5.1   Monitoring Phase for the Smart City

Fig. 4 describes the application of monitoring approach for the case study. In this application, we deploy a monitor next to each service. Each of these monitors

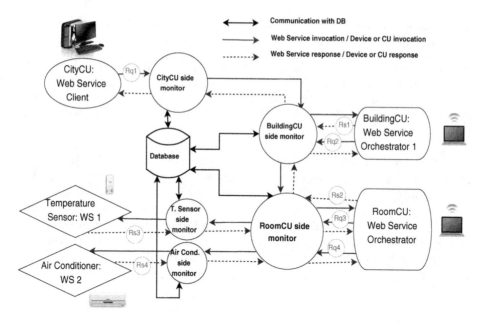

Fig. 4. Monitoring phase for the case study Smart City

saves the captured data in database. The presented scenario is an Air Conditioner reconfiguration. In fact, the administrator wants to limit the use of Air Conditioner: he defines a minimum temperature value which prohibits Air Conditioners to provide a lower temperature. First, the CityCU sends the minimum temperature to the BuildingCU (Rq 1). This request (Rq 1) is intercepted first time by the CityCU side monitor and a second time by the BuildingCU side monitor. Second, as BuildingCU is a Web Services orchestrator, it invokes RoomCU by sending the request (Rq 2). This is similarly intercepted twice.

The RoomCU invokes the Temperature Sensor (WS 1) by sending the request (Rq 3). The Temperature Sensor returns the value of the ambient temperature (Rs 3). Based on the value received by the Temperature Sensor, the RoomCU compares it with the minimum value received by the CityCU and decides to reconfigure the Air Conditioner. This decision will be transmitted to the Air Conditioner (WS 2) by the request (Rq 4). Once reconfigured, the Air Conditioner returns its new state to RoomCU (Rs 4). After that, the RoomCU transfers the data concerning its devices to BuildingCU through its response (Rs2). Then, the BuildingCU returns current status of devices and RoomCU belonging to the building, to the CityCU by the response (Rs 1).

5.2   Construction Phase of Architecture Graphical Representation for the Smart City

In this section, we apply the construction phase of architecture graphical representation for the case study Smart City at the run time. To illustrate the application of a normal behavior, we have executed the test scenario described by Fig. 4, 110 times. This gives us three graphs illustrated by the diagrams of Figs. 5, 6 and 7.

Web Services Graph. The Web Service graph shown in Fig. 5 has five nodes representing the five entities that participate in the scenario, namely: CityCU, the BuildingCU, the RoomCU, TemperatureSensor and the AirConditioner. In addition to the Web Service name, each node presents the IP address where Web service is deployed, the number of times it is invoked during the execution of 440 messages as well as the average number it is invoked during the execution of 440 messages, and finally the average time required by the Web Service to execute a request.

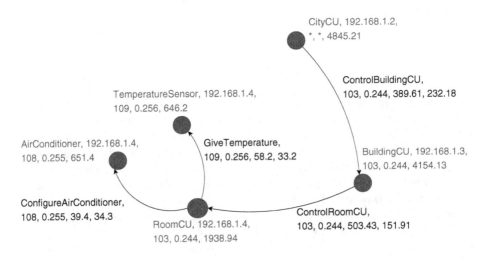

Fig. 5. Web Services Graph

Each link presents the invoked operation name like 'ConfigureAirConditioner', the number of times it is invoked during the execution of N messages, the average number the operation is invoked during the execution of N messages.

Machines Graph. The machines graph (Fig. 6), has three nodes. Each node represents a machine and presents its IP address. Each link shows source Web Service name, target Web Service name, the number of times that the target Web Service is invoked during the execution of 440 posts and the average number it is invoked during the execution of 440 posts.

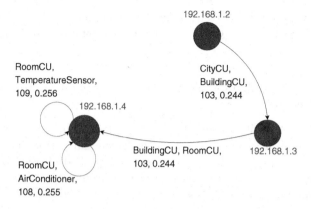

Fig. 6. Machine Graph

Interactions Bi-Graph. The interactions Bi-Graph (Fig. 7) summarizes the information introduced by the two previous graphs. We observe clearly the machines we used in our application and Web Services installed in each one.

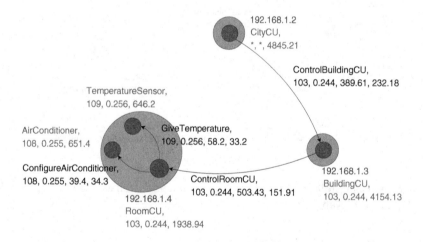

Fig. 7. Interactions Bi-Graph

6   Experiments

In our work, we use eclipse development environment that is based on Java EE. We profit of the benefits of the combined use of Tomcat Server and implementation of Axis2 Web Services. Thus, the Web Services orchestrators execute the described process by BPEL language. This allows to invoke atomic Web Services in order to obtain composed Web Service. The composed Web Service is equally presented in WSDL but deployed in Apache-Ode, which is deployment engine for orchestration process.

6.1   Application Behavior after Provoking a Communication Disturbance

The objective behind this experiment is to prove that using monitoring data and graphs representation, we were able to identify communication disturbances. In fact, we have executed the same test scenario described by Fig. 4, 110 times, causing a delay of 1000 ms during the execution of Web Service 'Air Conditioner'. Thus, we have obtained the Web services graph described in Fig. 8.

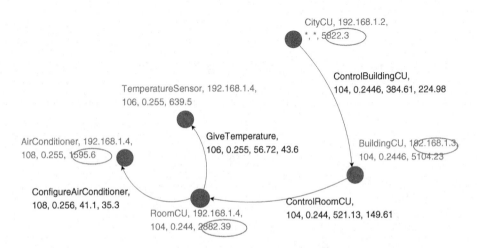

Fig. 8. Web Services Graph after provoking a communication disturbance

The delay is visible on the node 'AirConditioner' as well as 'RoomCU', 'BuildingCU' and the Web Service client 'CityCU'.

6.2   Monitoring Time Overhead

In this section we run the scenario: 'Normal behavior of the application', a number of times, to observe the additional time of monitor's. We use two machines with 4GB of RAM and Intel Pentium P6200 2.13GHz processor and one machine with 6GB of RAM and Intel Core i3 2330M 2.2GHz processor a 100 MB LAN network. We represent the evolution of the runtime system each 50 messages sent by the application client. The Fig. 9 shows that the execution time with the introduction of monitors is slightly higher than that measured without introducing them. In fact, the average execution time of 50 messages is 4350.32 ms without monitors and it is about 4845.21 ms after introducing monitors. For 500

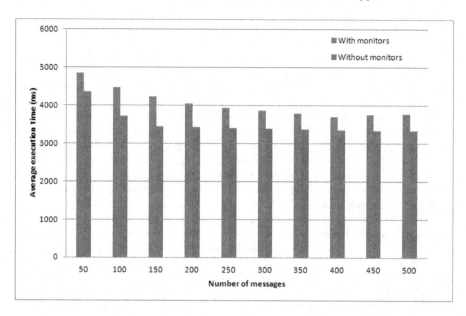

Fig. 9. The overhead introduced by monitoring module according to the number of messages

messages, we observe that the execution time varies between 3330.4 ms without monitoring and 3779 ms after introducing monitors. The monitoring operation takes 10% of total execution time.

7   Conclusion

We presented in this work an approach for monitoring service-oriented applications. Our challenge is to monitor the messages flow between the different entities of the application in order to build architecture graphical representation of the system at the run time. To achieve these objectives, we have used on a hand the support provided by Axis2 for engaging monitors in Web Services and clients. On the other hand, we have extended the monitors deployed on BPEL engine, Apache Ode, to monitor composed Web Services (BPEL process). Based on measurements collected by the monitors we have generated three types of graphs using the JUNG library. In fact this facilitates the task of the administrator by allowing it to manage interactions from several points of view. We have executed then a set of scenarios extracted from our case study Smart City. In the future, we plan to execute more experiments to validate more our approach and deploying our software layer for Smart City directly on smart devices.

Acknowledgment. The authors would like to thank University of Toulouse who permitted this research opportunity within the IDEX "chaire d'attractivite" delivered to Pr. Gene COOPERMAN.

References

1. An, K., Pradhan, S., Caglar, F., Gokhale, A.: A publish/subscribe middleware for dependable and real-time resource monitoring in the cloud. In: Proceedings of the Workshop on Secure and Dependable Middleware for Cloud Monitoring and Management, pp. 3:1–3:6, New York, USA (2012)
2. Apache. Axis2 architecture guide, 1.5.1 version (October 2009)
3. Bouassida Rodriguez, I., Chassot, C., Jmaiel, M.: Graph grammar-based transformation for context-aware architectures supporting group communication. Revue des Nouvelles Technologies de l'Information L(19), 29–42 (2010)
4. Erradi, A., Maheshwari, P., Tosic, V.: Ws-policy based monitoring of composite web services. In: Proceedings of the Fifth European Conference on Web Services, pp. 99–108, Washington, DC, USA (2007)
5. Giljong, Y., Eunseok, L.: Monitoring methodology using aspect oriented programming in functional based system. In: Proceedings of the 12th International Conference on Advanced Communication Technology, pp. 783–786, Piscataway, NJ, USA (2010)
6. Heijstek, W., Kuhne, T., Chaudron, M.R.V.: Experimental analysis of textual and graphical representations for software architecture design. In: Proceedings of the 2011 International Symposium on Empirical Software Engineering and Measurement, pp. 167–176 (2011)
7. IBM. BPEL4WS, Business Process Execution Language for Web Services Version 1.1 (2003)
8. Krichen, I., Loulou, I., Dhouib, H., Kacem, A.H.: P/s-com+: A formal approach to design correct publish/subscribe architectural styles. In: Proceedings of the 2012 IEEE 17th International Conference on Engineering of Complex Computer Systems, pp. 179–188 (2012)
9. Moser, O., Rosenberg, F., Dustdar, S.: Non-intrusive monitoring and service adaptation for ws-bpel. In: Proceedings of the 17th International Conference on World Wide Web, pp. 815–824 (2008)
10. Bouassida Rodriguez, I., Van Wambeke, N., Drira, K., Chassot, C., Jmaiel, M.: Multi-layer coordinated adaptation based on graph refinement for cooperative activities. In: Proceedings of the 4th International Conference on Self-organization and Adaptation of Computing and Communications, SSCC 2008, pp. 163–167, Glasgow, UK (July 2008)
11. Sancho, G., Bouassida Rodriguez, I., Villemur, T., Tazi, S.: What about collaboration in ubiquitous environments. In: 10th Annual International Conference on New Technologies of Distributed Systems, NOTERE 2010, pp. 143–150, Tozeur, Tunisia (2010)
12. Sun, M., Li, B., Zhang, P.: Monitoring bpel-based web service composition using aop. In: Proceedings of the 2009 Eigth IEEE/ACIS International Conference on Computer and Information Science, pp. 1172–1177 (2009)
13. Weinreich, R., Buchgeher, G.: Integrating requirements and design decisions in architecture representation. In: Babar, M.A., Gorton, I. (eds.) ECSA 2010. LNCS, vol. 6285, pp. 86–101. Springer, Heidelberg (2010)

Comparing Cuckoo Search, Bee Colony, Firefly Optimization, and Electromagnetism-Like Algorithms for Solving the Set Covering Problem

Ricardo Soto[1,2,3], Broderick Crawford[1,4,5], Cristian Galleguillos[1(✉)],
Jorge Barraza[1], Sebastián Lizama[1], Alexis Muñoz[1], José Vilches[1],
Sanjay Misra[6], and Fernando Paredes[7]

[1] Pontificia Universidad Católica de Valparaíso, Valparaíso, Chile
{ricardo.soto,broderick.crawford}@ucv.cl, cgalleguillosm@ieee.org,
{jorge.barraza.c,sebastian.lizama.a,alexis.munoz.c}@mail.pucv.cl,
josemiguelvilchesfierro@gmail.com
[2] Universidad Autónoma de Chile, Santiago, Chile
[3] Universidad Cientifica del Sur, Lima, Perú
[4] Universidad Central de Chile, Santiago, Chile
[5] Universidad San Sebastián, Santiago, Chile
[6] Atilim University, Ankara, Turkey
ssopam@gmail.com
[7] Escuela de Ingeniería Industrial, Universidad Diego Portales, Santiago, Chile
fernando.paredes@udp.cl

Abstract. The set covering problem is a classical model in the subject of combinatorial optimization for service allocation, that consists in finding a set of solutions for covering a range of needs at the lowest possible cost. In this paper, we report various approximate methods to solve this problem, such as Cuckoo Search, Bee Colony, Firefly Optimization, and Electromagnetism-Like Algorithms. We illustrate experimental results of these metaheuristics for solving a set of 65 non-unicost set covering problems from the Beasley's OR-Library.

Keywords: Combinatorial optimization · Set covering problem · Cuckoo search algorithm · Bee colony algorithm · Firefly optimization algorithm · Electromagnetism-Like algorithm

1   Introduction

The Set Covering Problem (SCP) is a classical problem in the subject of combinatorial optimization that consists in finding a set of solutions that cover a range of needs at the lowest possible cost. This problem is also a main model for several important applications such as line balancing, location or installation of services, simplification of Boolean expressions, and crew scheduling, among others [20,21].

© Springer International Publishing Switzerland 2015
O. Gervasi et al. (Eds.): ICCSA 2015, Part I, LNCS 9155, pp. 187–202, 2015.
DOI: 10.1007/978-3-319-21404-7_14

SCP belongs to the Karp's 21 NP-complete problems [14], as a result of this, various techniques has been reported to tackle it, as much to incomplete as to complete ones. The incomplete methods reports good results on large instances of this problem, such as Genetic Algorithm [4], Simulated Annealing [6], Tabu Search [9], among others [8]. In addition to this, complete techniques has been used, but those are not always successful and depend exclusively by search space size. The exact methods explores throughout the search space guaranteeing the global optimum, but requiring more resources while the search space is increasing, some examples are Branch-and-Bound and Branch-and-Cut [2], classical Greedy algorithm [12] and Lagragian relaxation-based heuristics [7,10].

In this paper, we present a comparison of Cuckoo Search, Artificial Bee Colony, Firefly Optimization, and Electromagnetism-Like Algorithms for solving the Set Covering Problem, these metaheuristics were implemented by the authors. As far as we know, this comparison has not been reported yet in the literature.

The remainder of this paper is organized as follows: In Section 2, we explain the problem. An overview of additional techniques employed in the metaheuristics are presented in Section 3. In Section 4, different metaheuristics for solving the SCP is presented. Finally, we provide the experimental results and conclusions.

2   Problem Description

The SCP aims to find a set of solutions to cover a set of needs at the lowest possible cost. Formally, we define the problem as follows:

Definition 1 (Set Covering Problem). *Let A a binary matrix of $m \times n$ size, when $a_{i,j}$ may contain a value from $\{0,1\}$ in the matrix. The $a_{i,j}$ is located in i^{th} row and j^{th} column, with $i = \{1, ..., m\}$ and $j = \{1, ..., n\}$. Through A, it is possible to say that a column j covers a row i contained in A if $a_{ij} = 1$.*

Let C a vector of length n, containing the costs of each column, where each $c_j \in C$ are non-negative real values.

The corresponding mathematical model is as follows:

$$min\,(z) = \sum_{j=1}^{n} c_j x_j \tag{1}$$

Subject to:

$$\sum_{j=1}^{n} a_{ij} x_j \geq 1 \quad \forall\, i \in \{1, ..., m\} \tag{2}$$

$$x_j = \{0, 1\} \quad \forall\, j \in \{1, ..., n\}$$

The SCP aims at finding a subset $x_j \in X$ of the $\{1, ..., n\}$ columns with the minimum cost, such that each row i existing in A, is covered by at least a j column existing in X.

3   Additional Techniques Employed

The following techniques have been used in the implemented metaheuristics, in order to improve their performance. The Pre-processing phase (explained in Section 3.1) is applied before the search procedure of the metaheuristics. The Heuristic Feasibility Operator (Section 3.2) is applied to fix the unfeasible solutions generated by the search process of the metaheuristics.

3.1   Preprocessing

An effective way to accelerate problem solving is to reduce the instance sizes by applying pre-processing phases. Then, to effectively tackle the SCPs we have employed *Column Domination* and *Column Inclusion* [19].

Column Domination: When a column c_j, whose rows I_j can be covered by another column with lowest ratio $ratio_j$, according to Eq. 3, this column j is named as dominated and it can be deleted. The *ratio* is computed as follows:

$$ratio_j = \frac{c_j}{\sum_{j=1}^{n} a_{ij} \, c_j} \quad \forall i \in I_j \tag{3}$$

Column Inclusion: When a row is covered by a unique column after column domination, this column must be included in the solution.

3.2   Heuristic Feasibility Operator

During the solution generation phase, may exists unfeasible solutions which do not satisfy the problem constraints. For instance, in the SCP, a new solution without some rows covered, clearly violates a subset of constraints. In order to provide feasible solutions, we have used an heuristic operator to fix the solutions.

We have employed a heuristic operator that achieves the generation of feasible solutions, and additionally eliminates column redundancy [4].

The heuristic operator procedure is depicted in Algorithm 1. Some definitions are explained below:

Let:

$I =$ The set of all rows,

$J =$ The set of all columns,

$\alpha_i =$ The set of columns that cover row i, $i \in I$,

$\beta_j =$ The set of rows covered by column j, $j \in J$,

$S =$ The set of columns in a solution,

$U =$ The set of uncovered rows,

$w_i =$ The number of columns that cover row i, $i \in I$.

Algorithm 1. Heuristic Feasibility Operator Algorithm

1: Initialize $w_i := |S \cap \alpha_i|, \forall i \in I$
2: Initialize $U := \{i|w_i = 0, \forall i \in I\}$
3: For each row $i \in U$(in ascendant order):
4:     Find the first column $j \in \alpha_i$ (in ascendant order) that minimizes $c_j/|U \cap b_j|$
5:     Add j to S
6:     Set $w_i := w_i + 1, \forall i \in b_j$
7:     Set $U := U - b_j$.
8: For each column $j \in S$ (in decendant order)
9:     if $w_i \geq 2, \forall i \in \beta_j$:
10:        Set $S := S - j$
11:        Set $w_i := w_i - 1, \forall i \in \beta_j$

To ensure all feasible solutions, we will calculate a ratio with the Eq. 3. An unfeasible solution is repaired by covering the columns of the solution that had the lower ratio. After this, a local optimization step is applied, where column redundancy is eliminated. A column is redundant when it can be deleted and the feasibility of the solution is not affected.

The recognition of the rows that are not covered is performed by the algorithm through a "greedy" heuristic. From line 3rd to 7th of Algorithm 1, the columns with lower ratios are added to the solution. Finally, the redundant columns with higher costs are deleted while the solution is being feasible.

4   Metaheuristics

Metaheuristics are search procedures which do not guarantee to find an optimal solution, but usually it is possible to get a solution with the enough quality in a considerable time. In general, metaheuristics are very simple and easy to implement. In contrast with heuristics, these techniques are developed to scape from local optima, redefining their search depending on the search evolution during the procedure. Metaheuristic techniques had been widely used to solve problems, specially with combinatorial optimization problems, where the search space is vast and a complete search could not be efficient.

The logic of these techniques is very similar, every metaheuristic has a start point, and from this, new solutions are generated until some criteria is reached. When the search process is stopped, some solution is given, being the best solution found (or optimal in best case). The metaheuristics techniques that we have used to solve the problem SCP are presented below.

4.1   Cuckoo Search Algorithm

Cuckoo Search (CS) [25] is inspired by the obligate brood parasitism of some cuckoo species whose main characteristic is to let their eggs in nests from other birds. CS was defined using three fundamental rules:

1. A cuckoo egg represents a solution to the problem and it is left in a randomly selected nest, a cuckoo only can left one egg at a time.
2. Nests holding the higher quality eggs will pass to the next generations.
3. The nest owner can discover a cuckoo egg with a probability $p_a \in [0, 1]$. If this occurs, the nest owner can left his nest and build other nest in other location. The number of total nests is a fixed value.

The generation of a new solution is performed by using Lévy flight as follows.

$$sol_i^{t+1} = sol_i^t + \alpha \oplus \text{Levy}(\beta) \tag{4}$$

where sol_i^{t+1} is the solution in iteration $t + 1$, and $\alpha \geq 0$ is the step size, which is associated to the range of values that the problem needs (scale value), being determined by upper and lower bounds [24].

$$\alpha = 0.01(U_b - L_b) \tag{5}$$

The step length in the generation of a new solution is given by the Lévy distribution.

$$\text{Lévy} \sim u = t^{-\beta}, \quad (1 < \beta < 3) \tag{6}$$

Algorithm 2 depicts the classic CS procedure. At the beginning, an initial population of n nests is generated. Then, a while loop manages the CS actions which are self-explanatory. The objective function of the SCP is employed to compute the fitness of each solution. Solutions are produced by using the Lévy flight distribution according to Eq. 4.

Algorithm 2. Cuckoo Search via Lévy Flights

Generate the first generation of n nests;
while $t < MaxGeneration$ **do**
    Obtain a random cuckoo/generate a solution by Lévy Flight distribution;
    Quantify its fitness F_j and then compare with old Fitness F_i;
    if $F_j > F_i$ **then**
        Substitute j as the new best solution;
    end if
    A fraction of worse nest $p(a)$ will be abandoned and new nests will be made;
    Maintain the better solutions (or nests with high quality);
    Rank the better solutions and find the best one;
end while
Postprocess results and visualization;

Since in Eq. 4 a real number between 0 and 1 is generated, it needs to be discretized, we have proceeded as follows:

$$x_j = \begin{cases} 1 & \text{if } r < x'_j \quad \text{or} \quad x'_j > U_b \\ 0 & \text{if } r \geq x'_j \quad \text{or} \quad x'_j < L_b \end{cases}$$

where x'_j holds the value to be discretized for variable x_j of the SCP solution, and r is a normal distributed random value. Then, a new solution is generated and next evaluated. If it is not feasible, it is repaired using the heuristic feasibility operator described in Section 3.2. Finally, the best solutions are memorized and the generation count is incremented or the process is stopped if the termination criteria has been met.

Additionally, as a pre-processing phase we have added the **Pre-process** explained in Section 3.1

4.2 Artificial Bee Colony Algorithm

The Artificial Bee Colony Algorithm (ABC) was developed by Karaboga [15,16] for solving numerical problems in optimization, it is inspired by the dance and feeding process behavior of real honey bee colonies.

For a theoretical background, we introduce the following concepts:

1. **Employed bees** are associated in particular to one food source, storing and sharing the information about their neighborhood.
2. **Unemployed bees** explores the space. They are classified into two subgroups.
   (a) **scout bees** seeks new food sources around the hive.
   (b) **onlooker** or **curious bees** remains in the hive, waiting for the information provided by the employed bees, with the objective of going out to search for a potential food source, based on the knowledge gained through the collected information.
3. **Colony** is the swarm which participates in the searching process, i.e. *employed bees + onlooker bees*. The scout bees are not counted, since they correspond to the state of the other bees.
4. **Food sources** correspond to the swarm size. Strictly in ABC, the initial population corresponds to the half of the colony. Since the relation between food sources and employed bees is always 1:1, therefore the half of the colony would be employed bees and remaining ones will be unemployed bees.

The ABC Algorithm is composed by 4 phases (Initialization, Employed bee, Onlooker bee and Scout bee), and they are executed in ordered way as they appear in the parenthesis, until a maximum cycle is reached.

Initialization phase: For every employed bee, a random solution is generated, with a drop counter initialized for each food source. This counter increases if the produced solution is worse or equal to the stored one. To evaluate each solution, the fitness value is calculated using Eq. 7.

$$fit_i = \begin{cases} \frac{1}{1+f_i} & if\ (f_i \geq 0) \\ 1 + abs\ (f_i) & if\ (f_i < 0) \end{cases} \tag{7}$$

Employed bee phase: The employed bee attempts to improve its solution, using the equation of motion depicted below.

$$X_i^j = X_i^j + \varphi \times \left(X_i^j - X_k^j \right) \tag{8}$$

Where X_i is the actual food source and the X_k is a randomly selected food source. These food sources must be different, then $i \neq k$. And j corresponds to the jth column.

Onlooker bee phase: The onlooker bee waits for the information to be shared from the other food sources. Each food source has a probability of being selected, depending on the potential for improvement. The equation used is not strict, as long as the use of priorities is respected at the time of selection of the food sources. After selecting a food source, the solution is updated with the equation of motion. If the new solution is better, it is replaced and the drop counter is reset, otherwise the counter is incremented by one.

Scout bee phase begins when the drop counter of a bee exceeds the defined limit. The employed bee becomes a scout bee and generates a new random solution. Once the new solution is generated and if it is feasible, the corresponding drop counter is started and the scout bee becomes an employed bee again.

Due to the nature of the problem, we have modified the classic ABC motion equation by Eq. 9, replacing the Eq. 8. The xor ($\oplus$) operator has been introduced in [17].

$$V_i^j = X_i^j \oplus \left[\varphi \left(X_i^j \oplus X_k^j \right) \right] \tag{9}$$

In order to improve the solution quality and increasing the solving speed, we have employed the pre-process (Section 3.1) and repairing function (Section 3.2). In addition, we have defined a dynamic parameter configuration after observing the *limit* (maximum tries to improve a food source) and *maximum cycles*(the stop criteria) parameters are associated with the problem size, we have improved the convergence and as a consequence the solving time.

We have determined that $m * 10$ (m being the number of rows of the SCP) plus a margin error equivalent to $2,000$ cycles, it is a good value for the maximum cycles, because after this quantity of iterations the best solution does not improve. Moreover, a suitable value for the limit [1] is $FN * n$, where FN is the number of food sources used in the implementation and n the amount of columns of the SCP.

4.3   Firefly Optimization Algorithm

Firefly Optimization Algorithm is a metaheuristic inspired by the flashing behavior of fireflies [22, 23], whose movements are mainly guided by the brightness that each one naturally emit.

Three main rules govern the implementation of this metaheuristic:

1. All fireflies are unisex, which means that they are attracted to other fireflies regardless of their sex.
2. The degree of attractiveness of a firefly is proportional to its brightness, and for any two flashing fireflies, the less brighter one will move towards the brighter one. More brightness means less distance between two fireflies. If there is no brighter one than a particular firefly, a random movement is performed.
3. Finally, the brightness of a firefly is determined by the value of the objective function. For a maximization problem, the brightness of each firefly is proportional to the value of the objective function. In case of minimization problem, brightness of each firefly is inversely proportional to the value of the objective function.

Below, we explain the employed algorithm including the applied modifications. Firstly, we have reduced the search space using the method explained in Section 3.1. Then the parameters are set, β, γ, number of fireflies, and maximum number of generation. Then, the initial population of n size is generated. For each solution the fitness function is calculated.

The fireflies are moved, this is done by using Eq. 10.

$$X'_p = X_p + \beta(r)(X_p - X_q) + \alpha(rand - \frac{1}{2}) \qquad (10)$$

Since X'_p is a real number, it needs to be discretized. For this, we have employed the following transfer function, which was the best performing in [11]:

$$S(x) = \left| \frac{2}{\pi} arctan(\frac{\pi}{2}x) \right| \qquad (11)$$

The generated solutions are evaluated in order to check whether it satisfies the constraints. If the constraints are not satisfied, the solution is repaired by using the heuristic feasibility operator described in Section 3.2. This process is repeated until the iteration counter have reached the *MaxGeneration* parameter.

4.4   Electromagnetism-Like Algorithm

The Electromagnetism-like algorithm (EM) was developed by Ilker Birbil and Shu-Cherng Fang [5]. EM is inspired in the electromagnetism theory where particles can be attracted or repealed by other ones. The magnitude of attraction or repulsion of a particle is determined by its charge. The higher is the charge of a particle, the stronger is its attraction. After charge calculations, the movement direction of each particle is determined, which will be subjected to the forces exerted among particles.

It consists in four phases: initialization of EM, local search procedure, calculation of the charge and force exerted on each particle and finally the movement of the particle.

Algorithm 3. Firefly Optimization Algorithm

Generate an initial population of n fireflies
Calculate light intensity I
Define absorption coefficient γ
while (t ¡ MaxGeneration) **do**
   for each $ff_i \in$ fireflies **do**
      for each $ff_j \in$ fireflies **do**
         if $(I_j > I_i)$
            move firefly i towards j
         end if
         Recalculate attractiveness
         Evaluate new solutions and update light intensity
      end for
   end for
   Rank fireflies and find the current best
end while

1. *Initialization:* The initialization procedure is randomly constructed defining a initial population of particles and belonging to a binary domain. Then, we compute the fitness to each particle and the best one is selected. The EM algorithm evaluates all the solutions w.r.t the objective function and selects the best one at each iteration, in such a way to obtain a result that can be used as a reference point to obtain best optimal solutions.

2. *Local Search:* Gathers local and relevant information for a given particle. Two input parameters are needed: LSITER defines the number of local iterations and δ represents the multiplier for the neighborhood search.

3. *Calculation of charge and force:* The calculation of charge and force is carried out via Eq. 12 and 13, respectively:

$$q^i = \exp\left(-n\frac{f\left(x^i\right) - f\left(x^{best}\right)}{\sum_{k=1}^{n}\left(f\left(x^k\right) - f\left(x^{best}\right)\right)}\right), \quad \forall i \in \{1, ..., n\} \qquad (12)$$

$$F_d^i = \sum_{k=1,k\neq i}^{n} \begin{cases} (x_d^k - x_d^i)\frac{q^i q^k}{\|x^i - x^k\|^2} & \text{if } f(x^k) < f(x^i) \\ (x_d^i - x_d^k)\frac{q^i q^k}{\|x^k - x^i\|^2} & \text{if } f(x^k) \geq f(x^i) \end{cases}, \forall i \qquad (13)$$

where q^i is the charge of particle x^i, n represents the dimension, x^{best} is the best particle found, and F_d^i is the force of component d of particle x^i. After charge computation, some particles will hold higher charges (better fitness) than others, which determines the attraction or repulsion magnitude exerted.

4. *Movement:* The movement is governed by Eq. 14, where λ is a random step length, uniformly distributed between 0 and 1. Finally, RNG is a vector controlling that movements do not escape from the domain of variables.

$$x^i = x^j + \lambda \frac{F^i}{\| F^i \|}(RNG) \quad i = 1, 2, ..., n \tag{14}$$

Because of Eq. 14 produces a real number, this number must be transformed to a binary one, due to the nature of the problem. To tackle this, we have employed the following transfer function, which was the best performing one in [18].

$$T(x) = \frac{1}{1 + e^{-2x}}$$

Finally, the resulting value is discretized via the standard method as follows:

$$x(t+1) = \begin{cases} 1 & \text{if } rand < T(x(t+1)) \\ 0 & \text{otherwise} \end{cases}$$

5   Experimental Results

We have tested 65 SCP non-unicost instances (organized in sets: 4, 5, 6, A, B, C, D, NRE, NRF, NRG, NRH) from OR-Library [3], all of them were executed 30 times.

Table 1 depicts detailed information about the tested instances, where "Density" is the percentage of non-zero entries in the SCP matrix.

Table 1. SCP instances from OR-Library

Instance group	No. of instances	Rows n	Columns m	Density	Optimal solution
4	10	200	1,000	2%	Known
5	10	200	2,000	2%	Known
6	5	200	1,000	5%	Known
A	5	300	3,000	2%	Known
B	5	300	3,000	5%	Known
C	5	400	4,000	2%	Known
D	5	400	4,000	5%	Known
NRE	5	500	5,000	10%	Unknown
NRF	5	500	5,000	20%	Unknown
NRG	5	1,000	10,000	2%	Unknown
NRH	5	1,000	10,000	5%	Unknown

Tables 3, 4 and 5 show the results obtained by the metaheuristics. Each instance was executed 30 times by each metaheuristic. The column "Best"

Table 2. Parameter configuration by used Metaheuristics

Metaheuristic	Population size	Max. Iterations	Specific Parameters
Cuckoo Search	50, for NR group 15 other case	5,000	$\beta = 1.5$, $\alpha = 0.01$, $p_a = 0.25$, $U_b = 1$, and $L_b = 0$
Artificial Bee Colony	150		Dynamic (See Section 4.2)
FireFly Optimization	20	7,000	$\gamma = 0.0002$ and $\beta_0 = 1$
Electromagnetism Like	25	MAXITER= 25	LSITER= 1000

Table 3. Computational results 4, 5 and 6 Group instances

Instance	Z_{opt}	Cuckoo Search		Artificial Bee Colony		Firefly Optimization		Electromagnetism Like	
		Best (Z)	RPD	Best (Z)	RPD	Best (Z)	RPD	Best (Z)	RPD
4.1	429	430	0.23	430	0.23	430	0.23	447	4.20
4.2	512	**512**	0.00	513	0.20	515	0.59	559	9.18
4.3	516	517	0.19	519	0.58	520	0.78	537	4.07
4.4	494	**494**	0.00	495	0.20	500	1.21	527	6.68
4.5	512	**512**	0.00	514	0.39	514	0.39	527	2.93
4.6	560	**560**	0.00	561	0.18	561	0.18	607	8.39
4.7	430	**430**	0.00	431	0.23	431	0.23	448	4.19
4.8	492	**492**	0.00	493	0.20	497	1.02	509	3.46
4.9	641	643	0.31	649	0.93	643	0.31	682	6.40
4.10	514	**514**	0.00	517	0.58	523	1.75	571	11.09
5.1	253	**253**	0.00	254	0.40	257	1.58	280	10.67
5.2	302	304	0.66	309	2.32	307	1.66	318	5.30
5.3	226	**226**	0.00	229	1.33	229	1.33	242	7.08
5.4	242	**242**	0.00	**242**	0.00	**242**	0.00	251	3.72
5.5	211	212	0.47	**211**	0.00	212	0.47	225	6.64
5.6	213	**213**	0.00	214	0.47	214	0.47	247	15.96
5.7	293	**293**	0.00	298	1.71	297	1.37	316	7.85
5.8	288	**288**	0.00	289	0.35	291	1.04	315	9.38
5.9	279	**279**	0.00	280	0.36	285	2.15	314	12.54
5.10	265	**265**	0.00	267	0.75	269	1.51	280	5.66
6.1	138	140	1.45	142	2.90	141	2.17	152	10.14
6.2	146	**146**	0.00	147	0.68	148	1.37	160	9.59
6.3	145	**145**	0.00	148	2.07	148	2.07	160	10.34
6.4	131	**131**	0.00	**131**	0.00	**131**	0.00	140	6.87
6.5	161	**161**	0.00	165	2.48	164	1.86	184	14.29
Optimal reached		19		3		2		0	

Table 4. Computational results A, B, C and D Group instances

Instance	Z_{opt}	Cuckoo Search		Artificial Bee Colony		Firefly Optimization		Electromagnetism Like	
		Best (Z)	RPD	Best (Z)	RPD	Best (Z)	RPD	Best (Z)	RPD
A.1	253	254	0.40	254	0.40	256	1.19	261	3.16
A.2	252	256	1.59	257	1.98	257	1.98	279	10.71
A.3	232	233	0.43	235	1.29	238	2.59	252	8.62
A.4	234	237	1.28	236	0.85	237	1.28	250	6.84
A.5	236	**236**	0.00	**236**	0.00	238	0.85	241	2.12
B.1	69	**69**	0.00	70	1.45	70	1.45	86	24.64
B.2	76	**76**	0.00	78	2.63	**76**	0.00	88	15.79
B.3	80	**80**	0.00	**80**	0.00	**80**	0.00	85	6.25
B.4	79	**79**	0.00	80	1.27	80	1.27	84	6.33
B.5	72	**72**	0.00	**72**	0.00	**72**	0.00	78	8.33
C.1	227	228	0.44	231	1.76	231	1.76	237	4.41
C.2	219	221	0.91	222	1.37	223	1.83	237	8.22
C.3	243	247	1.65	254	4.53	250	2.88	271	11.52
C.4	219	221	0.91	231	5.48	227	3.65	246	12.33
C.5	215	216	0.47	216	0.47	216	0.47	224	4.19
D.1	60	**60**	0.00	**60**	0.00	**60**	0.00	62	3.33
D.2	66	**66**	0.00	68	3.03	67	1.52	73	10.61
D.3	72	73	1.39	76	5.56	74	2.78	79	9.72
D.4	62	**62**	0.00	63	1.61	63	1.61	67	8.06
D.5	61	**61**	0.00	63	3.28	**61**	0.00	66	8.20
Optimal reached		10		4		5		0	

depicts the best results obtained by each metaheuristic and "RPD" shows the relative percentage deviation of the best solution, which is calculated as follows:

$$RDP = \frac{(Z - Z_{opt})}{Z_{opt}} \times 100$$

where Z is the best obtained, Z_{opt} is the optimal or best known solution. We have established the following parameters for each metaheuristic defined in Table 2. The experimental results exhibit the robustness of the used approaches, but specially with Cuckoo Search which is able to reach 34 of 65 global optimums, obtaining also reasonable good results by only increasing population size when the instance becomes harder. In general CS, ABC and FF have reported good results with a lower RPD fluctuating between 1.30% and 3.62%. A graphical comparison of four MHs can be seen in Figure 1.

Table 5. Computational results NRE, NRF, NRG and NRH Group instances

Instance	Z_{opt}	Cuckoo Search		Artificial Bee Colony		Firefly Optimization		Electromagnetism Like	
		Best (Z)	RPD	Best (Z)	RPD	Best (Z)	RPD	Best (Z)	RPD
NRE.1	29	**29**	0.00	**29**	0.00	31	6.90	30	3.45
NRE.2	30	31	3.33	32	6.67	31	3.33	35	16.67
NRE.3	27	28	3.70	29	7.41	29	7.41	34	25.93
NRE.4	28	30	7.14	29	3.57	31	10.71	33	17.86
NRE.5	28	**28**	0.00	29	3.57	**28**	0.00	30	7.14
NRF.1	14	**14**	0.00	**14**	0.00	16	14.29	17	21.43
NRF.2	15	**15**	0.00	16	6.67	**15**	0.00	18	20.00
NRF.3	14	15	7.14	16	14.29	16	14.29	17	21.43
NRF.4	14	15	7.14	15	7.14	16	14.29	17	21.43
NRF.5	13	14	7.69	15	15.38	14	7.69	16	23.08
NRG.1	176	**176**	0.00	183	3.98	**176**	0.00	194	10.23
NRG.2	154	156	1.30	162	5.19	155	0.65	176	14.29
NRG.3	166	169	1.81	174	4.82	168	1.20	184	10.84
NRG.4	168	170	1.19	175	4.17	171	1.79	196	16.67
NRG.5	168	170	1.19	179	6.55	171	1.79	198	17.86
NRH.1	63	64	1.59	70	11.11	65	3.17	70	11.11
NRH.2	63	64	1.59	69	9.52	66	4.76	71	12.70
NRH.3	59	62	5.08	66	11.86	63	6.78	68	15.25
NRH.4	58	59	1.72	64	10.34	61	5.17	70	20.69
NRH.5	55	56	1.82	60	9.09	56	1.82	69	25.45
Optimal reached		5		2		3		0	

Table 6. Average RPD for Group instances

Instance	Average RPD			
	Cuckoo Search	Artificial Bee Colony	Firefly Optimization	Electromagnetism Like
4	0.07	0.37	0.64	6.06
5	0.11	0.77	1.16	8.48
6	0.29	1.63	1.50	10.25
A	0.74	0.91	1.58	6.29
B	0.00	1.07	0.54	12.27
C	0.88	2.72	2.12	8.13
D	0.28	2.70	1.18	7.98
NRE	2.84	4.24	5.67	14.21
NRF	4.40	8.70	10.11	21.47
NRG	1.10	4.94	1.08	13.98
NRH	2.36	10.39	4.34	17.04
Average RPD	1.02	3.04	2.44	10.82

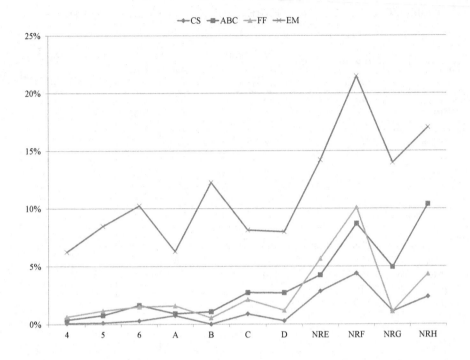

Fig. 1. Comparing average RPD by groups instances

6   Conclusions and Future Work

In this paper, we have presented four incomplete techniques for solving SCPs. The metaheuristics are quite simple to implement and can be adapted to binary domains. We have tested 65 non-unicost instances from the OR-Library where several global optimum values where reached (with RPD equal to 0). The results have also exhibited the rapid convergence and robustness of 3 of 4 proposed algorithms which are capable to obtain reasonable good results. In general, for every metaheuristic all best solutions were closed to each global optimum solution for the tested instances maintaining a lower RPD.

The integration of autonomous search [13] to the presented techniques would be another direction of research to follow, in order to autonomously self-tune their performance based on analysis of the performances of the solving process.

Acknowledgements. Cristian Galleguillos is supported by Postgraduate Grant PUCV 2015. Ricardo Soto is supported by Grant CONICYT / FONDECYT / INICIA-CION / 11130459. Broderick Crawford is supported by Grant CONICYT / FONDE-CYT /REGULAR / 1140897. Fernando Paredes is supported by Grant CONICYT / FONDECYT / REGULAR / 1130455.

References

1. Akay, B., Karaboga, D.: Parameter tuning for the artificial bee colony algorithm. In: Nguyen, N.T., Kowalczyk, R., Chen, S.-M. (eds.) ICCCI 2009. LNCS, vol. 5796, pp. 608–619. Springer, Heidelberg (2009)
2. Balas, E., Carrera, M.C.: Carnegie-Mellon University. Management Sciences Research Group. A Dynamic Subgradient-based Branch and Bound Procedure for Set Covering. Management sciences research report. Management Sciences Research Group, Graduate School of Industrial Administration, Carnegie Mellon University (1992)
3. Beasley, J.E.: http://www.brunel.ac.uk/mastjjb/jeb/info.html (last visited on January 30, 2015)
4. Beasley, J.E., Chu, P.C.: A genetic algorithm for the set covering problem. European Journal of Operational Research **94**(2), 392–404 (1996)
5. Birbil, S.I., Fang, S.-C.: An electromagnetism-like mechanism for global optimization. Journal of Global Optimization **25**(3), 263–282 (2003)
6. Brusco, M.J., Jacobs, L.W., Thompson, G.M.: A morphing procedure to supplement a simulated annealing heuristic for cost and coverage correlated setcovering problems. Annals of Operations Research **86**, 611–627 (1999)
7. Caprara, A., Fischetti, M., Toth, P.: A heuristic method for the set covering problem. Operations Research **47**, 730–743 (1995)
8. Caprara, A., Toth, P., Fischetti, M.: Algorithms for the set covering problem. Annals of Operations Research **98**(1–4), 353–371 (2000)
9. Caserta, M.: Tabu search-based metaheuristic algorithm for large-scale set covering problems. In: Doerner, K.F., Gendreau, M., Greistorfer, P., Gutjahr, W., Hartl, R.F., Reimann, M. (eds.) Metaheuristics. Operations Research/Computer Science Interfaces Series, vol. 39, pp. 43–63. Springer, US (2007)
10. Ceria, S., Nobili, P., Sassano, A.: A Lagrangian-based heuristic for large-scale set covering problems. Mathematical Programming **81**(2), 215–228 (1998)
11. Chandrasekaran, K., Simon, S.P., Padhy, N.P.: Binary real coded firefly algorithm for solving unit commitment problem. Information Sciences **249**, 67–84 (2013)
12. Chvatal, V.: A greedy heuristic for the set-covering problem. Mathematics of Operations Research **4**(3), 233–235 (1979)
13. Crawford, B., Castro, C., Monfroy, E., Soto, R., Palma, W., Paredes, F.: Dynamic selection of enumeration strategies for solving constraint satisfaction problems. Romanian Journal of Information Science and Technology **15**(2), 106–128 (2013)
14. Garey, M.R., Johnson, D.S.: Computers and Intractability: A Guide to the Theory of NP-Completeness. W. H. Freeman & Co., New York (1979)
15. Karaboga, D., Akay, B.: A survey: algorithms simulating bee swarm intelligence. Artificial Intelligence Review **31**(1–4), 61–85 (2009)
16. Karaboga, D., Basturk, B.: A powerful and efficient algorithm for numerical function optimization: artificial bee colony (abc) algorithm. Journal of Global Optimization **39**(3), 459–471 (2007)
17. Kiran, M.S., Gndz, M.: Xor-based artificial bee colony algorithm for binary optimization. Turkish Journal of Electrical Engineering and Computer Sciences **21**(suppl. 2), 2307–2328 (2013); cited By 2
18. Mirjalili, S., Hashim, S., Taherzadeh, G., Mirjalili, S.Z., Salehi, S.: A study of different transfer functions for binary version of particle swarm optimization. In: GEM 2011. CSREA Press (2011)

19. Pezzella, F., Faggioli, E.: Solving large set covering problems for crew scheduling. Top **5**(1), 41–59 (1997)
20. Vasko, F.J., Wilson, G.R.: Using a facility location algorithm to solve large set covering problems. Operations Research Letters **3**(2), 85–90 (1984)
21. Vasko, F.J., Wolf, F.E., Stott, K.L.: Optimal selection of ingot sizes via set covering. Oper. Res. **35**(3), 346–353 (1987)
22. Yang, X.S.: Nature-Inspired Metaheuristic Algorithms. Luniver Press (2008)
23. Yang, X.-S.: Firefly algorithms for multimodal optimization. In: Watanabe, O., Zeugmann, T. (eds.) SAGA 2009. LNCS, vol. 5792, pp. 169–178. Springer, Heidelberg (2009)
24. Yang, X.-S.: Bat algorithm and cuckoo search: a tutorial. In: Yang, X.-S. (ed.) Artificial Intelligence, Evolutionary Computing and Metaheuristics. SCI, vol. 427, pp. 421–434. Springer, Heidelberg (2013)
25. Yang, X.S., Deb, S.: Cuckoo Search via Levy Flights. ArXiv e-prints (March 2010)

Interactive Image Segmentation of Non-contiguous Classes Using Particle Competition and Cooperation

Fabricio Breve[1]([⊠]), Marcos G. Quiles[2], and Liang Zhao[3]

[1] São Paulo State University (UNESP), Rio Claro, SP, Brazil
fabricio@rc.unesp.br
http://www.rc.unesp.br/igce/demac/fbreve/
[2] Federal University of São Paulo (Unifesp), Rio Claro, SP, Brazil
quiles@unifesp.br
[3] University of São Paulo (USP), Rio Claro, SP, Brazil
zhao@usp.br
http://dcm.ffclrp.usp.br/~zhao/

Abstract. Semi-supervised learning methods employ both labeled and unlabeled data in their training process. Therefore, they are commonly applied to interactive image processing tasks, where a human specialist may label a few pixels from the image and the algorithm would automatically propagate them to the remaining pixels, classifying the entire image. The particle competition and cooperation model is a recently proposed graph-based model, which was developed to perform semi-supervised classification. It employs teams of particles walking in a undirected and unweighed graph in order to classify data items corresponding to graph nodes. Each team represents a class problem, they try to dominate the unlabeled nodes in their neighborhood, at the same time that they try to avoid invasion from other teams. In this paper, the particle competition and cooperation model is applied to the task of interactive image segmentation. Image pixels are converted to graph nodes. Nodes are connected if they represent pixels with visual similarities. Labeled pixels generate particles that propagate their labels to the unlabeled pixels. Computer simulations are performed on some real-world images to show the effectiveness of the proposed approach. Images are correctly segmented in regions of interest, including non-contiguous regions.

Keywords: Semi-supervised learning · Interactive image segmentation · Machine learning · Particle competition and cooperation

1 Introduction

Image segmentation is considered one of the most difficult tasks in image processing [23]. It is the process of dividing a digital image into parts (sets of pixels), identifying regions, objects or other relevant information [30]. Fully automatic methods are usually limited to simpler or specific types of images. Therefore,

© Springer International Publishing Switzerland 2015
O. Gervasi et al. (Eds.): ICCSA 2015, Part I, LNCS 9155, pp. 203–216, 2015.
DOI: 10.1007/978-3-319-21404-7_15

interactive image segmentation approaches, where some user input is used to help the segmentation process, are of increasing interest in the last decades [1,2,5,7,20,21,24,26–29,32].

Semi-supervised learning is an important class of machine learning classification techniques. They are usually applied to problems where unlabeled data is abundant, but the process of labeling them is expensive and/or time consuming, requiring the intense work of human specialists [19,34]. Differently from supervised and unsupervised learning approaches, semi-supervised learning algorithms employ both labeled and unlabeled data in their training process.

Many interactive image processing approaches are based on semi-supervised leaning. In such scenarios, a human specialist can quickly label some easier pixels, far from the boundaries, and the semi-supervised learning algorithm will then propagate the labels to the remaining pixels, according to the similarity between them, thus discovering regions or objects of interest.

Particle competition and cooperation is a graph-based semi-supervised learning method proposed recently [13]. It takes vector-based data sets and converts them into non-weighted undirected graphs. Each graph node corresponds to a data item, and edges are created between nodes corresponding to similar data items. Teams of particles are created, with each particle corresponding to labeled nodes. Cooperation takes place among particles in the same team, representing the same label; while competition takes place among particles playing for different teams, representing different labels. Each team tries to dominate unlabeled nodes near them, spreading their label, and preventing other teams from invading their territory. At the end of the iterative process, particles territory frontiers are usually close to the boundaries between the classes of the problem. Therefore, by labeling each unlabeled node after the team that dominated it, high classification accuracy is expected.

Most semi-supervised graph-based methods are similar [3,4,6,25,33,35], sharing the same regularization framework, differing only in the choice of a regularizer and a loss function [34]. The particle competition and cooperation method is different, it applies label spreading in a local fashion while most other methods spread labels globally. In this sense, while most semi-supervised graph-based methods have cubic computation complexity ($O(n^3)$), where n is data set size, the particle competition and cooperation method approaches $O(n)$ in most scenarios.

In spite of being a relatively new approach, particle competition and cooperation is already successfully extended and applied to some important machine learning problems, like overlapped community detection [11,12], learning with label noise [10,17,18], learning with concept drift [9,16], and combined active and semi-supervised learning [8,14,15]. Notice that in all those scenarios, the model was applied to vector-based and/or graph-based data, but not to images.

In this paper, the particle competition and cooperation model is employed to perform interactive image segmentation. The image is first converted into a graph, i.e., for each pixel in the image, a corresponding node is created in the graph. Each node is connected to its k-nearest neighbors. The distance between two pixels is given by their position and visual characteristics. Each labeled pixel also generates a particle that will be used to propagate its label to the unlabeled pixels nearby. Figures 1 and 2 illustrate this process. Computer simulations are performed on some real-world images and the results show that this is a promising approach.

The remaining of this paper is organized as follows. Section 2 presents an overview of the particle competition and cooperation method. Section 3, describes how the particle competition and cooperation model is applied to the task of interactive image segmentation. Section 4 shows some computer simulations on real-world images. Finally, some conclusions are drawn on Section 5.

2   Particle Competition and Cooperation Overview

The original particle competition and cooperation method for semi-supervised learning is detailed described in [13]. It employs non-weighted undirected graphs which are built based on the data set to be classified. For each data item, a corresponding graph node is created. Edges are created between nodes if they are close to each other in the original feature space. Usually, each node is connected to its k-nearest neighbors using the Euclidean distance.

For each labeled data item, a particle is created and its initial position is set to the corresponding graph node. The particles are divided into teams. Particles representing the same class label play for the same team, while particles representing different classes labels play for different teams. Particles on the same team cooperate among themselves to dominate the unlabeled nodes in their neighborhood. The different teams compete against each other. They try to avoid intrusion of enemy teams in the territory they are dominating.

Each node has a set of domination levels, a level for each team. When a particle visits a node, it increases its team domination level and it decreases the other teams domination levels on that node. Each particle has a strength level, which changes according to the domination level of its team in the node being visited. Each particle also holds a distance table, with the distance from their "home node" (the node where they started) to each node that they have visited at least once. Distance tables are updated dynamically as particles walk on the graph.

The particles walk in the graph and select the next node to be visited among the neighbors of the current node. A random-greedy rule is applied to choose which neighbor will be visited next. The particle stays on the chosen node only if its team is currently dominating that node, otherwise it is expelled and goes back to the previous node. At the end of the iterative process, each data item is labeled after the team (class) with the highest domination level on it.

Regarding computational complexity, the particle competition and cooperation model is usually $O(n)$ during the particle walking phase. The graph construction using brute force to find the k-nearest neighbors of each node is $O(n^2)$, as noted in [13], but one may use a faster method, like k-d trees to find nearest neighbors in logarithmic time [22].

Further details on the particle competition and cooperation approach are found in [13].

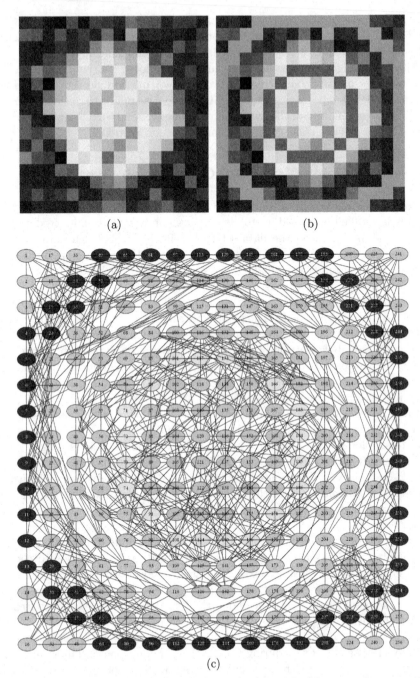

(a)   (b)

(c)

Fig. 1. Proposed Method Segmentation Example: (a) original image to be segmented (16x16 pixels); (b) original image with user labeling (green and red traces); and (c) graph generated after the original image, where each image pixel corresponds to a graph node. Labeled nodes are colored blue and yellow, and unlabeled nodes are colored grey. Each labeled node will have a particle assigned to it.

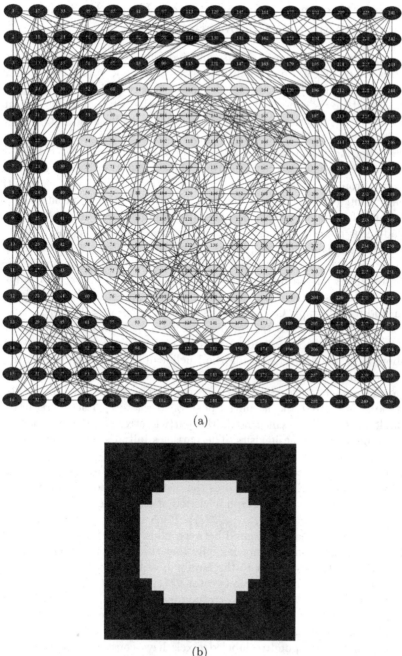

(a)

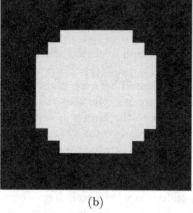

(b)

Fig. 2. Proposed Method Segmentation Example: (a) resulting graph after the segmentation process with nodes' colors representing the labels assigned to them; and (b) original image with the pixels colored after the resulting graph, where each color represents different class

3   Interactive Image Segmentation Using Particle Competition and Cooperation

Given a bidimensional image to be segmented with some pixels labeled by the user, the set of pixels are reorganized as $\mathcal{X} = \{x_1, x_2, \ldots, x_L, x_{L+1}, \ldots, x_N\}$, such that $\mathcal{X}_L = \{x_i\}_{i=1}^{L}$ is the labeled pixel subset and $\mathcal{X}_U = \{x_i\}_{i=L+1}^{N}$ is the unlabeled pixels subset. $\mathcal{L} = \{1, \ldots, C\}$ is the set containing the labels. $y : \mathcal{X} \rightarrow \mathcal{L}$ is the function associating each $x_i \in \chi$ to its label $y(x_i)$. For each unlabeled pixel $x_i \in \mathcal{X}_U$, the proposed model will estimate $y(x_i)$.

The following features are considered for each pixel: the three RGB (red, green, blue) components, the three HSV (hue, saturation, value) components (calculated from the RGB components using [31]), the average of each of these six components (RGB+HSV) in the pixel and its (up to) 8 adjacent pixels, and the average of standard deviation of the six components in the pixel and its (up to) 8 adjacent pixels. The horizontal and vertical positions of the pixel in the original image are also included in the feature set, totalling 20 features. For all measures with adjacent pixels, we consider the 8 adjacent pixels on the image, except for pixels on the image border, which have less (only 3 or 5) adjacent pixels. All features are normalized to have mean 0 and standard deviation 1.

An undirected graph $\mathbf{G} = (\mathbf{V}, \mathbf{E})$ is created, with $\mathbf{V} = \{v_1, v_2, \ldots, v_N\}$ as the set of nodes, and $\mathbf{E}$ as the set of edges (v_i, v_j). Each node v_i corresponds to a pixel x_i. Two nodes v_i and v_j are connected if v_j is among the k-nearest neighbors of v_i, or vice-versa, using the Euclidean distance between the x_i and x_j feature values. Otherwise, v_i and v_j are disconnected.

For each node $v_i \in \{v_1, v_2, \ldots, v_L\}$, corresponding to a labeled pixel $x_i \in \mathcal{X}_L$, a particle ρ_i is created and its initial position is set to v_i. Each particle ρ_j has a variable $\rho_j^\omega(t) \in [0, 1]$ which holds the particle strength, defining how much a particle impacts a node it visits. The particles initial strength are set to the maximum, $\rho_j^\omega(0) = 1$.

Each particle ρ_j also holds a distance table, which is dynamically updated as the particles walk. Each table stores the distance between the particle initial position and each node in the graph it visited at least once. The distance tables are defined as $\rho_j^{\mathbf{d}}(t) = \rho_j^{d_1}(t), \ldots, \rho_j^{d_N}(t)\}$. Each element $\rho_j^{d_i}(t) \in [0 \quad N-1]$ holds the distance in hops measured between node v_i and the initial node of ρ_j. Particles begin their journey knowing only that the distance to their initial node is zero. Other distances are set to the largest possible value $(n-1)$.

Each node v_i holds a domination vector $\mathbf{v}_i^\omega(\mathbf{t}) = \{v_i^{\omega_1}(t), v_i^{\omega_2}(t), \ldots, v_i^{\omega_C}(t)\}$, where each element $v_i^{\omega_c}(t) \in [0, 1]$ corresponds to the domination level from the team representing class c over the node v_i. The sum of the domination levels in each node is always constant, $\sum_{c=1}^{C} v_i^{\omega_c} = 1$.

Nodes which correspond to labeled pixels have constant domination levels, i.e., they are fully dominated by the corresponding team/class and this never changes. On the other hand, nodes which correspond to unlabeled pixels are variable. They start with all teams/classes domination levels set equally, but that changes as particles visits them. Thus, for each node v_i, the domination vector $\mathbf{v}_i^\omega$ is set as follows:

$$v_i^{\omega_c}(0) = \begin{cases} 1 & \text{if } x_i \text{ is labeled and } y(x_i) = c \\ 0 & \text{if } x_i \text{ is labeled and } y(x_i) \neq c \\ \frac{1}{C} & \text{if } x_i \text{ is unlabeled} \end{cases} \tag{1}$$

When a particle ρ_j visits a unlabeled node v_i, the node domination levels are updated as follows:

$$v_i^{\omega_c}(t+1) = \begin{cases} \max\{0, v_i^{\omega_c}(t) - \frac{0.1\rho_j^{\omega}(t)}{C-1}\} \\ \quad \text{if } c \neq \rho_j^c \\ v_i^{\omega_c}(t) + \sum_{r \neq c} v_i^{\omega_r}(t) - v_i^{\omega_r}(t+1) \\ \quad \text{if } c = \rho_j^c \end{cases}, \tag{2}$$

where ρ_j^c represents the class label of particle ρ_j. Each particle ρ_j changes the node its visiting v_i by increasing the domination level of its team ($v_i^{\omega_c}$, $c = \rho_j^c$) and, at the same time, decreasing the domination levels of other teams ($v_i^{\omega_c}$, $c \neq \rho_j^c$)). Notice that nodes which correspond to labeled pixels have constant domination levels, thus (2) is not applied to them.

A particle strength depends on the domination level of its team on the node it is visiting. Thus, at each iteration, the particle strength is updated as follows: $\rho_j^{\omega}(t) = v_i^{\omega_c}(t)$, where v_i is the node being visited, and $c = \rho_j^c$.

When a particle visits a node v_i, it updates its own distance table as follows:

$$\rho_j^{d_i}(t+1) = \begin{cases} \rho_j^{d_q}(t) + 1 & \text{if } \rho_j^{d_q}(t) + 1 < \rho_j^{d_i}(t) \\ \rho_j^{d_i}(t) & \text{otherwise} \end{cases}, \tag{3}$$

where $\rho_j^{d_q}(t)$ is the distance from the last visited node to the initial node of the particle ρ_j, and $\rho_j^{d_i}(t)$ is the distance from the node being visited to the initial node of the particle ρ_j. The distances are calculated dynamically. Particles do not know the graph structure. Unknown distances are calculated as the particles walk, and known distances are updated when particles naturally find shorter paths.

At each iteration, each particle ρ_j chooses a node v_i to visit among the neighbors of the node it currently is. The probability of choosing a neighbor node v_i is given by: a) the particle team domination on it, $v_i^{\omega_c}$, and b) the inverse of the node distance to the particle initial position, $\rho_j^{d_i}$, as follows:

$$p(v_i|\rho_j) = \frac{W_{qi}}{2\sum_{\mu=1}^{N} W_{q\mu}} + \frac{W_{qi}v_i^{\omega_c}(1+\rho_j^{d_i})^{-2}}{2\sum_{\mu=1}^{N} W_{q\mu}v_\mu^{\omega_c}(1+\rho_j^{d_\mu})^{-2}}, \tag{4}$$

where c is the class label of particle ρ_j, q is the index of the node the particle ρ_j currently is, $W_{qi} = 1$ if there is an edge between v_q and v_i, and $W_{qi} = 0$ otherwise. A particle stays on the chosen node only if, after applying (2), its team is dominating that node; otherwise, a shock happens and the particle goes back to the last node, and it stays there until the next iteration.

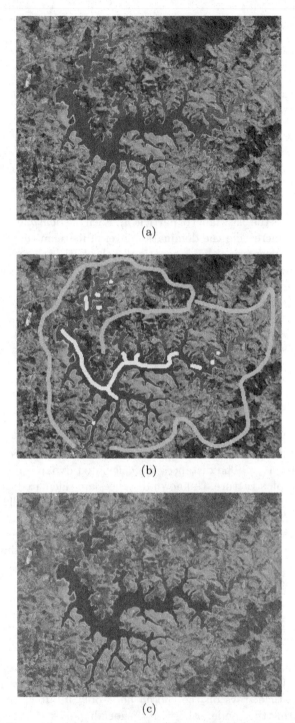

Fig. 3. Jaguari Reservoir: (a) original image, (b) image with user labeling (yellow and cyan traces), and (c) segmentation results by the proposed method

Fig. 4. Flowers: (a) original image, (b) image with user labeling (yellow and cyan traces), and (c) segmentation results by the proposed method

(a)

(b)

(c)

Fig. 5. Ducks: (a) original image, (b) image with user labeling (yellow and cyan traces), and (c) segmentation results by the proposed method

The average of the maximum domination levels of each node ($\langle v_i^{\omega_m} \rangle$, $m = \arg\max_c v_i^{\omega_c}$) is used as stop criterion. When this measure does not increase after a sufficient amount of iterations, the algorithm reached fair levels of stability, so it stops. The class of the dominating team is assigned to each unlabeled node.

4   Computer Simulations

In this section, some computer simulations using real-world images are presented to show the effectiveness of the proposed method. The parameter $k = 100$ is fixed, as it was empirically observed that it provides good results with the tested images.

The first experiment is performed on a Landsat satellite image from the Jaguari Reservoir in Brazil[1]. Figure 3a shows the original image with 1280×960 pixels. Figure 3b shows the image with some user labeling. Finally, Figure 3c shows the segmentation results by the proposed method, separating the water from the land. Notice that the algorithm actually resizes the image to 320×240 for faster processing.

The second experiment is performed on a picture of some flowers, shown in Figure 4a with 1536×1536 pixels. Figure 4b shows the image with some user labeling and Figure 4c shows the segmentation results by the proposed method, which correctly distinguishes between flowers and leafs. In this case, the algorithm resizes the picture to 256×256 for faster processing.

Finally, the third experiment is performed on a picture of some ducks, shown in Figure 5a with 3648×1824 pixels. Figure 5b shows the image with some user labeling on a single duck and a small portion of grass. The segmentation results by the proposed method are presented on Figure 4c. All the ducks were correctly identified and detached from the grass, even though only one of them had user labeled pixels. For this simulation, the algorithm resizes the picture to 304×152 for faster processing.

5   Conclusions

This paper has employed the semi-supervised learning graph-based model known as particle competition and cooperation to perform interactive image segmentation. In that model, particles walk in a unweighed and undirected graph created from the image to be segmented. Particles representing the same problem class play for the same team, thus they cooperate with each other to dominate the unlabeled pixels in their neighborhood. On the other hand, particles representing different problem classes play for different teams, thus they compete against each other to avoid invasion from enemy teams in the nodes they are dominating.

Computer simulations using some real-world images were performed. The proposed method was able to identify the objects of interest in all the proposed

---

[1] Available at the NASA Earth Observatory - http://earthobservatory.nasa.gov/IOTD/view.php?id=84564

scenarios, including non-contiguous classes, showing that this is a promising approach to interactive image segmentation.

As future work, we intend to extract different image features, and to refine the model to classify more types of images, including images from known repositories, so we may easily compare the results with those obtained by some state-of-the-art algorithms.

Acknowledgments. This work was supported by the São Paulo State Research Foundation (FAPESP) and the Brazilian National Research Council (CNPq).

References

1. Artan, Y.: Interactive image segmentation using machine learning techniques. In: 2011 Canadian Conference on Computer and Robot Vision (CRV), pp. 264–269 (May 2011)
2. Artan, Y., Yetik, I.: Improved random walker algorithm for image segmentation. In: 2010 IEEE Southwest Symposium on Image Analysis Interpretation (SSIAI), pp. 89–92 (May 2010)
3. Belkin, M., Matveeva, I., Niyogi, P.: Regularization and semi-supervised learning on large graphs. In: Shawe-Taylor, J., Singer, Y. (eds.) COLT 2004. LNCS (LNAI), vol. 3120, pp. 624–638. Springer, Heidelberg (2004)
4. Belkin, M., Niyogi, P., Sindhwani, V.: On manifold regularization. In: Proceedings of the Tenth International Workshop on Artificial Intelligence and Statistics, AISTAT 2005, pp. 17–24. Society for Artificial Intelligence and Statistics, New Jersey (2005)
5. Blake, A., Rother, C., Brown, M., Perez, P., Torr, P.: Interactive Image Segmentation Using an Adaptive GMMRF Model. In: Pajdla, T., Matas, J.G. (eds.) ECCV 2004. LNCS, vol. 3021, pp. 428–441. Springer, Heidelberg (2004). http://dx.doi.org/10.1007/978-3-540-24670-1_33
6. Blum, A., Chawla, S.: Learning from labeled and unlabeled data using graph mincuts. In: Proceedings of the Eighteenth International Conference on Machine Learning, pp. 19–26. Morgan Kaufmann, San Francisco (2001)
7. Boykov, Y., Jolly, M.P.: Interactive graph cuts for optimal boundary amp; region segmentation of objects in n-d images. In: Proceedings of the Eighth IEEE International Conference on Computer Vision, ICCV 2001, vol. 1, pp. 105–112 (2001)
8. Breve, F.: Active semi-supervised learning using particle competition and cooperation in networks. In: The 2013 International Joint Conference on Neural Networks (IJCNN), pp. 1–6 (August 2013)
9. Breve, F., Zhao, L.: Particle competition and cooperation in networks for semi-supervised learning with concept drift. In: The 2012 International Joint Conference on Neural Networks (IJCNN), pp. 1–6 (June 2012)
10. Breve, F., Zhao, L.: Particle competition and cooperation to prevent error propagation from mislabeled data in semi-supervised learning. In: 2012 Brazilian Symposium on Neural Networks (SBRN), pp. 79–84 (October 2012)
11. Breve, F., Zhao, L.: Fuzzy community structure detection by particle competition and cooperation. Soft Computing **17**(4), 659–673 (2013). http://dx.doi.org/10.1007/s00500-012-0924-3

12. Breve, F., Zhao, L., Quiles, M., Pedrycz, W., Liu, J.: Particle competition and cooperation for uncovering network overlap community structure. In: Liu, D., Zhang, H., Polycarpou, M., Alippi, C., He, H. (eds.) ISNN 2011, Part III. LNCS, vol. 6677, pp. 426–433. Springer, Heidelberg (2011). http://dx.doi.org/10.1007/978-3-642-21111-9_48

13. Breve, F., Zhao, L., Quiles, M., Pedrycz, W., Liu, J.: Particle competition and cooperation in networks for semi-supervised learning. IEEE Transactions on Knowledge and Data Engineering 24(9), 1686–1698 (2012)

14. Breve, F.A.: Combined active and semi-supervised learning using particle walking temporal dynamics. In: 2013 BRICS Congress on Computational Intelligence and 11th Brazilian Congress on Computational Intelligence (BRICS-CCI CBIC), pp. 15–20 (September 2013)

15. Breve, F.A.: Query rules study on active semi-supervised learning using particle competition and cooperation. In: Anais do Encontro Nacional de Inteligência Artificial e Computacional (ENIAC), pp. 134–140. São Carlos (2014)

16. Breve, F.A., Zhao, L.: Semi-supervised learning with concept drift using particle dynamics applied to network intrusion detection data. In: 2013 BRICS Congress on Computational Intelligence and 11th Brazilian Congress on Computational Intelligence (BRICS-CCI CBIC), pp. 335–340 (September 2013)

17. Breve, F.A., Zhao, L., Quiles, M.G.: Semi-supervised learning from imperfect data through particle cooperation and competition. In: The 2010 International Joint Conference on Neural Networks (IJCNN), pp. 1–8 (July 2010)

18. Breve, F.A., Zhao, L., Quiles, M.G.: Particle competition and cooperation for semi-supervised learning with label noise. Neurocomputing (2015) (article in Press)

19. Chapelle, O., Schölkopf, B., Zien, A. (eds.): Semi-Supervised Learning. Adaptive Computation and Machine Learning. The MIT Press, Cambridge (2006)

20. Ding, L., Yilmaz, A.: Interactive image segmentation using probabilistic hypergraphs. Pattern Recognition 43(5), 1863–1873 (2010). http://www.sciencedirect.com/science/article/pii/S0031320309004440

21. Ducournau, A., Bretto, A.: Random walks in directed hypergraphs and application to semi-supervised image segmentation. Computer Vision and Image Understanding 120, 91–102 (2014). http://www.sciencedirect.com/science/article/pii/S1077314213002038

22. Friedman, J.H., Bentley, J.L., Finkel, R.A.: An algorithm for finding best matches in logarithmic expected time. ACM Trans. Math. Softw. 3(3), 209–226 (1977). http://doi.acm.org/10.1145/355744.355745

23. Gonzalez, R.C., Woods, R.E.: Digital Image Processing, 3rd edn. Prentice-Hall Inc., Upper Saddle River (2008)

24. Grady, L.: Random walks for image segmentation. IEEE Transactions on Pattern Analysis and Machine Intelligence 28(11), 1768–1783 (2006)

25. Joachims, T.: Transductive learning via spectral graph partitioning. In: Proceedings of International Conference on Machine Learning, pp. 290–297. AAAI Press (2003)

26. Li, J., Bioucas-Dias, J., Plaza, A.: Semisupervised hyperspectral image segmentation using multinomial logistic regression with active learning. IEEE Transactions on Geoscience and Remote Sensing 48(11), 4085–4098 (2010)

27. Paiva, A., Tasdizen, T.: Fast semi-supervised image segmentation by novelty selection. In: 2010 IEEE International Conference on Acoustics Speech and Signal Processing (ICASSP), pp. 1054–1057 (March 2010)

28. Protiere, A., Sapiro, G.: Interactive image segmentation via adaptive weighted distances. IEEE Transactions on Image Processing 16(4), 1046–1057 (2007)

29. Rother, C., Kolmogorov, V., Blake, A.: grabcut: Interactive foreground extraction using iterated graph cuts. ACM Trans. Graph. **23**(3), 309–314 (2004). http://doi. acm.org/10.1145/1015706.1015720
30. Shapiro, L., Stockman, G.: Computer Vision. Prentice Hall (2001)
31. Smith, A.R.: Color gamut transform pairs. In: ACM Siggraph Computer Graphics, vol. 12, pp. 12–19. ACM (1978)
32. Xu, J., Chen, X., Huang, X.: Interactive image segmentation by semi-supervised learning ensemble. In: International Symposium on Knowledge Acquisition and Modeling, KAM 2008, pp. 645–648 (December 2008)
33. Zhou, D., Bousquet, O., Lal, T.N., Weston, J., Schölkopf, B.: Learning with local and global consistency. In: Advances in Neural Information Processing Systems, vol. 16, pp. 321–328. MIT Press (2004). http://www.kyb.tuebingen.mpg.de/bs/ people/weston/localglobal.pdf
34. Zhu, X.: Semi-supervised learning literature survey. Tech. Rep. 1530, Computer Sciences, University of Wisconsin-Madison (2005)
35. Zhu, X., Ghahramani, Z., Lafferty, J.: Semi-supervised learning using gaussian fields and harmonic functions. In: Proceedings of the Twentieth International Conference on Machine Learning, pp. 912–919 (2003)

The Evolution from a Web SPL of the e-Gov Domain to the Mobile Paradigm

Camilo Carromeu[1], Débora Maria Barroso Paiva[2], and Maria Istela Cagnin[2(✉)]

[1] Brazilian Agricultural Research Corporation - Embrapa Beef Cattle, Campo Grande, MS 79106-550, Brazil
[2] College of Computing, Federal University of Mato Grosso do Sul (UFMS), PO Box 549, Campo Grande, MS 79070-900, Brazil
istela@facom.ufms.br

Abstract. Lately, the demand for mobile applications development has increased significantly mainly due to growth of use of mobile devices and the need to port existing web applications. To reduce development's time and cost, Software Product Lines (SPLs) have also been used in the context of mobile applications. However, the existing SPLs do not worry about supporting the development of mobile applications corresponding to the existing Web applications, as it is desirable to have access to the information and main features of these applications in mobile devices. In face of this problem, this paper discusses the motivation and presents the evolution from a SPL in the e-Gov Web (e-Gov Web SPL) domain to a SPL in the mobile domain (e-Gov Mobile SPL) having in mind the need to supply market demand. The conducted evolution was supported by the PLUS approach (Product Line UML-Based Software Engineering) and by the features model. Furthermore, this work debates the main results obtained through some e-Gov Mobile SPL instantiations in the precision livestock domain.

Keywords: Software Product Line · Evolution · e-Gov · Web · Mobile

1 Introduction

The Portability non-functional requirement was defined in international rules [1, 2] published in the last years and refers to a set of sub-characteristics such as adaptability and replaceability. In practice, this requirement has followed the constant evolution of hardware devices, operational systems and infrastructure of computer networks and providing, among other things, the creation of applications that are executed in different environments and platforms. For example, nowadays, an application can easily run on desktop, web, tablets, smartphones and is presented according to the available resources, such as bandwidth and screen size.

In 2010, an SPL was developed for the e-Gov Web (e-Gov Web SPL) domain [3] aiming to obtain a flexible architecture that could reuse components developed from several previous experiences from the research group LEDES (Laboratory of Software Development from UFMS) and the PLEASE Lab (Laboratory for Precision

© Springer International Publishing Switzerland 2015
O. Gervasi et al. (Eds.): ICCSA 2015, Part I, LNCS 9155, pp. 217–231, 2015.
DOI: 10.1007/978-3-319-21404-7_16

Livestock, Environment and Software Engineering from Embrapa). This architecture made possible the development of several e-Gov Web applications which were used in different real life situations such as the Pandora[1] (a Web application which makes available to the user information and services from the information and management systems from the Embrapa Beef Cattle[2]).

With the availability of mobile devices, such as cellular devices and tablets, we observed that the applications developed based on the mentioned architecture needed to progress and adapt themselves to the new platforms, operational systems and mobile technologies. Thereby, mobile applications were developed corresponding to the Web applications, but without the e-Gov Web SPL support. From these mobile applications, we observed that a common set of services was implemented in their majority. Then, it was observed that the e-Gov Web SPL could evolve to a SPL for the mobile platform. To accomplish that, the common services in the developed mobile applications were abstracted as features[3] [4] and originated the e-Gov Mobile SPL, which is an evolution of the e-Gov Web SPL. The features are represented in a features model [4], which according to Sayyad et al. [5] allows visualization, reasoning, and configuration of SPLs, that can consist of hundreds (even thousands) of features, with complex dependencies and constraints that govern which features can or cannot live and interact with other features.

This paper aims to describe the evolutions which were made in the e-Gov Web SPL to obtain the e-Gov Mobile SPL having in mind the need to attend a new market demand which is to make available Web applications in mobile environments, since there is a lack of works that contemplate this demand [6,7,8]. The results of an instantiation of the e-Gov Mobile SPL are also presented aiming to show the application generation capacity for both Web platform and mobile platform.

This paper is organized as follows. Section 2 presents the related works. Section 3 describes the background to ease the understanding of concepts treated in the paper. Section 4 presents the evolution of the e-Gov Web SPL for the conception of the e-Gov Mobile SPL, as well as the resulting instance. Section 5 debates contributions, limitations and suggestions for future works.

2   Related Work

There are some works about mobiles SPLs in the literature and all of them use the features model for the variability modeling of the built SPLs. As the goal of the studied works is to develop an application in the mobile platform with the help of an SPL, all of them treated the variability of the business domain (also called vertical domain) and the variability of mobile domain (which can be considered as horizontal domain for it can be present in different vertical domains as in the case of context-awareness, location, battery level, security, communication, etc.).

---

[1] Brazilian National Institute of Industrial Property number: 14226-3.

[2] Institutional Site: https://www.embrapa.br/en/gado-de-corte

[3] Features are end-user visible characteristics of a system (Kang et al., 1990).

Quinton et al. [6] propose a SPL approach based in Model Driven Engineering, whose objective is to elaborate a variability modeling with the support of the features models of both system's business domain and the mobile devices domain to derive a software product taking into consideration both models. For this, the authors present a derivation process supported by a framework that performs the combination of both models (model merging) and the code generation. Additionally, the derivation process automatically analyzes restrictions for each derived product in relation to the available mobile devices.

Marinho et al. [7] present the SPL MobiLine, which supports the development of tour guide context aware mobile applications, that is, the software must be capable of adapting itself, during runtime, according to changes in its context (for example, user location, network conditions, etc.). The authors used the features model to document both tour guide domain related features and context aware mobility related features.

Mizouni et al. [8] propose a framework to build adaptive context aware and mobile applications, considering that context information can be related to the environment, users or device status. This framework is based on both variability modeling, with the support of the features model, and SPL. The adaptability modeling is done during the design phase, and the adaptability and context awareness are supported in execution time.

However, none of the studied SPLs considers the portability of the developed software system, as is the goal of the SPL approached in this work, that concerns about the development of a mobile application corresponding to a Web application and with the communication between them as well. That makes possible for the user to have access to the main information and to interact with the most important functionalities of the Web application in a mobile device. Next, we present some works about SPLs.

3   Background

Software Product Line (SPL) is one of the existing software reuse techniques and is of interest for this work because it provides the reuse of software products in the domain of e-Gov mobile applications corresponding to e-gov Web applications. The SPL was conceived from the adaptation of concepts of the Product Line Engineering (PLE), which is commonly applied in the engineering to reach mass customization with economy of efforts through the collective production (instead of individual) of multiple similar instances, but from distinct design and prototypes [9].

According to Clements and Northrop [10], an SPL is an set of intensive software systems that shares a set of managed and common features that meets the specific needs of a market segment in particular (domain) or company mission and is developed from a core set of assets in a pre-established way.

In a specific business domain, there are the mandatory characteristics (commonalities) and those that can be customized (variabilities). Both can be represented in variability models such as the features model. According to Pohl et al. [11], a variability is composed by a variation point and its variants. A variation point is the place where the variation may occur (payment methods, for example) and the variants are the existing possible solutions for a variation point (for example, money, credit card or banking billet). Thus, the management of variabilities aims to organize communalities and variabilities in a way to generate products with higher quality, reducing the use of organizational resources.

Still according to Pohl et al. [11], a SPL is built and instantiated from two processes named, respectively, Domain Engineering and Application Engineering. Basically, the Domain Engineering focuses on the variability management of all software artifacts from the SPL and Application Engineering is responsible for creating the final product from the SPL. In this work, the software artifacts from the presented e-Gov Mobile SPL are implemented with the support of application frameworks [12, 13] and application generators [14], whose definitions are presented as follows. These mechanisms together provide the instantiation of the e-Gov Mobile SPL without spending much effort by the application engineer.

The application frameworks are considered as semi-complete and reusable applications that, when specialized, produce personalized applications within a specific domain [12]. They are composed by a collection of abstract and concrete classes, and interfaces between them, representing the frozen spots and hot spots of the project of a subsystem [13].

Application generators are software systems that transform the specifications into an application. The specifications describe the problem or the task that the application has to perform, such that the generator can generate the source code. These specifications can be modelled in a graphic way, written in some intermediate language or even be created interactively, where the user selects desired features through the choices in a sequence of forms or menus [14].

4   SPL for e-Gov System Development

With the objective of reuse in different levels of abstraction, software patterns appeared in the 90´s [15, 16, 17] to try to capture acquired experience in the software development and synthesize it in the form of problem and solution. Besides providing the reuse of the solutions, patterns help to improve the communication among developers, which can conduct the discussions based on the usage of these patterns [17].

In this context, several experiences in the LEDES and in the PLEASE Lab in development of Web applications in the e-Gov domain resulted in the abstraction of analysis, architecture and codification patterns. Aiming to consolidate an agile development process for this domain, we decided to create a SPL, named e-Gov Web SPL.

Considering the abstracted patterns from different members of the SPL from e-Gov domain, the features model was created and the initial architecture of the e-Gov Web SPL was established, whose development was based in the PLUS (Product Line UML-Based Software Engineering) approach [18]. This approach is composed by two processes: Software Product Line Engineering and Application Engineering.

In the Software Product Line Engineering, the communalities and variabilities shown in the SPL features model, were mapped for the Titan Framework through classes, specifying the frozen-spots and the hot-spots, respectively, aiming to ease the instatiation of the line during the Application Engineering Process.

The e-Gov Web SPL is automatized by the Titan Framework and the application generator Titan Architect, as detailed by Carromeu et al. [3].

The use of the Titan Architect eliminates the need for programming, although the parametrization of variabilities is limited. The Titan Framework allows the configuration of business rules and parameters of the new application by modifying the markup language input and, in case it is not enough, it allows the programming of new software artifacts.

The Titan Framework architecture, illustrated in Fig. 1, is formed by a core and by the repository of components. The core is the implementation of all SPL similarities and, hence, it is unchangeable in the line instances. It is responsible for receiving as input the instance configuration files (XML and SQL) and generating an application in runtime. This framework is a gray-box, therefore, flexible and extensible.

The main features of the e-Gov Web SPL are the navigation sections and navigation actions which are related, respectively to the components and engines. The role-based access control, the audit log record, the search engine, the file system, the notification system, the manual generator, the backup system on demand, are other features inherent to its architecture.

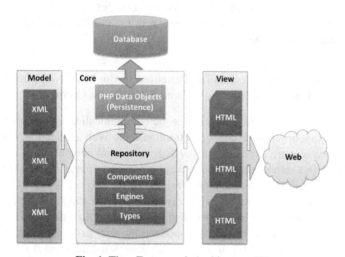

Fig. 1. Titan Framework Architecture [3]

5   Evolution of SPL for the Mobile Computing Paradigm

Since the creation of the e-Gov Web SPL, dozens of new Web applications were developed and others are in development at the moment. At the same time, the increasing use of mobile devices and the wide dissemination of wireless networks are increasingly stimulating the development of softwares in the mobile and ubiquitous computing paradigms. In this context, the e-Gov domain is one of the main application areas, once governments have been pushed to present higher efficiency when providing information and services to the society in a transparent and democratic way.

This way, it was identified a strong demand for mobile applications integrated with the developed Web systems, enabling the user access to the main information and functionalities of the system. For example, the SAC Gado de Corte [20] (Customer Services Center for Beef Cattle, in English), an mobile application with 1,700 questions and answers about the Brazilian chain of beef production; Suplementa Certo[4], a calculator that helps with decision making about feeding supplementation for beef cattle; and the Pandora Phone, that enables the access, by mobile devices, to some of the functionalities of the Pandora Web application, such as notifications, technology and knowledge bases, information about projects budgets (balance and statements) and the staff (employees and collaborators).

The experience with the development of several applications in the mobile platform, such as the mentioned above, enabled to initially identify codification patterns. For example, the code structure, in native Java, obeyed a recurring logic organization and demanded the same set of auxiliary classes (helpers).

Similarly, the implementation of the developed mobile applications required alterations in the Web applications (e-Gov Web SPL instances) with which they would have to synchronize data. In order to do that, it was necessary to implement exclusive service buses, related to communication and security, and decoupled from the repository components from the SPL as discussed below.

5.1   Discussion of Bus Services Implemented in the Mobile Applications: SAC Gado de Corte, Pandora Mobile and Suplementa Certo

One of the main services implemented was the communication, which is fundamental to keep the mobile application data consistent with the Web application. Furthermore, a mobile device is more exposed to risks than a personal computer and the data synchronization with a remote server is one way to prevent eventual losses. This communication has to be made through Web services, being also necessary to worry about security, performance and data integrity between the server and several clients.

Aiming to mitigate risks associated with **security** through **authentication** in the communication between client (mobile application) and server (Web application), it was specified a global scheme of authentication which can be used on any service available under the HTTP protocol, named Embrapa-Auth. It is a protocol that defines a scheme sufficiently secure, flexible, standardized and stateless to authenticate HTTP requests. The protocol uses headers and status codes from the HTTP protocol, favoring its usage in REST (Representational State Transfer) approaches. The protocol was developed to support an architecture where there are several users making HTTP requests to a Web service through applications running in several devices and platforms. The protocol is adaptable to the context, allowing up to three levels of authentication:

---

[4] Brazilian National Institute of Industrial Property number: 5120130013755.

- **Application authentication:** An authorized application is any program that possesses an identifier/token pair that authorizes it to make requests to the service. The credentials of the application have to, preferably, be embedded to the source code of the program and should not be editable by an ordinary user. This level of security allows to restrict that only some specific programs are able to make requests to the service.
- **Client authentication:** A client is a device that has a private identifier / key pair which authorizes it to make requests to a service. The authentication of a client allows to restrict the use of Web services by determined devices or specific systems. This security level allows implementations of quota policies; restriction of functionalities; sharing of keys by company, unity, department, team, etc.; authorization of third–party softwares; auditing; change of private key in case equipment loss or theft; cancellation of credentials at the end of contracts; user identification by a personal device; etc.
- **User authentication:** A user is somebody who owns an access credential (login/password pair).

Another implemented service refers to the **user registration** and **authentication of user by social network**. In the case of mobile devices with the Android operational system, it is possible to state that the application user has, necessarily, a Google account. This fact enables the user to use the account that is registered in the mobile device to make his/her registration in the mobile application. The biggest advantage of this approach is to offer the user a simplified and quick way to register without the requirement of filling out forms.

Another implemented service was to guarantee the local **data integrity** of mobile application in relation to the Web application.

It is easy to guarantee the data integrity when there is no creation of tuples in the mobile application database, as in the case of metadata entities (federative unit, country, marital status, etc.). In those cases, the mobile application assumes an architecture "**provider-consumer**", where the Web application will be the data provider and the mobile application will just "consume them". This way, the data synchronization controller shall be responsible for consulting the Web application and obtain new and updated tuples, so the data is synchronized from the Web application and cannot be locally modified by the mobile application.

However, often the mobile application must be responsible for the creation and modification of data, which will, at some point, be synchronized with the Web application data. In this case, it is necessary to worry about the fact that the same data can be modified in different devices and, when synchronizing with the Web application, it is necessary to **handle conflicts**. The adopted policy in this case is to always keep the most recent data.

The **disambiguation** service was implemented to solve the problem when tuples of entities created in mobile applications are sent to the Web application and synchronized in the mobile applications installed in other devices. In this case, there is no way to guarantee that it will be possible to consult the Web application to obtain a unique key, since, in many cases, the mobile application will be able to work without internet access. It is difficult, in this case, to guarantee that the sequential key be correctly interpreted by the mobile application, since there will be other mobile devices creating tuples in the same entity.

Through the disambiguation service, the Web application provides a unique key to the mobile application once it is installed, where it is stored locally on the device. This key is always concatenated with the entity key when the latter is a sequential key.

The **notification** service was also implemented to allow notifications launched by the Web application to be reflected in the mobile application. For this, services were created in the service bus that interact with the notification API from the Titan Framework and allow: i) to access the message list and related data (date, category and link), ii) to mark a notification as "read" and iii) to delete a notification. Moreover, the notifications API from Titan Framework was integrated with the notifications API from Android, named Google Cloud Messaging. This way, alerts issued on the Web application are received by mobile application users in a passive way, that is, without the need of the mobile application to be running in that moment.

Finally, the **image-processing** service was also considered because mobile applications that try to make the processing of big images locally are very susceptible to errors related to the lack of memory. Through this service, images loaded from the Web application are remotely processed and simply put in cache in the mobile application.

5.2   e-Gov Mobile SPL Conception

The implementation of the services presented in the previous section motivated the e-Gov Web SPL evolution to become an e-Gov Mobile SPL, being this evolution the focus of this work.

The e-Gov SPL Mobile features were identified from the services presented in Section 5.1, and are represented in the features model illustrated in Fig. 2. Next, for the incorporation of the raised features in the e-Gov Mobile SPL, there were used four evolution cycles of the Software Products Line Engineering process from the PLUS approach. This approach was used in the evolution of the e-Gov Web SPL to an e-Gov Mobile SPL since, beyond having being used in the creation of the e-Gov Web SPL, demonstrated to be effective during the held evolution.

Consequently, the execution of the evolution cycles are in compliance with the execution description of the PLUS approach commented in Section 4, that is, the features presented in Fig. 2 were mapped to the Titan Framework as frozen-spots or hotspots; and existing components in its repository were refactored to be able to interact with the incorporated services, as illustrated in Fig. 3.

The first evolution cycle of the e-Gov Web SPL started with the modification of its fixed core, that is, the artifacts which materialize communalities from the members of the family of products. In this core, there was added the service bus with the basic services represented by the feature **AuthenticationLayer**. This parameterizable implementation of the Embrapa-Auth protocol guarantees that any SPL instance can count on any combination of the authentication levels of this protocol represented in the features model from Fig. 2.

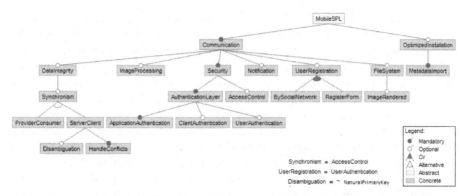

Fig. 2. Features model of the evolution from an e-Gov Web SPL to an e-Gov Mobile SPL

The feature **ProviderConsumer** was implemented to guarantee that all instances from the e-Gov SPL Mobile will have entities that will be treated in a provider-consumer architecture. A problem that was observed during the implementation of this feature was the overhead of the initial synchronization, that is, when the mobile application is installed. This can become a problem when there are millions of tuples that must be synchronized. In order to optimize the initial synchronization, this feature was implemented taking into consideration the sending of a binary database with all the metadata stored, that is, when the mobile application is installed in the mobile device, the installer already has a binary database with metadata tables.

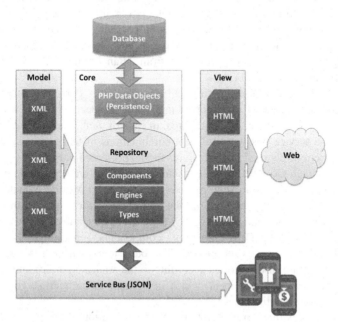

Fig. 3. Evolved Titan Framework Architecture for the development of mobile applications

Additionally, in this first evolution cycle, the Titan Framework has acquired its service bus, but lacking the corresponding maturity. The repository components were adapted to interact with the new layer of communication (**Communication** feature). For example, the classes that implement data types from the Titan Framework acquired the capacity to represent and understand values in the JSON (JavaScript Object Notation) format, which is a format more compact than XML and thus, chosen for the transmission of data between client and server.

It was also implemented, in this first cycle, the authentication layer (represented by the feature **AuthenticationLayer**) and its levels, represented by the features **ApplicationAuthentication**, **ClientAuthentication** and **UserAuthentication**. In the end, the service bus from Titan Framework was also capable to synchronize data in architectures provider-consumer and client-server, and to treat inherent conflicts.

After the above mentioned adaptations, it became evident the need for new improvements in the synchronization process, due to the already discussed problem, about the creation of tuples in entities without natural primary keys. Then, in the second evolution cycle, it was implemented the **Disambiguation** feature. This evolution required refactoring of classes from the Titan Framework that implement the model and control layers of the Web application, aiming to add to the artifacts of these layers, the capacity to deal with the disambiguation code. Another implemented feature in this cycle was the **ImageProcessing**, that is, the services were made available in the service bus to allow the mobile application to request the server for already processed images.

In the third evolution cycle, it was made the integration of the Titan Framework with the Google's APIs for Android. This integration allowed the user registration to be made in a simple and transparent way, using personal information already available in the device (**BySocialNetwork** feature). It was also added the previously discussed integration of the notification API from Titan Framework with the Google Cloud Messaging (Section 5.1).

Together with the implementation of all raised features corresponding to the service bus (Section 5.1), with the support of the PLUS approach, all the new e-Gov SPL instances are apt to use them, in other words, in each of them, the application engineer can activate the service bus with different combinations of features, according to the rules and restrictions detailed in the features model presented in Fig. 2.

To facilitate the e-Gov SPL Mobile instantiation, it was created a code generator in the fourth and last evolution cycle, named Titan Architect Mobile, which allows the application engineer to generate the mobile application code. For this, this generator scans the files in markup language (XML) from the Web application, responsible for the parametrization of the Titan Framework components, generating a Java code that composes the mobile application project. The basis of this project is also fixed and consists of a structure of directories according to the Android pattern and a collection of auxiliary classes, named helpers, which implement functionalities common to all members of the product line.

Based on the files that parameterize the Web application, the Titan Architect Mobile generates the following classes for each mobile application entity, corresponding to the Web application entities explicitly mapped (in a proper XML file) for the mobile platform:

- Contracts: classes responsible for the definition of the entity tables in the relational database of the mobile application. These classes also "know" how to create these tables charts when the application is installed.
- Models: classes that define the data model of the entity. When the code is generated, a transcription of the instance's XMLs is done (these XML files containing each entity attribute and its respective data type) for the creation of these classes in the mobile application.
- DAOs (Data Access Objects): classes that implement the persistence layer, making the interaction with the database.
- Adapters: classes with methods to convert values from the entities attributes. They are essential in the representation of classes objects as JSON (JavaScript Object Notation) objects, and for the conversion of values for the data record and recovery in the mobile application database.
- Tasks: classes, whose instances are called asynchronously by the mobile application and are responsible by the synchronization of the entity data with the Web application.
- Web services: they are classes that convert and send entities to the Web application and recover their information. They know the REST communication protocol and know how to treat errors when needed.

Moreover, the Titan Architect Mobile generates artifacts for the visualization layer (which includes classes called Views in the Android, and XML files) in a visual pattern that can be easily altered by the application engineer. Optionally, a binary database is created (corresponding to the **MetadataImport** feature) and it's loaded when the mobile application is installed, as previously mentioned.

Interventions in the source code obtained can be made by the developer to finish it and customize it, but not to optimize its performance. We highlight that the source code generated by Titan Architect Mobile will likely be more efficient than the source code manually implemented by a Java developer, who ignores or disregards the best practices of Android development [19]. We can say this because the implementation of the generator involved detailed study of these "best practices". Thus, since the source code generation of Titan Architect Mobile considers these assumptions, then we can infer that the generated source code will be efficient.

5.3   Instantiating the e-Gov Mobile SPL

In order to present the results of the conducted evolution, a Web application[5] (shown in Fig. 4) was developed with the support of the e-Gov SPL and then, a corresponding mobile application[6] (illustrated in Fig. 5) with the support of the e-Gov Mobile SPL, whose purpose is to provide a list of Brazilian embassies for overseas travelers. The development of both applications was carried out following the Application Engineering Process of the PLUS approach.

---

[5] Available at: http://titan.cnpgc.embrapa.br/sample/travel/

[6] Available at: https://play.google.com/store/apps/details?id=com.carromeu.titan.sample.travel

Fig. 4. Web application of Brazilian embassies for overseas travelers

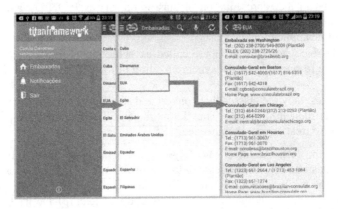

Fig. 5. Mobile application of Brazilian embassies for overseas travelers

Thereby, firstly, an e-Gov Web SPL was instantiated. For this, it was followed a Titan Framework step-by-step creation guide for Web applications [21], which involves the instantiation of the base code (the default code that all applications have, but can be modified by the developer) present in the framework's public repository and its configuration. Next, the list of Brazilian embassies was obtained in the Brazilian's Ministry of External Relations (commonly called Itamaraty) public website[7], and imported into the Web application's database, in a table with columns corresponding to the quantity and types of the extracted data. It was created a new CRUD (create, read, update and delete) section in the Web application to enable the representation and manipulation of data from the table. Once the section was created, it was just necessary to enable its integration to the service bus through the parametrization of the corresponding component and available in the Titan Framework repository. Lastly, it was activated in the Web instance the integration with the Google Plus API, allowing users to register and have access to the system using their Google accounts.

---

[7] Available at: http://www.itamaraty.gov.br/o-ministerio/o-brasil-no-exterior. Accessed: June, 2014.

The second step was the creation of the mobile application. For this, the Titan Architect Mobile was executed, generating the Contract, Model, DAO, Adapter, Task, Web services classes, XMLs and classes for the visualization layer as well, corresponding to the CRUD section of representation and manipulation for the list of embassies. The classes were generated in the packages and folders of the base code, including in it the functionality of synchronizing the embassies list in the mobile device. Finally, graphic alterations were made in the already functional mobile application, so that it became correspondent to the web application visual identity.

It is noted that a single application engineer developed both applications (Web and Mobile) and spent about 16 hours, being half of that time used in the creation of graphic elements. The optimization of development time was possible due to automatic code generation, which also helped to guarantee the final quality of the developed product.

Other real world solutions, cited below, were obtained based on the same instantiation process aforementioned and are therefore, members of the e-Gov Mobile SPL products line:

- Livestock Management Platform: this software platform is composed by a Web application (Titan instance) and a mobile application, whose code was generated from the instantiation of the e-gov Mobile SPL. This solution allows beef cattle producers to manage the herd on their rural properties. The mobile application makes integral use of the services mentioned in Section 5.1, once it adopts a policy of asynchronous synchronization, where the user can interact with the mobile application while offline, creating tuples, identified making use of the disambiguation technique, that will be synced afterwards with the Web application using the service bus authenticated in application and client levels, provided in the Embrapa-Auth protocol.

- Sanitary Calendar: the objective of this application is to notify the beef cattle producer about sanitary managements and vaccinations that the producer will have to apply in the herd. For this, the mobile application makes use of the Android notification system, named Google Cloud Messaging, which allows alerts to be launched to the mobile device. With that, from the record of business rules in the Web application, such as the correct age to apply specific vaccines in the animals, and the registration of animal lots by the producer, the mobile application emits notices that allow the user to schedule coming managements and vaccinations that should be done.

- Summary of Bulls from the Geneplus Program: the Geneplus program (http://geneplus.cnpgc.embrapa.br/) is specialized service for animal genetic improvement available for the cattle breeder. One of the results of this program is a catalog that is annually updated with certified bulls (reproducers). This way, the developed mobile application aims to replace the catalog's distribution format that, until then, was printed. Then, this application's database was populated with about 70,000 animals, which compose the catalog. This mobile application remains synchronized with the corresponding Web application, providing the user with up to date information from the catalog, as well as other information like a search engine for bulls with specific characteristics.

6   Conclusion

This paper presented the changes made in the e-Gov SPL to obtain an e-Gov Mobile SPL. In particular, a service bus that contemplates the communication between Web applications and their corresponding mobile applications was implemented in the Titan Framework (some of the existing components in its repository were refactored in order to be able to interact with the incorporated services); and, a code generator to create mobile applications (Titan Architect Mobile) was implemented. The main motivation on conducting this work was the need to migrate several web systems, previously developed with the support of the e-Gov Web SPL, to a mobile platform, besides allowing the communication between the applications in both platforms (that is, Web and mobile).

From this work, it is possible to reduce time and effort during development of Android applications and their integration with a web application that acts as a centralized database; and to ensure the adoption of best practices following Google's performance guidelines by mobile applications developed with Titan Architect Mobile.

As the main restrictions for this work, it can be pointed out the small number of instantiations performed until the moment of the writing of this paper with the support of the e-Gov Mobile SPL, what prevented the conduction of an empirical study. However, the results obtained so far indicate considerable reduction in development time of mobile applications corresponding to Web applications obtained through the e-Gov Web SPL and, consequently, cost reductions and increased productivity for application engineers. Moreover, the generated source code specifically works with the patterns adopted by Titan, that is, it must necessarily use the Embrapa-Auth, for example.

Main future works arising from this paper may include: release of other services in the e-Gov SPL Mobile which will be identified with the usage of product line; addition of support to the code generation for iOS and Windows Phone in the Titan Architect Mobile; the development of an expressive number of mobile applications through the instantiation of the SPL treated in this paper, aiming to obtain statistically significant results and the conduction of controlled experiments that would allow to verify the effectiveness of the e-Gov SPL Mobile.

References

1. ISO. ISO/IEC 25010:2011 Systems and software engineering - Systems and software Quality Requirements and Evaluation (SQuaRE) - System and software quality models (2011)
2. ISO. ISO/IEC 9126:2001 Software engineering - Product quality - Part 1: Quality model (2001)
3. Carromeu, C., Paiva, D.M.B., Machado, M.I.C., Rubinsztjn, H.K.S., Breitman, K., Turine, M.A.S.: Component-based architecture for e-Gov web systems development. In: 17th IEEE International Conference and Workshops on Engineering of Computer-Based Systems, Oxford (2010)

4. Kang, K.C., Cohen, S.G., Hess, J.A., Novak, W.E., Peterson, A.S.: Feature-Oriented Domain Analysis (FODA): Feasibility Study, Technical Report CMU/SEI-90-TR-21, November 1990
5. Sayyad, A.S., Menzies, T., Ammar, H.: On the value of user preferences in search-based software engineering: a case study in software product lines. In: 35th International Conference on Software Engineering (ICSE 2013), San Francisco, CA, USA, pp. 492–501 (2013)
6. Quinton, C., Mosser, S., Parra, C., Duchien, L.: Using multiple feature models to design applications for mobile phones. In: 15th International Software Product Line Conference (SPLC 2011), Munich, Germany, vol. 2, pp. 1–8 (2011)
7. Marinho, F.G., Andrade, R.M.C., Werner, C., Viana, W., Maia, M.E.F., Rocha, L.S., Teixeira, E., Filho, J.B.F., Dantas, V.L.L., Lima, F., Aguiar, S.: MobiLine: A Nested Software Product Line for the domain of mobile and context-aware applications. Science of Computer Programming 78(12), 2381–2398 (2013)
8. Mizouni, R., Matarb, M.A., Mahmoudb, Z.A., Alzahmib, S., Salahc, A.: A framework for context-aware self-adaptive mobile applications SPL. Expert Systems with Applications 41(16), 7549–7564 (2014)
9. Weiss, D.M., Lai, C.T.R.: Software Product-Line Engineering: A Family-based Software Development Process. Addison-Wesley Professional, Boston (1999)
10. Clements, P., Northrop, L.: Software Product Lines: Practices and Patterns. Addison-Wesley, Boston (2002)
11. Pohl, K., Bockle, G., van der Linden, F.J.: Software Product Line Engineering: Foundations, Principles and Techniques, 1st edn. Springer-Verlag, Secaucus (2005)
12. Foote, B., Johnson, R.E.: Designing reusable classes. Journal of Object Oriented Programming 1(2), 22–35 (1988)
13. Sommerville, I.: Software Engineering, 9th edn. Addison-Wesley (2010)
14. Cleaveland, J.C.: Building application generators. IEEE Software 5(4), 25–33 (1988)
15. Buschmann, F., Meunier, R., Rohnert, H., Sommerland, P., Stal, M.: Pattern-oriented software architecture - a system of patterns. Wiley & Sons (1996)
16. Coplien, J.O.: Software Design Patterns: Common Questions And Answers. The patterns handbook: techniques, strategies, and applications. Cambridge University Press, New York (1998)
17. Gamma, E., Helm, R., Johnson, R., Vlissides, J.: Design Patterns - Elements Of Reusable Object-Oriented Software. Addison-Wesley (1995)
18. Gomma, H.: Designing Software Product Lines with UML. Addison-Wesley (2005)
19. Android. Best Practices. http://developer.android.com/guide/prac-tices/index.html
20. Souza, D.C.G., Righes, B., Rodrigues Filho, J.R., Queiroz, H.P., Carromeu, C.: Mobile service for citizen: SAC mobile. In: VIII Scientific Meeting of the Embrapa Beef Cattle, Campo Grande, MS, Brazil, pp. 120–121 (2012) (in portuguese)
21. Carromeu, C.: Titan Framework Cookbook, September (2014) (in portuguese) http://cloud.cnpgc.embrapa.br/titan/documentacao/

Advanced Induction Variable Elimination
for the Matrix Multiplication Task

Jerzy Respondek[✉]

Faculty of Automatic Control, Electronics and Computer Science,
Institute of Computer Science, Silesian University of Technology, Gliwice, Poland
jerzy.respondek@polsl.pl

Abstract. The main objective of this article is to make use of the induction variable elimination in the matrix multiplication task. The main obstacle to this aim is iterating through a matrix column, because it requires jumping over tables. As a solution to this trouble we propose a shifting window in a form of a table of auxiliary double pointers. The ready-to-use C++ source code is presented. Finally, we performed thorough time execution tests of the new C++ matrix multiplication algorithm. Those tests proved the high efficiency of the proposed optimization.

Keywords: C++ · Iterators · Linear algebra · Matrix multiplication · Pointers · Programming languages · Smart pointers

1 Introduction

The C++ programming language is based on the C programming language and enables to create the object-oriented software with the speed of the C programs. The C++ was designed by Bjarne Stroustrup [15,16]. The C programming language was designed by Kernighan, Ritchie [8] as a highly effective tool for operating systems designing. Apart from the operating systems, the majority of the severe contemporary software is still programmed in the C++ language. The high efficiency of the software coded in C++ follows, to a large extent, from C++ pointers. The pointers appear also in other general purpose languages but in the C++ programs they are usually the main part of the algorithms code. The pointer-oriented C++ code closes the programming style to the direct assembler programming causing its unmatched effectiveness.

The main objective of this article is to make use of the induction variable elimination together with the auxiliary pointer table in one of the fundamental numerical algorithms, i.e. in the matrix multiplication. The main obstacle to this aim is iterating through a matrix column, because it requires jumping over rows. Details can be found in Sect. 5.1. As a solution to this, we propose a shifting window in a form of a table of auxiliary double pointers.

O. Gervasi et al. (Eds.): ICCSA 2015, Part I, LNCS 9155, pp. 232–241, 2015.
DOI: 10.1007/978-3-319-21404-7_17

2   Theoretical Background

2.1   Matrices from the Mathematical Point of View

The sole matrix definition can be found e.g. in the monograph Bellman [5] p.37 and other works [1-4, 13-14]. The $m \times n$-dimensional matrix A we will denote as $A = \left[a_{ij} \right]_{\substack{i=1,..,m \\ j=1,..,n}}$. Thus let us first refer to the formal definition of the matrix multiplication.

Matrix Multiplication Definition (Bellman [5] pp.39)

Let *us be given two real matrices* $A = \left[a_{ij} \right]_{\substack{i=1,..,m \\ j=1,..,n}}$, $B = \left[b_{ij} \right]_{\substack{i=1,..,n \\ j=1,..,l}}$. *As a result of the multiplication of the* A *matrix by the* B *matrix we define the matrix* $C = \left[c_{ij} \right]_{\substack{i=1,..,m \\ j=1,..,l}}$ *with entries expressed by* (1)*:*

$$c_{ij} = \sum_{k=1}^{n} a_{ik} b_{kj}, \quad i = 1,...,m, \quad j = 1,...,l \tag{1}$$

2.2   Double Pointer vs. the Flat Representation of Matrices

The simplest matrix representation is the double pointer structure. It consists of a series of separate tables with an additional, auxiliary table of pointers. This topic is presented in the most general case, for an arbitrary matrix dimension, in the monograph [17]. This idea of storing the matrices is presented on Fig. 1.

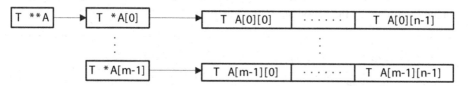

Fig. 1. Representation of the matrix as the series of separate tables

In this method the allocator lays out each row of the matrix separately. This representation is more flexible e.g. can represent diagonal matrices using less space, often it is used for sparse representations, etc. The primary deficiency of the double pointer representation, with separate rows, is the potential lack of locality. It causes weak cache utilization and gives great variation in execution times depending upon how the allocator lays out the rows of the matrix. Thus double pointer representation typically has performance problems when arrays are dense and compared to the standard flat representation.

In this article we use the flat matrix representation, which is common in contemporary (Java, ML) and earlier (e.g. Pascal) languages. In this structure all the matrix data is stored in a single one dimensional table, ensuring its locality in a coherent memory location. To give it double dimensional behavior, with convenient random access to any element of the matrix in the form $A[i][j]$, we use additional table of pointers. Such a representation is utilized by the reputable monograph [10, p. 21], devoted to the implementation of numerical methods in C. This idea is presented on Fig. 2.

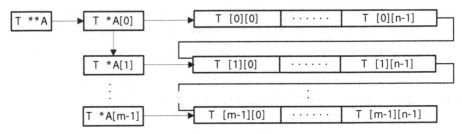

Fig. 2. Flat representation of the matrix

The proper C++ type definition code has the following form:

```
struct TMatrix
{
  T **aux;
  int m,n;
};
```

where *aux* is the double pointer indicating the auxiliary pointer table matrix data while *m* and *n* are the number of matrix rows and columns, respectively. The task of storing and deleting in a memory a $m \times n$-dimensional matrix A of the data type T can be performed by the following C++ code, respectively:

```
T *data=new T [m*n]; A.aux=new T* [m];
for(int i=0;i<m;i++)
  { A.aux[i]=data; data+=n; }

delete [] A.aux[0]; delete [] A.aux;
```

Such a way of a computer matrix representation has two other important advantages:

- The dimensions of the allocated matrix can be dynamically determined during the program execution. Thus there is no need to reserve an additional amount of memory during the code compilation; we use exactly as much memory as we need.
- We avoid the potential lack of locality preserving convenient access to a random element.

3   The C++ Pointer-Based Programming Paradigm Fundamentals

The pointers in C++ play a much greater role than in other programming languages. The fundamental C reference book [8] devotes a broad separate Chap. 5 to the pointer notion. The most important advantage of the pointers is the high efficiency of the sequential table operations coded with their use. Below we show, by the exemplary task of summing the table elements, its two opposite implementations: by the table indexing and by the pointer iteration.

A)
```
float tab[1000],s=0.0;

for(int i=0;i<100;i++)
   s += tab[i] ;
```
B)
```
float tab[1000],*p=tab,s=0.0;

for(int i=0;i<100;i++)
   s += *p++ ;
```

We can conclude the following reasons of the pointer-based programming execution time boost:

- In the table-based implementation (cell A) we have to determine the memory address for each summed table element in each loop iteration again and again in compliance with the expression: `&tab[i] = tab + i*`**`sizeof(float)`**. It is worth to notice that in the table implementation it is necessary to perform an additional integer multiplication in each loop iteration.
- In the pointer-based implementation (cell B), in order to sequentially sum up the table elements the only task we have to perform in each loop iteration (apart from the sole element addition) is to shift the pointer value to the next table element by a constant equal to **`sizeof(float)`**.

We illustrated the idea of pointer-based table access compared with the classic one in Fig. 3.

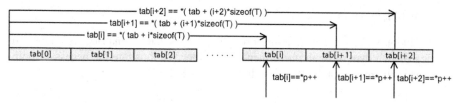

Fig. 3. The classic table access vs. pointer based table access

The main objective of this article is to apply the C++ pointer programming paradigm in order to obtain highly efficient matrix multiplication.

4   The Classic Matrix Multiplication Algorithm

The matrix multiplication definition (1) leads directly to the following, well known, function:

```
template <class T>
void Classic_Matrix_Multiply(Matrix<T> &A,
                                Matrix<T> &B,Matrix<T> &C)
{
  T temp;
  for(int i=0;i<A.m;i++)
    for(int j=0;j<B.n;j++)
    {
      temp=0.0;
      for(int k=0;k<A.n;k++)
        temp += A.aux[i][k] * B.aux[k][j];
      C.data[i][j]=temp;
    }
}
```

It can be noticed that at each iteration of the most nested loop, without the induction variable optimization, we have to calculate anew the memory address of the proper elements of the multiplied matrices, namely, the addresses of the elements *A.aux[i][k]* and *B.aux[k][j]*. It is highly ineffective. In the next item we explain how to achieve this optimization.

5   The Advanced Induction Variable Elimination Implementation

5.1   Problem Analysis

The problem of the pointer iteration through two dimensional matrices is a more sophisticated one than through one dimensional table. We illustrate the necessary pointer paths in the matrix A by B multiplication in Fig. 4.

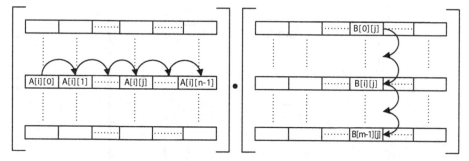

Fig. 4. The pointer iteration paths in the matrix multiplication process

Let us take into account the most nested loops of the classic algorithm:

```
temp += A.aux[i][k] * B.aux[k][j]
```

We can observe the following:

- The *A.aux[i][k]* expression at higher optimization levels will be automatically optimized by the compiler with the induction variable elimination technique. This follows from the fact, that its first part *A.aux[i]* refers to the most outside loop, and next in the most nested loop we pick its *k-th* element.
- The problem significantly complicates when it comes to the iteration along a matrix column in the expression *B.aux[k][j]* which needs to iterate over a series of rows. This part will not be automatically optimized by the variable elimination, because the first expression's part *B.aux[k]* refers to the most nested loop while its most nested part *B.aux[k][j]* pertains to the outside loop.

The effective implementation of the induction variable elimination is the crucial part of this problem and we present it in the next Chapter.

5.2   The Main Idea of the Proposed Induction Variable Elimination Implementation

To implement the induction variable elimination optimization for the iterating along the matrix columns we propose the following solutions:

- Introduction of an auxiliary pointer table *p1B*. In that table we store the addresses of the current column elements, initializing them by the addresses of the first column cells of the *B* matrix (code line 11 in item 5.3). Next, we sequentially increment each pointer value in the pointer table *p1B* (26th code line). In this way we get a shifting memory window, enveloping a single matrix column. This idea is clearly illustrated in Fig. 5.

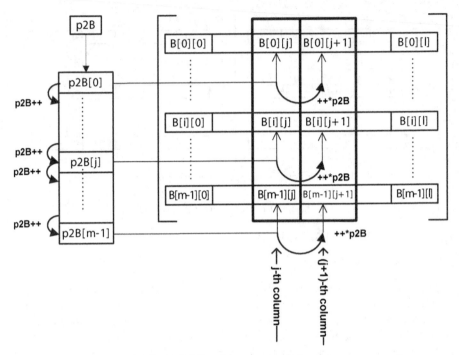

Fig. 5. Shifting the memory window

The most nested loop code (line 22) of the proposed algorithm deserves a closer look. It adds the subsequent scalar terms of the form $a_{ik}b_{kj}$, due to the definition (1). It has the following form:

```
temp += *p2A++ * **p2B++;
```
(2)

The sub-expression: < * **p2B++> performs 3 tasks at once:

— gets the value of the current B matrix cell (sub-expression **p2B),
— multiplies it by the value of the proper cell of the A matrix,
— finally post-increments the value of the *p2B* pointer, shifting it to the next matrix cell.

• Changing the order of calculations of the matrix product. The classic matrix multiplication algorithm (Chap. 4) calculates the elements of the $C = AB$ matrix in the natural row-by-row order. In the advanced multiplication algorithm we changed the calculation order of the $C = AB$ matrix to the column-by-column order. It enabled to minimize the number of necessary memory window shifting moves (26th code line) by moving this operation to the least nested loop.

It is worth to notice that the code (2) requires no integer multiplication to determine the addresses of the a_{ik}, b_{kj} matrix cells. Instead we apply the induction variable optimization by the use of the C++-oriented code.

5.3   The Final C++ Matrix Multiplication Algorithm Code

The innovations proposed in the previous item lead to the following, final C++ code for the fast matrix multiplication:

```
1 : T **p1B,**p1C;
2 :
3 : template <class T> // C=AB
4 : void Fast_Matrix_Multiply(Matrix<T> &A,
                              Matrix<T> &B,Matrix<T> &C)
5 : {
6 :    int m1=A.m,n1=A.n,n2=B.n;
7 :    int i,j,k;
8 :    T temp;
9 :    T **p1A,*p2A,**p2B=p1B,**p3B=B.aux,
       **p2C=p1C,**p3C=C.aux;
10:
11:    for(k=0;k<n1;k++)  *p2B++=*p3B++;
12:    for(k=0;k<m1;k++)  *p2C++=*p3C++;
13:
14:    for(j=0;j<n;j++)
15:    {
16:      p1A=A.aux,p2C=Cp1;
17:      for(i=0;i<m1;i++)
18:      {
19:        p2A=*p1A++,p2B=p1B;
20:        temp=0.0;
21:        for(k=0;k<n1;k++)
22:          temp += *p2A++ * **p2B++;
23:        **p2C++ = temp;
24:      }
25:      p2B=p1B,p2C=p1C;
26:      for(k=0;k<n1;k++)  ++*p2B++;
27:      for(k=0;k<m1;k++)  ++*p2C++;
28:    }
29: }
```

The meaning of the variables is as follows:

— i, j, k   – loop working variables,
— temp   – the variable storing the temporary value of the scalar product,
— m1, n1, n2   – number of the respective matrix rows and columns,
— A, B   – matrices to be multiplied,
— C   – result of the matrix multiplication, with respect to the formula
              $C = AB$,
— pxA ,pxB, pxC – auxiliary pointers.

6   The Benchmarks

In this Sect. we present the results of the performance tests we performed to verify the robustness of the proposed fast C++ matrix multiplication algorithm 5.3. The test was performed on the Intel™ Core i7 workstation under the Visual Studio™ Ultimate 2010 suite with the control of the Windows 7 64bit operating system. We performed the benchmarks for square matrices with dimensions ranging from 1×1 up to 512×512.

On the graph we present the received execution time shortage of the fast algorithm with respect to the classic one.

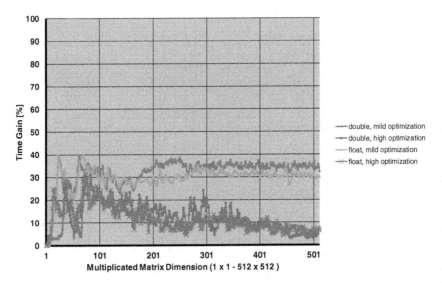

Fig. 6. The benchmark tests

We can observe the following:

- At mild optimization level we can see the constant, equal to approximately one third, performance gain.
- At the high optimization level (with the *maximizing speed* option set on) we still observe meaningful performance boost with comparison to the classical algorithm, despite automatically performed by the compiler the induction variable optimization. The highest performance gain we obtained for the matrix dimensions up to 200 which, in most cases, the user has to deal with.

7   Summary

The main novelty of this article is that we proposed a fast C++ algorithm for the matrix multiplication which supports the compiler in the full utilization of the induction variable elimination optimization method by introducing the auxiliary double pointer shifting window.

To sum up, we hope that many researchers and engineers will find the proposed algorithms implementations useful in their work.

References

1. Aceto, L.: Some applications of the Pascal matrix to the study of numerical methods for differential equations. Boll. del. Unione Matem. Ital. B **8**(3), 639–651 (2005)
2. Aceto, L., Magherini, C., Weinmuller, E.B.: Matrix methods for radial Schrödinger eigenproblems defined on a semi-infinite domain. Appl. Math. Comp. **255**, 179–188 (2015)
3. Augustyn, D.R., Warchal, L.: Cloud service solving N–body problem based on Windows Azure platform. Comm. in Comput. and Inf. Sci. **79**, 84–95 (2010)
4. Augustyn, D.R.: Query-condition-aware histograms in selectivity estimation method. Adv. in Intel. and Soft Comput. **103**, 437–446 (2011)
5. Bellman, R.: Introduction to Matrix Analysis. Society for Industrial Mathematics, New York (1987)
6. Coppersmith, Winograd S.: Matrix multiplication via arithmetic progressions. J. Symb. Comput. **9**, 251–280 (1990)
7. Cormen, T.H., Leiserson, Ch.E., Rivest, R.L., Stein, C.: Introduction to Algorithms, 2nd Edn. McGraw–Hill (2001)
8. Kernighan, B.W., Ritchie, D.M.: The C Programming Language. Prentice-Hall, New Jersey (1978)
9. Kincaid, D.R., Cheney, E.W.: Numerical Analysis: Mathematics of Scientific Computing, 3rd edn. Brooks Cole, California (2001)
10. Press, W.H., Teukolsky, S.A., Vetterling, W.T., Flannery, B.P.: Numerical Recipes in C, 2nd edn. Cambridge University Press, Cambridge (1992)
11. Respondek, J.: On the confluent Vandermonde matrix calculation algorithm. Appl. Math. Lett. **24**, 103–106 (2011)
12. Respondek, J.: Numerical recipes for the high efficient inverse of the confluent Vandermonde matrices. Appl. Math. Comp. **218**(5), 2044–2054 (2011)
13. Sakthivel, R., Ganesh, R., Anthoni, S.M.: Approximate controllability of fractional nonlinear differential inclusions. Appl. Math. and Comp. **225**, 708–717 (2013)
14. Strassen, V.: Gaussian elimination is not optimal. Numer. Math. **13**, 354–356 (1969)
15. Stroustrup, B.: The C++ Programming Language, 3rd edn. AT&T Labs, New Jersey (2000)
16. Stroustrup, B.: The Design and Evolution of C++, 9th edn. Addison-Wesley, Massachusetts (1994)
17. Waite, W.M., Goos, G.: Compiler Construction, Monographs in Computer Science, 2nd edn. Springer Verlag, New York (1983)

Name Entity Recognition for Malay Texts Using Cross-Lingual Annotation Projection Approach

Norshuhani Zamin[1(✉)] and Zainab Abu Bakar[2]

[1] Faculty of Science and Information Technology, Universiti Teknologi PETRONAS,
32610, Bandar Seri Iskandar, Perak, Malaysia
norshuhani@petronas.com.my
[2] Faculty of Computer and Mathematical Sciences, Universiti Teknologi MARA, 40000,
Shah Alam, Selangor, Malaysia
zainabcs@salam.uitm.edu.my

Abstract. Cross-lingual annotation projection methods can benefit from rich-resourced languages to improve the performance of Natural Language Processing (NLP) tasks in less-resourced languages. In this research, Malay is experimented as the less-resourced language and English is experimented as the rich-resourced language. The research is proposed to reduce the deadlock in Malay computational linguistic research due to the shortage of Malay tools and annotated corpus by exploiting state-of-the-art English tools. This paper proposes an alignment method known as MEWA (Malay-English Word Aligner) that integrates a Dice Coefficient and bigram string similarity measure with little supervision to automatically recognize three common named entities – person (PER), organization (ORG) and location (LOC). Firstly, the test collection of Malay journalistic articles describing on Indonesian terrorism is established in three volumes – 646, 5413 and 10002 words. Secondly, a comparative study between selected state-of-the-art tools is conducted to evaluate the performance of the tools against the test collection. Thirdly, MEWA is experimented to automatically induced annotations using the test collection and the identified English tool. A total of 93% accuracy rate is achieved in a series of NE annotation projection experiment.

1 Introduction

Information overload is now a reality. The overload of documents has created the potential for there to be a vast amount of valuable information buried in those texts. This has become a motivation for the research - to explore potential methods for discovering relevant knowledge in a collection without the user having to read everything. Learning algorithms and methods typically require large amounts of supervised training data in order to produce accurate results. Large annotated corpora have been constructed for popular languages such as English, French and Italian. Building large, clean, well-balanced, annotated corpora requires significant infrastructure and many hours of dedicated effort by expert linguists. Hence, constructing similar large corpora for less-studied languages is regularly not practical.

© Springer International Publishing Switzerland 2015
O. Gervasi et al. (Eds.): ICCSA 2015, Part I, LNCS 9155, pp. 242–256, 2015.
DOI: 10.1007/978-3-319-21404-7_18

This research is closely related to the study of Natural Language Processing (NLP) in the area of Artificial Intelligence [1]. NLP research has been going on for the last thirty years. With the sufficient power and memory capacity of today's computers, most of NLP work has turn into real applications. information extraction (IE) is one of the sub-topics in NLP that has been moving rapidly since 2000. IE takes unstructured natural language texts to produce structured or fixed-format data as output. Structured data is an important form of data for analysis purpose [2]. It shows the relationship of entities, for instance, the relationship between people, companies and locations. IE output helps to ease the searching, for instance, given a location, other entities such as the companies and types of business can be discovered straightforward. Table 1 shows an example of IE output for text input *John and Jane joined IBM New York in 2006.*

Table 1. Example of an IE Output – A Structured Data

Person	Organization	Location	Date
John	*IBM*	*New York*	*2006*
Jane			

This research is conducted to propose a method to leverage pre-existing resources into a less-resourced language to perform some significant tasks in IE i.e. the tagging of Named Entity (NE) annotation on a natural language corpus. A tool to demonstrate the effectiveness of the method is developed using C++. To achieve the objectives, Malay is chosen to be in the experiments as it has been one of the less-resourced languages in the literature. As English is well-studied with satisfactory achievements in much computational linguistic research, it is used as the pre-existing resource to annotate the Malay corpus. Bridging two languages is a process known as cross-lingual annotation projection. It has been a popular framework for transferring linguistic annotations from one language to another while exploiting the translational equivalences present in parallel corpora. The idea was originated by Yarowsky et al. [3].

2   Problems

Cross-lingual annotation projection for Named Entity Recognition (NER) is relatively a new research area in Malay. Thus, three prominent problems have been identified which daunt the NER research in Malay:

1. Malay has little annotated data that is publicly available. Examples of private data include the Malay Practical Grammar Corpus [4], the Dewan Bahasa Pustaka Database Corpus[1], the Malay Corpus by Unit Terjemahan Melalui Komputer from the University Science of Malaysia [5] and, more recently, the Malay

---

[1] http://www.dbp.gov.my

LEXicon (MALEX) [6]. The freely available Malay Concordance Project Corpus[2] is a collection of 3,000,000 words extracted from classical Malay texts, ones that are not related to this research domain. Limited access to such important sources of linguistic knowledge is a major hurdle in Malay NLP research.

2. Text-processing tools are available for only a few rich-resourced languages, such as English, German and Chinese. As for example, existing English POS-taggers, there are many advanced versions that have reached almost up to 98% accuracy with almost no room for improvement [7, 8]. This presents significant opportunities to leverage these pre-existing resources from a rich-resourced language, such as English, to avoid building new text-processing tools from scratch for a less-resourced language. The idea to exploit the linguistic information from one language to another is referred as cross-lingual annotation projection. Reuse of resources helps to reduce costs and overheads in system development.

3. There is a gap between supervised and unsupervised research for less-resourced languages. Christodoulopoulos et al. [9] uncovered further evidence for this analysis. The study showed that the best accuracy i.e. the F1 score recorded for unsupervised POS taggers, for less-studied languages, was only 76.1%. This research contributes into bridging the gap.

3   Named Entity Recognition for Malay

Named entities are prominent entities in text belonging to predefined categories such as names of persons, organizations, locations, expressions of time, monetary values, etc. Named Entity Recognition (NER) involves the identification of certain occurrences of words or expressions in unstructured texts and the classification of them into a set of predefined categories of interest. NER is often implemented as a pre-processing tool for a text mining application. Among the popular applications of NER are document retrieval and question answering. Research in NER was initiated in 1988 from the Message Understanding Conference (MUC) funded by Defense Advanced Research Projects Agency (DARPA) in United States of America. MUC continued looking for more NER applications until 1998 and the effort was taken over by the Automatic Content Extraction (ACE) until 2008 and handed over to the Knowledge Based Population (KBP) on 2009 onwards [10].

Mikheev et al. [11] stated that a textual document which was annotated with NE information can be searched more accurately than unannotated document or raw text. NER supports semantic searching i.e allowing the user to perform semantic querying, thus going beyond keyword searching. Similar to POS tagging, NER deals with metonymy too. For example the *White House* is recognized as the organization in the expression *The White House is located in the Washington D.C.* but it is recognized as a person (staff of the White House) in the expression *The White House will be announcing the decision around noon today.*

---

[2] http://mcp.anu.edu.au/

In IR research, NE annotation allows the search for all texts mentioning the *company* with the name *Hong Leong* such as the notable *Hong Leong Bank* and *Hong Leong Group*, both Malaysian establishments. Theoretically, NER is expected to ignore documents about unrelated entities of the same name. However, in Malaysia, there exists a person with the name *Hong Leong*, particularly the Chinese citizens. Classification of entities might be a huge challenge without additional information. In this case, an NER system would have a trouble to classify the entity *Hong Leong* into either a name of a *company* or a *person*. An example of how NER work is shown in Fig. 1.

Kofi Atta Annan is a Ghanaian diplomat who served as the seventh Secretary General of the United Nations from January 1, 1997, serving two five-year terms. Annan was the co-recipient of the Nobel Prize in October 2001. Kofi Annan was born on April 8, 1938, to Victoria and Henry Reginald Annan in Kumasi, Ghana. He is a twin, an occurrence that is regarded as special in Ghanaian culture. Efua Atta, his twin sister, shares the same middle name, which means 'twin'. As with most Akan named, his first name indicates the day of the week he was born.

Legend:
PERSON, ORGANIZATION, LOCATION, DATE.

Fig. 1. An Example of NER Application

Unfortunately, not much research has been published on NER for Malay. Sharum et al. [12] proposed a free indexing method to recognize person names in Malay text. The research manually analyzed the structures of Malaysians' names commonly presented in Malay news articles. They extracted common titles and names found as shown in Table 2 and build the name indexer.

Table 2. List of Extracted Common Titles and Names in Malay Texts

Type	List
Abbreviated Title	Sdr, Sdri, En, Tn, Pn, Hj, Hjh, YBhg, Sen, Prof, PM, Dr, Drs, Ir, YB, YAB, Kol, Brig, Jen, Kapt, Sjn, DYMM, DYTM *etc.*
Title	Tunku, Tengku, Tun, Toh, Tan Sri, Datok, Dato', Datu, Datuk, Datin, Seri Paduka, Senator, Profesor, Kolonel, Brigidier, Jeneral, Komisioner, Mejar, Kapten, Sarjan, Leftenan, Lans, Koperal, Inspektor, Laksamana, Cif *etc.*
First Name	Ahmad, Muhd, Mohd, Muhamad, Muhammad,. Mohamed, Mohammed, Md, Mat, M., Syed, Abd., Abdul, A., Siti, Noor, Nor, Nur, Nurul, Ku, Wan, Nik, Che, Chong, Lee, Tan *etc.*
Final Name	Singh, Kaur
Middle Name (Kinship)	Bin, B. Son of (Malay) Binti, Bt. Daughter of (Malay) Anak, Ak., AL., A/L Son of (non-Malay) AP., A/P Daughter of (non-Malay)

An experiment conducted on 117 news articles produced 68% accuracy rate in identifying person's names. The algorithm worked poorly on business news articles but showed an increased performance rate in political news articles. This was due to the nature of political news that involved name of authorities, VIPs or honorable persons. More recently, [13] proposed a rule-based NER for Malay texts which has improved the previous work [12] by using contextual rules for the identification of organization and location entities as shown in Table 3. The research also compiled a dictionary look-up that consist of all possible entries found in Malay text for each of the identified contextual rules.

Table 3. An Example of Contextual Rules for Malay Text

Feature	Example
Location Prefix	Jalan, Bukit, Kampung
Preposition that usually followed by location	di, ke
Organization Prefix	Syarikat, Kelab, Persatuan
Organization Suffix	Sdn. Bhd.
Person Prefix	Tan, Lim
Person Middle	Bin, Binti, A/P, A/L, anak
Person Title	Dato Paduka, Tun, Datuk, Tan Sri
Preposition that usually followed by person	Oleh

4   Cross-Lingual Annotation Projection or NER

Annotating the corpus manually is a laborious and expensive task. This task is called corpus annotation in linguistic. Some research do provides annotation standards and guidelines for natural language annotation in various domains such as the research by Galescu and Blaylock [14] and Roberts et al. [15]. Annotations can be of different nature, such as prosodic, semantic or historical annotation. The most common form of annotated corpora is the grammatically tagged such as the POS tagged corpus. However, annotation is not only limited to text but also images, multimedia, database queries and even procedures [16]. An example of automatic image annotation is demonstrated by [17]. Annotation projection is a task within parallel corpora where information from Source Language (SL) is projected or mapped onto the Target Language (TL). Parallel corpora or bitext are extensively studied in the machine translation field where the aligned phrases and words are used to create translation models.

The scope of this research is focusing on projecting the annotation information from the SL (English) to the TL (Malay) within a non-pre-aligned bilingual corpora or parallel corpus. This task is also referred as cross-language annotation projection [18] and tools through the annotation projection process. The challenge is to develop an algorithm to automatically align the sentences from two languages of different structures and morphology, a rich-resourced language (English) and a less-resourced language (Malay) and thus to project the English annotations across the parallel corpus with higher accuracy. Word-level and sentence-level alignment are commonly used to bridge the gap between the two corpus. Some annotation projection algorithms might need to filter the noisy annotation results.

Cross-lingual annotation projection aims at overcoming this resource bottleneck by scaling monolingual resources and tools to multilingual level, a word alignment as a bridge to transfer the annotations from the resourced-rich to resourced-poor languages [19]. This finding has motivated the study to leverage pre-existing resources from a rich-resourced language, such as English, to less-resourced language as in this case study, Malay. This provides a significant opportunity for the study not to build new domain specific Malay text processing tools from scratch. A survey of the literature revealed that reuse of resources helps to reduce costs and overheads in system development ([20, 21, 22, 23]. Moreover, it has been noticed by many scholars through all the previous work on annotation projection across languages that we should support the sharing of knowledge resources. The principle of reuse of resources helps to avoid the waste in time and effort in repeated construction of linguistic resources [24, 25].

A cross-lingual annotation projection process is visualized as three dimensional in Fig. 2 by Frank [30]. The *Resource Axis* (representing SL) holds the annotations following the corpus annotation standards. It could also hold multiple annotations such as shown in the figure – *syntax (syn), word sense, semantic (sem) and anaphora resolution (anaph)*. At the *Language Axis* lies the projection annotation algorithm to project the annotations to the *Processing Axis* (representing TL) cum the target axis as the newly created resource.

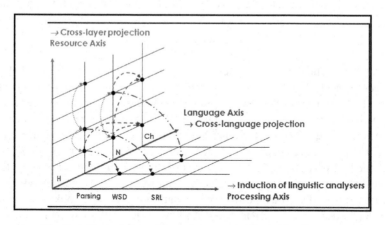

Fig. 2. A General Cross-Lingual Annotation Projection Process

In cross-lingual annotation projection, the annotations from the SL are projected to the TL via a word alignment algorithm as a bridge. The annotation projection is commonly performed better in a pre-aligned parallel corpus. This can be achieved using some existing tools which will be discussed in detail in the next sub topics. Annotation projection using pre-aligned parallel corpus is demonstrated successfully in projecting coreference resolution in English-Portuguese parallel corpus [26], relation detection in English-Korean parallel corpus [18], dependency analysis in English-Swahili parallel corpus [27], semantic roles in English-German parallel corpus [28] and syntactic relations in English-Romanian parallel corpus [29].

5   Methodology

The underlying idea in this paper is the alignment between the two languages. A perfect mapping ensures perfect projection of annotations. We refer the proposed method as Malay-English Word Aligner or MEWA. MEWA is a hybrid method using heuristic and probabilistic approach that aligns English to its corresponding Malay word, thus projects the Named Entities (NE) of the English word which was assigned by the off-the-shelf NE tagger to the newly aligned Malay word. In detailed, MEWA uses a probabilistic approach i.e. N-gram scoring method for two characters which is commonly referred as bigram and is integrated with a heuristic approach, Dice Coefficient function [31] in order to calculate the probability distribution of letter sequences between Malay and English texts. This is a hybrid method which never been applied in any research involving the Malay corpus.

Dice Coefficient is a heuristic alignment method which differs from statistical alignment. It uses specific associative measures rather than pure statistical measures. The function applies basic heuristic rule – a word pair are chosen as the aligned words if the pair has the highest co-occurrence score. It is a simple and intuitive estimation approach for word alignment [32]. The Dice Coefficient Function is shown in Eq. (1).

$$S_{Dice} = \frac{2\sum_{i=1}^{d} P_i Q_i}{\sum_{i=1}^{d} P_i^2 + \sum_{i=1}^{d} Q_i^2} \tag{1}$$

The Dice Coefficient is calculated by counting the number sentences where the words co-occur ($P_i Q_i$), the number of occurrences of the English word (P_i) and the number of occurrences of the Malay word (Q_i). The measure is always between 0 and 1. The use of the Dice Coefficient function in bitext alignment research is inspired by the research of Dien [33]. On the other hand, bigram is a simple probabilistic method to measure the string closeness of two different texts and often generates good results. Bigram method performs well on languages of different structures and is widely implemented in text-mining research including Yarowsky's text projection research, [3]. A bigram is also referred as the first-order Markov model as it looks one token into the past and a trigram is a second order Markov model while N-gram is the Nth order Markov model [34]. N-gram has been widely demonstrated in early research for word prediction [35, 36]. The successful of an N-gram model is measured by its perplexity. A better model has the lowest perplexity value.

Hence, a significant improvement to integrate the two methods (Dice Coefficient + Character Bigram) is proposed and experimented on Malay language. A collection of news articles on Indonesian terrorism written in Malay are chosen as the platform to test the algorithm. The bigram scoring method is employed in the Dice Coefficient function in order to calculate the probability distribution of letter sequences between Malay and English texts, without using any training data or morphological analyzer. It is a variation of string similarity function introduced by Sørensen [37] and is referred as Sørensen Coefficient. The two input strings are internally tokenized into character bigram. The formula is shown in Eq. (2).

$$S_{\text{Sørensen}} = \frac{2n_t}{n_x + n_y} \tag{2}$$

where n_t is the number of character bigrams found in both strings, n_x is the number of bigrams in string x and n_y is the number of bigrams in string y. For example, to calculate the similarity between *night* and *light*, we would find the set of bigrams in each word as follows:

{ni,ig,gh,ht}

{li,ig,gh,ht}

Here, each set has four elements and the intersection of these two sets has two elements i.e. *gh* and *ht*. Hence, putting in the values, we get $S_{\text{Sørensen}} = (2 \times 3) / (4+4) = 0.75$. All variations of Dice Coefficient return probabilistic value between 0 and 1. The higher the number, the greater the similarity of the two compared strings. Sørensen-Dice Coefficient supports for lexical similarity, robustness to changes of word order, and language independence [38]. The overall framework is shown in Fig. 3.

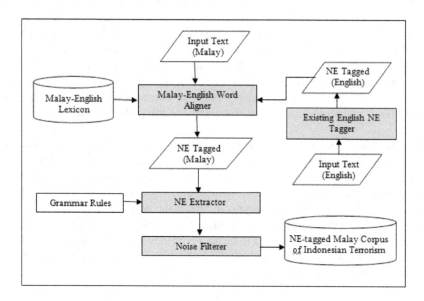

Fig. 3. Proposed Framework

In this framework, the Malay-English Lexicon is not only compiling the list of lexemes and their translations but also the translations for each person, organization and location and their variations. In this case, all the proper names appearing in the corpus are also stored as they are in our lexicon except for those that may have more than one reference. Some examples are shown in Fig. 4.

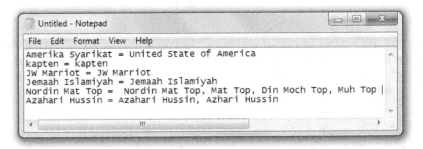

Fig. 4. Lexicon Entries for the Malay NEs

The mapping for NEs is simple and straight forward as it does not require much the use of lexemes. MEWA works similarly as described in the previous section. Let's exemplify a Malay statement *"John dan Jane telah menyertai IBM di New York pada 2006"* which has its translation in English as *"John and Jane joined IBM in New York in 2006"*. Assume that the English translation has been tagged with particular NEs using a readily available NER English tagger as to produce annotated text as *"John/PER and Jane/PER joined IBM/ORG in New York/LOC in 2006/DATE"*. Proper names in the English corpus are identified using grammar rules as the common pattern. Rules are constructed manually based on the translated version of the terrorism corpus and also by using the projected POS tag for proper names. Feldman & Sanger [39] have also provided certain rules for common English expressions and some of them are incorporated and merged with the manually constructed rules in this study.

A Noise Filterer module is included in the model to filter out any low-confidence prediction so that only useful names remain. This is due to the possibility of false positives in the dictionary matching scheme where a proper name recognized by the Entity Extractor may actually be a non-name which is also known as noise. The occurrences of false positives will negatively affect the accuracy and reliability of the system. A Noise Filterer module by Minkov et al. [40] is used in this study. This module consists of two metrics: Predicted Frequency (PF) and Inverse Document Frequency (IDF). The PF metric calculates the frequency of consistent occurrences of a word as a proper name in the corpus as shown in Eq. (3).

$$PF(w) = \frac{cpf(w)}{ctf(w)} \tag{3}$$

where *cpf(w)* is the number of times that a word *w* is identified as a name and *ctf(w)* is the number of times it appears in the entire test corpus. The IDF metric is calculated as in Eq. 4.

$$IDF(w) = \frac{\log(\frac{N+0.5}{df(w)})}{\log(N+1)} \tag{4}$$

where $df(w)$ is the number of articles that contain the word w and N is the total number of articles in the corpus. *PFIDF* is a measure which combines these two metrics multiplicatively, giving a single probability of a word being a name and showing how common it is in the entire corpus; the measure is in Eq. 5. A word with low *PFIDF* score is considered ambiguous in the corpus and is excluded from it.

$$PFIDF(w) = PF(w) \times IDF(w) \tag{5}$$

6 Experimental Results

Two experiments were conducted: 1) the comparative study between three off-the-shelf English NE Taggers and 2) the evaluation of the framework's performance using different volume of test collection against human results. The common evaluation metric for IR is used i.e. Precision (P), Recall (R) and F1-Score (F1).

In the first experiment, the performance of our English terrorism corpus is compared with three most advanced open-source NER Taggers (i.e. UIUC, LingPipe and Stanford) over their capacity to recognize PER, ORG and LOC entities in the 25 selected news articles. These NER Taggers are selected for the comparison because they were trained on similar form of data (news articles) and on top of that, they are publicly available. Out of these three, the best NER tagger is chosen based on how accurate it is compared to human annotations. The result is shown in Table 4.

Table 4. A Comparative Study between Stanford, LingPipe and UIUC NE Tagger on English Terrorism Corpus

NE Tagger	PER			ORG			LOC		
	P	R	F1	P	R	F1	P	R	F1
Stanford	0.76	0.59	0.66	0.59	0.42	0.49	0.89	0.79	0.84
LingPipe	0.67	0.56	0.61	0.62	0.61	0.62	0.73	0.72	0.72
UIUC	0.97	0.77	0.86	0.93	0.84	0.88	0.86	0.82	0.84

UIUC NER Tagger produced the best results due to its Wikipedia-based gazetteers that generates NE list by extracting information from Wikipedia[3]. Wikipedia is an open-system and free encyclopedia that can be updated independently and regularly by its huge numbers of collaborators worldwide, hence the vast amount of data available. Another attractive property of this knowledge base is its capacity to list out spelling variations of one entity and link all of the related pages in one basic entry.

---

[3] http:// www.wikipedia.org

However, this approach may not be suitable to our research since Wikipedia informa-tion in Malay language is limited with low update rate. Hence, to tailor our lexicon to the domain of this research, manual English – Malay word entries extracted from the online dictionary[4] is conducted. The lexicon is interchangeably used with dictionary look-up and gazetteer in most of text mining research.

The second experiment is divided into two sub-experiments, each involving two different test collections.

1.   Experiment A: To test the framework on a large scale corpus consists of 5413 words (First collection).
2.   Experiment B: To test the framework on a larger scale corpus consists of 10002 words (Second collection).

The performances of our algorithm on the datasets are presented respectively in Table 5.

Table 5. Performance Evaluation Results

Words	Correct Tags	Wrong Tags	Missed Tags	P	R	F1
5413	494	34	35	0.94	0.88	0.90
10002	2062	157	132	0.93	0.93	0.93

The accuracy and reliability of an NE depends on how well it matches the equiva-lent entity in the manually annotated dataset. The addition of grammar rules to per-form together with word aligner has improved its performance to 90% in Experiment A and 93% in Experiment B. The tests have shown that the value of F1 has increased by the average of 7.2% using the hybrid method. Surprisingly, the results remain un-changed for all the P, R and F1 in the larger test collection when MEWA is incorpo-rated with grammar rules but the increment when compared to MEWA alone in the same test collection is a relatively significant improvement. It is proven that incorpo-rating rules can enhance the identification of proper entities because the statistical algorithm alone may have missed some of those entities. The performance improves when more in depth rules with fixed variable strings are added (e.g. *district*, *hotel*, *new*, *province*, *north of*, *agency*, etc.). The results also show that 25% of the mistakes in the identification process are caused by spelling variations in the news articles and lexicon, no lexeme match in the lexicon, and errors from the existing tagger itself. In addition, misidentification rate increases when words of different sizes such as *Ame-rika Syarikat'* and its English translation, *United States of America* are mapped. Final-ly, the error in the output comes from UIUC NE Tagger itself when it incorrectly tags capitalized nouns and fails to tag some of the entities. The examples of wrong tagging by the UIUC NE Tagger using our dataset are presented in Table 6. Words that are incorrectly tagged by the hybrid MEWA are underlined.

---

[4] http://kamus.lamanmini.com/index.php

Table 6. UIUC's Tagger Tagging Errors

Error	Examples
Missed Tagging	*Example 1:* MISC Southeast Asian) extremist group inspired by <u>al-Qaeda</u> and blamed for several attacks in (LOC Indonesia).
	Example 2: They are conducting military training in remote areas of the forests of (LOC Aceh), probably because (LOC Aceh) is now peaceful," he said, as quoted by the agency of <u>Antara</u> news today.
Wrong Tagging	*Example 1:* More than 100 heavy-armed police officers took part in the raid just before midnight last night in the forest areas of the (MISC <u>Aceh Besar</u>) district, about 70 kilometers north of (PER <u>Banda Aceh</u>).
	Example 2: Police made the arrests after fighting for an hour late yesterday in the mountainous area of (PER <u>Jalin</u>), said the Provincial Police Chief, Major General (PER Aditya Warman).

7   Conclusions

This paper introduces a new method for an automated NE tagger for Malay using parallel data from a rich-resourced language. As far as the research is concerned, no previous work has studied the alignment of a Malay corpus with an English corpus by projecting tags using hybrid algorithms. The bitext alignment method appears to be a powerful unsupervised learning algorithm mapping two dissimilar languages at minimum computational cost. Additionally, unsupervised methods heavily reduce the labour required to annotate a Malay corpus, and generate quick results. Three state-of-the-art NE Taggers, i.e. the Stanford, the LingPipe and the UIUC have been evaluated on the English parallel corpus. The performance was decisively low for both the Stanford and the LingPipe NE Taggers. The most probable reason for the poor outcomes are the different genre and domain used, ones that were never tested previously on these taggers. However, the use of a gazetteer extracted from Wikipedia has shown an increased performance in the UIUC NER Tagger that it is chosen as the NER resource for English. A hybrid NER Tagger for a Malay journalistic corpus of Indonesian terrorism referred as MEWA that employs an unsupervised learning method has been developed - a statistical technique using the classical Dice Coefficient function with bigram scoring and a rule-based method combined with a set of knowledge engineered rules. Usage was made of a Malay-English lexicon as an add-on to overcome the costly lemmatization effort needed for the Malay language.

References

1. Cowie, J., Wills, Y.: Information Extraction: A Handbook of Natural Language Processing. Marcel Dekker, New York (2000)
2. Bird, S., Klein, E., Loper, E.: Natural Language Processing with Python, 1st edn. O'Reilly Bookstore (2009)
3. Yarowsky, D., Ngai, G., Wicentowski, R.: Inducing multilingual text analysis tools via robust projection across aligned corpora. In: Proceedings of the Human Language Technology Research, pp. 1–8 (2001)
4. Abdullah, I.H., Ahmad, Z., Ghani, R.A., Jalaludin, N.H., Aman, I.: A Practical Grammar of Malay–A Corpus based Algorithm to the Description of Malay: Extending the Possibilities for Endless and Lifelong Language Learning. National University of Singapore (2004)
5. Ranaivo, M.B.: Computational analysis of affixed words in malay language. In: Proceedings of the 8th International Symposium on Malay/Indonesian Linguistics, Penang, Malaysia (2004)
6. Don, Z.M.: Processing Natural Malay Texts: A Data-driven approach. Trames **1**, 90–103 (2010)
7. Indurkhya, N., Damerau, F.J.: Handbook of Natural Language Processing, 2nd edn. Chapman & Hall / CRC Press (2010)
8. Tsuruoka, Y., Tateishi, Y., Kim, J.-D., Ohta, T., McNaught, J., Ananiadou, S., Tsujii, J.: Developing a robust part-of-speech tagger for biomedical text. In: Bozanis, P., Houstis, E.N. (eds.) PCI 2005. LNCS, vol. 3746, pp. 382–392. Springer, Heidelberg (2005)
9. Christodoulopoulus, C., Goldwater, S., Steedman, M.: Two decades of unsupervised POS induction: how far have we come. In: Proceedings of Empirical Methods in Natural Language Processing (2010)
10. Grishman, R.: Lecture Notes on Information Extraction (2013). http://cs/nu.edu/grishman/tarragona.pdf
11. Mikheev, A., Moens, M., Grover, C.: Named entity recognition without gazetteers. In: Proceedings of the 9th European Chapter of the Association for Computational Linguistics, pp. 1–8. Association for Computational Linguistics (1999)
12. Sharum, M.Y., Abdullah, M.T., Sulaiman, M.N., Murad, M.A.A., Hamzah, Z.A.Z.: Name extraction for unstructured malay text. In: IEEE Symposium on Computers & Informatics (ISCI), pp. 787–791. IEEE (2011)
13. Alfred, R., Leong, L.C., On, C.K., Anthony, P.: Malay Named Entity Recognition Based on Rule-Based Approach. International Journal of Machine Learning and Computing **4**(3), June 2014
14. Galescu, L., Blaylock, N.: A corpus of clinical narratives annotated with temporal information. In: Proceedings of the 2nd ACM SIGHIT International Health Informatics Symposium, pp. 715–720. ACM (2012)
15. Roberts, A., Gaizauskas, R., Hepple, M., Demetriou, G., Guo, Y., Roberts, I.: Building a Semantically Annotated Corpus of Clinical Texts. Journal of Biomedical Informatics **42**(5), 950–966 (2009)
16. Katz, B.: Annotating the world wide web using natural language. In: Proceedings of the 5th RIAO Conference on Computer Assisted Information Searching on the Internet (RIAO 1997), pp. 136–59 (1997)
17. Manaf, S.A., Nordin, M.J.: Review on statistical approaches for automatic image annotation. In: International Conference on Electrical Engineering and Informatics, 2009. ICEEI 2009, vol. 1, pp. 56–61 (2009)

18. Kim, S., Jeong, M., Lee, J., Lee, G.G.: A cross-lingual annotation projection algorithm for relation detection. In: Proceedings of the 23rd International Conference on Computational Linguistics, pp. 564–571. Association for Computational Linguistics (2010)
19. Spreyer, K., Frank, A.: Projection-based acquisition of a temporal labeller. In: Proceedings of IJCNLP 2008 (2008)
20. Mayobre, G.: Using code reusability analysis to identify reusable components from the software related to an application domain. In: Proceedings of the 4th Annual Workshop on Software Reuse, pp. 1–14 (1991)
21. Bollinger, T.B., Pfleeger, S.L.: Economics of Software Reuse: Issues and Alternatives. Information and Software Technology 32(10), 643–652 (1990)
22. Barnes, B.H., Bollinger, T.B.: Making Reuse Cost-Effective. IEEE Software 8(1), 13–24 (1991)
23. Kim, Y., Stohr, E.A.: Software Reuse: Survey and Research Directions. Journal of Management Information Systems, 113–147 (1998)
24. Brill, E., Lin, J., Banko, M., Dumais, S., Ng, A.: Data-intensive question answering. In: Proceedings of the Tenth Text Retrieval Conference (TREC 2001) (2001)
25. Banko, M., Brill, E.: Mitigating the paucity-of-data problem: exploring the effect of training corpus size on classifier performance for natural language processing. In: Proceedings of the First International Conference on Human Language Technology Research, pp. 1–5. Association for Computational Linguistics (2001)
26. de Souza, J.G.C., Orăsan, C.: Can projected chains in parallel corpora help coreference resolution? In: Hendrickx, I., Lalitha Devi, S., Branco, A., Mitkov, R. (eds.) DAARC 2011. LNCS, vol. 7099, pp. 59–69. Springer, Heidelberg (2011)
27. De Pauw, G., Wagacha, P.W., De Schryver, G.M.: The SAWA corpus: a parallel corpus english-swahili. In: Proceedings of the First Workshop on Language Technologies for African Languages, pp. 9–16. Association for Computational Linguistics (2009)
28. Padó, M., Lapata, M.: Cross-linguistic projection of role-semantic information. In: Proceedings of the Conference on Human Language Technology and Empirical Methods in Natural Language Processing, pp. 859–866. Association for Computational Linguistics (2005)
29. Mititelu, V.B., Ion, R.: Cross-Language Transfer of Syntactic Relations Using Parallel Corpora. Cross-Language Knowledge Induction Workshop, Romania (2005)
30. Frank, A.: Network of Linguistic Annotation: The Linguist Web [Power Point Slides]. University of Heidelberg, Heidelberg (2007)
31. Dice, L.R.: Measures of the Amount of Ecologic Association between Species. Ecology 26(3), 297–302 (1945)
32. Moore, R.C.: Improving IBM word-alignment model 1. In: Proceedings of the 42nd Annual Meeting on Association for Computational Linguistics (518). Association for Computational Linguistics (2004)
33. Dien, D.I.N.H.: Building an Annotated English-Vietnamese Parallel Corpus. MKS: A Journal of Southeast Asian Linguistics and Languages 35, 21–36 (2005)
34. Jurafsky, D., Martin, J.H.: Speech and Language Processing: An Introduction to Natural Language Processing, Computational Linguistics and Speech. Prentice Hall (2000)
35. Jurafsky, D., Bates, R., Coccaro, N., Martin, R., Meteer, M., Ries, K., Ess-Dykema, V.: Automatic detection of discourse structure for speech recognition and understanding. In: Proceedings of the 1997 IEEE Workshop on Automatic Speech Recognition and Understanding, 1997, pp. 88–95. IEEE (1997)
36. Jurafsky, D., Wooters, C., Tajchman, G., Segal, J., Stolcke, A., Foster, E., Morgan, N.: The Berkeley Restaurant Project. ICSLP 94, 2139–2142 (1994)

37. Sørensen, T.: A Method of Establishing Groups of Equal Amplitude in Plant Sociology based on Similarity of Species and its Application to Analyses of the Vegetation on Danish Commons. Biol. Skr. **5**, 1–34 (1948)
38. Kondrak, G., Marcu, D., Knight, K.: Cognates can improve statistical translation models. In: Proceedings of the 2003 Conference of the North American Chapter of the Association for Computational Linguistics on Human Language Technology: Companion Volume of the Proceedings of HLT-NAACL 2003–Short Papers, vol. 2, pp. 46–48. Association for Computational Linguistics, May 2003
39. Feldman, R., Sanger, J.: The Text Mining Handbook: Advanced Algorithm in Analyzing Unstructured Data. Cambridge University Press (2006)
40. Minkov, E., Wang, R., Cohen, W.: Extracting personal names from emails: applying named entity recognition to informal text. In: Proceedings of the Human Language Technology and Conference on Empirical Methods in Natural Language Processing, pp. 443–450 (2005). doi:10.3115/1220575.1220631

The Maximum Similarity Partitioning Problem and Its Application in the Transcriptome Reconstruction and Quantification Problem

Alex Z. Zaccaron[1]([✉]), Said S. Adi[1], Carlos H.A. Higa[1], Eloi Araujo[1],
and Burton H. Bluhm[2]

[1] Faculdade de Computação, Universidade Federal de Mato Grosso do Sul,
Campo Grande, Brazil
alex.zaccaron@gmail.com
[2] Plant Pathology Department, University of Arkansas, Fayetteville, USA
bbluhm@uark.edu

Abstract. Reconstruct and quantify the RNA molecules in a cell at a given moment is an important problem in molecular biology that allows one to know which genes are being expressed and at which intensity level. Such problem is known as Transcriptome Reconstruction and Quantification Problem (TRQP). Although several approaches were already designed that solve the TRQP, none of them model it as a combinatorial optimization problem. In order to narrow this gap, we present here a new combinatorial optimization problem called Maximum Similarity Partitioning Problem (MSPP) that models the TRQP. In addition, we prove that the MSPP is NP-complete in the strong sense and present a greedy heuristic for it.

Keywords: Partition · Similarity · Transcriptome · Reconstruction and quantification

1 Introduction

With the advent of Next Generation Sequencing (NGS) technologies, new interesting problems involving strings arise at the same time that traditional ones become more difficult. One of these problems, called Transcriptome Reconstruction and Quantification Problem (TRQP), consists in determining the RNA molecules expressed in a cell at a given moment. More precisely, given a set of reads resulting from an RNA-Seq experiment, the purpose is to assemble these reads into transcribed RNA sequences and to quantify such sequences.

Problems like the TRQP are a valuable source of combinatorial optimization problems over strings, and as such, require efficient algorithms to cope with

This work has been supported by FUNDECT-Brasil/MS (process number: 23/200, 500/2014. FUNDECT number: 185/2014).

O. Gervasi et al. (Eds.): ICCSA 2015, Part I, LNCS 9155, pp. 257–266, 2015.
DOI: 10.1007/978-3-319-21404-7_19

them. Although some approaches for the TRQP have already been designed [1–9], to the best of our knowledge, none of them address it as a combinatorial optimization problem.

In this work we describe a new combinatorial optimization problem involving strings, called Maximum Similarity Partitioning Problem (MSPP), that models the TRQP. The MSPP consistis in finding a partition of an ordered subset of segments as similar as possible with other subsets of ordered segments. Besides introducing the MSPP, we show that it is NP-complete in the strong sense by a reduction from the 3-Partition Problem and we propose a greedy heuristic for it. Such heuristic is based on another problem involving strings called Segment Alignment Problem [10].

This paper is organized as follows. In Section 2 we present some basic concepts and the MSPP is defined. In Section 3 we prove that the MSPP is NP-complete in the strong sense. In Section 4 we propose a greedy heuristic for it and in Section 5 is presented some experimental results. Lastly, the paper is finalized in Section 6 with some concluding remarks and future research directions.

2   Basic Concepts

Let s be a *string* over an alphabet Σ, with $|s|$ denoting the length of s. A *substring* $c = s[i..j]$ of s is a sequence of $j - i + 1$ consecutive characters of s starting at the i-th character and ending at the j-th character of this sequence. We denote by $first(c)$ the position of the first symbol of c in s and by $last(c)$ the position of the last symbol of c in s. Given a set $C = \{c_1, c_2, ..., c_k\}$ of substrings of s, we say that C is an *ordered set of substrings* if $\forall c_i \in C$ 1) $first(c_i) < first(c_{i+1})$ or 2) $first(c_i) = first(c_{i+1})$ and $last(c_i) < last(c_{i+1})$. We also say that two substrings $c' = s[l..m]$ and $c'' = s[o..p]$ of s *overlap* each other if $o \le l \le p$ or $l \le o \le m$. If $m < o$, it is said that c' *precedes* c'', and such relation is denoted by $c' \prec c''$. A subset $\Gamma = \{c_i, c_j, ..., c_q\}$ of C is called a *chain* of C if $c_i \prec c_j \prec ... \prec c_q$, and we denote by $\Gamma^\bullet = c_i \bullet c_j \bullet ... \bullet c_q$ the string resulting from the concatenation of all elements of Γ.

Given two sequences s and t over an alphabet Σ that does not include the space character ('−'), an *alignment* of s and t is a $2 \times l$ matrix A ($|s|, |t| \le l \le |s| + |t|$) such that the first and second row of A contains the characters of s and t, respectively, in the same order that they appear in their sequences and interspersed with $l - |s|$ and $l - |t|$ spaces. A *score function* $\omega : \bar{\Sigma} \times \bar{\Sigma} \to \mathbb{R}$, where $\bar{\Sigma} = \Sigma \cup \{'−'\}$, assigns a real number for each column of the alignment, and it is used to determine the *score of the alignment*, which corresponds to the sum of the values of its columns. We call *similarity* of s and t, denoted by $sim_\omega(s, t)$, the highest score among all possible alignments of s and t [11].

Strings and substrings, as well as numbers, can be given as input to an algorithm that solves a specific problem. We call an *instance* $I = (i_1, i_2, ..., i_n)$ of a computational problem P the input $i_1, i_2, ..., i_n$ necessary for an algorithm solve P. Given an instance I of a computational problem, the function Length(I) represents the number of symbols necessary to represent I and the function

$\text{Max}(I)$ corresponds to the magnitude of the greatest number in I or zero in case that there are no numbers in I [12]. For instance, if $I = (A)$, where $A = \{a_1, a_2, ..., a_n\}$ is a set of n (decimal) numbers in $\mathbb{Z}^+$, $\text{Length}(I) \approx \sum_{i=1}^{n} \log_{10} a_i$ and $\text{Max}(I) = \max\{a_i : a_i \in A\}$. An algorithm L is said to be *pseudo-polynomial* if there exists a polinomial q such that L can be executed in time bounded above by $q(\text{Length}(I), \text{Max}(I))$. A decision problem Π is called *NP-complete in the strong sense* if $\Pi \in \text{NP}$ and there exists a polynomial p over the integers such that Π_p is NP-complete, where Π_p is the problem Π restricted to instances I that satisfy $\text{Max}(I) \leq p(\text{Length}(I))$ [12]. Informally, a problem is NP-complete in the strong sense if it is NP-complete and there is no pseudo-polynomial algorithm that solves it, unless P = NP.

Algorithms that solve problems having sets of strings as input are generally applied in text processing as well as in other fields like molecular biology. In this context, the DNA, represented computationally by a sequence over the alphabet $\Sigma = \{A, C, G, T\}$, is one of the main subjects of study. The genetic code of a cell is codified in specific regions of its DNA, called *genes*. These regions are expressed producing RNA molecules that are essential for the maintenance and correct behavior of the organism. An RNA molecule transcribed from the DNA is called a *transcript* and the set of all transcripts in a cell at a given moment is called the *transcriptome* of the cell. Commonly in eukaryotes, the genes are interspersed with sequences that have unknown function, called *introns*, while the regions that have a function are called *exons*. During the gene expression the introns are spliced out, while the exons remain in the transcript. Eukaryotic genes are likely to undergo an event called *alternative splicing*, which consists in the splicing of one or more exons together with the introns, resulting in a transcript with a different function. A common way to analyze the transcriptome of a cell is following the steps of an RNA-Seq experiment that consists, basically, in the extraction of the transcripts from the cell followed by a process called reverse transcription, that converts them into DNA sequences. These sequences, known as complementary DNA (cDNA), are thus sequenced by NGS technologies, producing short fragments called *reads* [13]. Given a set of reads R, the *Transcriptome Reconstruction and Quantification Problem* (TRQP) is defined as the problem of determining the set $T = \{t_1, t_2, ..., t_i\}$ of transcripts and, for each $t_j \in T$, determining how many copies of t_j exist [3].

In this work, we model the TRQP as a combinatorial optimization problem called Maximum Similarity Partitioning Problem (MSPP) defined bellow.

Maximum Similarity Partitioning Problem (MSPP): *given two ordered sets S and T of strings over an alphabet Σ such that the strings in T do not overlap each other, a score function $\omega : \bar{\Sigma} \times \bar{\Sigma} \mapsto \mathbb{R}$ and a bound K, is there a partition $P = \{S_1, S_2, ..., S_k\}$ of S in k parts, where each part corresponds to a chain of S, and k ordered subsets $T_1, T_2, ..., T_k$ of T such that $\sum_{i=1}^{k} sim_\omega(S_i^\bullet, T_i^\bullet) \geq K$?*

Informally, the MSPP consists in finding a partition of S in k chains (of S) and k subsets of T such that each chain S_i of S is very similar to the ordered subset T_i of T.

The relation between the MSPP and the TRQP is established considering the set S as the set of reads R and the set T as a set of exons of a gene g that is able to be expressed. A solution of the MSPP consists of k pairs (S_i, T_i), where each T_i corresponds to a transcript of g. The number of distinct pairs that contain the same T_i represents the quantification of such transcript.

3   Proving that the MSPP is NP-complete in the Strong Sense

In this section we prove that the MSPP is strongly NP-complete by a reduction of the 3-Partition Problem which is proven to be NP-complete in the strong sense [12]. This problem is defined as follows:

3-Partition Problem: *given a set $A = \{a_1, a_2, ..., a_{3m}\}$ of $3m$ positive integers and a bound $B \in \mathbb{Z}^+$ such that $B/4 < a_i < B/2$ and $\sum_{i=1}^{3m} a_i = mB$, is there a partition of A in m parts A_1, A_2, ..., A_m such that $\sum_{a_j \in A_i} a_j = B$?*

The reduction in this paper follows the conditions of the pseudo-polynomial transformation presented by Garey and Johnson in [12]. Given an instance I of the 3-Partition Problem, we define an instance $f(I)$ of the MSPP satisfying the following conditions:

1. I is a "yes" instance if and only if $f(I)$ is a "yes" instance;
2. f is determined in pseudo-polynomial time;
3. there is a polynomial p such that $p(\text{Length}(f(I))) \geq \text{Length}(I)$;
4. there is a polynomial q such that $\text{Max}(f(I)) \leq q(\text{Max}(I), \text{Length}(I))$.

Proof. From an instance $I = (A, B)$ of the 3-Partition Problem we define an instance $f(I) = (S, T, \omega, K)$ of the MSPP where $S = \{a'_1, a'_2, ..., a'_{3m}\}$ and $T = \{B'\}$ are sets of strings over the alphabet $\Sigma = \{1\}$ such that $|a'_i| = a_i$ for $a_i \in A$ and $|B'| = B$. We also set $K = 0$ and $\omega(\alpha, \beta) = 0$ if $\alpha = \beta$ or -1 if $\alpha \neq \beta$. Observe that once there is no string s from which the substrings in S and T are derived from, we consider S and T trivially ordered sets of non-overlapping substrings. This means that any subset of S is a chain of S and any subset of T is a chain of T. Suppose that I is a "yes" instance of the 3-Partition Problem. Then there exists a partition of A into m parts $A_1, A_2, ..., A_m$ such that $\sum_{a_j \in A_i} a_j = B$. As a consequence, there will be a partition of S into m parts $S_1, S_2, ..., S_m$ such that $\sum_{a'_j \in S_i} |a'_j| = |B'|$. Thus it is true that $|S_i^\bullet| = |B'|$ and an optimal alignment of $S_i^\bullet$ and B' consists of $|B'| = |S_i^\bullet|$ pairs ('1', '1') whose score is $sim_\omega(S_i^\bullet, B') = |B'| \cdot \omega('1', '1') = 0$. Consequently, $\sum_{i=1}^k sim_\omega(S_i^\bullet, B') \geq K$, which means that $f(I)$ is also a "yes" instance of the MSPP. On the other hand, if $f(I)$ is a "yes" instance of the MSPP then there exists a partition of S into k parts $S_1, S_2, ..., S_k$ and k subsets $T_1, T_2, ..., T_k$ of T such that $\sum_{i=1}^k sim_\omega(S_i^\bullet, T_i^\bullet) \geq K$. It is possible to obtain the set $A = \{A_1 \cup A_2 \cup ... \cup A_m\}$ and the bound B of the 3-Partition Problem counting the characters '1' in $S = \{S_1 \cup S_2 \cup ... \cup S_k\}$ and setting $B = |T^\bullet|$. Considering that $\sum_{a'_j \in S_i} |a'_j| = |B'|$ and $\sum_{a_j \in A_i} a_j = B$, I is a "yes" instance of the 3-Partition Problem. Since $f(I)$ is a "yes" instance of the MSPP if and only if I is a "yes" instance of the 3-Partition, the first condition of the pseudo-polynomial transformation is verified.

In order to determine the time to compute $f(I)$, one must analyze the time necessary to create each element of an instance of the MSPP. The values of K and ω are defined in constant time and therefore can be ignored. However, the time to define the sets S and T depends on the magnitude of the numbers in A and the bound B. One straightforward way to define the sets S and T is schematized in Algorithm 1. Assuming that a character '1' can be created in constant time and that a string a' of characters '1' can be created in time $O(|a'|)$, the set T can be defined in time $O(B) = O(\mathrm{Max}(I))$ and the set S in time $O(\mathrm{Max}(I) \cdot 3m)$. Therefore, $f(I)$ can be defined in time $O(\mathrm{Max}(I) + \mathrm{Max}(I) \cdot 3m) = O(\mathrm{Max}(I) \cdot m)$.

Algorithm 1. Create sets S and T.

Data: set $A = \{a_1, a_2, ..., a_{3m}\}$ and bound B.
Result: set $S = \{a'_1, a'_2, ..., a'_{3m}\}$ and $T = \{B'\}$.
1 create a string B' of B characters '1';
2 **for** $i \leftarrow 1$ **to** $3m$ **do**
3 $\quad \lfloor$ create a string a'_i of a_i characters '1';

Assuming that it is necessary at least one symbol to represent a number, we can say that $|A| \leq \mathrm{Length}(I)$ and $m \leq \mathrm{Length}(I)$. From this, the time to define $f(I)$ can be rewritten as $O(\mathrm{Max}(I) \cdot \mathrm{Length}(I))$ which is pseudo-polynomial, satisfying the second condition of the pseudo-polynomial transformation.

It is easy to see that $\mathrm{Length}(f(I)) \geq \mathrm{Length}(I)$ since strings of '1' are created for each number in A, increasing the number of symbols necessary to represent $f(I)$. Therefore, the third condition of the pseudo-polynomial transformation is also verified.

The instance $f(I)$ of the MSPP is formed by the strings in S and T, the score function ω which can be 0 or -1 and the number $K = 0$. As one can see, the magnitude of the greatest number in $f(I)$ is $|-1| = 1$, which satisfies the fourth and last condition of the pseudo-polynomial transformation.

With all the conditions of the pseudo-polynomial transformation verified, the MSPP is proven to be NP-complete in the strong sense.   □

4   A Greedy Heuristic for the MSPP

Although strongly NP-complete problems cannot be solved by pseudo-polynomial (and consequently polynomial) time algorithms unless P = NP, they can be addressed with the design of heuristics and approximation algorithms. Here we propose a greedy heuristic for the MSPP based on the Segment Alignment Problem (PASG) which is defined as follows [10].

Segment Alignment Problem (PASG): *given two sequences s and t over an alphabet Σ, two ordered sets of substrings $B = \{b_1, b_2, ..., b_u\}$ and $C = \{c_1, c_2, ..., c_v\}$ from s and t, respectively, and a score function $\omega : \bar{\Sigma} \times \bar{\Sigma} \to \mathbb{R}$, determine a chain $\Gamma_B = \{b_p, b_q, ..., b_r\}$ of B and a chain $\Gamma_C = \{c_w, c_x, ..., c_y\}$ of C such that $sim_\omega(\Gamma_B^\bullet, \Gamma_C^\bullet)$ is maximum.*

Basically, the PASG consists in finding a chain of B and a chain of C such that the similarity of the sequences resulting from the concatenation of B and C is maximum. In [10], de Lima showed that the PASG can be solved in polynomial time by means of a dynamic programming algorithm that runs in time $O(|B|^2 \cdot |C|^2)$.

The main idea of the heuristic proposed here for the MSPP consists in using the solution of the PASG proposed by de Lima in [10] to find a non-empty chain Γ_S of S and a non-empty chain Γ_T of T such that $sim_\omega(\Gamma_S^\bullet, \Gamma_T^\bullet)$ is maximum. Setting Γ_S as S_i and Γ_T as T_i, the elements in Γ_S will thus be removed from S and the process is repeated until S becomes empty. More precisely, at each iteration i, $1 \leq i \leq k$, the algorithm that solves the PASG takes as input the sets $S' = \{S\} \setminus \{S_{i-1} \cup S_{i-2} \cup ... \cup S_1\}$ and T, and gives as output the part S_i of S and the subset T_i of T. At the end of the k-th iteration, the partition $P = \{S_1, S_2, ..., S_k\}$ of S and the subsets $T_1, T_2, ..., T_k$ of T will be defined. Such approach is illustrated in Algorithm 2, where Sol_PASG() corresponds to the algorithm that solves the PASG.

Algorithm 2. Heuristic MSPP.

Data: sets S and T.
Result: partition P of S and k subsets of T.
1 $S' \leftarrow S$;
2 $i \leftarrow 1$;
3 **while** $S' \neq \emptyset$ **do**
4 $\quad (S_i, T_i) \leftarrow$ Sol_PASG(S', T);
5 $\quad P \leftarrow P \cup S_i$;
6 $\quad S' \leftarrow S' \setminus S_i$;
7 $\quad i \leftarrow i + 1$;
8 **return** $(S_1, T_1), (S_2, T_2), ..., (S_k, T_k)$;

Undoubtedly, the complexity of the line 4 will dominate the running time of the Algorithm 2. Since the PASG can be solved in time $O(|B|^2 \cdot |C|^2)$, the line 4 is executed in time $O(|S|^2 \cdot |T|^2)$. It will be repeated until S is empty, so at most $|S|$ times. Therefore, the heuristic can be executed in time $O(|S| \cdot |S|^2 \cdot |T|^2) = O(|S|^3 \cdot |T|^2)$.

5   Experimental Results

In order to assess the performance of the greedy heuristic presented here, we simulated an RNA-Seq experiment from 14 *Homo sapiens* genes using the FLUX simulator [14]. Regarding its options, we disabled the PCR amplification, since it can change the real quantification of the transcripts. Besides that, we set the read length option to be big enough in order to allow the sequencing of the whole fragments in the library. Otherwise, pieces of fragments may not be sequenced, resulting in loss of information for the assembly of the transcripts. From the genes, 42 isoforms (average of 3 per gene) were created corresponding to alternative splicing events of exon skipping. This means that each isoform is a distinct combination of exons from the gene. For each isoform we simulated a number of copies between 14 and 295, which corresponds to its real quantification.

Because that there are no standard metrics to assess transcriptome assembly [15,16], we used the four metrics suggested in [16], namely: accuracy, completeness, contiguity and variant resolution.These metrics are calculated with the following equations, where N is the set of reference transcripts.

$$\text{Accuracy} = 100 \cdot \frac{\sum\limits_{i=1}^{M} A_i}{\sum\limits_{i=1}^{M} L_i}, \tag{1}$$

where L_i is the length of an alignment between a reconstructed transcript T_i and a reference transcript R_i, and A_i is the number of correct bases in T_i.

$$\text{Completeness} = 100 \cdot \frac{\sum\limits_{i=1}^{N} l(C_i \geq \delta)}{N}, \tag{2}$$

where l can be zero or one, depending on whether C_i, which is the percentage of a reconstructed transcript T_i covered by the reference transcripts, is greater than a threshold δ.

$$\text{Contiguity} = 100 \cdot \frac{\sum\limits_{i=1}^{N} l(C_i \geq \delta)}{N}, \tag{3}$$

where l can be zero or one, depending on whether C_i, which is the percentage of a reconstructed transcript T_i covered by a single reference transcript, is greater than a threshold δ.

$$\text{Variant resolution} = 100 \cdot \frac{\sum\limits_{i=1}^{N} \frac{max((C_i - E_i), 0)}{V_i}}{N}, \tag{4}$$

where C_i and E_i are the number of correctly and incorrectly transcripts reconstructed of a gene i, respectively, and V_i is the number of isoforms of i.

These metrics correspond, respectively, to the percentage of correctly assembled bases, coverage of the reference transcripts by the assembled transcripts, coverage of the reference transcripts by a single assembled transcript, the percentage of transcript variants assembled and transcripts that have parts of two or more genes. For the experimental tests presented here, we set the threshold δ in (2) and (3) equals to 80%.

For each gene, the heuristic was run using the PASG solution implemented by [10], taking as input the exons coordinates of the gene and the mapping coordinates of the simulated reads to the respective gene, given by FLUX. The spliced reads (i.e. the reads that span multiple exons) were split into individual ones, that is, if a read maps to two different exons, then it is split into two individual reads, where each maps to one exon.

The results showed that the accuracy, completeness and contiguity are 100%. This can be explained by the fact that the simulated reads are from the reference genome itself and that the coordinates of the exons are known. On the other hand, the variant resolution is only 6%. This low value is due to the large number of isofoms incorrectly assembled (Table 1). From the 42 isoforms in the reference set, the heuristic reconstructed 197, from which 28 are true positives.

Comparing with the reference set, the quantification of the 28 isoforms correctly reconstructed has an average error rate of 26.6%, meaning that, on average, the quantification of each isoform is 26.6% above or below the real number. From the 28 isoforms correctly reconstructed, the error rate of 15 are below 10%, indicating that its quantification is very close to the expected value.

Table 1. Reconstruction and quantification results using the greedy heuristic for the MSPP. The second column represents the number of simulated isoforms of the respective gene. The third and fourth columns corresponds to the number of correctly and incorrectly isoforms reconstructed. The fifth column shows the average error rate of the quantification of the isoforms correctly reconstructed.

Gene	isoforms	correctly reconstructed	incorrectly reconstructed	quantification error rate
chmp2a	3	2	10	85.3%
ssp2	4	3	13	42.0%
ooep	3	3	0	4.5%
pes1	3	1	26	69.7%
fteb	4	3	12	19.7%
znf584	2	2	8	31.3%
slc22a4	4	2	23	25.2%
tec	4	2	17	49.4%
znf622	2	2	5	3.1%
tbc1d10a	4	1	17	0%
rps5	3	3	5	9.2%
samd9l	1	1	0	0%
tnni2	5	1	32	53.4%
zbtb45	2	2	1	1.4%

6   Conclusion

We presented a new combinatorial optimization problem called MSPP and showed how it models the TRQP. In addition, we proved that the MSPP is NP-complete in the strong sense and proposed a greedy heuristic for it, which was implemented and tested.

Experimental results showed that the greedy heuristic for the MSPP does not solve the TRPQ effectively, mainly because of the large number of transcripts incorrectly reconstructed. However, the quantification of the transcripts correctly reconstructed was close to its real number of copies.

As future work, the greedy heuristic for the MSPP could be improved in order to make use of spliced reads information, and compare its results with state-of-the-art tools. Additionally, new heuristics for the MSPP could be developed and tested, as well as approximation algorithms, if any.

References

1. Li, J.J., Jiang, C.R., Brown, J.B., Huang, H., Bicke, P.J.: Sparse linear modeling of next-generation mRNA sequencing (RNA-Seq) data for isoform discovery and abundance estimation. Proceedings of the National Academy of Sciences **108**(50), 19 867–19 872 (2011)
2. Schulz, M.H., Zerbino, D.R., Vingron, M., Birney, E.: Oases: robust de novo RNA-Seq assembly across the dynamic range of expression levels. Bioinformatics **28**(8), 1086–1092 (2012)
3. Trapnell, C., Williams, B.A., Pertea, G., Mortazavi, A., Kwan, G., van Baren, M.J., Salzberg, S.L., Wold, B.J., Pachter, L.: Transcript assembly and quantification by RNA-Seq reveals unannotated transcripts and isoform switching during cell differentiation. Nature biotechnology **28**(5), 511–515 (2010)
4. Guttman, M., Garber, M., Levin, J.Z., Donaghey, J., Robinson, J., Adiconis, X., Fan, L., Koziol, M.J., Gnirke, A., Nusbaum, C., et al.: Ab initio reconstruction of cell type-specific transcriptomes in mouse reveals the conserved multi-exonic structure of lincRNAs. Nature biotechnology **28**(5), 503–510 (2010)
5. Grabherr, M.G., Haas, B.J., Yassour, M., Levin, J.Z., Thompson, D.A., Amit, I., Adiconis, X., Fan, L., Raychowdhury, R., Zeng, Q., et al.: Full-length transcriptome assembly from RNA-Seq data without a reference genome. Nature biotechnology **29**(7), 644–652 (2011)
6. Li, W., Feng, J., Jiang, T.: IsoLasso: a LASSO regression approach to RNA-Seq based transcriptome assembly. Journal of Computational Biology **18**(11), 1693–1707 (2011)
7. Mezlini, A.M., Smith, E.J., Fiume, M., Buske, O., Savich, G.L., Shah, S., Aparicio, S., Chiang, D.Y., Goldenberg, A., Brudno, M.: iReckon: Simultaneous isoform discovery and abundance estimation from RNA-seq data. Genome research **23**(3), 519–529 (2013)
8. Behr, J., Kahles, A., Zhong, Y., Sreedharan, V.T., Drewe, P., R atsch, G.: MITIE: Simultaneous RNA-Seq-based transcript identification and quantification in multiple samples. Bioinformatics **29**(20), 2529–2538 (2013)
9. Martin, J., Bruno, V.M., Fang, Z., Meng, X., Blow, M., Zhang, T., Sherlock, G., Snyder, M., Wang, Z.: Rnnotator: an automated de novo transcriptome assembly pipeline from stranded RNA-Seq reads. BMC genomics **11**(1), 663 (2010)

10. de Lima, L.I.S.: O problema do alinhamento de segmentos: Master's thesis, Universidade Federal de Mato Grosso do Sul, October (2013) (in portuguese)
11. Pevzner, P.: Computational molecular biology: an algorithmic approach. MIT press (2000)
12. Garey, M., Johnson, D.: Computers and Intractability: A Guide to the Theory of NP-Completeness. Series of books in the mathematical sciences. W.H. Freeman (1979)
13. Wang, Z., Gerstein, M., Snyder, M.: RNA-Seq: a revolutionary tool for transcriptomics. Nature Reviews Genetics 10(1), 57–63 (2009)
14. Griebel, T., Zacher, B., Ribeca, P., Raineri, E., Lacroix, V., Guigó, R., Sammeth, M.: Modelling and simulating generic RNA-Seq experiments with the flux simulator. Nucleic acids research 40(20), 10073–10083 (2012)
15. Lu, B., Zeng, Z., Shi, T.: Comparative study of de novo assembly and genome-guided assembly strategies for transcriptome reconstruction based on RNA-Seq. Science China Life Sciences 56(2), 143–155 (2013)
16. Martin, J.A., Wang, Z.: Next-generation transcriptome assembly. Nature Reviews Genetics 12(10), 671–682 (2011)

Largest Empty Square Queries
in Rectilinear Polygons

Michael Gester$^{(\boxtimes)}$, Nicolai Hähnle, and Jan Schneider

Research Institute for Discrete Mathematics, University of Bonn,
Lennéstr. 2, 53113 Bonn, Germany
{gester,haehnle,schneid}@or.uni-bonn.de

Abstract. Given a rectilinear polygon P and a point $p \in P$, what is a largest axis-parallel square in P that contains p? This question arises in VLSI design from physical limitations of manufacturing processes. Related problems with disks instead of squares and point sets instead of polygons have been studied previously.

We present an efficient algorithm to preprocess P in time $O(n)$ for simple polygons or $O(n \log n)$ if holes are allowed. The resulting data structure of size $O(n)$ can be used to answer largest square queries for any point in P in time $O(\log n)$. Given a set of points Q instead of a rectilinear polygon, the same algorithm can be used to find a largest square containing a given query point but not containing any point in Q in its interior.

Keywords: Voronoi diagram · Computational geometry · Rectilinear polygon · VLSI design

1 Introduction

As the physical features in very large scale integrated (VLSI) circuit designs become ever smaller, it becomes increasingly challenging to manufacture these features correctly. Foundries establish a growing number of *design rules* which must be satisfied by circuit designs to reduce the expected number of manufacturing defects as well as electrical problems such as coupling [12]. Computer aided design and design automation tools for VLSI must be able to verify that those design rules are satisfied and must take them into account when generating physical designs. Design rules can be quite complex. For example, the minimum distance between shapes may depend on the so-called *width* of a point p in a shape P:

Definition 1. *For a compact set $P \subset \mathbb{R}^2$ and $p \in P$ we define the* width *of p in P, which is denoted by $\mathrm{w}(p, P)$, as the maximum edge length of an axis-parallel square Q with $p \in Q \subseteq P$.*

We develop an efficient method for querying the widths of points even in very complex rectilinear polygons, which are the shapes of interest in VLSI design.

© Springer International Publishing Switzerland 2015
O. Gervasi et al. (Eds.): ICCSA 2015, Part I, LNCS 9155, pp. 267–282, 2015.
DOI: 10.1007/978-3-319-21404-7_20

More formally, we say that an *empty square* is a square whose interior is disjoint from the boundary of P and consider the *largest empty square query problem* in rectilinear polygons: Preprocess a rectilinear polygon P such that for any given point $p \in P$ a largest axis-parallel square Q with $p \in Q \subseteq P$ can be reported efficiently.

Every rectilinear polygon P can be written as a finite union of axis-parallel rectangles $R_1, \ldots, R_k$ such that every maximal square in P is contained in one of the R_j. Computing $w(p, P)$ is then simply a matter of finding the rectangles containing p and selecting one that yields the largest square. However, p may be covered by a number of rectangles that is linear in the number of vertices of P, thereby leading to an unacceptably large query time.

Our contribution is twofold. We consider the decomposition of rectilinear polygons into *width classes*, that is, regions on which $w(\cdot, P)$ is constant, and show that the decomposition has linear description complexity. Furthermore, we present an efficient algorithm for computing this decomposition in time $O(n)$ for simple polygons or $O(n \log n)$ if holes are accounted for, where n is the number of polygon vertices. Given this decomposition, $w(p, P)$ can be queried in time $O(\log n)$ using standard point location techniques.

Related problems like querying a largest square whose interior avoids a set of points or axis-parallel line segments can be solved with the presented techniques as well. In these cases, the sets that must be avoided by the square can be modeled as degenerated holes in a sufficiently large polygon.

2   Related Work

A well-known class of problems in computational geometry closely related to the largest empty square query problem are *largest empty circle* problems. In its original form, as proposed by Shamos and Hoey [25], a finite set of points in the plane is given and one is searching for the largest circle centered within the convex hull of the points and not containing any of the points in its interior. The authors show how to solve this problem in $O(n \log n)$ time using Voronoi diagrams.

The problem of finding a largest empty axis-parallel square can be similarly solved in $O(n \log n)$ time using L_∞ Voronoi diagrams [15,19]. If pairwise non-intersecting but possibly touching line segments are given rather than discrete points, then the same runtime can be achieved by using the L_∞ Voronoi diagram of line segments [21]. This result also applies to not necessarily rectilinear polygons, as their border can be represented as a collection of line segments.

Kaplan and Sharir [17] consider a problem similar to ours where $w(p, P)$ is defined in terms of the radius of the largest disk that contains p but whose interior avoids P, where P is a set of n points in the plane. They prove that $w(p, P)$ can be queried in time $O(\log n)$ after a preprocessing step that takes $O(n \log^2 n)$ time and $O(n \log n)$ space.

A related problem, where P is a simple polygon and $w(p, P)$ is the radius of the largest disk inside P containing p, is studied by Augustine et al. [4].

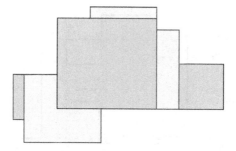

Fig. 1. A partition of a rectilinear polygon into its width classes

They provide a solution where $w(p, P)$ can be queried in time $O(\log n)$ using $O(n \log^2 n)$ preprocessing time and $O(n \log n)$ space.

Kaplan et al. [16] provide a method that solves the problem of finding a rectangle of maximum area which avoids a given point set in $O(\log n)$ query time, given a preprocessing time of $O(n\alpha(n) \log^4 n)$, where $\alpha(n)$ is the inverse Ackermann function, $O(n\alpha(n) \log^3 n)$ space, and a query time of $O(\log^4 n)$.

Boissonnat et al. [8] show that it is possible to preprocess a convex polygon P in $O(n)$ time and space, such that given as a query a set of k points, the largest disk inside of P enclosing all k points may be computed in $O(k \log n)$ time and $O(n + k)$ space. Dumitrescu and Juang [11] study the number of maximal empty axis-parallel rectangles (or boxes, in higher dimension) among a randomly chosen point set.

3   Preliminaries

3.1   Basic Definitions

In the following, we consider a rectilinear polygon P which may contain rectilinear holes. The polygon's interior is denoted by P° and its border by $\partial P := P \backslash P^\circ$. The border can be seen as the union of vertical and horizontal line segments connecting the vertices of the polygon.

The idea of the preprocessing phase is to represent P as the interior-disjoint union of subsets on which w is constant. An example for such a partition is shown in Fig. 1. We call these subsets width classes.

Definition 2. *For a rectilinear polygon $P \subseteq \mathbb{R}^2$ and a number $s \in \mathbb{R}$ we define the width class of size s as the set $C_s := \{p \in P \mid w(p, P) = s\}$. A decomposition of P into interior-disjoint rectangles with the property that $w(\cdot, P)$ is constant within the interior of each rectangle is called a width decomposition of P with respect to w.*

The following chapters provide an algorithm that efficiently computes these sets and a proof stating that their complexity is sufficiently small such that

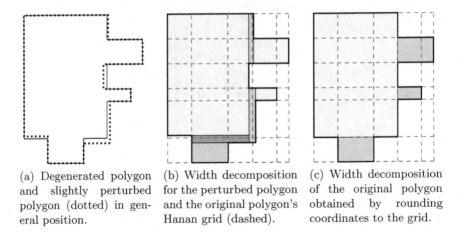

(a) Degenerated polygon and slightly perturbed polygon (dotted) in general position.

(b) Width decomposition for the perturbed polygon and the original polygon's Hanan grid (dashed).

(c) Width decomposition of the original polygon obtained by rounding coordinates to the grid.

Fig. 2. How to transform an arbitrary polygon into a polygon in general position.

common point location methods can be used to obtain a logarithmic query time. The queries will not just return the number $w(p, P)$, but also a square with that edge length containing p. Such squares are called width representatives.

Definition 3. *For a rectilinear polygon $P \subseteq \mathbb{R}^2$ and a point $p \in P$, an axis-parallel square Q is called* width representative *for p if its edge length is $w(p, P)$ and $p \in Q \subset P$.*

Our main result can be stated as follows.

Theorem 1. *For a rectilinear polygon P with n vertices, the largest empty square query problem can be solved with preprocessing time $O(n \log n)$, space $O(n)$, and query time $O(\log n)$. If P is simple, then a preprocessing time of $O(n)$ suffices.*

3.2 General Position

To simplify our proofs, we assume that P is in general position in the following sense. We assume that no two horizontal segments of ∂P have the same y-coordinate and no two vertical segments of ∂P have the same x-coordinate. Let $c_1, c_2, \ldots, c_n$ be the x-coordinates of vertical segments and the y-coordinates of horizontal segments of ∂P (ordered arbitrarily). We assume that for any four distinct indices $i, j, k, l \in \{1, 2, \ldots, n\}$ we have $c_i - c_j \neq c_k - c_l$. In other words, no difference between two coordinates occurs more than once.

While many polygons in practice do not have these properties, they can be enforced by slightly perturbing the input data. After applying our algorithm to the perturbed polygon, we can round the borders of the width classes to the next x- and y-coordinates occurring in vertices from the original polygon. This works because the shapes of width classes vary continuously with changes of the coordinates in P. This procedure is illustrated in Fig. 2. For more details on standard perturbation techniques for geometric algorithms see [24]. Instead of perturbing the data, the algorithm can also be adapted such that the general position assumption is not necessary.

4 Relation between Width Classes and Voronoi Diagrams

4.1 Properties of the L_∞ Voronoi Diagram

Our solution makes use of the L_∞ *Voronoi diagram* of P.

Definition 4. *The L_∞ bisector of two line segments s_1, s_2 is the set $\{x \in \mathbb{R}^2 \mid d_\infty(s_1, x) = d_\infty(s_2, x)\}$. The L_∞ Voronoi diagram of P is the set of all points $p \in P$ for which there exist at least two different border segments s_1, s_2 of P with*

$$d_\infty(p, s_1) = d_\infty(p, s_2) = \min_{q \in \partial P} d_\infty(p, q). \tag{1}$$

The L_∞ Voronoi diagram can be computed in time $O(n \log n)$ for arbitrary polygons [21] and in time $O(n)$ for simple polygons [9].

From now on, we omit the term L_∞ when there is no risk for confusion. It is easy to see that the Voronoi diagram is a subset of the union of all bisectors between line segments of the polygon. Therefore, bisectors define the structure of Voronoi diagrams and can be viewed as their basic modules.

Bisectors of line segments may contain two-dimensional parts, see Fig. 3a. To avoid this, we use lines with slope ± 1 as bisectors for touching segments instead of original L_∞ bisectors (see again Fig. 3a). We call them *refined bisectors*. We are only interested in the parts of the bisectors within P. By our general position assumption, no border segment of an inscribed square in P can have proper intersection with two non-touching segments of ∂P. This implies that two-dimensional bisector parts between non-touching segments do not appear within P. For more details see [5, Chapter 7] and [21].

Using the refined bisectors to define the Voronoi diagram (more exactly, intersecting the Voronoi diagram with the union of all refined bisectors between any two border segments) results in a *refined Voronoi diagram* which we denote as $\mathcal{V}_\infty(P)$. In the following, we only consider this Voronoi diagram which consists of line segments only. We call points on $\mathcal{V}_\infty(P)$ which lie on ∂P or have at least three incident segments in $\mathcal{V}_\infty(P)$ *Voronoi vertices* and the parts between these vertices *Voronoi edges*.

We associate with every point p lying on $\mathcal{V}_\infty(P)$ the unique square $Q(p)$ centered at p and touching the nearest segments of ∂P. The following simple but important lemma is the key relation between the partition of a polygon into its width classes and the Voronoi diagram which we use for our algorithm.

Lemma 1. *All width representatives for points in P are of the form $Q(q)$ for some q lying on $\mathcal{V}_\infty(P)$.*

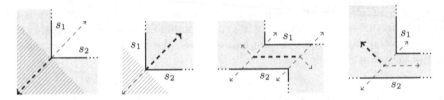

Fig. 3. Each figure shows a section of a polygon (gray) and refined bisectors of its segments (dashed lines). Dashed regions are two-dimensional parts belonging to the original, non-refined bisectors. Parts contributing to the Voronoi edge of s_1 and s_2 are thicker.

Proof. Each width representative Q for some point p touches ∂P in at least two non-touching segments, otherwise Q could be enlarged within P while still containing p, a contradiction to the definition of width representatives. Therefore, the center of Q lies on $\mathcal{V}_\infty(P)$. □

Before we show how a width decomposition can be obtained efficiently by using this fact we need to prove some structural properties of $\mathcal{V}_\infty(P)$.

Lemma 2. *All Voronoi edges of $\mathcal{V}_\infty(P)$ are single horizontal, vertical or diagonal segments.*

Proof. First note that for each Voronoi edge e there exist two segments s_1, s_2 of ∂P such that each point on e has the same d_∞-distance to s_1 and s_2, and there is no segment of ∂P with smaller distance (by definition of the Voronoi diagram, note that this property still holds with refined bisectors).

If s_1 and s_2 are touching, then the refined bisector of the segments is a diagonal line, see Fig. 3a and Fig. 3b. Thus the Voronoi edge, being a connected subset of the bisector, is a diagonal segment.

If s_1 and s_2 are non-touching and parallel (say both horizontal, w.l.o.g.), then by the general position assumptions the segments have different y-coordinates. The part of the bisector inside of P consists of at most three pieces, and both endpoints of the horizontal piece intersect bisectors induced by vertices incident to s_1 and s_2 (see Fig. 3c). The Voronoi edge of s_1 and s_2 clearly cannot cross these bisectors and thus consists of the horizontal segment only.

If s_1 is horizontal and s_2 is vertical (or vice versa), then the part of the bisector inside of P again consists of at most three pieces (see Fig. 3d). Here only one diagonal part of the bisector contributes to the Voronoi edge, because the Voronoi edge cannot cross the bisector induced by the polygon vertex which causes the break in the bisector of s_1 and s_2. □

Lemma 2 implies that $\mathcal{V}_\infty(P)$ can be interpreted as a planar straight-line graph $G = (V, E_{\mathrm{orth}} \,\dot\cup\, E_{\mathrm{diag}})$, where E_{orth} is the set of horizontal and vertical edges and E_{diag} is the set of diagonal edges. In the following we identify vertices and edges with their embeddings in the plane. We now collect some statements about the structure of G for later use.

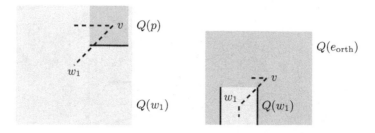

Fig. 4. Contradictions in the proof of Lemma 6 (G dashed)

The next two lemmas follow easily from the definition of $\mathcal{V}_\infty(P)$ and our general position assumption, see also the proof of Lemma 2 and Fig. 3.

Lemma 3. *For any edge $e = \{v, w\} \in E_{orth}$ and any point q on e, ∂P intersects both border segments of $Q(q)$ which are parallel to e, at least one of them in its interior. If $q \notin \{v, w\}$, then both are intersected in their interiors.*

Lemma 4. *For any edge $e \in E_{diag}$ and any distinct $v', w' \in e$ we have either $Q(v') \subsetneq Q(w')$ or $Q(w') \subsetneq Q(v')$.*

Lemma 5. *For all $v \in V$ we have $|\delta(v) \cap E_{orth}| = \begin{cases} 0, & \text{if } v \in \partial P \\ 1, & \text{if } v \in P \setminus \partial P \end{cases}$.*

Proof. If $v \in \partial P$, then v is a vertex of P and the only Voronoi edge containing v is diagonal (see Fig. 3a and 3b), proving $|\delta(v) \cap E_{orth}| = 0$.

If v is a Voronoi vertex in $P \setminus \partial P$, then by the general position assumptions there is exactly one border segment of $Q(v)$ whose interior does not intersect ∂P, thus we can move $Q(v)$ in direction of this segment while touching two parallel border segments of ∂P, implying the existence of an incident horizontal or vertical Voronoi edge. By the general position assumptions we cannot have two such incident edges, so $|\delta(v) \cap E_{orth}| = 1$.   □

Lemma 6. *Let $e_{orth}, e_{diag} \in E(G)$ be incident edges forming a 45° angle, where $e_{diag} = \{v, w_1\}$ is the diagonal and $e_{orth} = \{v, w_2\}$ the horizontal or vertical edge. Then $w_1 \in \partial P$.*

Proof. Suppose e_{orth} is horizontal and e_{diag} leaves the right endpoint of e_{orth} to the lower left, all other cases are symmetric. By Lemma 4, we have either $Q(v) \subsetneq Q(w_1)$ or $Q(w_1) \subsetneq Q(v)$.

In the first case (Fig. 4a), there exists a point p in the interior of e_{orth} such that the lower border segment of $Q(p)$ (the black line in Fig. 4a) is contained in the interior of $Q(w_1)$, but by Lemma 3 this segment also intersects ∂P, a contradiction.

For the second case (Fig. 4b), suppose $w_1 \notin \partial P$, then by Lemma 5 w_1 is incident to an edge $e'_{orth} \in E_{orth}$. By Lemma 3, ∂P intersects both segments

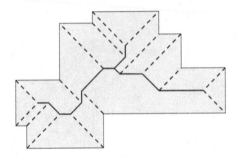

Fig. 5. Rectilinear polygon P, refined Voronoi diagram $\mathcal{V}_\infty(P)$, and Voronoi core G_{VC} (without dashed lines)

of $Q(w_1)$ parallel to e'_{orth} (the black lines), at least one in its interior, so ∂P intersects the interior of $\bigcup_{q \in Q} Q(q)$. This is also a contradiction.   □

4.2   Voronoi Core

We now define the *Voronoi core* G_{VC} as the embedded planar graph that is obtained from G by deleting all vertices lying on ∂P (the leaves of G) and their incident edges (Fig. 5). This graph is a subset of the *medial axis* [7] which, for rectilinear polygons, coincides with the *straight skeleton* [2]. The medial axis has applications in pattern recognition [10,23], solid modeling [27], and mesh generation [14]. The straight skeleton is used in computer graphics [26], graph drawing [6], and for roof construction [1,2]. The roots of straight skeletons used for roof construction actually date back to the 19th century [22, p.86-122] as pointed out in [3].

Corollary 1. *For each edge $e \in E_{orth}$, all incident edges in G_{VC} are diagonal and form $135°$ angles with e.*

Proof. Incident edges must be diagonal by Lemma 5 and cannot form $45°$ angles with e because G_{VC} does not contain edges ending in ∂P (see Lemma 6), thus they must form $135°$ angles as claimed.   □

By Corollary 1, we can provide E_{orth} with a natural notion of diagonal neighborhood. For an edge $e \in E_{orth}$ we define $n_{\nearrow}(e)$, the *upper right neighbor* of e, as the edge $f \in E_{orth}$ that is reached from the top or right end of e, respectively, when following the diagonal edge in the top right direction. If no such diagonal edge exists, we set $n_{\nearrow}(e) := \varnothing$. In the same sense the *upper left neighbor* $n_{\nwarrow}(e)$, the *lower right neighbor* $n_{\searrow}(e)$, and the *lower left neighbor* $n_{\swarrow}(e)$ are defined.

4.3   Edge Rectangles and Width Classes

We need the following stronger version of Lemma 1, which is a direct implication of Lemma 1 and Lemma 4.

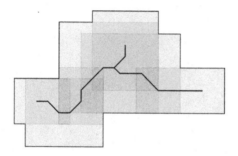

Fig. 6. Rectilinear polygon with Voronoi core and overlapping edge rectangles

Lemma 7. *All width representatives for points in P are of the form $Q(q)$ for some point q lying on an edge of E_{orth}.*

For $e \in E_{\text{orth}}$ the union of all $Q(q)$ for points q on e is a rectangle which we denote as $Q(e)$ and which we call *edge rectangle* (Fig. 6). We define $\text{w}(e)$ as the width (i.e. the smaller edge length) of $Q(e)$ and $C_e := C_{\text{w}(e)}$ (see Definition 2) for $e \in E_{\text{orth}}$. We can now extend Lemma 7 as follows.

Lemma 8. *For all $p \in P$ we have*

$$\text{w}(p, P) = \max\{\text{w}(e) \mid e \in E_{orth} \wedge p \in Q(e)\} \tag{2}$$

and each width class can be written as

$$C_e = Q(e) \setminus \bigcup_{\substack{f \in E_{orth} \\ \text{w}(f) > \text{w}(e)}} Q(f). \tag{3}$$

for some $e \in E_{orth}$.

Proof. The first part follows directly from Lemma 7. For the second part, note that by the general position assumptions the widths of all edge rectangles are pairwise distinct, so for each width class C_s we have exactly one edge e with $\text{w}(e) = s$. Subtracting all edge rectangles with greater width from $Q(e)$ results in $C_s = C_e$. □

The lemma implies that, given the Voronoi core for a polygon, its width classes can be computed as differences of edge rectangles. However, using the formula for C_e from Lemma 8 may result in quadratic total runtime. It turns out that only a *constant* number of edge rectangles need to be subtracted from $Q(e)$. Furthermore, while those may be arbitrarily far away from e in terms of the Voronoi core, we will give an algorithm that allows them to be found efficiently.

For any $e \in E_{\text{orth}}$ we define $P_{\nearrow}(e)$ to be the undirected path in G_{VC} that contains all edges of the sequence $(n_{\nearrow}(e), n_{\nearrow}(n_{\nearrow}(e)), (n_{\nearrow}(n_{\nearrow}(n_{\nearrow}(e)))), \ldots)$, all connecting diagonal edges between them, and the diagonal edge connecting e

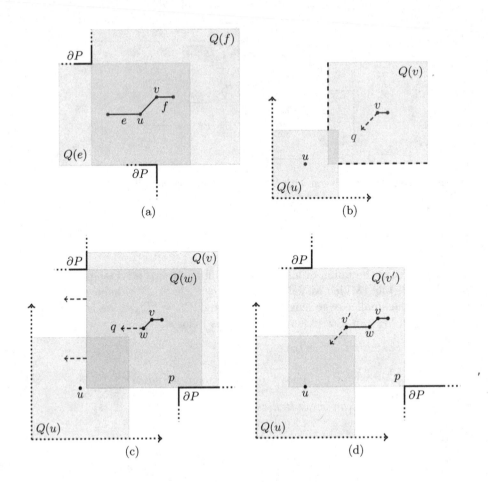

Fig. 7. Notation in the proof of Theorem 2

and $n_{\nearrow}(e)$. $P_{\nwarrow}(e)$, $P_{\searrow}(e)$ and $P_{\swarrow}(e)$ are defined analogously, and we call them *the diagonal paths of e* in the following.

We further define $e_{\nearrow}$ to be the first edge f on $P_{\nearrow}(e)$ with $w(f) > w(e)$, or $e_{\nearrow} := \varnothing$ if no such edge exists. Let $e_{\nwarrow}$, $e_{\searrow}$ and $e_{\swarrow}$ be defined analogously. These edges are called *diagonal edge pointers of e*.

We are now ready to prove the following description of width classes which involves only a constant number of edge rectangles.

Theorem 2. *For each $e \in E_{orth}$ we have*

$$C_e = Q(e) \setminus (Q(e_{\nearrow}) \cup Q(e_{\searrow}) \cup Q(e_{\nwarrow}) \cup Q(e_{\swarrow})). \tag{4}$$

Proof. Let $f \in E_{orth}$ be arbitrary with $w(f) > w(e)$ and $Q(e) \cap Q(f) \neq \varnothing$ maximal, i.e. there is no $g \in E_{orth}$ with $w(g) > w(e)$ and $Q(e) \cap Q(g) \supsetneq Q(e) \cap Q(f)$. If there is no such f, then we clearly have $C_e = Q(e)$, finishing the proof.

We show that $f \in \{e_{\nearrow}, e_{\nwarrow}, e_{\searrow}, e_{\swarrow}\}$, proving the theorem by using Lemma 8. We first claim that $Q(f)$ contains at least one vertex of $Q(e)$. Assume that it doesn't. Then by $w(f) > w(e)$, $Q(f)$ intersects at least one border segment $s \subset \partial Q(e)$ parallel to e, and it does so only in the interior of s. There is a point $q \in e$ such that the square Q of edge length $w(e)$ centered in q satisfies $Q \cap s \subset Q(f)$. Consequently, there is a larger square $Q' \supset Q$ which is entirely contained in $Q(e) \cup Q(f)$, and hence $w(q, P) > w(e)$, which is a contradiction. Hence we can assume w.l.o.g. that $Q(f)$ contains the upper right vertex of $Q(e)$; the other cases are symmetric.

Let v be the left or lower vertex of f and u the right or upper vertex of e. We will slide a point q starting from v along the Voronoi core until reaching u by using only edges in lower or left direction, implying that $f \in E(P_{\nearrow}(e))$.

First consider the case that ∂P does not intersect the right border of $Q(u)$. Then by Lemma 3 applied to e and u, e must be horizontal and for either $\bar{f} := n_{\nearrow}(e)$ or $\bar{f} := n_{\searrow}(e)$ we must have $w(\bar{f}) > w(e)$. Figure 7a illustrates this case. Moreover, $Q(e) \cap Q(f) \subseteq Q(e) \cap Q(\bar{f})$ must hold, otherwise the interior of $Q(f)$ would intersect ∂P, a contradiction. By the maximality assumption for f we must have $Q(e) \cap Q(f) = Q(e) \cap Q(\bar{f})$ which implies, by general position, $f = \bar{f} \in \{e_{\nearrow}, e_{\searrow}\}$ as claimed. Similarly, if ∂P does not intersect the upper border of $Q(u)$, then e must be vertical and $f \in \{e_{\nearrow}, e_{\searrow}\}$ as claimed. So in the following we may assume that ∂P intersects both right and upper border of $Q(u)$. This implies that for each p on G_{VC} with $Q(p)$ containing the upper right vertex of $Q(u)$ the interior of $Q(p)$ cannot intersect one of the dotted lines in Fig. 7b, thus we have $Q(e) \cap Q(p) = Q(u) \cap Q(p)$.

By the maximality of f the interiors of the parts of the left and lower borders of $Q(v)$ which are not contained in $Q(u)$ (dashed in Fig. 7b) intersect a vertical and a horizontal segment of ∂P, respectively, and the bisector of these two segments induces a diagonal edge in G_{VC} from v to the lower left. We move q to the other vertex w of this edge. Note that $Q(u) \cap Q(q)$ remains unaffected during this process. Figure 7c provides an illustration.

By Lemma 5, w is incident to an edge $f' \in E_{orth}$, and by Corollary 1 f' proceeds in lower or left direction. Suppose $f' \neq e$ and f' is horizontal as in Fig. 7c (the vertical case is symmetric), then we can move q along f' until either the left border of $Q(q)$ intersects ∂P, which happens at the latest at the dotted line, or the lower right corner of $Q(q)$ meets the polygon vertex denoted as p in Fig. 7c. In both cases, we arrive at a Voronoi vertex v' incident to a diagonal edge in lower left direction. Note that $Q(e) \cap Q(q) = Q(u) \cap Q(q)$ increases for this movement, thus we must have $w(f') \leq w(e)$ by the maximality of f.

This procedure can be iterated, always following edges in lower left direction. Note that at any time $Q(q)$ contains the upper right corner of $Q(u)$ and the interior of $Q(q)$ cannot intersect the dotted lines in Fig. 7b–7d, thus q is in upper right direction of u until, after a finite number of steps, we must have

$q = u$. Also note that $Q(e) \cap Q(q) = Q(u) \cap Q(q)$ never shrinks, so for all traversed horizontal or vertical edges $g \neq f$ we have $w(g) \leq w(e)$, implying that $f = e_\nearrow$ as claimed.   □

5   The Algorithm

According to Theorem 2, the computation of the width classes can be reduced to finding the diagonal edge pointers for each edge in E_{orth} and subtracting four rectangles from one other rectangle, where the latter part can be trivially done in constant time.

5.1   The All Next Greater Numbers Problem

The determination of all edge pointers on a maximal diagonal path (that means, a diagonal path of some edge which is not a proper subset of a diagonal path of some other edge) in one direction can be abstracted to the following problem.

Problem 1 (All Next Greater Numbers Problem). Let $x_1, x_2, \ldots, x_n$ be a sequence of real numbers. For each index $i \in \{1, 2, \ldots, n\}$ find the smallest index $j > i$ with $x_j > x_i$, if existing.

Here the indices correspond to the edges on the diagonal path, ordered in direction of the pointers to be determined, and the real numbers correspond to the widths of the edge rectangles. This problem can be solved in $O(n)$ time with Algorithm 1.

Input   : A sequence of real numbers $x_1, x_2, \ldots, x_n$.
Output: A function next : $\{1, 2, \ldots, n\} \to \{1, 2, \ldots, n\}$ such that next$[i]$ is the
           smallest index greater than i with $x_{\mathrm{next}[i]} > x_i$, if it exists, and
           next$[i] = -1$ else.

1 prev$[i] \leftarrow -1$, next$[i] \leftarrow -1$ $\forall i \in \{1, 2, \ldots, n\}$
2 **for** $i \leftarrow 2$ **to** n **do**
3 $\qquad j \leftarrow i - 1$
4 $\qquad$ **while** $j \geq 1$ *and* $x_j < x_i$ **do**
5 $\qquad\qquad$ next$[j] \leftarrow i$
6 $\qquad\qquad$ $j \leftarrow$ prev$[j]$
7 $\qquad$ **end**
8 $\qquad$ **if** $j \geq 1$ *and* $x_j \geq x_i$ **then**
9 $\qquad\qquad$ prev$[i] \leftarrow j$
10 $\qquad$ **end**
11 **end**
12 **return** next

Algorithm 1. Solving the All Next Greater Numbers Problem

Theorem 3. *Algorithm 1 works correctly and runs in $O(n)$ time.*

Proof. For the correctness we show that the following conditions hold after executing line 9 of the algorithm:

1. $\forall j \leq i :$ prev$[j]$ is the largest index $m < j$ with $x_m \geq x_j$ (or -1 if no such index exists)
2. $\forall j \leq i :$ next$[j]$ is the smallest index m with $j < m \leq i$ and $x_m > x_j$ (or -1 if no such index exists)

For $i = 2$ there are two possible cases: If $x_1 < x_2$ we enter the while loop and set next$[1] \leftarrow 2$ in line 5, otherwise we set prev$[2] \leftarrow 1$ in line 9, thus for the first iteration ($i = 2$) both conditions are clearly satisfied after line 9.

Now let $i > 2$ and suppose the conditions hold for $i - 1$. We show that they also hold for i. If $x_{i-1} \geq x_i$, then in iteration i we only have to set prev$[i] \leftarrow i-1$ which is done correctly in line 9 of iteration i.

Otherwise, if $x_{i-1} < x_i$, we have to set next$[j] \leftarrow i$ for each index j with $x_j \geq x'_j \, \forall j < j' < i$ and $x_j < x_i$. Because condition 1 holds for $i - 1$, we traverse exactly those indices in the while loop and set next$[j] \leftarrow i$ correctly. After finishing the while loop, j is either -1 or the largest index smaller than i such that $x_j \geq x_i$, in which case we correctly set prev$[i] \leftarrow j$ in line 9. In summary, both conditions are satisfied after line 9 of iteration i. Condition 2 for $i = n$ proves the correctness of the algorithm.

For the runtime observe that $x_j < x_i$ holds when arriving at the body of the while loop for some j. Thus in any later iteration $i' > i$ we have prev$[i'] \neq j$ by condition 1. Therefore, the body of the loop is never reached for the same j again, bounding the total number of iterations of the while loop by n. Consequently, the total runtime of the algorithm is $O(n)$. □

5.2 Computing a Width Decomposition

This completes the preparations for the computation of a width decomposition. Algorithm 2 presents the method we propose.

The first step is to compute the Voronoi core (line 2). This step takes $O(n \log n)$ time for polygons with holes [21] and $O(n)$ time for simple polygons [9]. This is the only part of the algorithm that may require super-linear time. We assume the Voronoi core is stored in an appropriate data structure for planar straight-line graphs such as *doubly connected edge list* [20] or *quad edge data structure* [13]. In fact, any structure will suffice that allows to access the incident edges of a vertex in constant time.

Subsequently, for each edge in E_{orth} all diagonal edge pointers on maximal diagonal paths containing the edge are computed (lines 4–9). Note that we set the pointers only once for each maximal diagonal path, therefore this step takes linear time in total by using Algorithm 1.

Finally, we traverse all edges in E_{orth} again and construct the width decomposition using Theorem 2 (line 12). The algorithm's correctness and runtime follows immediately from this discussion.

Input : A rectilinear polygon P with n vertices.
Output: A width decomposition W of P, containing $O(n)$ rectangles.

1   $W \leftarrow \varnothing$
2   Compute Voronoi core $\mathcal{V}_\infty(P)$
3   **foreach** $e \in E_{orth}$ **do**
4   **if** $e_\nearrow$ *or* $e_\nearrow$ *not set* **then**
5   Set $e'_\nearrow$ and $e'_\nearrow$ for all edges $e' \in E_{orth} \cap (E(P_\nearrow(e)) \cup \{e\} \cup E(P_\nearrow(e)))$
6   **end**
7   **if** $e_\nwarrow$ *or* $e_\searrow$ *not set* **then**
8   Set $e'_\nwarrow$ or $e'_\searrow$ for all edges $e' \in E_{orth} \cap (E(P_\nwarrow(e)) \cup \{e\} \cup E(P_\searrow(e)))$
9   **end**
10 **end**
11 **foreach** $e \in E_{orth}$ **do**
12   Add constant number of interior-disjoint rectangles covering the closure of
      $Q(e) \setminus (Q(e_\nearrow) \cup Q(e_\searrow) \cup Q(e_\nwarrow) \cup Q(e_\nearrow))$ to W
13 **end**
14 **return** W

Algorithm 2. Computing a width decomposition

Theorem 4. *Given a rectilinear polygon P, a width decomposition of P containing $O(n)$ rectangles can be computed in $O(n \log n)$ time. If P is simple, the runtime reduces to $O(n)$.*

A corresponding edge rectangle can easily be stored together with every rectangle in such a data structure. Combining it with standard point location techniques, e.g. [18], yields our main result, Theorem 1.

For practical implementations the complicated linear time algorithm for computing the Voronoi core of a simple polygon may be replaced by a much simpler randomized algorithm running in $O(n \log^* n)$ expected time [9].

6   Conclusion

In this paper we consider the following problem: Given a rectilinear polygon P and a point $p \in P$, what is a largest axis-parallel square in P that contains p? We present an efficient algorithm which computes in $O(n \log n)$ time a data structure of size $O(n)$ such that largest square queries for any point in P can be answered in time $O(\log n)$. For simple polygons the preprocessing time decreases to $O(n)$. Our algorithm is based on the L_∞ Voronoi diagram.

An interesting open question is how generalizations of the query problem can be solved efficiently, for example for arbitrary polygons or more general query objects.

Acknowledgments. The authors would like to thank Dr. Dirk Müller for valuable input and discussions.

References

1. Ahn, H.-K., Bae, S.W., Knauer, C., Lee, M., Shin, C.-S., Vigneron, A.: Realistic roofs over a rectilinear polygon. Computational Geometry: Theory and Applications **46**(9), 1042–1055 (2013)
2. Aichholzer, O., Aurenhammer, F., Alberts, D., Gärtner, B.: A novel type of skeleton for polygons. Journal of Universal Computer Science **1**(12), 752–761 (1995)
3. Aichholzer, O., Cheng, H., Devadoss, S.L., Hackl, T., Huber, S., Li, B., Risteski, A.: What makes a tree a straight skeleton? In: CCCG, pp. 253–258 (2012)
4. Augustine, J., Das, S., Maheshwari, A., Nandy, S.C., Roy, S., Sarvattomananda, S.: Localized geometric query problems. Computational Geometry **46**(3), 340–357 (2013)
5. Aurenhammer, F., Klein, R., Lee, D.-T.: Voronoi Diagrams and Delaunay Triangulations. World Scientific (2013)
6. Bagheri, A., Razzazi, M.: Drawing free trees inside simple polygons using polygon skeleton. Computing and Informatics **23**(3), 239–254 (2012)
7. Blum, H., et al.: A transformation for extracting new descriptors of shape. Models for the perception of speech and visual form **19**(5), 362–380 (1967)
8. Boissonnat, J.-D., Czyzowicz, J., Devillers, O., Yvinec, M.: Circular separability of polygons. Algorithmica **30**(1), 67–82 (2001)
9. Chin, F., Snoeyink, J., Wang, C.A.: Finding the medial axis of a simple polygon in linear time. Discrete & Computational Geometry **21**(3), 405–420 (1999)
10. Duda, R., Hart, P.: Pattern classification and scene analysis (1973)
11. Dumitrescu, A., Jiang, M.: Maximal empty boxes amidst random points. In: Gupta, A., Jansen, K., Rolim, J., Servedio, R. (eds.) APPROX 2012 and RANDOM 2012. LNCS, vol. 7408, pp. 529–540. Springer, Heidelberg (2012)
12. Gester, M., Müller, D., Nieberg, T., Panten, C., Schulte, C., Vygen, J.: BonnRoute: Algorithms and data structures for fast and good VLSI routing. ACM Transactions on Design Automation of Electronic Systems (TODAES) **18**(2), 32:1–32:24 (2013). Preliminary version in the Proceedings of the 49th Annual Design Automation Conference, pp. 459–464
13. Guibas, L., Stolfi, J.: Primitives for the manipulation of general subdivisions and the computation of Voronoi. ACM Transactions on Graphics (TOG) **4**(2), 74–123 (1985)
14. Gürsoy, H.N., Patrikalakis, N.M.: An automatic coarse and fine surface mesh generation scheme based on medial axis transform: Part I algorithms. Engineering with computers **8**(3), 121–137 (1992)
15. Hwang, F.K.: An $O(n \log n)$ algorithm for rectilinear minimal spanning trees. Journal of the ACM **26**(2), 177–182 (1979)
16. Kaplan, H., Mozes, S., Nussbaum, Y., Sharir, M.: Submatrix maximum queries in monge matrices and monge partial matrices, and their applications. In: Proceedings of the Twenty-Third Annual ACM-SIAM Symposium on Discrete Algorithms, SODA 2012, pp. 338–355. SIAM (2012)
17. Kaplan, H., Sharir, M.: Finding the maximal empty disk containing a query point. In: Proceedings of the 2012 Symposium on Computational Geometry, pp. 287–292. ACM (2012)
18. Kirkpatrick, D.: Optimal search in planar subdivisions. SIAM Journal on Computing **12**(1), 28–35 (1983)
19. Lee, D.-T., Wong, C.K.: Voronoi diagrams in $L_1(L_\infty)$ metrics with 2-dimensional storage applications. SIAM Journal on Computing **9**(1), 200–211 (1980)

20. Muller, D.E., Preparata, F.P.: Finding the intersection of two convex polyhedra. Theoretical Computer Science **7**(2), 217–236 (1978)
21. Papadopoulou, E., Lee, D.-T.: The L_∞ Voronoi diagram of segments and VLSI applications. International Journal of Computational Geometry & Applications **11**(05), 503–528 (2001)
22. Peschka, G.: Kotirte Ebenen und deren Anwendung. Verlag Buschak & Irrgang, Brünn (1877)
23. Rosenfeld, A.: Axial representations of shape. Computer Vision, Graphics, and Image Processing **33**(2), 156–173 (1986)
24. Seidel, R.: The nature and meaning of perturbations in geometric computing. Discrete & Computational Geometry **19**(1), 1–17 (1998)
25. Shamos, M.I., Hoey, D.: Closest-point problems. In: 16th Annual Symposium on Foundations of Computer Science, pp. 151–162 (1975)
26. Tanase, M., Veltkamp, R.C.: Polygon decomposition based on the straight line skeleton. In: Proceedings of the Nineteenth Annual Symposium on Computational Geometry, pp. 58–67. ACM (2003)
27. Vermeer, P.J.: Two-dimensional MAT to boundary conversion. In: Proceedings on the Second ACM Symposium on Solid Modeling and Applications, pp. 493–494. ACM (1993)

Variable Size Block Matching Trajectories for Human Action Recognition

Fábio L.M. de Oliveira[⊠] and Marcelo B. Vieira

Universidade Federal de Juiz de Fora, Juiz de Fora, MG, Brazil
{fabio,marcelo.bernardes}@ice.ufjf.br

Abstract. In the context of the human action recognition problem, we propose a tensor descriptor based on sparse trajectories extracted via Variable Size Block Matching. Compared to other action recognition descriptors, our method runs fast and yields a compact descriptor, due to its simplicity and the coarse representation of movement provided by block matching. We validate our method using the KTH dataset, showing improvements over a previous block matching based descriptor. The recognition rates are comparable to those of state-of-the-art methods with the additional feature of having frame rates close to real-time computation.

Keywords: Human action recognition · Variable size block matching · Self-descriptor · Tensor descriptor

1 Introduction

Human Action Recognition (HAR) has been a rapid growing research field over recent years. Along with other computer vision themes, it has been boosted especially by the massive popularization of video recording devices, such as phone cameras and surveillance equipment. These devices generate an amount of video data too big to have its semantic value or significance examined by people. Hence the need of an automated system capable of extracting this information from videos.

Many, if not all, of the methods for this purpose rely on detecting and tracking movement in sequences of frames. That is the case with optical flow, 3-D gradients, spatio-temporal interest points, block matching, amongst others. The assumption is that movement can be detected, described, and used to categorize different actions portrayed in a video.

In the case of block matching, which is the focus of this paper, this motion information has been largely used for video compression since the introduction of the method by Jain and Jain [9]. It consists in dividing a frame into rectangular blocks and searching for corresponding blocks on the following frame. These corresponding blocks are the ones that are most similar regarding brightness or color. This results in a set of displacement vectors, one for each block. The original authors [9] described it as a "piecewise translation".

© Springer International Publishing Switzerland 2015
O. Gervasi et al. (Eds.): ICCSA 2015, Part I, LNCS 9155, pp. 283–297, 2015.
DOI: 10.1007/978-3-319-21404-7_21

In this work, we propose a video tensor descriptor based on trajectories obtained via multiple variable size block matchings [26]. This work is an improvement over our latest works [6]. Compared to other well known action recognition descriptors, like ones based on dense trajectories or 3-D gradients, our method is fast and yields a compact descriptor, due to its coarser representation of the movement. In fact, the method runs fast enough to be potentially incorporated into a real-time application and the final descriptor file size for the whole dataset rarely exceed that of a single video.

The following subsections show related works and an overview of the method. In Sec. 2 we go into details for each stage of the method. Section 3 shows the experimental environment and Sec. 4 the results and discussion.

1.1   Related Work

Since block matching was introduced by Jain and Jain [9] and the later introduction of its variable block size variant [3,26], it has appeared in a number of publications concerning video coding [2,3,5,13,24,29] and was part of specifications of codecs like H.263 [15] and H.264 [4,23], amongst others. In this work, we employ block matching as a motion flow under the assumption that it generates quality and compact descriptors, as it has been so broadly explored in the video coding and compression contexts.

Aside from video compression, block matching has been used for shot boundaries detection by Amel et al. [1], and video registration by Hafiane et al. [7].

As a crucial component of the block matching routine, several search strategies have been proposed in order to speed up the process of finding the best matches. Some examples include: Three Step Search [16], Four Step Search (4SS) [25], Simple and Efficient Search [18], Diamond Search [34], and others. In our implementation, we chose 4SS as it is a simple to implement, steepest descent based strategy that runs approximately 10 times faster than the exhaustive approach.

In the action recognition context, techniques like optical flow [21] and 3D gradients [14] have been shown to produce high recognition rates. The same applies for the use of trajectories. Both Wang et al. [31,33] and Jain et al. [10] combine dense sampled trajectories and other descriptors, like Motion Boundary Histogram (MBH) and Histogram of Optical Flow (HOF) to achieve very high recognition rates on several datasets. In this work, we use only block matching information to generate the descriptor. With this, we obtain a computationally inexpensive descriptor, both in terms of time and space, with recognition rates comparable to state-of-the-art.

To condense the motion information extracted, a histogram is commonly used [11,14,20,32]. Histograms are interesting for video description as they are simple structures which encode a compact representation of the motion information. In Mota et al. [20], the final descriptor is an orientation tensor generated from a Histogram of Oriented Gradients (HOG). Tensors are robust mathematical tools and good aggregators. They can capture the local orientation and uncertainties of motion, and thus, could carry more useful information than a histogram. In this work, the final video descriptor is an orientation tensor that accumulates information from all trajectories.

1.2   Overview

The general workflow of a human action recognition system consists of three main tasks: motion extraction, motion representation, and action classification. Our focus is on the first two tasks, that is, computing a descriptor from block matching information. A modified Variable Size Block Matching Algorithm (VSBMA) [26] extracts trajectories from a set of frames, and thus is responsible for the first task. For the motion representation task, the vectors composing these trajectories are accumulated into a histogram of directions and then an orientation tensor is coded from it. These steps are repeated for all the frames of a video, in order to generate the video descriptor. Figure 1 illustrates this sequence of steps.

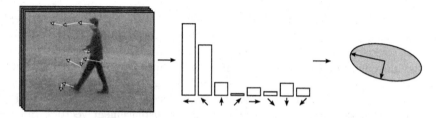

Fig. 1. Technique overview. From left to right: trajectories generated from the block matching routine, histogram of directions, and orientation tensor. The ellipse is merely an illustration since the tensor dimension is greater than 2.

2   Proposed Method

2.1   Variable Size Block Matching

The main algorithm consists in dividing a so called "reference frame" into blocks of pixels and finding for each one a corresponding block in a so called "target frame" which minimizes a dissimilarity (or error) function. For each block, the algorithm outputs a displacement vector between the coordinates of a reference block and its corresponding target.

This dissimilarity function is generally based on pixel intensities within the analyzed blocks. Although others can be found in related literature [3,9,13], In our implementation the function of choice was the Sum of Absolute Differences (SAD). In previous works and preliminary tests, we found that employing other error functions such as Mean Absolute Differences (MAD) or Sum of Squared Differences (SSD) had very slight or no impact at all regarding the action recognition context.

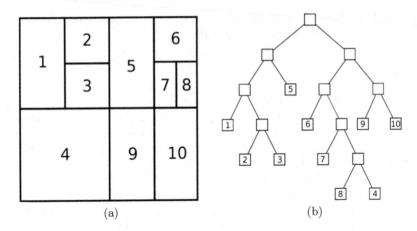

(a)   (b)

Fig. 2. Binary tree image segmentation from [3]. (a) Frame segmentation scheme. (b) Tree structure corresponding to (a).

When matching blocks, the blocks evaluated in the target frame are restricted to a close vicinity of the block coordinates in the reference frame. This vicinity is called a search window. The search window is established under the assumption that the movement of objects between frames is somewhat smooth, so it is not necessary to search the whole target frame. This greatly reduces the cost of the search, especially for higher resolution sequences. Note also that the size of the search window limits the magnitude of the displacement vectors. In our implementation we established a 15×15 pixel search window. This means that the largest vector that could result from a match would be $(\pm 7, \pm 7)$.

Another noteworthy element of the algorithm is the search strategy. Even with a search window, evaluating all blocks within it is still too big an effort. To reduce this cost, a number of strategies have been proposed over the years [16, 18, 22, 25, 34]. For this work, we chose Four Step Search (4SS) [25] as the search strategy. 4SS is a steepest descent strategy that compares at most 27 blocks to find the match, as opposed to evaluating all the blocks of the search window, which requires 225 comparisons.

All these elements are also present in a conventional Block Matching Algorithm (BMA). What differentiates VSBMA from BMA is that the sizes of the blocks in VSBMA change during the matching routine. All blocks have an initial size, but blocks that have a minimum matching error above a fixed error threshold are divided into smaller blocks and matched again. This process repeats until the error is below the threshold or the size of the blocks reach a fixed minimum size of 4×4 pixels. Just like first proposed in [3], we split the blocks into two smaller blocks, alternating between horizontal and vertical partitions. This is appropriately represented by a binary tree (Fig. 2), in which the leaves are blocks of varying sizes. As in picture segmentation [8], the goal is to have blocks that encompass homogeneous regions, possibly representing objects in the scene.

Fig. 3. VSBMA displacement vectors. Hotter colors indicate bigger blocks.

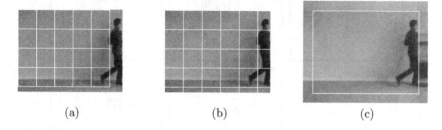

(a)   (b)   (c)

Fig. 4. Treatment for out-of-bounds block coordinates. (a) No treatment. Blocks are marked in white and regions close to the right and bottom of the image are not considered. (b) Fill solution. Gaps in said regions are occupied with blocks with different sizes. (c) Frame extension. The white rectangle now highlights the original frame. Intensity values for pixels outside of this rectangle are the same as those in the border.

Figure 3 shows a frame with displacement vectors computed through VSBMA drawn over it. Hotter colored vectors correspond to bigger blocks and colder colored vectors correspond to smaller blocks. The size of the vectors are proportional to their norms. This example suggests that the motion of more homogeneous regions of the image can be represented by a single vector, while more detailed regions need more vectors in order to properly represent its motion.

Boundary Treatment and Trajectory Extration. Each video frame is subdivided into non-overlapping blocks with an initial block size. When the video resolution is not a multiple of the block size, blocks may encompass regions out of the bounds of a frame, and thus, are not considered, as in Fig. 4. In this case, a border treatment becomes necessary to completely cover a frame. To solve this issue we use the initial segmentation depicted in Fig. 4, positioning as many whole blocks as possible within the frame and then completing the remainder of the frame with rectangular blocks, with initial sizes different from other blocks.

Even with this approach of filling the frame with blocks, another issue remains. Since all blocks are confined within the bounds of the frame, objects leaving the scene result in extraneous vectors with high error values. The frame is then extended as shown in Fig. 4 to overcome this problem. The intensity value for an out-of-bounds pixel is the same as its closest neighbor in the borders of the image, creating a stretching effect. Note that this solution also supplants the previous one, as blocks that do not fit in the original frame, fit in the extended frame. This is also coherent with block matching implementations specified in H.263 and H.264 (HEVC) [15, 30].

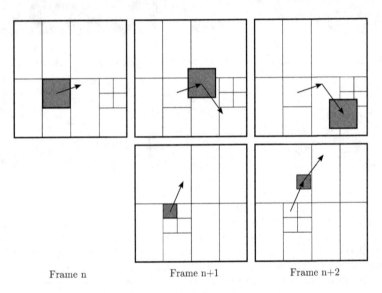

Frame n   Frame n+1   Frame n+2

Fig. 5. Example trajectories starting on frames n (top) and $n + 1$ (bottom). On the upper sequence, an object is tracked from frame n to frame $n+2$. Segmentation remains as on frame n. On the lower sequence, another object is tracked from frame $n + 1$ to frame $n + 3$ (not depicted). In this case, the segmentation used is the one resulting from matches between frames $n + 1$ and $n + 2$.

In order to extract trajectories, a set of frames, instead of just a pair, is analyzed. When doing so, VSBMA is used between the first and second frames of the set, and the following matches are carried out as conventional block matching, using the target of a previous match as reference for the following one. This way, the frame segmentation occurs only once and stays the same for the rest of the round of computations. This procedure is repeated in such way that all frames of the video serve as starting point for a trajectory. Figure 5 shows trajectories starting on two consecutive frames presenting different behaviours. Notice that the same region of the frame may be segmented in two different ways, depending on the reference frame, resulting in different trajectories. This is the case when objects appear or disappear during the sequence, enter or leave the camera's field of view, or if the sequence has distinctive shots or cuts. For such cases, especially regarding scene cuts, the trajectories carry erroneous motion information, since the assumption of continuous motion is toppled. By overlapping multiple trajectories, these errors have less impact on the overall accumulated motion information.

The use of these sparse trajectories computed via block matching is an attempt to obtain quality results, as found in other trajectory based works [10, 31, 32], but at a lower computational cost. Not only block matching serves as a low-cost kind of flow, compared to gradients or optical flow, but it also provides a coarser representation of objects in the scene compared to dense or keypoint sampling. This results in less trajectories carrying possibly the same meaningful motion information as a set of dense trajectories.

2.2   Generating the Descriptor

Histogram of Directions. The result of block matching is a displacement vector $d(i,j) = (d_x, d_y)$ for each block, where (i,j) are the block indexes. These vectors are converted to equivalent polar coordinates $c(i,j) = (\theta, r)$ with $\theta = \tan^{-1}(\frac{d_y}{d_x})$, $\theta \in [0, 2\pi]$ and $r = \parallel d(i,j) \parallel$. All displacement vectors of all trajectories starting in the reference frame are used to form a histogram of directions.

A motion direction histogram is used as a compact representation of the motion vector field obtained from each frame. It is defined as the column vector $h_f = (h_1, h_2, \ldots, h_{n_\theta})^T$, where n_θ is the number of cells for the θ coordinate. We use a uniform subdivision of the angle intervals. Each interval is populated as the following equation:

$$h_l = \sum_{i,j} r(i,j) \cdot \omega(i,j) \ , \tag{1}$$

where $l = 1, 2, \ldots, n_\theta$ and $\omega(i,j)$ is a vector weighting factor, which is a Gaussian function with $\sigma = 0.01$ in our experiments.

Consequently, a histogram is computed for each video frame, represented by a vector h_f with n_θ elements. It encodes all displacements of all blocks forming trajectories starting at the frame and spreading throughout a fixed number of frames ahead.

Tensor Descriptor. An orientation tensor is a representation of local orientation which takes the form of a $n \times n$ real symmetric matrix for n-dimensional signals [12]. Given a vector $v \in \mathbb{R}^n$, it can be represented by the tensor $\mathbf{T} = vv^T$. Then, we use the orientation tensor to represent the histogram $h_f \in \mathbb{R}^{n_\theta}$. The frame tensor, $\mathbf{T}_f \in \mathbb{R}^{n_\theta \times n_\theta}$, is given by:

$$\mathbf{T}_f = h_f \cdot h_f^T \ . \tag{2}$$

Individually, these frame tensors have the same information as h_f, but several tensors can be combined to find component covariances.

Orientation Tensor. The average motion of consecutive frames can be expressed using a series of tensors, given by

$$\mathbf{T} = \sum_{f=1}^{n_f} \frac{\mathbf{T}_f}{\| \mathbf{T}_f \|_2} ,$$

using all video frames. By normalizing $\mathbf{T}$ with a L_2 norm, we are able to compare different video clips or snapshots regardless their length or image resolution. Since $\mathbf{T}$ is a symmetric matrix, it can be stored with $d = \frac{n_\theta (n_\theta+1)}{2}$ elements.

If the motions captured in the histograms are too different from each other, we obtain an isotropic tensor which does not hold useful information. But, if the accumulation results in an anisotropic tensor, it carries meaningful average motion information of the frame sequence [20].

3 Experiments

The method is validated using the KTH dataset [28], which contains 600 videos of six human actions: walking, running, jogging, boxing, hand waving and hand clapping. These actions are performed by 25 people in 4 different scenarios: outdoors, outdoors with scale variation, outdoors with different clothes, and indoors. In this work, contrary to the protocol suggested in [28], the dataset is not split in sequences. The videos have a resolution of 160×120 pixels and 25fps frame rate.

We use a SVM classifier to evaluate our descriptor on KTH. All tests were run on an Intel®Core™2 Quad Q9550 2.83GHz with 4GB memory running a single thread per video. This is an important remark, as the speed of the method could be even further improved running more parallel threads or in an up to date system.

Table 1 shows the parameter values used on our experiments. Throughout descriptor computation and classification, there are other parameters to be tuned, such as descriptor size (related to number of histogram bins), standard deviation σ of histogram Gaussian weighing, other error functions and search strategies, search window size, and minimum block size. However, we chose to focus on the three main parameters: initial square block width, VSBM error threshold and trajectory size. We have conducted experiments with 7 initial block sizes, 4 threshold values, and 7 trajectory sizes, to a total of $7 \times 4 \times 7 = 196$ different parameter settings. For all the other parameters, we rely on values found in the literature and previous experience with some of the tools used in this work.

The values for block sizes were chosen based on the dataset resolution of 160×120 pixels, H.264 (HEVC) specification of block sizes ranging from 8×8 up to 64×64 [30], and recognition rates from preliminary experiments that

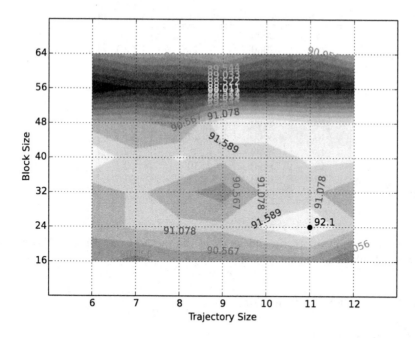

Fig. 6. Contour plot showing classifier accuracy for each parameter setting regarding block and trajectory sizes. Lighter colors indicate higher accuracy values. The highest value is marked as a black dot.

Table 1. Parameter values for experiments with VSBMA

Parameter	Values
Initial block width (pixels)	16, 24, 32, 40, 48, 56, 64
VSBM error threshold	2000, 4000, 8000, 16000
Trajectory size (frames)	6, 7, 8, 9, 10, 11, 12

showed a decrease in quality for larger blocks. As for threshold values, the choice was based on the maximum matching error value ($block\_width \times block\_height \times \#channels \times 255$), and on the apparently decreasing power law distribution of error values suggested also by preliminary experiments.

To assess and compare the quality of our descriptor in the action recognition context, we measure the recognition rate, which is the output of a SVM classifier, which takes the descriptors for the whole database. We follow the same classification protocol as [28]. The classifier produces 6 recognition rates for each parameter combination, 3 using a triangular kernel, and 3 using a Gaussian kernel. From these results, we take the highest ones achieved and present them in

Table 2. Summary of best recognition rates

Block Size	Threshold	Trajectory Size	Accuracy
24	2000	10	91.7
24	2000	11	92.1
40	4000	8	91.7
40	4000	9	91.2
40	4000	11	91.7
40	8000	6	92.1
40	8000	8	91.7
40	8000	12	91.7
48	8000	9	92.1
48	8000	10	92.1
48	8000	11	91.7
48	8000	12	91.7

Sec. 4. In order to evaluate the efficiency of the method, we consider the frame rate at which block matching operates.

4   Results and Discussion

4.1   Recognition Rates

Table 2 shows a summary of the results obtained. Notice the predominance of block sizes 40 and 48, and of thresholds 4000 and 8000. This can be attributed to the segmentation process including all relevant motion information from the cases with smaller block sizes into the cases with bigger block sizes. This is also true for the cases with block size 56 and 64, except that these block sizes do not produce better results. In these cases, the initial segmentation includes too many of out-of-bounds pixels. Consider, for example, a block size of 56 and the dataset resolution of 160×120. Horizontally, 3 blocks fit within the image with only 2 columns of pixels out-of-bounds, but vertically only 2 blocks fit perfectly, and the third one has 48 rows of out-of-bounds pixels. Although not amongst the best results, the recognition rates of these parameter settings reach upwards of 90.7%.

By observing the accuracy results to come to these conclusions, we are assuming that these block matching settings have no interaction or confounding with any other parameters of the process of generating and classifying the descriptor.

Figure 6 shows a contour plot of the same data. In lighter shades we can see the higher accuracy cases. This visualization allows for a quick recognition of what may be the optimal parameter setting, or at least delineate a relation between the parameters in order to achieve good results. It also shows the effect previously mentioned, where block size 56 has noticeably poorer results.

State-of-the-art comparison: We can see a clear improvement over our previous block matching based descriptor [6], especially considering that the best result

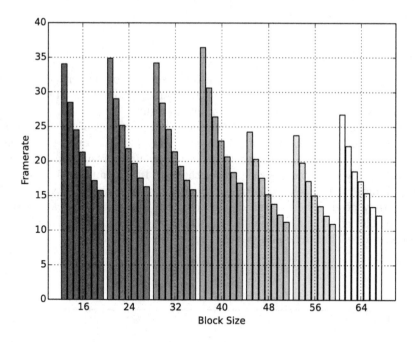

Fig. 7. Bar graph showing running speed for each parameter setting. Each group of bars represents a block size and bars within a group represent a different trajectory size, increasing from left to right.

Table 3. Comparison with state-of-the-art and previous works for KTH dataset

Authors	Recognition Rate
Klaser et al. (2008) [14]	91.0
Liu et al. (2009) [17]	93.8
Mota et al. (2013) [19]	93.2
Sad et al. (2013) [27]	93.3
Wang et al. (2013) [33]	**95.3**
Figueiredo et al. (2014) [6]	87.7
This work	**92.1**

in that work was obtained with the most computationally expensive method proposed, MSMV. Although still below the state-of-the-art recognition rates like 93.2% in [20] and 95.0% in [33], the results obtained using block matching are comparable. The best recognition rates from the literature are obtained using combinations of video characteristics [10, 20, 32]. This often leads to a very demanding process, in terms of computational effort. In our work, we use only motion information extracted with VSBMA, achieving real time computation capability in some parameter settings.

4.2 Frame Rates

Figure 7 shows a bar graph of the frame rates measured in our experiments. Bars are grouped by block size and each bar within a group shows the frame rate for a different trajectory size, increasing from left to right.

By comparing different groups, we can see that the impact of the trajectory size is approximately the same on all cases, reducing frame rates in roughly 50% between trajectory sizes 6 and 12. Like the accuracy results, frame rates are noticeably lower for block sizes greater than 40. In our implementation, the intensity values for pixels beyond image dimensions are computed by demand, whenever they need to be evaluated. For the cases with bigger blocks, greater portions of blocks are out-of-bounds, thus leading to an increased number of pixel intensities calculations in order to compare two blocks.

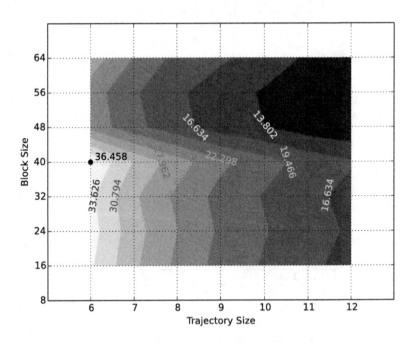

Fig. 8. Contour plot showing running speed for each parameter setting on VSBMA. Lighter colors indicate higher framerate values.

Figure 8 shows a contour plot of average frame rates for the same experiments as above. Lighter shades show higher frame rates. This graph highlights one important aspect: the execution speed is not a trade-off versus recognition accuracy. Note that the best recognition rates were obtained with block sizes 40 and 48, and trajectory sizes of 6, 9, 10, and 11. With the exception of the case with block size 40 and trajectory size 6, none of these parameter combinations are within the region of highest frame rates, but they are not within the region of lower frame rates either. In fact, the two contour plots represent very distinct surfaces, with no apparent correlation between them.

5   Conclusion

We presented a tensor self-descriptor based on variable block matching trajectories. The trajectories are accumulated into orientation histograms which in turn are coded into orientation tensors.

This is a work within the context of Human Action Recognition. It aims to provide a baseline for further developments regarding sparse trajectories and block matching in this context. We view our approach as a promising work, since it yields results close to those of state-of-the-art methods at low computational costs.

Future works may include several improvements, both on block matching and on its use for action recognition. Better exploration of the parameters, adaptive threshold values, adaptive trajectory sizes, block merging operations, and different block geometry are a few examples of improvements that can be made on the block matching end. As for human action recognition, the integration of other, more complex, techniques and datasets is going to be the next improvement of this work.

Acknowledgments. Authors would like to thank FAPEMIG, CAPES and UFJF for funding.

References

1. Amel, A.M., Abdessalem, B.A., Abdellatif, M.: Video shot boundary detection using motion activity descriptor. Journal of Telecommunications **2**(1), 54–59 (2010)
2. Calzone, S., Chen, K., Chuang, C.C., Divakaran, A., Dube, S., Hurd, L., Kari, J., Liang, G., Lin, F.H., Muller, J., Rising, H.K.: Video compression by mean-corrected motion compensation of partial quadtrees. IEEE Transactions on Circuits and Systems for Video Technology **7**(1), 86–96 (1997)
3. Chan, M., Yu, Y., Constantinides, A.: Variable size block matching motion compensation with applications to video coding. IEEE in Proceedings **137**(4), 205–212 (1990). http://dl.acm.org/citation.cfm?id=646271.685624
4. Chen, C.Y., Chien, S.Y., Huang, Y.W., Chen, T.C., Wang, T.C., Chen, L.G.: Analysis and architecture design of variable block-size motion estimation for h.264/avc. IEEE Transactions on Circuits and Systems I: Regular Papers **53**(3), 578–593 (2006)

5. Choi, S.J., Woods, J.: Motion-compensated 3-d subband coding of video. IEEE Transactions on Image Processing **8**(2), 155–167 (1999)
6. Figueiredo, A.M.O., Maia, H.A., Oliveira, F.L.M., Mota, V.F., Vieira, M.B.: A video tensor self-descriptor based on block matching. In: Murgante, B., Misra, S., Rocha, A.M.A.C., Torre, C., Rocha, J.G., Falcão, M.I., Taniar, D., Apduhan, B.O., Gervasi, O. (eds.) ICCSA 2014, Part VI. LNCS, vol. 8584, pp. 401–414. Springer, Heidelberg (2014)
7. Hafiane, A., Palaniappan, K., Seetharaman, G.: Uav-video registration using block-based features. In: IEEE International Geoscience and Remote Sensing Symposium (IGARSS), vol. 2, pp. 1104–1107 (2008)
8. Horowitz, S.L., Pavlidis, T.: Picture segmentation by a tree traversal algorithm. J. ACM **23**(2), 368–388 (1976). http://doi.acm.org/10.1145/321941.321956
9. Jain, J.R., Jain, A.K.: Displacement measurement and its application in interframe image coding. IEEE Transactions on Communications COM **29**(12), 1799–1808 (1981)
10. Jain, M., Jegou, H., Bouthemy, P.: Better exploiting motion for better action recognition. In: 2013 IEEE Conference on Computer Vision and Pattern Recognition (CVPR), pp. 2555–2562, June 2013
11. Ji, Y., Shimada, A., Taniguchi, R.I.: A compact 3d descriptor in roi for human action recognition. In: IEEE TENCON, pp. 454–459 (2010)
12. Johansson, B., Farnebäck, G.: A theoretical comparison of different orientation tensors. In: Proceedings of the SSAB Symposium on Image Analysis, pp. 69–73 (2002)
13. Kim, J.W., Lee, S.U.: Hierarchical variable block size motion estimation technique for motion sequence coding. Optical Engineering **33**(8), 2553–2561 (1994)
14. Kläser, A., Marszałek, M., Schmid, C.: A spatio-temporal descriptor based on 3d-gradients. In: British Machine Vision Conference (BMVC), pp. 995–1004, September 2008
15. Ku, C.W., Lin, G.S., Chen, L.G., Lee, Y.P.: Architecture design of motion estimation for itu-t h.263, vol. 3024, pp. 482–493 (1997). http://dx.doi.org/10.1117/12.263260
16. Li, R., Zeng, B., Liou, M.L.: A new three-step search algorithm for block motion estimation. IEEE Transactions on Circuits and Systems for Video Technology **4**(4), 438–442 (1994)
17. Liu, J., Luo, J., Shah, M.: Recognizing realistic actions from videos in the wild. In: IEEE Conference on Computer Vision and Pattern Recognition (CVPR), pp. 1996–2003. IEEE (2009)
18. Lu, J., Liou, M.L.: A simple and efficient search algorithm for block-matching motion estimation. IEEE Transactions on Circuits and Systems for Video Technology **7**(2), 429–433 (1997)
19. Mota, V.F., Souza, J.I., de A. Araújo, A., Vieira, M.B.: Combining orientation tensors for human action recognition. In: Conference on Graphics, Patterns and Images (SIBGRAPI), pp. 328–333. IEEE (2013)
20. Mota, V.F., Perez, E.D.A., Maciel, L.M., Vieira, M.B., Gosselin, P.H.: A tensor motion descriptor based on histograms of gradients and optical flow. Pattern Recognition Letters **31**, 85–91 (2013)
21. Mota, V.F., Perez, E.D.A., Vieira, M.B., Maciel, L., Precioso, F., Gosselin, P.H.: A tensor based on optical flow for global description of motion in videos. In: Conference on Graphics, Patterns and Images (SIBGRAPI), pp. 298–301. IEEE (2012)
22. Nie, Y., Ma, K.K.: Adaptive rood pattern search for fast block-matching motion estimation. IEEE Transactions on Image Processing **11**(12), 1442–1449 (2002)

23. Muralidhar, P., Rama Rao, C.B., Ranjith Kumar, I.: Efficient architecture for variable block size motion estimation of h.264 video encoder. In: International Conference on Solid-State and Integrated Circuit (ICSIC), vol. 32, p. 6 (2012)
24. Pirsch, P., Demassieux, N., Gehrke, W.: Vlsi architectures for video compression-a survey. Proceedings of the IEEE **83**(2), 220–246 (1995)
25. Po, L.M., Ma, W.C.: A novel four-step search algorithm for fast block motion estimation. IEEE Transactions on Circuits and Systems for Video Technology **6**(3), 313–317 (1996)
26. Puri, A., Hang, H., Schilling, D.: Interframe coding with variable block-size motion compensation. In: IEEE Global Telecommunication Conference, pp. 65–69 (1987)
27. Sad, D., Mota, V.F., Maciel, L.M., Vieira, M.B., Araújo, A.d.A.: A tensor motion descriptor based on multiple gradient estimators. In: Conference on Graphics, Patterns and Images (SIBGRAPI), pp. 70–74. IEEE (2013)
28. Schuldt, C., Laptev, I., Caputo, B.: Recognizing human actions: a local svm approach. In: Proceedings of the 17th International Conference on Pattern Recognition (ICPR), vol. 3, pp. 32–36. IEEE (2004)
29. Sullivan, G., Baker, R.: Rate-distortion optimized motion compensation for video compression using fixed or variable size blocks. In: Global Telecommunications Conference, GLOBECOM 1991. 'Countdown to the New Millennium. Featuring a Mini-Theme on: Personal Communications Services, vol. 1, pp. 85–90, December 1991
30. Sullivan, G., Ohm, J., Han, W.J., Wiegand, T.: Overview of the high efficiency video coding (hevc) standard. IEEE Transactions on Circuits and Systems for Video Technology **22**(12), 1649–1668 (2012)
31. Wang, H., Klaser, A., Schmid, C., Liu, C.L.: Action recognition by dense trajectories. In: 2011 IEEE Conference on Computer Vision and Pattern Recognition (CVPR), pp. 3169–3176, June 2011
32. Wang, H., Kläser, A., Schmid, C., Liu, C.L.: Dense trajectories and motion boundary descriptors for action recognition. International Journal of Computer Vision **103**(1), 60–79 (2013)
33. Wang, H., Schmid, C., et al.: Action recognition with improved trajectories. In: International Conference on Computer Vision (2013)
34. Zhu, S., Ma, K.K.: A new diamond search algorithm for fast block-matching motion estimation. IEEE Transactions on Image Processing **9**(2), 287–290 (2000)

An Impact Study of Business Process Models for Requirements Elicitation in XP

Hugo Ordóñez[1], Andrés Felipe Escobar Villada[1], Diana Lorena Velandia Vanegas[1], Carlos Cobos[2], Armando Ordóñez[3(✉)], and Rocio Segovia[1]

[1] Departamento de Ingenieria de Sistemas, Universidad de san Buenaventura, Buenaventura, Colombia
{haordonez,ersegovia}@usbcali.edu.co,
{anfelesvillada,dilove0122}@gmail.com
[2] Departamento de Ingeniería de Sistemas, Universidad del Cauca, Cauca, Colombia
ccobos@unicauca.edu.co
[3] Intelligent Management Systems Group, Foundation University of Popayan, Popayan, Colombia
armando.ordonez@docente.fup.edu.co

Abstract. Many communication problems may appear during requirements elicitation causing that final products do not accomplish client expectations. This paper analyzes the impact of using business processes management notation (BPMN) instead of user stories during requirements analysis in agile methodologies. For analyzing the effectiveness of our approach, we compare the use of user stories vs. BP models in eleven software projects during requirements elicitation phase. Experiments evidence that BPMN models improve quality and quantity of information collected during requirements elicitation and ease that clients specify clearly their needs and business goals.

Keywords: Software · Requirements elicitation · XP (eXtreme Programming) · BPMN (Business Process Management Notation) · User stories

1 Introduction

Software development is an important part of global capital, because it leverages business development and knowledge management [1]. Furthermore, software industry is a white industry that does not pollute and generates well paid jobs [2]. In this scenario, competitive software tools must integrate appropriate methodologies to each project goals [3], Most of the software methodologies must deal with projects with tight deadlines, volatile requirements, and new technologies [4]. One of these methodologies is XP, which is focused on enhancing interpersonal relationships as a key point for successful software development, promoting teamwork, learning developers and good working environment; based on continuous feedback between the client and the development team [5].

In XP, system requirements are gathered in meetings with clients and stakeholders using user stories. User stories describe the functionality of the software to build and customer's needs. However, these user stories are written in natural language, which

O. Gervasi et al. (Eds.): ICCSA 2015, Part I, LNCS 9155, pp. 298–312, 2015.
DOI: 10.1007/978-3-319-21404-7_22

may guide to misinterpretation, given the ambiguity and uncertainty of the collected information [6, 7].

Often in meetings between developers and clients, the dialogue may be not productive enough, causing that requirements are not identified clearly and do not cover all client requirements [8]. Recent studies show that the graphical representation of requirements can contribute to a clearer understanding of the system requirements. This may reduce the ambiguities in the requirements specification. The representations are clear and provide a good overview of the system ("a picture is worth a thousand words") (Pohl et al. 2013). In this paper we evaluate the use of business process (BP) for requirements elicitation. BP can be defined as a set of coordinated activities that aim for achieving a common business goal [9]. In our work, BP are represented using BPMN (Business Process Management Notation) (BPMI, 2006) standard.

For analysis purposes, two strategies (user stories and BPMN models) were applied to eleven software development projects in requirement analysis phase. For each project, a group of analysts evaluated the advantages of the two options for requirements elicitation. This article is organized as follows: in section two some of the most representative works on the subject of research are presented, in section three the proposed method is shown, section four puts forward the validation and results of the proposed technique, finally the conclusions and future work are depicted in section five.

2   Related Work

As previously mentioned, our approach uses user stories. While not all agile methodologies have focus on user stories for requirements elicitation, some of them use these user stories. Consequently, in the present review we analyze some projects using user stories in agile methodologies.

Agile methodologies aims for enhancing interpersonal relationships as a key to success in software development; some of these methodologies promote the use of user stories [6] and constant communication between clients and developers. There are several suggested templates for user stories, however, there is no consensus. In general terms, templates may include one or more of the following parts: name, description, tasks description, and estimate effort in days. Besides, it is recommended that one user story exist for each major functionality; in addition, it is suggested that each developer works in one or two stories each month [10].

Jaqueira et al. affirm that the greatest weakness of user stories is that managers, clients and developers describe non-essential elements instead of describing essential functionalities [11]. The user story should only describe external behavior of the system, which can be understood by the client [12]. As an alternative to the textual requirements elicitation, some studies have used visual models that help to understand to the client how the system works, equally these models also facilitate communication, understanding, problem detection and exploration of scenarios.

Zheng et al. present the adaptation of an agile methodology for implementing a Business Process Management Systems [15], the authors analyze the overall project implementation as well as the impact in the organization. This work uses Scrum methodology and a team formed by developers and customers. The team defines requirements and test deliverables at the end of each iteration. At the end of each iteration, assessment of deliverables is performed. With this assessment, business analysts begin to implement BPMN processes.

Avner et al. compare use case templates vs. graphical notation of BPMN business process that represents the system requirements [16]. This study shows that business analysts and a clients have different visions of processes inside the organization. The experiment concluded that the graphical notation enables faster cognitive understanding. Moreover, clients and analysts understanding improves when use cases templates are used first before BPMN models.

Existing approaches evidence that the use of natural language between clients and developers may produce that that requirements or priorities are not well defined, creating delays in development review and redefinition of the requirements for each user story. Furthermore during user stories creation some information considered "obvious" may be lost. Some approaches propose using BPMN for requirements elicitation but they do analyze the impact on agile methodologies.

3   Our Proposal

Our proposal replaces user stories by BP models during requirements elicitation in agile development methodologies such as XP. BP models allows client and developers to adopt a simple and standard notation to facilitate requirements capture. BP models represent may be created between development team and clients. For a description of the proposal, each element of the user story is taken and the correspondence is established within a model BP, in the BPMN notation. For this, the following considerations are taken into account:

3.1   Title of the Story

Title describes the main objective is clearly propounded. It also defines the interactions between users and the system. In addition the title of the story contains descriptions of all functions (each interaction) of the System. In BPMN the title of the story is represented by the name of the business process model.

3.2   Description of the Story

It describes the sequence of activities identified that make up the flow model for the development process. It is then, the step by step to follow and the description of the activities. In BP, is defined by the name and description attached to each one of the activities or tasks that make up the logical sequence of the BP model.

3.3   Historical Review

The historical revision allows to trace changes made to the requirements and establish responsible and change dates. These changes and its traceability are represented in BP models through the versioning process.

3.4   Reference Documents

Reference documents are a helpful way to elicit new requirements or as an input or output products. In the BP, reference documents are represented as data objects, however, if the document is not part of the process it is represented as an annotation in BP.

3.5 Actors

These are actors perform the sequence of actions that the system must perform to meet each of the set requirements. In the BP, actors are represented by the roles within each process. Roles are defined by Pools and lanes. A pool represents the main participants in a process, and can be divided according to the organizations they belong. A pool contains one or more lanes. The lanes are groups of tasks by area or participant involved in the process.

3.6 Dependencies

Within the user stories some dependencies may be established. For example, some user stories must be previous to some activities, this dependence allows establishing order constraints between stories. In the BP, these dependencies are determined by sending messages between processes and creating threads to indicate the sequence of activities between different roles and tasks. The sequence of actions is structured in terms of steps expressed consistently and sequentially generating the complete specification of the process through the control flow. Consequently, from the description of the individual processes and the use of messaging between processes, pools and lanes may be established dependencies of processes and responsibility of the actors involved in them. Figure 1 provides a graphic representation of the relationship between each of the elements of a user story and the elements within a business process model.

USER STORY	BPMN NOTATION	SYMBOL
Title of the story	Name of Process	
Historical Review	Process versioning	
Reference Documents	Data Objects, but if the document is not part of the process but to elicit reference is treated as annotation.	
Actors	Roles. Task Allocation via Lanes.	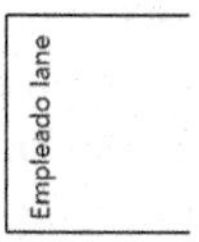
Dependencies	Related threads or processes. Sending messages between processes.	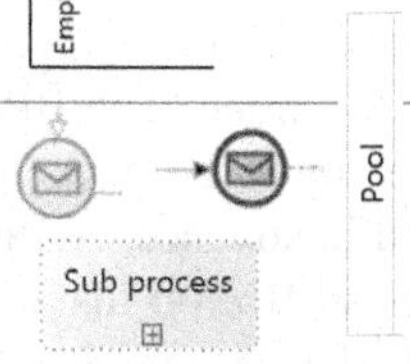
Description of the Story	Sequence of Activities	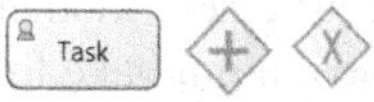

Fig. 1. User Story vs. BP notation

3.7   Adaptation of BP in the XP Methodology

BP generated models are integrated into the later stages of the XP methodology. This section shows how this integration is performed. XP proposes a dynamic life cycle supported on short development cycles (called iterations) with functional deliverables at the end of each cycle [13]. In each iteration a full cycle of analysis, design, development and testing is done (see Figure 2). Next the integration of BP models in the XP methodology is described.

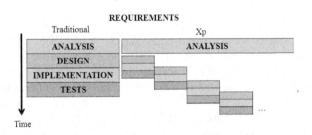

Fig. 2. Stages of development Traditional methodology vs. XP

Exploration: In this phase, the client defines what he needs by writing "user stories" at a high level. Developers estimate the development time of each activity described in the story [14]. This estimation may vary later when the requirements are analyzed in more detail. With the BP model, more details are specified, because this is created in the interview with the client or stakeholder and the BP model is created. Therefore, in this phase, low detail level is maintained as methodology proposes. Functionalities in BP models should be programmed by developers in less that three weeks. If time is greater, the process should be divided into two or more processes.

Planning: In this phase the stakeholder group and the developers agree the order in which user stories should be implemented, and, associated with them, its deliverables. These deliverables are specified in a document called the Delivery plan, or Release Plan. Furthermore at this stage, when developers develop small test programs called "spikes" to validate the client functionality, spikes are created in order to reduce risks and effort estimation. These spikes are used exclusively during requirements phase, and subsequently discarded for further phases; however developing of such spikes may be time consuming. The BP model specifies the order, priority and dependency of the functionality to be developed, therefore, the implementation and deliverables order is defined in advance. Additionally, in this phase, there are some BP tools such as: BonitaSoft and BizAgi that automatically generate small prototypes from defined process models. In these prototypes the developer does not invest more time than the necessary for requirement modelling, he also validates the requirement immediately, reducing the time it might take for validation.

Iterations: in this phase each user story is translated into specific tasks scheduling software functionality, generating deliverables. Because user stories are not defined in sufficient detail in the exploration phase, then a process of analysis is performed with the client to specify items as constraints or data necessary for the required features. With the BP data details and restrictions are already defined from the exploration phase, this allows estimation of time for the developer to be more accurate, therefore at this stage the process with the client is more a validation process and not about specification.

Deployment: this phase includes Implementation of functionalities and testing, these test may require modifications or adjustments, these changes are known as "fine tuning". Table 1 summarizes the use of User Stories vs. BP models.

Table 1. BP models vs. user stories in XP methodology

User Stories	BP Models
During exploration some aspects has are not defined such as conditions and data. It requires "spikes"	More detail to facilitate the estimation of time and risks, it becomes possible to avoid the generation of spikes
Release Plan based on the prioritization of the stories, agreed between developers and customers.	Release Plan based on prioritized BP models modeled with stakeholder.
They are written by the client, in their own language	Modeled in BPMN, understandable graphic language.
Minimum Detail, programmers make an estimate of the time that it will take to be development	Greater detail, it facilitates estimation, i.e. it needs more time for modeling, but reduces time in following phases.
If calculated effort is higher to 3 weeks by story, so it is divided in 2 or more stories. If one story take less than a week, it is combined other one.	If the estimation is more than 3 weeks, it is divided into two or more processes. The division is done using threads, avoiding the most complex model.
The customer order and group user stories and defines the dependencies.	The customer defines priority and the model gives dependencies over pools and threads.
Each user story is translated into specific tasks.	The sequence and priority of tasks are already defined in the BP flow.
Spikes help to estimate the risk	Some BPMN tools generate quick prototypes which permit validate in the moment when a model is defined.

4   Hypothesis

Some hypotheses were defined in our research:

- H1: requirements elicitation using BPMN improves domain understanding compared to user stories
- H2: The use of BP models after performing user stories in requirements elicitation increases domain understanding.

- H3: Requirements elicitation using user stories after performing the BPMN model increases the understanding of domination by those involved.
- H4: The generation of release plan from BPMN process model specifies in greater detail the deliverables, unlike those that use user stories.
- H5: The activities of plan iterations from BPMN process model specified better, instead of the use of user stories.
- H6: The percentage of implemented tasks in the first iteration of release plan that accomplish customer requirements, modeled with BPMN is greater than the percentage of tasks implemented from user stories

The aim of the first hypotheses (H1 to H3) is to identify which technique improves the understanding between the client and developers. The last three hypotheses aimed to determine if the use of BP models for requirements elicitation, instead of the user story improves the phases release execution plan and plan iterations of the XP methodology.

For the hypothesis validation, an experimental approach was used; this validation is based in the survey by sampling proposed by Liu et al. [17]. For performing the survey, each technique (user stories and BP) where tested separately for requirements elicitation. Then a comparative analysis of the results with each technique is done. Finally the BP based approach was tested during planning and iterations phases

4.1   Experimentation

Both techniques (BP and user stories) were applied for requirements elicitation in 11 projects in analysis phase, with the participation of 58 analysts (software engineers) in the software building department of SENA - Latin American Minor Species (LCMS) Valley Regional and 25 organizations stakeholder from different companies.

To begin the research, participants were trained in XP methodology, user stories creation and BPMN Once analyst groups for each project were trained, two groups were created for the purpose of applying the two techniques in each project. In these projects, both techniques were applied to compare the quantity and quality of information collected. In each of the techniques, the analysts interacted with stakeholders involved in each project through focus groups that collect data using a semi-structured group interviews [18]. The main purpose of the focus group is to arise concerns and reactions in stakeholder and analysts. Moreover, the focus group focuses on the user story that is being generated or the BP model being defined.

Table 2 shows the projects and participants for each technique. For requirements elicitation with user stories we worked with a template based in the work of Quasaimeh [13], for the elicitation of requirements under BP models, BonitaSoft tool was used (see Figure 3). As an example we take one of the user stories and one BP model defined in project A. This project is focused on staff evaluation. The A project aims to create a tool that allows the evaluation of employees performance when a contract ends. This system will help determine the performance of the employees based on a predetermined performance factors. It also allows the user to register and establish assessment criteria and elements to be considered during the assessment.

Table 2. Number of projects and Stakeholders by Hypothesis

Hypothesis	Projects	Number of Analysts		Number of Stakeholder
		User stories	BP Models	
H1	A	2	2	2
	B	2	2	2
	C	4	3	3
	D	4	3	3
H2	E	6	6	2
	F	6	6	5
	G	6	6	4
H3	H	4	4	3
	I	4	4	3
	J	4	4	3
	K	4	4	3
H4	H	6	6	3
	I	6	6	3
	J	6	6	3
	K	4	4	3
H5	H	6	6	3
	I	6	6	3
	J	6	6	3
	K	4	4	3

This in order to generate performance reports and identify weaknesses and strengths of their employees. Table 3 shows the user story created for one of the requirements (R1) Project A. The requirement (R1) consists of recording the performance evaluation of an employee by the manager of human resources. This registration process is performed after the close of each employee and is input to the renewal of the contract each year or semester.

Table 3. Template of user story for Project A, R1

Title: Project A – R1. Register staff evaluation			
Historical review			
Version.	Change description	Author	Date
Reference document			
Document number:	title:		
Word Format – staff evaluation template			
GENERAL INFORMATION			
Actor:	Human capital manager		
Dependencies:	Staff must be registered.		
Priority:	Low	Medium	High	X
Description of the story			
Human talent manager can record performance evaluation of staff. To do so, he must enter the employee's id and if it is registered with a valid contract, the last active evaluation will be displayed... the system must register the scores per criterion, the date of the assessment and the overall score			
Observations: None			

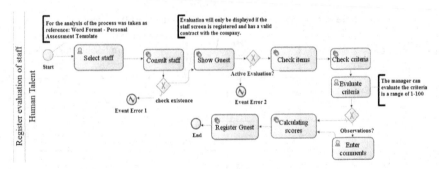

Fig. 3. Example of model for project A

Figure 3 shows the BP model created for requirements (R1) of Project A. This process consists of the evaluation record of the organization's staff. To do this the manager should select human talent through the identification, the employee to whom the assessment will be applied; in this case the system must consult if the employee exists and has a valid contract with the company. Assessment that will be applied, provided such active displays. If the assessment is active for being applied, so the elements and the evaluation criteria can be evaluated in a range of 1-100 points. If necessary observations may be recorded. Eventually the system will calculate and display the scores and the assessment will be recorded.

Upon completion of the requirements elicitation stage, a survey for each applied technique was performed. The survey allows evaluating how each analyst experienced information capture. For more detailed and accurate information on the surveys, the sample was defined considering different groups of analysts working on development projects, in which the hypotheses were validated. The end of the sample is to avoid partial or biased results. For example, Table 4 shows the survey to the evaluation process in user stories.

Table 4. Survey for user stories application

Question
Does the template used to describe user stories allow you to define a concise and complete customer requirements?
Does the template used to describe the user story allowed you to establish if the requirement outlined by the user was required for the system or the organization?
Does the template used to describe user stories allow you to define whether the requirements were consistent? This means, if it was established that the requirements were not contradictory?
When the review of user stories was performed written by another analyst, did you find that these were not ambiguous and that its interpretation was that the stakeholder had said?
Are the requirements documented in the user stories are verifiable, i.e. could do tests to validate their compliance once implemented?
With the template used to write user stories could set the priority or urgency of the requirement according to the objectives of the stakeholder?
It is clear the language used by the stakeholder when the system requirements were specified?
Highlight of the percentage of the total system information that could capture using user stories.
It was adequate the disposition of the customer to specify their requirements?

Later a survey for Stakeholders was also applied. This survey has the following purposes: 1) To measure the percentage of needs and objectives that the stakeholder consider that are specified correctly with the two applied techniques, 2) to determine how the analyst is dealing with customers and/or users communication in order to unify criteria, 3) to identify how stakeholder are included in the process of requirements elicitation and how the analyst makes use of information collected in the interview with the stakeholders.

To validate the hypothesis H4- H6 both groups used the requirements elicitation techniques and subsequently they created iterations plan and release plan both for user stories and BP models approaches. For generating the plan from user stories and BP models, it was used a template generated from Valkenhoef et al. [19]. The release plan in the first iteration includes: priority, that may be low, medium or high, estimated effort at a range of 1-10, where 10 is the largest effort, delivery date. Each process or user story results in specific programming tasks.

Each module of the project was developed by a couple of developers who from the user story or BP model generated activities for each iteration. At the end of this first iteration, an interview with the developers and stakeholder was performed to verify the status of each task and acceptance by the user. The interview allowed the validation of hypotheses H1 to H6 in the following items: 1) Measure the percentage of requirements that were implemented according to the needs of the stakeholder, after the first iteration with each of the techniques applied, 2) determine whether release plan generation from BP model specifies in greater detail the deliveries, unlike that with the use of user stories, 3) the activities of the iterations plan from the BP model specify more completeness, unlike when using user stories. Table 5 shows the questions asked in the survey to the evaluation.

Table 5. Survey for evaluation of release plan

Does the technique provides enough detail assigned for software development?
The selected technique allows you to know restrictions in some requirements during development time?
Did you take into account other sources of requirements different to the assigned technique? Which one?
Does the assigned technique allow you identify other requirements during development?
Does the release plan modified structurally at the end of the first iteration?
The esteemed effort in the release plan is consistent with what you experienced during development?
Does the assigned technique allowed you to establish a precise number or tasks? During development these tasks were increased / decreased?
Does the estimated delivery date of each user story / BP correctly established?

5   Discussion of Results

5.1   Validation of the Hypotheses

H1: According to information collected in the survey, requirements documentation in natural language creates ambiguities and different interpretations by the analyst when other person review the description, furthermore it was not possible to establish whether the requirements contradicted each other, for example when multiple stakeholder were involved in the process of requirements elicitation. Moreover, in the survey of analysts, the percentage of the total system information that could be captured with this technique was between 50% and 80%, indicating that the requirements are not defined completely in this phase. However it was established that the client's participation was 75% favoring validation of requirements. In addition language used by clients to specify his needs was clear, but could be improved.

A second survey with the second analysts group who applied requirements identification using BP models in projects was done. This survey allowed to establish the utility of the use of BP models to represent the client's needs. According to information collected in the survey, documentation requirements through model BP generates more clarity and a single interpretation by the analyst when reviewing models made by another analyst, equally, the BP model, allowed to identify consistency between the requirements using relationships / dependencies between processes. It also clearly defines the assignment of responsibilities to each role or an actor through the Lanes. From the results of this survey it was highlighted that 87% of analysts indicated that the requirements modeled through BP allowed to validate information with stakeholder, verifying that the sequence of activities was adequate, and besides if the information requested in each activity was complete.

H2: eliciting requirements through BP after performing user stories increases domain understanding by those involved. A group of analysts used user stories for requirements elicitation in 3 projects. It was established that 73.33% of the analysts found that user stories written by another analyst were ambiguous and that its interpretation was not the same of the stakeholder. In these surveys it was established that 56% of the analysts said that they captured between 50 and 70% of the requirements. After application of user stories, the same group use the BP model to raise the requirements, at this point it was found that 95% of analysts affirm that BP Model allow them to define a concise and complete requirements client. Now, from the previous survey increased from 22% of analysts who say that when reviewing a model BP has made another analyst team are not ambiguous and its interpretation was that the same that stakeholder had expressed. The rate of requirements capture was between 78% and 89% with this technique, demonstrating that BP model allows to increase by 19% the capture and identification of requirements.

H3: eliciting requirements with user stories after making BP models increases domain understanding by those involved. A group of analysts applied first BP for requirements elicitation in 4 projects, through a survey, I was established that for 80% of the analysts, to understand the notation of the model was easy, considering they pass first through a training process. Now for the stakeholder, it was found that 83.33% said that the notation was easy to understand, because analyst generates diagrams with stakeholders in real time, which allows both parties to resolve doubts immediately. The validation of this hypothesis also found that 73.33% of analysts indicated that when reviewing a model of BP was made by another analyst, no ambiguities arise and that his interpretation was that the same that the stakeholder said. After applying the BP model, the same group used user stories for requirements elicitation. In this process it was found that 83% of analysts used models that BP had done before, to write the story. In addition, it was noted that 85% of analysts indicated that when they conducted the review of user stories written by another analyst, these were not ambiguous and that its interpretation was that the stakeholder had said. The results therefore validate the four hypotheses allowing to evidence that the BP model can increase by 19%, the number of identified requirements, besides the ambiguity in the requirements specification is decreased by 25%. This demonstrates that using BP models increases the level of understanding between the different actors in the process of building software in XP.

H4 – H5 – H5: Generation Release Plan: with the projects worked in the previous stage, it was evaluated the result the first iteration. The results are listed in Table 6. The results of the first iteration may identify that the technical proposal (BP) reaches 98.33% in the tasks approved by the stakeholder and implemented by the development team, which exceeds 23% of the requirements elicitation technique with user stories in the first iteration, which achieves 75.85 of the Release Plan.

Table 6. Execution results of the first iteration (P: Project T: Technique used, TT: Total specified tasks, NT: Number of implemented tasks, NS: Number of tasks approved by the stakeholder, NI: Number of approved and implemented tasks

P	T	TT	NT	%	NS	%	NI
H	BP	15	15	100	15	100%	100%
I	BP	18	17	94,44	17	94,44	100%
J	BP	15	15	100	14	93,33%	93,33%
K	BP	14	12	85,77	12	85,71%	100%
H	HU	16	13	81,25	13	81,25%	92,86%
I	HU	19	14	73,68	16	84,21%	100%
J	HU	11	9	81,81	9	75,00%	90 %
K	HU	12	8	66,66	6	54,55%	75%

Additionally at this stage, it was used some metrics that allowed to establish the value of client needs. With these metrics, the client will know if in each iteration functionalities are being implemented according to the needs and according to the established time without making many changes between. For this, a set of different metrics of different related aspects (see Table 8). It can be seen that with the use of BP model, the averaged percentage of requirements that were not approved by the client at the end of the first iteration was 20% lower than the average of the requirements with user stories.

Table 7. Validation Metrics for hypotheses H4-H6

Metrics evaluated by iteration	User story	BP Model
Average requirements completed in iteration	12	14,75
Average incorporated changes and added requirements on the initial scope of the project	4	1
Average completed requirements for total iteration requirements.	75,85	98,43
Average conditions not approved by the client	26,25	6,63

Results obtained when evaluating the hypothesis H4 –H6 demonstrate that proposed technique allows 23% more fully developed requirements by iteration, consequently the functionality defined by the client is widely covered, which significantly reduces the time of software development, because the client does not need to go back to the redefinition of the requirements that were not approved. Furthermore clear and understandable communication between the development team, client and stakeholder in the identification and definition of the model using BP reduces changes in the identified requirements. The reduction in changes avoid delays and allows the development times stays in line with client needs.

5.2   Results of Stakeholder and Customers Surveys

Using BP models for requirements elicitation in an agile methodologies such as XP instead of using user stories, according to the data provides some advantages: improved communication between the analyst and the stakeholder, allowing to establish requirements more clearly. The BP models allow to model the processes in a unified and standardized way, representing the sequence of the process just as it does the stakeholder in your organization. Furthermore, understand a model through this notation by the client does not require much time and additional training. The stakeholder process and describes the BP model and may validate at real time, allowing clarify some requirements that are not understood at the time of the interview.   In addition it was found in the study that 87% of stakeholders and customers (see Figure 5) indicated that the notation used by the analyst to specify their requirements were clear and for easier to understand compared to the use of user stories. With BP models, customers can actively participate in the validation requirement, due to the fact that they can add or define clearly the sequence of activities to develop, also assign users to specific tasks as well as the data required in each task.

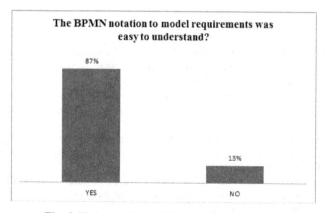

Fig. 4. Understanding of BP model by stakeholder

6   Conclusions and Future Work

Errors in requirements elicitation stage are caused mainly by the ambiguity of the communication between users and the development team. Using BP models for requirements elicitation improved the communication between the analyst, stakeholder and clients. BP notation used was easier to understand that natural language. Furthermore, using BP models was much more productive and easy to understand by 80% of analysts, which represents a major improvement in this activity. Analysts also emphasize that BP models can be more productive, because they elicit a greater number of requirements compared to user stories. BP models improve interpretation and understanding, upon review of the requirements that have been elicited by other analyst team. This is because the BP models are defined a common and standardized language allowing to clearly identify the sequence of activities and what the user expects the system.

In relation to the support among the techniques of elicitation of requirements, we can say that the user story itself did not allow fully clarify requirements compared to BP models. The description of requirements in user stories created from BP models increased understanding in 11.67 % due to the fact that BP models allowed to make clearer the redaction of the user story by analysts. However, it is convenient to use two techniques for requirements elicitation, this is because this will add an additional activity that does not contribute more to the process would and this will affect one of the agile manifest that supports on simplicity. The generation and implementation of release plan from BP model is faster and generates a higher percentage of implemented and approved requirements at each iteration, this allows that the development to be best fit to the client needs. We can conclude that BPs is appropriate to give a top-level view of the processes supported by one or more organizations; on the other hand, user stories may be used in XP for describing the user-centric requirements of the system under design and development. That means that BPs and user stories should be better considered as complementary techniques.

References

1. Paternoster, N., Giardino, C., Unterkalmsteiner, M., Gorschek, T., Abrahamsson, P.: Software development in startup companies: A systematic mapping study. Information and Software Technology **56**, 1200–1218 (2014)
2. Nguyen-Duc, A., Cruzes, D.S., Conradi, R.: The impact of global dispersion on coordination, team performance and software quality – A systematic literature review. Information and Software Technology **57**, 277–294 (2015)
3. Christensen, H.B., Hansen, K.M., Kyng, M., Manikas, K.: Analysis and design of software ecosystem architectures – Towards the 4S telemedicine ecosystem. Information and Software Technology **56**, 1476–1492 (2014)
4. Losada, B., Urretavizcaya, M., Fernández-Castro, I.: A guide to agile development of interactive software with a "User Objectives"-driven methodology. Science of Computer Programming **78**, 2268–2281 (2013)
5. Thiyagarajan, P.S., Verma, S.: A Closer Look at Extreme Programming (XP) with an On-site-Offshore Model to Develop Software Projects Using XP Methodology. In: Berkling, K., Joseph, M., Meyer, B., Nordio, M. (eds.) Software Engineering Approaches for Offshore and Outsourced Development. LNBIP, vol. 16, pp. 166–180. Springer, Heidelberg (2009)
6. Blom, M.: Is Scrum and XP suitable for CSE Development? Procedia Computer Science **1**, 1511–1517 (2010)
7. Domann, J., Hartmann, S., Burkhardt, M., Barge, A., Albayrak, S.: An Agile Method for Multiagent Software Engineering. Procedia Computer Science **32**, 928–934 (2014)
8. Wood, S., Michaelides, G., Thomson, C.: Successful extreme programming: Fidelity to the methodology or good teamworking? Information and Software Technology **55**, 660–672 (2013)
9. Chinosi, M., Trombetta, A.: Computer Standards & Interfaces BPMN : An introduction to the standard. Computer Standards & Interfaces **34**, 124–134 (2012)
10. Tong, K.I.: Chapter 8 Managing Software Projects with User Stories. In: Essential Skills for Agile Development, pp. 217–252 (2010)
11. Jaqueira, M.L.A., Aranha, E., Alencar, F., Castro, J.: Using i Models to Enrich User Stories Objectives of the research. iStar. In: Proceeding ICSE 2002 Proceedings of the 24th International Conference on Software Engineering, pp. 695–696 (2013)
12. Layman, L., Williams, L., Damian, D., Bures, H.: Essential communication practices for Extreme Programming in a global software development team. Information and Software Technology **48**, 781–794 (2006)
13. Qasaimeh, M.: Extending Extreme Programming User Stories to Meet ISO 9001 Formality Requirements. Software Engineering and Applications, pp. 626–638 (2011)
14. Kent Beck, A.-W.: Extreme Programming Explained: Embrace Change, 2nd edn., pp. 178–200. Addison Wesley (2012)
15. Zheng, G.: Implementing a business process management system applying Agile development methodology: A real-world case study. Thesis. Erasmus Universiteit Rotterdam (2012)
16. Ottensooser, A., Fekete, A., Reijers, H.A., Mendling, J., Menictas, C.: Making sense of business process descriptions: An experimental comparison of graphical and textual notations. Journal of Systems and Software **85**(3), pp. 596–606 (2012)
17. Liu, Y., Tian, G.-L.: A variant of the parallel model for sample surveys with sensitive characteristics. Computational Statistics & Data Analysis **67**, 115–135 (2013)
18. Singh, G.N., Priyanka, K., et al.: Estimation of population mean using imputation techniques in sample surveys. Journal of the Korean Statistical Society **39**(1), 67–74 (2010)
19. Van Valkenhoef, G.T., Tervonen, T., de Brock, B., Postmus, D.: Quantitative release planning in extreme programming. Information and Software Technology **53**(11), 1227–1235 (2011)

Moving Meshes to Fit Large Deformations Based on Centroidal Voronoi Tessellation (CVT)

Witalij Wambold[1,2], Günter Bärwolff[3]([✉]), and Hartmut Schwandt[3]

[1] Component Development, Volkswagen AG, Letter box 7359,
38231 Salzgitter, Germany
[2] Institute of Mathematics, TU Berlin, Berlin, Germany
[3] Institute of Mathematics, Technische Universität Berlin,
Straße des 17. Juni 136, 10623 Berlin, Germany
`baerwolf@math.tu-berlin.de`

Abstract. The essential criterion for stability and fast convergence of CFD-solvers (CFD - computational fluid dynamics) is a good quality of the mesh. Based on results of [30] in this paper we use the so-called centroidal Voronoi tessellation (CVT) not only for mesh generation and optimization. The CVT is applied to develop a new mesh motion method. The CVT provides an optimal distribution of generating points with respect to a cell density function. For a uniform cell density function the CVT results in high-quality isotropic meshes. The non-uniform cases lead to a trade-off between isotropy and fulfilling cell density function constraints. The idea of the proposed approach is to start with the CVT-mesh and apply for each time step of transient simulation the so-called Lloyd's method in order to correct the mesh as a response to the boundary motion. This leads to the motion of the whole mesh as a reaction to movement. Furthermore, each step of Lloyd's method provides a further optimization of the underlying mesh, thus the mesh remains close to the CVT-mesh. Experience has shown that it is usually sufficient to apply a few iterations of the Lloyd's method per time step in order to achieve high-quality meshes during the whole transient simulation. In comparison to previous methods our method provides high-quality and nearly isotropic meshes even for large deformations of computational domains.

Keywords: Mesh motion · Centroidal Voronoi tessellation · Finite Volume method

1 Introduction

Currently there are numerous areas of applications in which the shape of the solution domain is variable for every time step of the simulation. Examples for such cases are prescribed boundary motion in pumps, internal combustion engines, free-rising bubbles in water as well as wind turbine simulations. One problem of such simulations is the propagation of the displacement at the surfaces into the volume mesh. It is well known that the essential criterion for stability and fast

© Springer International Publishing Switzerland 2015
O. Gervasi et al. (Eds.): ICCSA 2015, Part I, LNCS 9155, pp. 313–328, 2015.
DOI: 10.1007/978-3-319-21404-7_23

convergence of CFD-solvers is a good quality of the mesh. Maintaining these criteria for the internal mesh is a quite difficult task if the domain suffers from large deformations. In this paper we present a method based on centroidal Voronoi tessellation that provides a high-quality mesh even for large boundary motions. The algorithm is implemented as an extension to the OpenFOAM® framework [15].

2   Previous Work

Several methods for mesh deformation have been presented in the literature during the past decades. In [30] we discussed the properties, advantages and disadvantages for example of Laplacian smoothing [23], the spring analogy method of [2–4,6,10,28], a finite element method proposed in [11], interpolation of the boundary displacements to the interior mesh by radial basis functions proposed by [25], and a new method based on a disk relaxation algorithm recently developed by Xuan Zhou [29].

Regardless of all methods described above, each is just suitable for certain degrees of deformations, because these approaches describe the mesh motion simply through relocation of mesh vertices. The cell topology, however, remains unchanged.

3   Contribution

Up to now the CVT has been used primarily for mesh generation and optimization [7,13,17,18]. CVT provides an optimal distribution of generating points with respect to a given cell density function. For a uniform cell density function the CVT results in high-quality isotropic meshes. The non-uniform cases lead to a trade-off between isotropy and fulfilling of the cell density function constraints.

The idea of the approach is to start with a CVT-mesh and to apply for each time step of transient simulation the so-called Lloyd's method [19,20,22] to correct the mesh with respect to the boundary motion. This leads to the motion of the whole mesh if the boundary is moved. Furthermore, each step of Lloyd's method provides a further optimization of the underlying mesh, thus the mesh remains close to the CVT-mesh. In order to create the initial CVT-mesh from a given arbitrary mesh, we also apply Lloyd's method. An integral part of our work is to develop an efficient CVT implementation that allows a mesh generation close to CVT at each simulation step. Our code is written in C++ as extension to the OpenFOAM® framework. This package enables the developer through, a convenient class hierarchy to extend the built-in code without great effort.

Compared to previous approaches our technique provides nearly isotropic polyhedral meshes even for large boundary deformation. Another advantage of our approach is the operation on already existing cells, so the interpolation of fields between two different meshes is avoided.

The algorithm affects solely the cell topology and thus allows to keep the field affiliation to each cell. Only the face fluxes must be calculated for a new generated CVT-mesh. Moreover, our algorithm can be run in parallel, because the computation of each Voronoi cell is carried out independently from other cells.

4 Theoretical Background of the CVT

We restrict our treatment of the Voronoi tessellation to 3D-space. Given an open set $\Omega \subseteq \mathbb{R}^3$, the set $\{V_i\}_{i=1}^n$ is called a tessellation of Ω if $V_i \cap V_j = \emptyset$ for $i \neq j$ and $\cup_{i=1}^n \overline{V_i} = \overline{\Omega}$. Let $\|\cdot\|$ denote the Euclidean norm on $\mathbb{R}^3$. Given a set of points $\{x_i\}_{i=1}^n$ belonging to $\overline{\Omega}$, the Voronoi region $\hat{V}_i$ corresponding to the point x_i is defined by

$$\hat{V}_i = \{x \in \Omega \ \mid \ \|x_i - x\| < \|x_j - x\| \text{ for } j = 1, \ldots, n \text{ with } j \neq i\}. \quad (1)$$

The points $\{x_i\}_{i=1}^n$ are called generating points or generators. The set $\{\hat{V}_i\}_{i=1}^n$ is a Voronoi tessellation of Ω, and each $\hat{V}_i$ corresponds to the Voronoi region of the generator x_i. Given a region $V \subseteq \mathbb{R}^3$ and a cell density function ρ, defined on V, the centre of mass x^* of V is defined by

$$x^* = \int_V x\rho(x)dx / \int_V \rho(x)dx. \quad (2)$$

A centroidal Voronoi region V_i^* is a Voronoi region $\hat{V}_i$ with the property:

$$x_i = x_i^*. \quad (3)$$

A Voronoi tessellation where all regions satisfy the condition in (3) is called centroidal Voronoi tessellation.

One of the algorithms for CVT computation is the Lloyd's method. This is an iterative algorithm consisting of the following simple steps: starting from an initial Voronoi tessellation corresponding to an old set of generators, a new set of generators is defined by the centres of mass of the old Voronoi regions. A mathematical scheme of the Lloyd's method is given in algorithm 1. In the sequel we give a short mathematical argumentation for Lloyd's method. For a detailed mathematical treatment we refer to [20]. Let us define the set

$$\mathcal{M} := \left\{ (x_1, x_2, \cdots, x_n)^T \ \mid \ x_i \in \Omega \text{ for } i = 1, \cdots, n \right\}. \quad (4)$$

Then we know that considering (1) for a fixed boundary $\partial \Omega$ each V_i depends on x_i and a neighbourhood of x_i. Therefore we can say that, V_i depends on $X := (x_1, x_2, \cdots, x_n)^T \in \mathcal{M}$. Then for each step of the Lloyd's iteration we have:

$$x_i^{(k+1)} = \frac{\int\limits_{V_i(X^{(k)})} x\rho(x)dx}{\int\limits_{V_i(X^{(k)})} \rho(x)dx}. \quad (5)$$

Algorithm 1. Lloyd algorithm

For a given domain Ω with cell density function ρ defined on Ω and the initial set
of generators $\{x_i\}_{i=1}^n$ perform following iterations:

for k=1,2,...,nIterations **do**

Construct the Voronoi tessellation $\{V_i^{(k-1)}\}_{i=1}^n$ of Ω with generators
$\{x_i^{(k-1)}\}_{i=1}^n$.

Take the centres of mass of $\{V_i^{(k)}\}_{i=1}^n$ as the new set of generators $\{x_i^{(k)}\}_{i=1}^n$.

Break if some stopping criterion is met.

end for

Further we define the map

$$T : \mathcal{M} \mapsto \mathcal{M}, \quad X^{(k)} \mapsto X^{(k+1)}, \quad \text{such that } T_i(X^{(k)}) := x_i^{(k+1)}. \tag{6}$$

Considering (2) and (3) we obtain in case of CVT the following equality:

$$X^{(k)} = T(X^{(k)}) \text{ or } X = T(X). \tag{7}$$

In view of (7) Lloyd's method may be viewed as a fixed point iteration. This
shows that, Lloyd's method has a linear convergence rate. And according to [19]
this convergence rate decreases as the number of generators gets large. Some
accelerating techniques like the Lloyd-Newton method are given in [19]. Apart
from this fact Lloyd's method is very well suited for the mesh motion, because
the time step of the simulation should be set smaller while the number of cells
gets larger. This is done in order to bound the Courant number by one.

Unfortunately, and despite great advantages of the CVT, its computation
becomes difficult for complicated domain boundaries. The most widely used
approach to construct the Voronoi tessellation in 3D-space is the computation
of a dual data structure referred as Delaunay triangulation. See for example
[8,16,18]. We propose an alternative approach for the generation of Voronoi
tessellations.

5   Centroidal Voronoi Generator

There are numerous preprocessors for the efficient generation of tetrahedral
meshes. We presume that the fluid domain is already meshed. We are interested
in construction of the CVT from a given tetrahedral or polyhedral mesh. Now
we use Lloyd's algorithm as described above. The sum of the absolut values of
the differences between old generators and new generators serves as convergence
criterion for Lloyd's method. One step of Lloyd's method includes the following
tasks:

1. Calculate the cell density distribution in the whole domain.
2. Compute the new generators with respect to the cell density function.
3. Compute the tessellation of the cuboid containing the underlying domain.

4. Clip the cuboid-mesh with the given domain boundaries.

In general there does not exist a pure CVT for the domains with non-flat boundaries. We are also forced to distinguish between internal cells and patch cells. We think with patch on the set of the faces, which compound the domain boundary. In general the mesh patches can get or loose some faces during the simulation. For this purpose, we use additional bounding surfaces, which bound the domain. So we have the boundary part of the volume mesh (the set of the patch faces) as well as the bounding surfaces. The boundary motion means in our work the motion of the bounding surfaces. This allows us to keep the topology of the boundary representation fixed.

In general we can handle both parametric surfaces and triangulated surfaces as boundary representation of the domain. Here we focus on managing triangulated surfaces. In this work we use two different techniques to handle the boundary cells described in subsection 5.2. The next subsection shows how the cell density function can be calculated.

5.1   Computation of the Cell Density Function

As described above we start with a discretized model. This means that the user has defined an appropriate mesh edge size for each edge of the model. Further we refer to the mesh edge size as feature size. Qiang Du [18] proposed to take for the cell density function the inverse 5th power of the feature size. Some experiments have confirmed that this strategy provides the best description of the cell density function. Let e_{ij} and n_i be edges and number of edges of the Voronoi cell V_i, then we define its feature size as

$$s_i = \frac{\sum_{j=1}^{n_i} \|e_{ij}\|}{n_i}. \tag{8}$$

The cell density ρ_i of V_i is defined by:

$$\rho_i = \frac{1}{s_i^5}. \tag{9}$$

In order to determine the distribution of the cell density in the interior of the domain we first compute the distribution of the feature size in the whole domain, and than we get the cell density using the equation (9). For the distribution of the feature size in the interior we solve the Laplace equation:

$$\nabla \cdot (d^2 \nabla s) = 0, \tag{10}$$

where d square denotes the diffusion coefficient for the feature size. In order to solve the equation (10) we employ the Finite-Volume method using the mesh from the previous Lloyd's iteration. So we get the feature edge size on the cell centres of the mass. After that we use linear interpolation in order to get the feature edge size on the vertices of the mesh.

An appropriate mesh grading can be achieved by variation of the diffusion coefficient. It turned out that, the inverse distance from the boundaries often provides an optimal mesh grading. The algorithm for computing of the distance to the nearest patch was implemented in OpenFOAM® [15]. With the cell density function on the cell centres we interpolate it to the mesh vertices.

5.2   Computation of Generators

For the internal cells we take the centres of mass as the new generators. The centre of mass for a polyhedron V_i with respect to the cell density function ρ can be computed as follows. First, V_i is decomposed into n tetrahedra as proposed by [18]. This straightforward decomposition technique is very efficient and always feasible for convex polyhedra. Let x_j and m_j be the centres of mass and the masses of the tetrahedra forming the polyhedron V_i. Than the centre of mass y_i of V_i can be computed from

$$y_i = \int_{V_i} y\rho(y)dy \bigg/ \int_{V_i} \rho(y)dy = \sum_{j=1}^{n} x_j m_j \bigg/ \sum_{j=1}^{n} m_j.$$

The centre of mass and the mass of each tetrahedron can be computed by any quadrature rule. We now distinguish between so-called constrained centroidal Voronoi tessellation (CCVT) and unconstrained (only clipped) CVT. In our work we use both methods depending on the given geometry.

CCVT. By the CCVT we will understand a CVT, where generators of the patch cells are constrained to be located on the surface. The CCVT for surfaces in $\mathbb{R}^n$ was originally proposed by Qiang Du, see [21]. In his further work [18] he suggests to project the centre of mass of the patch cell onto the bounding surface. Unfortunately, this procedure does not have a unique solution. The projection of sphere centre to the sphere surface, for example, leads to an infinite number of results. This could occur if the centre of mass is located on the centre of the osculating circle of the surface. In order to avoid this problem we use another method. Depending on the number of patch faces of the patch cell we perform an appropriate step:

1. One patch face: The centre of mass is taken as the generator of the appropriate patch face. We decompose the underlying face into a number of triangles and use the same technique as for a polyhedron mentioned above.
2. Two patch faces: When both faces contain a common edge we reduce the problem to a one dimensional problem. Hence, the generator is computed as the centre of mass lying on the common edge. If both patch faces do not contain a common edge, we perform the step 1 for the patch face with the centre nearest to the centre of mass of the cell.
3. At least three patch faces: If we can find three patch faces, which contain a common vertex, we set the generator to the common vertex. In case of

several common vertices, we take the vertex with the nearest distance to the centre of mass of the cell. If there are no common vertices, then we search for the faces with common edges, and perform step 2 for the common edge, which centre located nearest to the centre of mass of the cell. If we can find neither a common vertex nor a common edge, we perform step 1 for the patch face with the centre nearest to the centre of mass of the cell.

It is clear, that the generator produced by the described technique does not reside exactly on the surface, but is located very close to it. The technique described here is very simple and leads always to a unique solution.

Clipped CVT. Another technique to manage patch cells should be described in the sequel. It is not necessary that, the generators are constrained to the boundary. It is also possible to simply clip the boundary cells by the bounding surface. In such case the generator can be inside as well as outside of the domain. This has some disadvantages:

- The domain boundary gets new cells or loses cells during the simulation. As a consequence small patch faces arise on the boundary. This reduces the mesh quality.
- In comparison to CCVT, there is no optimization of the patch cells.
- For large deformations of the domain boundary per simulation step it can happen, that cells lie completely outside of the domain.
- The volume of the discretisation domain changes slightly at each step of the simulation.

A major advantage of the clipped CVT compared to CCVT is the exact fulfilment of the cell density function constraints. In case of the CCVT the cells stick to the surface and can not leave this surface. If we have too many cells on the surface and not enough in the interior, we can not fulfil the required cell density function constraints.

5.3   Tessellation of the Cuboid

With the new generators, we can proceed with the tessellation of the underlying domain. First, we compute the Voronoi tessellation of the cuboid containing the whole domain and after that we clip this mesh with domain boundaries. For the computation of the CVT of the cuboid we opted to use the voro++ library [24]. This library deals directly with Voronoi cells and computes a Voronoi tessellation by the cuts with perpendicular-bisector planes. The cells are handled and saved independently by this library. We use only one function from this library to compute the plane cuts. Now we describe the computation of a new Voronoi cell.

- First, we initialize each Voronoi cell as cuboid so that the whole domain is contained in it. The dimensions of the cuboid are cumputed as dimensions of the domain plus offset. Each neighbouring generator creates with own generator the corresponding perpendicular-bisector. Our task is to find the correct indices of the generators, which perpendicular-bisectors would contain a face of the new Voronoi cell. In order to determine these indices we make use of the connectivity information of the previous mesh.

- We cut the initial cuboid-cell using the perpendicular-bisectors created by generator indices of the direct neighbours of the previous mesh.
- We use multiple levels of the old neighbouring indices, because each cell of the new mesh can get new neighbours if the mesh is moved. By neighbours we understand here all cells, which contain at least one common vertex with the current cell.
- After each cut the cell gets smaller. So we use the maximal radius of the own cell in order to determine whether the perpendicular-bisector created by the relevant neighbouring index cuts the cell. This can significantly speed up the cutting routine.
- We stop the recursive routine at a level where the previous level of the old neighbouring indices did not create the perpendicular-bisector, which intersects the cell. The experiments have shown that this technique works well and is very efficient.

After performing the described procedure for all cells, we achieve a decomposition of the cuboid which has the same dimensions as the cuboid used for the initial cell. It turned out that for each Voronoi cell it is usually sufficient to visit two levels of the neighbours from the previous mesh. Therefore, we get for each cell an average computational amount t_c, that only depends on the number of the visited levels of the neighbours. That means that t_c is independent of the whole number of the cells n. Therefore, for the n cells the decomposition of the whole domain takes the expected time $t_c \cdot n \approx \mathcal{O}(n)$.

5.4   Clipping with Boundaries

Once we have the decomposition of a cuboid covering the whole domain, we start clipping the underlying cells with bounding surfaces. Usually we have a lot of surfaces bounding the domain. Suppose we have found for each surface at least one intersected cell. So we can perform the cuts for remaining cells recursively using neighbourhood relations saved during tessellation of the cuboid. The tessellated cuboid is shown in Fig 1a. Fig. 1b shows the first intersected cell and its neighbours. Using the example in Fig. 1b we are explaining our procedure. During the cutting of each cell (magenta cell in Fig. 1b) the next intersected cell (blue cells in Fig. 1b) is determined by the face containing the edge which is currently being intersected. So we push the neighbour index into the so-called FIFO queue and search for the next intersected edge of the cell. Once the cut of the current cell is completed we pop out the cell index from the queue and compute the intersection for this cell. This process is terminated when all edges, that intersect the current surface, have been visited.

As mentioned above for clipping of the underlying mesh with each surface we need at least one cell intersecting this surface. For that we make use of the

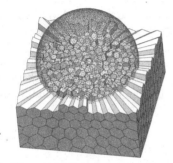

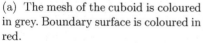

(a) The mesh of the cuboid is coloured in grey. Boundary surface is coloured in red.

(b) Boundary surface is coloured in red. The first intersected cell is coloured in magenta and its neighbours are coloured in blue.

Fig. 1. Clipping with domain boundaries

patch cells from the previous mesh. For each patch we choose an appropriate cell and compute intersections for each edge of this cell. Now we still need to know which part of the clipped cell remains in the domain and which must be removed. That can be done by means of the surface normal.

A similar approach for recursive cutting of the boundary cells with the corresponding surface was proposed by [8]. In comparison to this approach we handle the Voronoi cells directly instead of using the dual data structure known as the Delaunay triangulation.

5.5   Merging to the Global Mesh

After some iterations of Lloyd's method we achieve a mesh close to CVT within the limit of the predefined tolerance. Since all cells are computed independently, we need to merge these to a global mesh. Hence, we need to create the global vertices and faces from those local entities. The global faces can be created easily using the fact that each global face belongs to exactly two local faces. In order to construct the global vertices we march through the local faces and correlate the vertices for both corresponding faces. We mark the visited local vertices of both adjacent cells with new created global labels. Only the non-visited local vertices create new global vertex labels.

This technique works as long as the two corresponding local faces are equal. As a result of rounding errors there are neighbours with non-equal corresponding faces. Hence, the number of vertices contained in both faces is not equal. In such cases these faces are corrected. We compute correlations for both faces and remove the redundant vertex from the face with greater number of vertices. Since each vertex belongs to at least three faces the procedure also modifies other faces from the handled cell. We also have to check the modified local faces for equality to the corresponding local faces. In case of inequality, we correct the underlying faces too. Each removal of a vertex leads to non-flat faces. Therefore the position of the affected vertices is computed by the least squares method, i.e. each affected vertex forms the smallest distance to the original planes containing it. Although the described method causes an increasing number of arithmetic operations the whole computational effort increases only very slightly, because only very few cells are affected. The proposed technique works reliably and is substantiated by a series of test cases.

6   Mesh Motion Cases

This section shows three simple examples with prescribed boundary motion. We emphasize that, the generated meshes have a very high quality and fulfil the quality-check criteria of OpenFOAM® at each time step of the simulations.

Fig. 2 shows a cuboid which top wall is moved down during the simulation. The cell density function is uniform. The simulation starts with a tetrahedral mesh (Fig. 2a). After some iterations of the Lloyd's method we achieve a mesh close to CVT (Fig. 2b). The cells partition the whole domain perfectly, because the generators are not constrained to lie on the boundaries (CVT technique). It can be observed that, the cell density of the cells increases for smaller volumes, because the number of the cells remains constant during the simulation. In the last step (Fig. 2e) the simulation arrives at a single thin layer of cells.

Fig. 3 illustrates a volume between two concentric spheres. The radius of the inner sphere is increased during the simulation. The cell density function is computed as the inverse third power of the sphere radius. Here we used a very fine triangle surface mesh, in order to reach a very small gap between both spheres. See Fig. 3f.

The next example, illustrated in Fig. 4, shows a cylinder with an enclosed sphere. The cell density function is explicit defined on the boundary and computed in the interior by Laplacian Smoothing as described in subsection 5.1. The generators are fixed on the parametric surfaces. This keeps the volume of the interior domain constant, because the boundary-mesh remains unchanged during any affine transformations of the enclosed sphere surface.

7   Validation of Usability Within the Finite-Volume Method

Obviously employing of the centroidal Voronoi meshes for the Finite-Volume method (FVM) can be very beneficial. As already mentioned above, the proposed mesh motion approach do not affect number of cells during the simulation. We just move the cells like particles through space. In case of mesh motion the physical phenomena are described by so-called arbitrary Lagrangian-Eulerian Formulation. For further details please refer to [12]. For an incompressible and divergence-free flow the momentum equation and continuity equation is of the

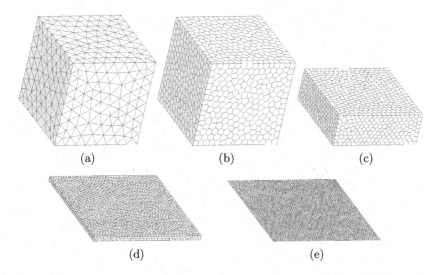

Fig. 2. Cuboid, top wall moved down (consequent steps (a)-(e))

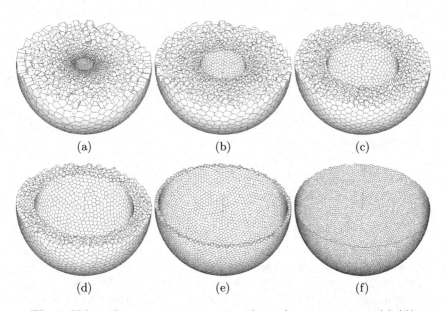

Fig. 3. Volume between two concentric spheres (consequent steps (a)-(f))

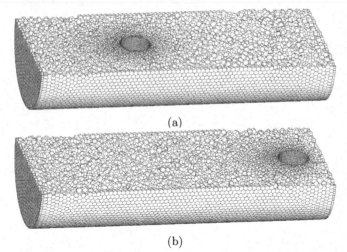

(a)

(b)

Fig. 4. A cylinder with enclosed sphere (consequent steps (a)-(b))

form:

$$\left.\frac{\partial U}{\partial t}\right|_{\chi} + \nabla \cdot ((U - U_m)U) - \nabla \cdot (\nu \nabla U) = -\nabla p \tag{11}$$

$$\nabla \cdot U = 0 \tag{12}$$

Where U, U_m, ν and p denote respectively the velocity, the mesh velocity, the kinematic viscosity and the pressure. In order to solve the system of the differential equations (11)-(12) we use the so-called Semi-implicit Method for Pressure Linked Equations (SIMPLE). For further details we refer to [1,9].

For a validation of the developed mesh motion solver we need a simple model, which has an analytical solution. We decided to simulate the free-falling sphere in a viscous fluid. Such a case can be also validated by a steady state solution. For the purpose of comparisons we have to transform the steady state solution to the reference coordinate system of the transient case. For both simulations the following parameters were used:

- $v = 0.01\ m/s$ - sphere relative velocity
- $r = 0.001\ m$ - sphere radius
- $\eta = 0.000885$ - dynamic viscosity
- $R = 0.05\ m$ - radius of cylinder
- $H = 0.4\ m$ - length of cylinder

Before the start of the transient simulation we have to create a quasi-CVT mesh. After round 60 iteration of the Lloyd's method we got a quasi CVT-mesh, which is shown in the Fig. 5a. Considering the Fig. 5b we can note that, the cell density in the downwind region is significantly less than the cell density in the upwind region. This phenomenon will be explained in the section 8.

Fig. 6 shows the simulation results of both the steady state and the transient cases. Except for small differences, we get a good agreement of both fields. In

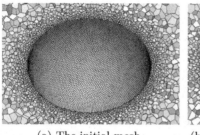 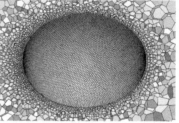

(a) The initial mesh   (b) The mesh at the time 0.3585

Fig. 5. The quasi-CVT mesh in the area of the sphere. The whole mesh contains 775380 Voronoi cells. The x-axis points to the right.

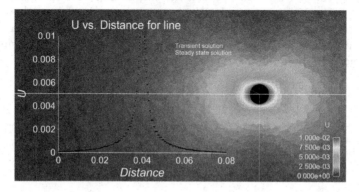

Fig. 6. Velocity field of the steady state and the transient simulations of free-falling sphere in a viscous fluid. The solutions are plotted over the distance on the line (0.16, 0., 0.)-(0.24, 0., 0.). The zero point on the x-axis corresponds to the point (0.16, 0., 0.) in the 3D-space. The steady state case is computed with 9000 iterations of the simpleFoam. The transient case is simulated up to 0.3585 seconds.

order to make a quantitative comparison, we computed the drag force on the sphere:

$$F_D = \oint_S p \, n \, dS - \oint_S \tau \cdot n \, dS, \tag{13}$$

where p, n and τ denote pressure, outward-pointing normal to face vector and shear stress tensor. Using the formula (13) for both cases we get following results:

$$F_D^{st} \approx 1.7451e - 7N \quad \text{(with 9000 simpleFoam steps)} \tag{14}$$

$$F_D^{tr} \approx 1.7358e - 7N \quad \text{(transient up to 0.3585 sec.)} \tag{15}$$

$$|(F_D^{tr} - F_D^{st})/F_D^{tr}| \approx 0.53\% \tag{16}$$

We recognise that the deviation in the drag force is within the acceptable range. Furthermore, for the considered geometry the drag force can be calculated by means of the Stokes' Law:

$$F_D = 6\pi\eta v r\lambda \tag{17}$$

$$\lambda = \lambda_R\lambda_H = (1 + 2.1\,r/R)\,(1 + 3.3\,r/H), \tag{18}$$

where λ - Ladenburg-correctors for finite vessel dimensions with $(r \ll R, r \ll H)$. Using the equations (15), (17) and (18) we get the following deviation in the drag force:

$$F_D^{Stoke} \approx 1.7526e - 07N. \tag{19}$$

$$|(F_D^{tr} - F_D^{Stoke})\,/F_D^{tr}\,| \approx 0.96\% \tag{20}$$

Summarising the above we can say that the proposed mesh motion method can be used by Finite-Volume methods. We observed the mesh quality values during the whole transient simulation. The maxima of the critical mesh values are shown in table 1 and compared to a polyhedral mesh of a commercial preprocessing tool. Table 1 indicates that the CVT provides better quality meshes than classical

Table 1. Mesh quality results

	Max aspect ratio	Max non-orthogonality	Max skewness
CVT mesh	2.33210739	15.20106280	3.14209776
Commercial preprocessor	6.23371	55.1742	1.7585

mesh generators. The mesh skewness in case of the CVT is higher, because we got higher gradients of the cell density function in the area of the sphere. The mesh skewness can be improved varying the mesh diffusion coefficient as defined in the equation (10).

8   Conclusions and Future Work

Up to now the CVT has been used for mesh generation and optimization. This paper shows the possibility to use the CVT for mesh motion. We use the term mesh motion, because the number of cells remains constant during the simulation. Previous approaches relating to retopologization usually lead to a change of the number of cells, see [26].

The computational effort of the developed mesh motion method is comparable with one iteration of the `pimpleDyMFoam` solver (See [15]). We are currently working on the development of an efficient solution for a treatment of the non-convex boundaries. To our knowledge, in all previous works, the non-convex boundaries lead to non-convex star-shaped Voronoi cells, see [8,14]. A further decomposition of such cells leads to convex cells, but these decomposed cells are not centroidal Voronoi cells.

As already mentioned in section 7 we get non-symmetric cell density distribution in the area of the moving sphere. This is due to the fact that, our cell motion approach is exclusively based on the CVT. The cells move because we try to reconstruct the CVT mesh. For that we just perform a few steps of Lloyd's method. In case of the sphere motion there are no motion condition for the laterally placed cells. The cells on the sphere surface just float through the surrounding cells. This leads to a high cell density in the back of the sphere. This problem can be solved by adding an additional displacement to the newly computed generator, which can be determined by the Laplacian of the boundary displacement. Using this strategy we have already performed some experiments, which shows that the cell density remains constant in the area of the sphere.

It seems interesting to use error estimators for the construction of appropriate cell density distribution. But here we are with Finite-Volume methods not in such a good situation as in the case of Finite-Element methods. But some results of [5] should be helpful.

References

1. de Foy, B., Dawes, W.: Unstructured pressure-correction solver based on a consistent discretization of the Poisson equation. International journal for numerical methods in fluids **34**, 463–478 (1999)
2. Farhat, C., Degand, C., Koobus, B., Lesoinne, M.: Torsional springs for twodimensional dynamic unstructured fluid meshes. Computer Methods in Applied Mechanics and Engineering **163**(1–4), 231–245 (1998)
3. Bottassoa, C.L., Detomib, D., Serra, R.: The ball-vertex method: a new simple spring analogy method for unstructured dynamic meshes. Computer Methods in Applied Mechanics and Engineering **194**(39–41), 4244–4264 (2005)
4. Degand, C., Farhat, C.: A three-dimensional torsional spring analogy method for unstructured dynamic meshes. Computers & Structures **80**(3–4), 305–316 (2002)
5. Eymard, R., Herard, J.-M.: Finite Volumes for Complex Applications V. Wiley (2008)
6. Zeng, D., Ethier, C.R.: A semi-torsional spring analogy model for updating unstructured meshes in 3D moving domains. Finite Elements in Analysis and Design **41**(11–12), 1118–1139 (2005)
7. Wang, D., Qiang, D.: Mesh optimization based on the centroidal voronoi tessellation. International Journal of Numerical Analysis and Modeling **2**, 100–113 (2005)
8. Yan, D.-M., Wang, W., Levy, B., Liu, Y.: Efficient Computation of Clipped Voronoi Diagram for Mesh Generation. Computer-Aided Design **45**, 843–852 (2013)
9. Lien, F.-S.: A pressure-based unstructured grid method for all-speed flows. International journal for numerical methods in fluids **33**, 355–375 (1999)
10. Markou, G.A., Mouroutis, Z.S., Charmpis, D.C., Papadrakakis, M.: The ortho-semi-torsional (OST) spring analogy method for 3D mesh moving boundary problems. Computer Methods in Applied Mechanics and Engineering **196**(4–6), 747–765 (2007)
11. Jasak, H., Tukovic, Z.: Automatic mesh motion for the unstructured finite volume method (November 2006)
12. Donea, J., Huerta, A., Ponthot, J.Ph., Rodriguez-Ferran, A.: In: Encyclopedia of Computational Mechanics, Chapter 14, Arbitrary Lagrangian-Eulerian Methods (2004)

13. Chen, L.: Mesh smoothing schemes based on optimal delaunay triangulations. Math Department, The Pennsylvania State University, State College
14. Ebeida, M.S., Mitchell, S.A.: Uniform random Voronoi meshes. In: Proceedings of the 20th International Meshing Roundtable, Paris, France, pp. 273–290. Sandia National Laboratories, Albuquerque (2011)
15. OpenFOAM C++ Documentation. http://foam.sourceforge.net/docs/cpp/
16. Alliez, P., Cohen-Steiner, D., Yvinec, M., Desbrun, M.: Variational Tetrahedral Meshing. ACM Transactions on Graphics. Proceedings of ACM SIGGRAPH 2005 **24**, 617–625 (2005)
17. Qiang, D., Wang, D.: Anisotropic centroidal voronoi tessellations and their applications. SIAM Journal on Scientific Computing **26**(3), 737–761 (2005)
18. Qiang, D., Wang, D.: Tetrahedral mesh generation and optimization based on centroidal Voronoi tessellation. International journal for numerical methods in engineering **56**, 1355–1373 (2003)
19. Qiang, D., Emelianenko, M.: Acceleration schemes for computing centroidal Voronoi tessellations. Numerical linear algebra with applications **0**, 1–19 (2005)
20. Qiang, D., Emelianenko, M., Lili, J.: Convergence of the Lloyd algorithm for computing centroidal voronoi tessellations. SIAM Journal Numerical Analysis **44**(1), 102–119 (2006)
21. Qiang, D., Gunzburger, M.D., Lili, J.: Constrained Centroidal Voronoi Tessellations For Surfaces. SIAM Journal on Scientific Computing **24**(5), 1488–1506 (2003)
22. Qiang, D., Faber, V., Gunzburger, M.: Centroidal Voronoi Tessellations: Applications and Algorithms. SIAM REVIEW **41**(4), 637–676 (1999)
23. Löhner, R., Yang, C.: Improved ALE mesh velocities for moving bodies. Communications in Numerical Methods in Engineering **12**, 599–608 (1996)
24. Rycroft, C.H.: Voro++: a three-dimensional Voronoi cell library in C++ (2009)
25. Jakobsson, S., Amoignon, O.: Mesh deformation using radial basis functions for gradient-based aerodynamic shape optimization. Computers & Fluids **36**, 1119–1136 (2007)
26. Menon, S., Schmidt, D.P.: Conservative interpolation on unstructured polyhedral meshes: An extension of the supermesh approach to cell-centered finite-volume variables. Computer Methods in Applied Mechanics and Engineering **200**, 2797–2804 (2011)
27. Arabi, S., Camarero, R., Guibault, F.: Unstructured meshes for large body motion using mapping operators. Mathematics and computers in simulation **106**, 26–43 (2014)
28. Zhang, X., Zhou, D., Bao, Y.: Mesh motion approach based on spring analogy method for unstructured meshes. Journal of Shanghai Jiaotong University **15**, 138–146 (2010)
29. Zhou, X., Li, S.: A new mesh deformation method based on disk relaxation algorithm with pre-displacement and post-smoothing. Journal of Computational Physics **235**, 199–215 (2013)
30. Wambold, W., Bärwolff, G.: New mesh motion solver for large deformations based on CVT. Procedia Engineering **82**, 390–402 (2014)

Deployment of Collaborative Softwares as a Service in a Private Cloud to a Software Factory

Guilherme Fay Vergara$^{(\boxtimes)}$, Edna Dias Canedo,
and Sergio Antônio Andrade de Freitas

Faculdade UnB Gama, Universidade de Brasília (UnB), Caixa Postal 8114,
Gama, DF 72.405-610, Brazil
gfv.unb@gmail.com, {ednacanedo,sergiofreitas}@unb.br

Abstract. This paper presents a proposal of deploying secure communication services in the cloud for software factory university UNB (University of Brasília - Brazil). The deployment of these services will be conducted in a private cloud, allocated in the CESPE (Centro de Seleção e de Promoção de Eventos) servers. The main service that will be available is the Expresso, which is a system maintained by SERPRO (Serviço Federal de Processamento de Dados). These services increase the productivity of the factory members and increase their collaboration in projects developed internally

Keywords: Cloud computing · Deployment · Private cloud · Software factory · Owncloud

1 Introduction

The development of several new computational technologies, as distributed computing, pervasive computing, grid computing, internet and programming languages, made possible the study and increase of new areas of computing. The computing resources sale on demand, according to customer's need is an old necessity that was limited to large institutions able to get computational resources of computing powerful enterprises.

Due to the development of new technologies in the broader fields of computing and the widespread use of the Internet, has provided web applications that can be accessed independently of the location by the internet. The development of virtualization technology enable sales on demand, in a scalable way, of computing resources and infrastructure, which are able to support web applications, arising this way cloud computing, generating an increasing tendency of applications that can be accessed efficiently, independent of the location.

The technology of cloud computing aims to provide services on demand, being billed or not by usage, as well as other basic services. Prior trends to cloud computing were limited to a certain class of users or focused on making available a specific demand for IT resources [1]. This technology tends to comply with wide goals, being used not only by big companies that would outsource all its IT services to another company, but also for user who wants to host their personal documents on the

O. Gervasi et al. (Eds.): ICCSA 2015, Part I, LNCS 9155, pp. 329–344, 2015.
DOI: 10.1007/978-3-319-21404-7_24

Internet. This type of technology allows not only the use of storage resources and processing, but all computer services.

In cloud computing, resources are provided as a service, allowing users to access without knowing the technology used. Thus, the users and companies began to access services on demand, independent of location, which increased the amount of services available [2]. With this, users are moving their data and applications to the cloud and can access them easily from any location.

Cloud computing emerges from the need to build less complex IT infrastructures compared to traditional, where users have to perform installation, configuration and upgrade of software systems, also infrastructure assets are inclined to become obsolete quickly. Therefore, the use of computational platforms of others is a smart solution for users dealing with IT infrastructure. Cloud computing is a distributed computing model that derives characteristics of grid computing, with regard to the provision of information on demand to multiple concurrent users [2]. A cloud service provider offers cloud applications without the user having to worry about where the services are hosted or how they are offered. Slices of the computational power of the nodes of the network are offered, reducing costs to purvey own infrastructure to provide services. Resources are assigned only during the period of use, reducing power consumption when utilization is no longer needed.

Virtualization technology [6] provides the foundation for many cloud solutions. Moreover, in many solutions environments, where users are able to choose their virtualized resources, such as programming languages, operating system and other custom services are offered. The main benefits are reduction of the costs of infrastructure investment, operating costs and scalability for the provision of services on demand.

Cloud computing is an area that is increasingly growing and attracting diverse audiences. Ever more organizations has adopted cloud computing based solutions. The objectives of this paper can be summarized by making a study of existing technologies in cloud computing with application to a software factory and contribute for a collaborative environment for software factory.

This document is organized as followed. Section 2 provides a review of the concepts of cloud computing. Section 3 presents a set of possible implementations of solutions for factory software. Section 4 presents an implementation proposal for the software factory. Section 5 shows deployment. Section 6 shows some results of this deployment. Finally, Section 7 shows conclusion.

2 Cloud Computing

Cloud computing refers to the use, through the Internet, of diverse applications as if they were installed in the user's computer, independently of platform and location. Several formal definitions for cloud computing have been proposed by industry and academy. We adopt the following definition: "Cloud computing is a model for enabling convenient, on-demand network access to a shared pool of configurable computing resources (e.g., networks, servers, storage, applications, and services) that can be rapidly provisioned and released with minimal management effort or service provider interaction" [3].

Cloud computing is being progressively adopted in different business scenarios in order to obtain flexible and reliable computing environments, with several supporting solutions available in the market. Being based on diverse technologies (e.g., virtualization, utility computing, grid computing, and service-oriented architectures) and constituting a completely new computational paradigm, cloud computing requires high-level management routines. Such management activities include: (a) service provider selection, (b) virtualization technology selection, (c) virtual resources allocation, and (d) monitoring and auditing in order to guarantee Service Level Agreements (SLA).

A solution of cloud computing is composed of several elements, as clients, data center and distributed servers, as shown in Figure 1. These elements form the three parts of a solution cloud [3] [6]. Each element has a purpose and has a specific role in delivering a working application based on cloud.

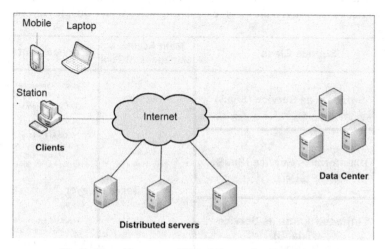

Fig. 1. Three Elements of Cloud Computing Solution [6]

Cloud computing architecture is based on layers. Each layer deals with a particular aspect of making application resources available. Basically, there are two main layers, namely, a lower and a higher resource layer. The lower layer comprises the physical infrastructure and is responsible for the virtualization of storage and computational resources. The higher layer provides specific services, such as: software as service, platform as service and infrastructure as service. These layers may have their own management and monitoring system, independent of each other, thus improving flexibility, reuse and scalability. Figure 2 presents the cloud computing architectural layers [4] [5].

Software as a Service (SaaS): Provides all the functions of a traditional application, but provides access to specific applications through Internet. The SaaS model reduces concerns with application servers, operating systems, storage, application development, etc. Hence, developers may focus on innovation, and not on infrastructure, leading to faster software systems development. SaaS systems reduce costs since no software licenses are required to access the applications. Instead, users access services on demand. Since the software is mostly Web based, SaaS allows better integration among the business units of a given organization or even among different software services. Examples of SaaS include [7] Google Docs and CRM.

Plataform as a Service (PaaS): Is the middle component of the service layer in the cloud. It offers users software and services that do not require downloads or installations. PaaS provides an infrastructure with a high level of integration in order to implement and test cloud applications. The user does not manage the infrastructure (including network, servers, operating systems and storage), but he controls deployed applications and, possibly, their configurations [4]. PaaS provides an operating system, programming languages and application programming environments. Therefore, it enables more efficient software systems implementation, as it includes tools for development and collaboration among developers. Examples of SaaS [7] include: Azure Services Platform (Azure), Force.com, EngineYard and Google App Engine.

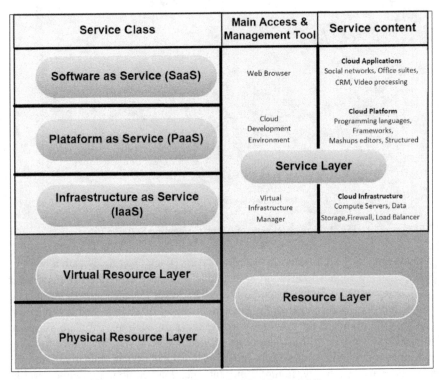

Fig. 2. Cloud Computing Architecture [4]

Infrastructure as a Service (IaaS): Is the portion of the architecture responsible for providing the infrastructure necessary for PaaS and SaaS. Its main objective is to make resources such as servers, network and storage more readily accessible by including applications and operating systems. Thus, it offers basic infrastructure on-demand services. IaaS has a unique interface for infrastructure management, an

Application Programming Interface (API) for interactions with hosts, switches, and routers, and the capability of adding new equipment in a simple and transparent manner. In general, the user does not manage the underlying hardware in the cloud infrastructure, but he controls the operating systems, storage and deployed applications. The main benefit provided by IaaS is the pay-per-use business model [4]. Examples of IaaS [7] include: Amazon Elastic Cloud Computing (EC2) and Eucalyptus.

According to the intended access methods and availability of cloud computing environments, there are different models of deployment; they include private cloud, public cloud, community cloud, and hybrid cloud [9].

Table 1. Models of deployment of cloud services [9]

Cloud Model	Description
Private	In this model, the cloud infrastructure is exclusively used by a specific organization. The cloud may be local or remote, and managed by the company itself or by a third party. There are policies for accessing cloud services. The techniques employed to enforce such private model may be implemented by means of network management, service provider configuration, authorization and authentication technologies or a combination of these.
Public	Infrastructure is made available to the public at large and can be accessed by any user that knows the service location. In this model, no access restrictions can be applied and no authorization and authentication techniques can be used.
Community	Several organizations may share the cloud services. These services are supported by a specific community with similar interests such as mission, security requirements and policies, or considerations about flexibility. A cloud environment operating according to this model may exist locally or remotely and is normally managed by a commission that represents the community or by a third party.
Hybrid	Involves the composition of two or more clouds. These can be private, community or public clouds which are linked by a proprietary or standard technology that provides portability of data and applications among the composing clouds.

3   Solutions for Deployment

This section aims to investigate what are the possible options available in cloud computing to the university software factory.

3.1   IAAS

IAAS solutions primarily aim to provide virtual machines with all the features of cloud for software factory.

Microsoft Windows AZURE: Windows Azure, which was released on February 1, 2010. Is a cloud computing platform and infrastructure, created by Microsoft, for building, deploying and managing applications and services through a global network of Microsoft-managed datacenters. It provides both PaaS and IaaS services and supports many different programming languages, tools and frameworks, including both Microsoft-specific and third party software and systems [11].

OpenStack: OpenStack is an open source software able to manage components of multiple virtualized infrastructures, like the operational system manages the components of a computers, OpenStack is called Cloud Operating System, to fulfill the same role in larger scale.

OpenStack is a collection of open source software projects that companies and service providers can use to configure and operate their infrastructure computing and cloud storage. Rackspace and NASA (the U.S. space agency) were the main contributors to the initial project. Rackspace provided a platform "Cloud Files" to implement the storage object OpenStack, while NASA entered the "Nebula" for implementing the computational side.

Elastic Utility Computing Architecture Linking Your Programs To Useful Systems (Eucalyptus): The Eucalyptus project [10] is an open source infrastructure that provides a compatible interface with Amazon EC2, S3, Elastic Block Store (EBS) and allows users to create an infrastructure and experience the cloud computing interface. The Eucalyptus architecture is simple, flexible, modular and contains a hierarchical design which reflects the common features of the environment.

Eucalyptus aims to assist research and development of technologies for cloud computing and has the following features: compatible with EC2, simple installation and deployment management using clusters tools, presents a set of allocation policies interface extensible cloud overlapping functionality that requires no modification on Linux environment, tools for managing and assisting the management of the system and users and the ability to configure multiple clusters, each with private internal network addresses in a single cloud.

3.2   PAAS

The PaaS solutions offer a platform for users to simply and quickly put their programs into production, thus providing an environment for quickly testing.

Tsuru: Tsuru is an open source and polyglot platform for cloud computing developed by globo.com since 2012 and it has began to be offered in a preliminary version in 2013 [12]. Like other platforms, the Tsuru helps the development of web applications without the any charge of a server environment. Tsuru uses Juju orchestration of services and takes advantage of the attractive features of its architecture. Supported programming development languages include Go, Java, Python and Ruby.

Heroku: Heroku [13] is a platform as a cloud service with support for several pro-gramming languages. Heroku was acquired by Salesforce.com in 2010. Heroku is one of the first cloud platforms and it has been in development since June 2007, when it supported only the Ruby programming language, but since then added support for Java, Node.js, Scala, Python and Perl. The base operating system is Debian or in the latest Ubuntu based on Debian.

3.3   SAAS

Owncloud: The Owncloud is a free and open source web application for data syn-chronization, file sharing and remote storage of documents written in scripting languages PHP and Java Script. The Owncloud is very similar to the widely used Dropbox, with the primary difference being that ownCloud is free and open-source, thereby giving to anyone the option to install on your own private server, with no limits on storage space (except for hard disk capacity).

Expresso: Expresso [14] is a complete communication solution that brings together email, calendar, address book, instant messaging and workflow in a single environ-ment. Because it is a custom version of the E-GroupWare, its development is also based entirely on free software.

4   Proposed Deployment

From the studies of cloud computing solutions, a proposal was created based on ease of deployment X relevance X time to deployment. For this paper, two software (SaaS) were chosen to contribute in collaborative software factory. Such software should help to ensure that members of the factory could share project documents, tasks, shared calendars and other tools for project management. An important factor to be considered during the time of adoption of a cloud service is the security. This security issue has attracted several discussions with the Brazilian Federal Government to the extent of having been issued a presidential decree (Decree No. 8.135, of November 4, 2013). The first article of this decree is as follows:

"Article 1 - Data communications direct, independent federal government and foundations shall be conducted by telecommunications networks and services of in-formation technology provided by agencies or entities of the federal public adminis-tration, including public enterprises and joint stock companies of the Union and its subsidiaries."

This article clearly shows the concern of the Government, which began to be wide-ly commented after being publicize cases of espionage on the emails of the President of Brazil.

4.1   IAAS Proposal

The first option selected for the software factory was in IaaS solutions, i.e., to provide a software factory the entire necessary infrastructure in a transparent and scalable way. The software factory does not provide today's servers and storage required to keep, such hardware resources are still going through the bidding process. By aiming to solve this problem, the main solution is that use virtual machines until the factory have their own means of keeping it going. For the provisioning of virtual machines, the XEN Hypervisor [15] was chosen as a solution, because it has a large use in the market besides being open source and has already been studied previously. Figure 3 presents the installation of two Linux virtual machines on the Windows client. One of the machines is a Linux server and the other one with an Ubuntu GUI.

The interesting aspect that can be seen in the Figure 4 is that we can control independently and completely the memory, the disk and the network; this way, we can have an idea if the VM was well provisioned. But, in the course of the project, we came across a pleasant surprise; CESPE (Center of Selection and Promotion Events), which is one of the partners of the factory, provides virtual machines using the Windows solution, Microsoft Azure and then making XEN relevant only for research. These machines will be used for the allocation of software offered as a service (SaaS) to the factory.

4.2   PaaS Proposal

As a software factory, it is very interesting that the factory is able to produce prototypes as quickly as possible, because, as soon as the customer has a prototype, reduces the risks of software that does not add useful features to the client and sooner he will have a preview of the software that will be delivered. It can be very interesting to the factory because it will have a platform where you can quickly put the software into production and can show their customers.

For this type of service, both OpenShift and Tsuru tools were analyzed, but neither locally, although OpenShift proved to be a powerful tool for such functions. OpenShift supports major language currently used and is a leading solution Open Source.

Fig. 3. Xen Client

4.3   SAAS Proposal

Lastly, we evaluated SaaS solutions and at the first view, these solutions did not prove to be relevant compared to IaaS and PaaS, because the Internet is full of them, such as Dropbox and Google Docs; but, one of the main reasons that lead to not use such software is the security that they do not provide. Because it is a software factory, is extremely important that their projects are not opened at all to other users. It is directly related to the decree mentioned before, SERPRO (Federal Data Processing Service) as the primary IT Company of the federal government, which was designed to provide such services and especially covering the second paragraph.

"The agencies and entities of the Union referred to header of this article should adopt the email services and its additional features offered by agencies and entities of the federal public administration."

With this scope, Expresso [14] comes in as an email platform and other features that SERPRO is the main developer and it has an internal allocation of its resources, thereby increasing the security of the application. As previously mentioned, the factory needs solutions that add value to collaborative work. As a solution to this problem, two solutions for SaaS are mentioned:

1.   Expresso

Expresso, as mentioned in the previous section, is one complete communication solution, which includes email, calendar, address book, instant messaging and workflow in a single environment; in Figure 3, the log in of Expresso is shown. This solution greatly facilitates the work of the members of the software factory because they can set up meetings and schedule them on a shared agenda, facilitating the allocation of free time between them. Another very interesting service to be used by the factory is the video call, where project members can have meetings without leaving home, facilitating meetings and expediting meetings, which no longer need to be physically occurred. We cannot forget of course the principal Expresso service, which is email, which will be much more secure than allocated in the internet environments, such as Gmail, Hotmail, etc. In Figure 5, the main screen of Expresso is shown.

Fig. 4. Expresso Log in

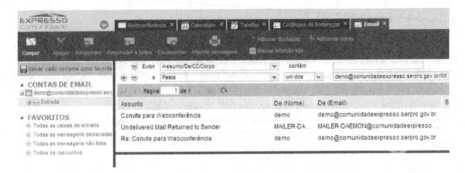

Fig. 5. Main Screen of Expresso

Looking at the top of Figure 5, we can see all five tabs that represent the key features of Expresso. First, the selected tab shows the functionality of email with multiple filter options, favorites, etc. Beside the email tab, we have the address book, where all the user's contacts are saved. The next tab shows a very interesting feature that is the task manager where for example a team manager can delegate tasks to any of his members in a simple and fast way. Perhaps, after the email feature, the second most used feature used by the teams is the calendar functionality, because the managers can easily manage the agenda of each of the participants, and facilitate the allocation of meetings, deadlines, etc.

Lastly, we have the functionality of web conferencing, where members of a meeting for example, may join a video conference by simply accepting a request via email, facilitating the occurrence of meetings distributed teams or with difficulty time to face meetings.

2. Owncloud

Besides the implementation of the email service, is intended to provide to members of the factory an archive, where it will be possible to share important documents in a safer environment then other solutions, because the files are storage in the factory servers. The software chosen was the ownCloud, which is a similar solution as the Dropbox; ownCloud is free open source software and has great community support. OwnCloud has as main feature the ability for the users to store their files, so that it can access on any computer with the client, both desktop and smartphones.

Looking at Figure 6, it is possible to notice the options allocation of ownCloud; the first file is an example of code written in C ++. This type of file can be viewed on its own web interface, already accepted by application and a similar visualization to the IDE for this extension and can it be edited directly from the browser, thus making it easier for programmers in the factory. The same applies for all other languages shown in Figure 6 and also to txt files, printing, and others. By clicking on the images, a pop-up appears with the selected image, facilitating the visualization of it. Lastly we have the songs that can be played directly from the internet.

Fig. 6. Main Screen of Owncloud

All these files can be downloaded in two ways: First, when passed the mouse over the file you click on download, and the download is done. On the other hand, the second way, that is by downloading the client ownCloud, so you keep all your files updated.

5   Deployment

The implementation of Expresso and ownCloud in the software factory is a partnership between three institutions SERPRO, UNB and CESPE. The CESPE assumes the role of infrastructure provider, providing virtual machines and the entire necessary hardware infrastructure for deploying SaaS subsequently this work will be assuming the role of supplier of service provider, providing Expresso to the factory and its customers. This interaction can be best represented by Figure 7.

Fig. 7. Roles in cloud computing in this context

5.1   Process

The deployment of services is a process that will take all the time devoted to a future work. Figure 8 proposes an initial process that can be appropriately adjusted within the first stage of this process.

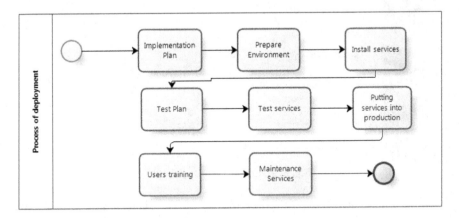

Fig. 8. Process of deployment

5.2   Implementation Plan

This will be the beginning of the process for the services implementation; this phase is intended to study more "low level" of that material (storage, servers) and will be needed for deployment. In this case, we need a better study of the Expresso and the ownCloud to clarify what are the elements that need to be installed and configured.

5.3   Prepare Environment

At this stage, we already have a good overview of the elements necessary for the installation of the Expresso and the ownCloud. CESPE will help during this phase preparing and configuring all virtual machines.

5.4   Install Services

This phase is intended to install/configure two services presented; this phase is perhaps one of the most time- consuming. Since there is not yet a broad understanding of the process of the installation of services, always unexpected can occur and end up taking a long time. In the case of Expresso, we have a greater ease, since they can have SEPRO aid.

5.5   Test Plan

After the services are already installed, it is planned to test the software; one can create a document specifying at this stage all the tests that will be performed, as well as expected results and comparing them with the results in the next stage.

5.6   Test services

After services are fully installed, you must run the tests planned in the previous step for problems that may have occurred in the installation, as a wrong connection with database, causing when people try to login or save your files. In this case you should review if the database has been properly installed and configured, and test again until all tests of test plan are completed.

5.7   Putting Services into Production

This phase along with the installation is perhaps the most difficult and time consuming. This is because one must ensure that the services are running correctly, and ensure that clients are using them. It should always be close to customers looking to receive feedback on potential improvements and problems.

5.8   Users Training

Right after the services have been put into production, users need a little coaching on how to use the tools. Since users are mostly students of software engineering, this phase will be very short and easy. However, as the services needed for future maintenance is also included in this phase the transfer of knowledge for any member of the factory, so it can continue the operation of the service. At this stage, we will have the help of SERPRO also, which may give workshops on the operation of the Expresso, thus arousing the interest of members of the factory proposing future functionalities for the tool.

5.9   Maintenance Services

As most software evolves and presents problems, there is always the need for continued maintenance and upgrades thereof. The members walked in the previous training will first take the maintenance services.

6   Deployment Results

After having defined the Expresso and the ownCloud as software as services to be deployed, it moved to the implementation step. Firstly CESPE released 3 virtual machines with Debian 7 operating system and Intel Xeon x5660 2.8 GHZ 4 cores and 4GB memory, two machines with 100GB HD and the third one with 1TB . In the first virtual machine, we installed all the software's and firstly was installed the Expresso and the OwnCloud, in the second one we installed one LDAP (Lightweight Directory Access Protocol) to control all the active users in this cloud, and in the third one the Postgres database, this organization can be seen in Figure 8.

Today, 5 months after the installation, Expresso proved to be a powerful communication suite and the OwnCloud surprises by its functionality and interface very similar to the already known Dropbox. However, later we felt the need to install two new applications that were not previously thought, the first was the gitlab, a web-based Git repository manager, and the second was the Redmine for project management. LDAP proved an important tool to control users since we had to control all users of all installed applications in the cloud and can add and remove new users easily.

We opted for a centralized database for both applications, thereby centralizing all information; thus, it is safer to keep the data because you can restrict the access to this virtual machine, for example by closing all ports in the firewall, leaving only port 5432 open, hindering unauthorized access.

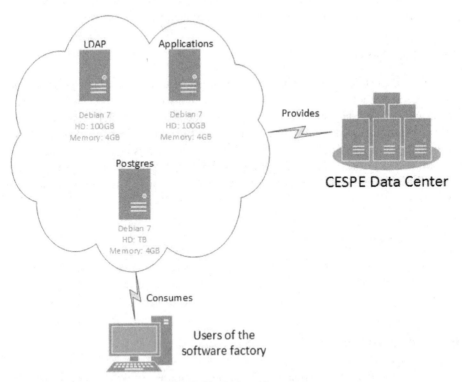

Fig. 9. Physical architecture

7   Conclusion

With each passing day cloud computing has been present in our lives, not only for personal use, but also increasingly in professional use. As much as cloud computing brings many benefits, it also brings many challenges. One of the major challenges is in relation to data security, especially after the scandals in Brazil and in the world in relation to privacy.

This paper provided an overview of tools that implement the three main types of cloud architecture; additionally, a proposal for the implementation of two of the tools analyzed, one ownCloud, which deploys a system of file sharing, and Expresso as email service.

Using these services is expected that the collaborative development of the factory becomes more facilitated and reliable with the exchange of files and emails, which will be more secure than allocated in the internet environments, such as Gmail and Dropbox.

References

1. Miller. Cloud Computing Web-Based Applications, That Change the Way You Work and Collaborate online. Que Publishing, Pearson Education, Canada (2008)
2. Zhou, M., Zhang, R., Zeng, D., Qian, W.: "Services in the cloud computing era: a survey," Software Engineering Institute. In: 4th International Universal Communication. Symposium (IUCS), pp. 40–46. IEEE, Shanghai (2010)
3. Uppoor, S., Flouris, M., Bilas, A.: Cloud-based synchronization of distributed file system hierarchies. In: IEEE International Conference on Cluster Computing Workshops and Posters (CLUSTER WORKSHOPS), pp. 1–4 (2010)
4. Jing, X., Jian-Jun, Z.: A Brief Survey on the Security Model of Cloud Computing. In: Ninth International Symposium on Distributed Computing and Applications to Business, Engineering and Science (DCABES), pp. 475–478. IEEE, Hong Kong, August 2010
5. Zhou, M., Zhang, R., Zeng, D., Qian, W.: "Services in the cloud computing era: a survey," Software Engineering Institute. In: 4th International Universal Communication of Symposium (IUCS). IEEE, Shanghai, pp. 40–46 (2010)
6. Velve, A.T., Elsenpeter, T.J.: Cloud Computing – Computação em Nuvem - Uma Abordagem Prática, Mei, G.E. (translator.), pp. 352–359. Alta Books (2011)
7. Jian-Jun, Z., Jing, X.: A Brief Survey on the Security odel of Cloud Computing. In: Ninth International Symposium on Distributed Computing and Applications to Business, Engineering and Science, pp. 475–478 (2010)
8. Marinos, A., Briscoe, G.: Community Cloud Computing. In: Jaatun, M.G., Zhao, G., Rong, C. (eds.) Cloud Computing. LNCS, vol. 5931, pp. 472–484. Springer, Heidelberg (2009)
9. Mell, P., Grance, T.: The NIST definition of cloud computing. Technical report, National Institute of Standards and Technology (August 2009)

10. Liu, S., Liang, Y., Brooks, M.: Eucalyptus: a web service-enabled einfrastructure. In: Conference of the Centre for Advanced Studies on Collaborative Research (CASCON 2007), pp. 1–11, October 2007
11. Windows Azure - The Official Microsoft. http://azure.microsoft.com/en-us
12. Tsuru-A Open Source PAAS from Globo.com. http://www.tsuru.io/
13. Heroku. https://devcenter.heroku.com/
14. Expresso http://www.expressolivre.org

A Meta-Information Extractor for Interrogative Sentences

Cleyton Souza[1]([✉]), Joaquim Maia[1], Luiz Silva[1], Jonathas Magalhães[2], Heitor Barros[2], Evandro Costa[3], and Joseana Fechine[2]

[1] Federal Institute of Education, Science and Technology of Paraíba - IFPB, Monteiro, PB, Brazil
cleyton.caetano.souza@gmail.com
[2] Federal University of Campina Grande - UFCG, Campina Grande, PB, Brazil
[3] Federal University of Alagoas - UFAL, Maceió, AL, Brazil

Abstract. The development of tools for Information Retrieval Systems or Expertise Finding Systems has a common task: the understanding of the information need. Since the information need is usually expressed through natural language, the computational processing of the information need involves several NLP techniques. Fortunately, there is a vast set of tools for English, but others languages have been marginalized. Thus, in this paper, we present a Web Service that offers to clients NLP treatment for interrogative sentences written in Brazilian Portuguese. The Web Service receives the question as input e returns its meta-information. The main differential of our proposal is that we offer a full analysis of the question text using a single function. We evaluate a feature of the Web Service named Category labeler, responsible for automatic discover the subject of the question, and we found that it has a true positive rate higher than 50% (α=10%).

Keywords: Natural Language Processing · Brazilian portuguese · NLP · Web services

1 Introduction

In current society, information plays a major role. The lack of information is one of the main issues faced by people and organizations [3]. In this context, inside computing, we have many areas that the main goal is to solve information needs. The Information Retrieval (IR) area, for instance, handles with the searching for documents that can solve an information need. Expertise Finding Systems (EFS) are applied to recommend experts to offer supporting to a task or an issue.

In all these cases, understanding the information need is an essential step. Since the information need is usually expressed through natural language, several techniques from Natural Language Processing (NLP) area are frequently applied in order to computationally represent the requests [1]. Nowadays, there are many tools that offer support to the application of these techniques [7]. However, there is still a gap in literature for NLP tools for other languages besides English,

© Springer International Publishing Switzerland 2015
O. Gervasi et al. (Eds.): ICCSA 2015, Part I, LNCS 9155, pp. 345–354, 2015.
DOI: 10.1007/978-3-319-21404-7_25

e.g., Portuguese. In addition, each tool usually provides access to only a small set of techniques, being necessary to combine many solutions in order to achieve a more complex and complete analysis.

In this paper, we present a NLP Web Service that offers only one function; specifically destined for the analysis of interrogative sentences written in Brazilian Portuguese. The input of the Web Service should be a question and the output will be the meta-information related to the question, such as topic, type, vector model representation, etc. We are not proposing a novel NLP technique, but making available a tool that could be used for many ends, as will be fully demonstrated later.

This work is organized as follows: Section 2 presents the Related Work; Section 3 details our proposal; Section 4 shows results related to the evaluation of a couple of our meta-information extractors; and Section 5 ends the paper with Conclusions and Future Work.

2   Related Work

We are not the first work about providing NLP techniques via Web Service for a non-English language neither for Portuguese. Ogrodniczuk and Przepirkowski [7] summarize and compare eight frameworks for NLP in a more complete review than we aim providing in this section. Our related work analysis concentrates in NLP Web Services for non-English languages.

In [6], it is described a text handler that analyzes free text and outputs sentence boundaries, among other basic patters. Their tool is meant for Catalan and Spanish, although authors say that it could be expanded for English. According to Martinez et al. [6], even basic functionalities are not equally available for every language. In addition, the authors propose a Web Service Description Language (WSDL) interface for text handling tools.

Erygit [4] presented a NLP platform, namely ITU Turkish NLP Web Service, that provides the state of art NLP tools for Turkish language. The users may communicate with the platform via three channels: (1) via a Web interface, (2) by file upload, and (3) by using a Web API. According to Erygit [4], the importance of providing such tools this way comes from the difficulty of sharing NLP resources through different people with distinct purposes and with varying level of computer background.

Prokopidis et al. [8] provide an overview of NLP technologies available from the Institute for Language and Speech Processing (ILSP). This NLP suite is exclusive for the Greek language and comprises a series of processing units based on both machine learning and rule-based approaches. According to the authors, given the large number of linguistic services and tools already developed by various organizations, it is imperative made a NLP suite like theirs available as a Web Service that can be combined with services provided by other teams in larger processing workflows.

In [2], it is reported the development of a cluster of Web Services for NLP for Portuguese named LXService. This cluster includes a Sentence Chunker, a Tokenizer, POS tagger, Nominal featurizer, Nominal lemmatizer, Verbal featurizer and lemmatizer, Verbal conjugator and Nominal inflector; all available through four Web Services named: LX-Inflector, LX-Lemmatizer, LX-Conjugator and LX-Suite. These tools provide the basic of NLP functions that can be further combined and expanded by other client applications.

Branco et al. [2] is the closest work to ours, however, we consider that our work is an interesting complement to theirs. Specifically, Branco et al.s work provides many basic functions, while we are focusing in providing a more complex NLP analysis; and specialized in interrogative sentences. This specialization results in, at least, two additional functions: (1) a Question-category labeler and (2) a Question-type labeler. Regarding the complex analysis, our aim is making in a single operation a full analysis of the question text in order to extract its meta-information, as it will be detailed in Section 3.

In addition to this tools, we want highlight the Linguateca Project [5]. Its goal is being a resource center of tools for the computational processing of Portuguese. They make the access to these existing tools easier through the aggregation of their information in their web site. The importance of such space is huge, either for experienced researchers in computing, and for students with little or none background in computing.

3   Proposal

Our proposal consists in a Web Service that extracts the following meta-information from the questions using a single operation. Although, it is possible also invoke the Web Services separately. In the following, we describe what consists these meta-information.

- **Vector-model Representation** – It represents the question text using the vector-model. In this representation, each coordinate of the vector represents a stem present in the question text and the value of the coordinate is the product between the TF and IDF. To build this vector, we used Lucene API. It provides many NLP basic functions like tokenization, stop-words removal and stemming. In addition, it has analyzers available for several languages, including Portuguese.
- **Category Label** – Each question has a subject. The category label single represents the question subject. The list of possible categories is pre-defined. Each category is also represented by a vector. We calculate the cosine similarity between the question vector and all category vectors. The one with higher similarity is assigned as Category of the question.
- **Type Label** – Questions can be classified according its intention. Morris et al. (2010) proposed a classification into one of eight types: recommendation, opinion, factual knowledge, rhetorical, invitation, favor, social connection and offer. We represent the question using some descriptors. Next, we use a Naive Bayes to automatically classify the question into one of these eight types.

- **Category Similarities** – It represents an array of tuples with the Category label plus the cosine similarity. Before we label the question with a category, we test the similarity between the question-vector and all category-vector in order to find the higher similarity. All these data is persisted in an Array.
- **Type Probabilities** – It represents an array of tuples with the Type label plus the probability calculated by the Naive Bayes model. The output of the Naive Bayes is a table with probabilities. The meta-information consists in this table.
- **Expressions** – It represents an array of tuples with Entities mentioned in the question text plus a categorization of the entity (e.g., place, date, people, product, etc.). We use a NER tool for this extraction. However, we are still testing the efficacy of different tools with Brazilian Portuguese. Currently, we are working with Stanford NER and Open NLP API.
- **Interrogative Particle** – It represents the interrogative particle of the question. In English would be equivalent to what, who, where, how, etc. Before apply the stop-words removal, we basically search in the question text for one or more of the Portuguese interrogative particle.

3.1 Design

Figure 1 presents the components of the proposed Web Service. These components are discussed later.

- **Information Extraction Services** (IES) Layer – This layer contains the functions used to extract the information from the question. These functions were also separately implemented as Web Service, but can be indirectly invoked through our proposal. This separation leads to high cohesion and low coupling; making the maintaining and reuse of the code easier and also allowing other applications to single handling with them. How the functions were implemented is described in the next section.
- **Management** Layer – The role of this layer is to manage all others Web Services. In addition, it works like a Facade that provides sequential access to all Web Services functions. Basically, this layer has a single Web Service that defines the order in which each Web Service of the IES Layer will be invoked and also invoke them. The input of this Web Service should be a question and the output will be the meta-information related to the question.
- **Client** Layer – This layer is responsible for invoke the functions from the Management Layer to get the meta-information from the question. Due the dynamic nature of Web Services, these clients could be implemented in any platform or language and could even be others Web Services.

3.2 Service Workflow

In this section, we briefly detail the sequential process of execution of the Service Manager. Figure 2 is an Activity Diagram of the Service Manager, which is the Web Service responsible for invoke the others.

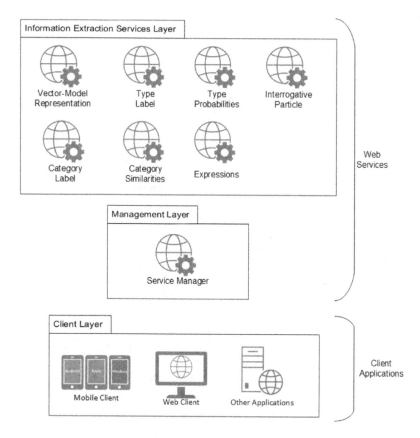

Fig. 1. Components of the Proposed Web Service

First, the Service Managers receives the question as input. Next, it starts the process of call the remaining services in chain. The first service to be call returns the Vector-Model representation of the question. It is the first because some of the others Services depend of this information like the Category Labeler. Next, the remaining Web Services are executed in parallel. Each one returns a different meta-information. These data are combined and structured before the Service Manager returns it to the client. It is interesting to notice that the Category Similarities Service is invoked before the Category Label. Since the function of this service is to estimate the probability of the question fit in each category, it is imperative that it be executed first, to guarantee a minimum of efficiency. The same can be said to the Type Label Service and Type Probabilities Services.

In the end, after all services being executed, their results are checked for inconsistencies and the meta-information structured in returned in a XML format. An example of such file is presented in Figure 3.

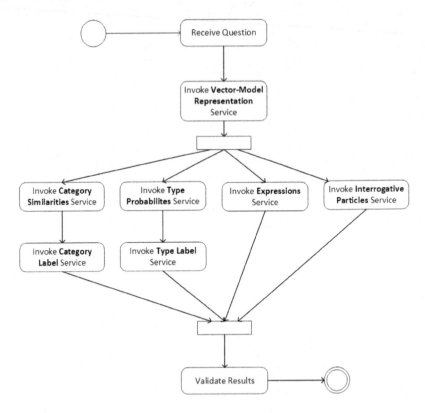

Fig. 2. Activity Diagram of the Proposal

4   Evaluation and Results

As our proposal has many functions and some of these functions are very basic, we concentrate our evaluation in the more complex features: the Category Labeler, the Type Labeler and Named Entity Recognizer.

4.1   The Category Labeler

Methodology. To evaluate the category labeler function, we performed an experiment using two datasets. The datasets are composed by a set of questions labeled with at least one tag, but it can has multiple, and the goal of the experiment is to confirm if the category labeler is able to correctly assign one of these tags.

We split each dataset in a set for training (30%) and other for testing (70%). The questions in the training dataset were used to build the vectors for each category. A category vector is built by the merging among all questions in this category of the training dataset. Next, we compared each question vector of the test dataset with the category vectors using the cosine similarity. The highest similarity was used to assign a category for the question.

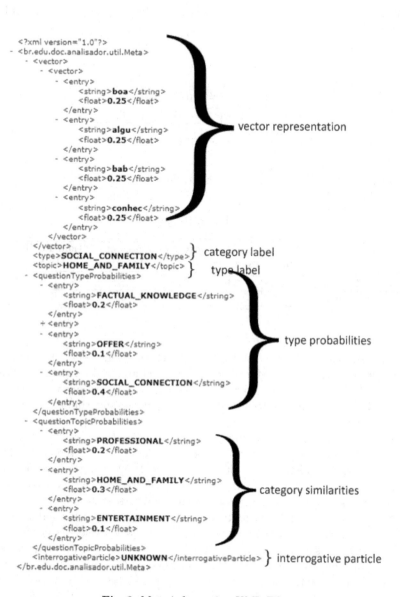

```
<?xml version="1.0"?>
- <br.edu.doc.analisador.util.Meta>
  - <vector>
    - <vector>
      - <entry>
          <string>boa</string>
          <float>0.25</float>
        </entry>
      - <entry>
          <string>algu</string>
          <float>0.25</float>
        </entry>
      - <entry>
          <string>bab</string>
          <float>0.25</float>
        </entry>
      - <entry>
          <string>conhec</string>
          <float>0.25</float>
        </entry>
      </entry>
    </vector>
  </vector>
  <type>SOCIAL_CONNECTION</type>
  <topic>HOME_AND_FAMILY</topic>
  - <questionTypeProbabilities>
    - <entry>
        <string>FACTUAL_KNOWLEDGE</string>
        <float>0.2</float>
      </entry>
    + <entry>
    - <entry>
        <string>OFFER</string>
        <float>0.1</float>
      </entry>
    - <entry>
        <string>SOCIAL_CONNECTION</string>
        <float>0.4</float>
      </entry>
    </questionTypeProbabilities>
  - <questionTopicProbabilities>
    - <entry>
        <string>PROFESSIONAL</string>
        <float>0.2</float>
      </entry>
    - <entry>
        <string>HOME_AND_FAMILY</string>
        <float>0.3</float>
      </entry>
    - <entry>
        <string>ENTERTAINMENT</string>
        <float>0.1</float>
      </entry>
    </questionTopicProbabilities>
  <interrogativeParticle>UNKNOWN</interrogativeParticle>
</br.edu.doc.analisador.util.Meta>
```

} vector representation

} category label

} type label

} type probabilities

} category similarities

} interrogative particle

Fig. 3. Meta-information XML File

The first dataset was built with questions from Stack Overflow in Portuguese[1]. The other dataset was built with questions from Forro Square[2], a regional CQA site for questions about So Joo, a traditional party in Brazil that happens in June. Table 1 details both datasets. Both datasets are available for download[3].

Table 1. Datasets detailing

Dataset	# Questions	# Tags
Stack Overflow	501	60
Forro Square	65	8

Results. Next, in Table 2, we summarize the results for both datasets.

Table 2. Results

Dataset	True Positive	False Positive	Left Out	Amount of Questions	True Positive Rate
Stack Overflow	267	223	11	501	53%
Forro Square	42	19	4	65	64%

Discussion. As can be seen in Table 2, the Category Labeler achieved a True Positive rate of 53% for the Stack Overflow dataset and 64% for the Forro Square dataset. The left out column represents the number of questions which the Category Labeler was not able to assign a category due cosine similarity equal to 0 in all comparisons.

We used a one-tailed-right binomial test to confirm these results statistically. The test showed that the Category labeler statistically achieved a rate higher than 50% with a confidence level of 95% (p-value 0.01241) in the classification of questions from the Forro Square dataset. Since this dataset has far less questions than the Stack Overflow, it is not too difficult achieve such results. However, it is easier to observe some interesting tendencies. For instance, as we said a question can have multiple tags. In this scenario, if the Category labeler assigns any of its tags we counted as a hit. However, as the tags were assigned firstly by users, we are assuming that they are correct, but in a small dataset like that, we observed the over-use of some tags and also, sometimes, the incorrect use. As our training is based on assigned tags it is expected that questions from the same category share a common vocabulary, but the over-use and incorrect assignment of tags makes it harder.

---

[1] http://pt.stackoverflow.com/

[2] http://forrosquare.lsd.ufcg.edu.br/index.php

[3] https://docs.google.com/file/d/0B4ZS_d4fhCZVdVdpZEw2LW04enc/

Regarding the Stack Overflow dataset, the category labeler also achieved a True Positive rate statistically higher than 50%, but for a confidence level of 90% ($p-value$=0.07637). These results are quite impressive; when we analyze that the Stack Overflow dataset has over 300 hundred different tags. Higher the amount of tags more difficult is for the Category Labeler correctly assign one, since its decision is sole based on the cosine similarity comparison. In addition, the tags from the Stack Overflow dataset had close subjects. For instance, there was a tag called mysql, but there was also tags like foreign key, database and table-database[4]. So, the possibility of a common vocabulary makes more difficult the task of the Category labeler. However, this was not enough for interfere with the results, since the results for both datasets were so similar.

Threats to Validity. We identify two types of threats to these results regarding External and Construct validity.

External validity refers to the generalizability of the results. Regarding external validity, although we evaluated the Category Labeler using dataset from two different domains (with two groups of unrelated tags) and we found similar results with both of them, we cant assume that this will repeat through different domains. However, we are planning another study case using more generic categories like those used on Quora[5] or Yahoo! Answers[6]. May seem odd do not use Yahoo! Answers on this study, but, in the time when we were planning this evaluation, unfortunately, Yahoo! Answers API was discontinued[7].

Construct validity defines how well a test or experiment measures up to its claims and relates with problems in the design of the experiment. Regarding construct validity, we build two datasets to use in the experiment, however we do not consider treat our control the dataset data. Thus, there is no intended balance in the number of questions with each tag or accuracy of assigned tags taken as ground truth.

5 Conclusion and Future Work

In this paper, we present a Web Service for NLP processing of interrogative sentences written in Brazilian Portuguese. This tool realizes with a single function a full analysis over the question text and returns a XML containing all the meta-information extracted. In [9], we have been using such analysis for Expertise Finding and Hint Recommendation. However, we believe that there are many other applications like information retrieval and information extraction, for instance.

We validate so far one of the features of our proposal named Category labeler. A Web Service responsible for assign a label to categorize the subject of the

---

[4] This tags were translated to english, but they are in portuguese in the dataset.
[5] http://www.quora.com/
[6] http//answers.yahoo.com
[7] https://developer.yahoo.com/answers/

question. The experiment with two wide different datasets showed that this feature has a true positive rate of over 50%. The interesting part is that the Category Labeler reached such results in a dataset with more than 60 possible tags for a question.

As future work, we planning evaluate other features of our proposal that are still in development, such as the Named Entity Recognizer and the Type labeler. In addition, we are planning developing other tools as clients of this Web Service, like a Chatter Bot able to offer answers or a Information Extraction System, that look in documents the answer for questions asked in natural language.

Acknowledgments. We want to thanks the Laboratory of Distributed Systems (LSD) from the Federal University of Campina Grande (UFCG) for give in the dataset of the Forro Square.

References

1. Baeza-Yates, R., Ribeiro-Neto, B.: Modern Information Retrieval: The Concepts and Technology Behind Search. Addison Wesley (2011)
2. Branco, A., Costa, F., Martins, P., Nunes, F., Silva, J., Silveira, S.: Lx-service: web services of language technology for portuguese. In: LREC 2008 (2008)
3. Castells, M.: The Rise of the Network Society: The Information Age: Economy, Society, and Culture. Information Age Series. Wiley (2010)
4. Eryiğit, G.: Itu turkish nlp web service. In: Proceedings of the Demonstrations at the 14th Conference of the European Chapter of the Association for Computational Linguistics, pp. 1–4. Association for Computational Linguistics (2014)
5. Linguateca. Processamento computacional do português, January 2003. http://www.linguateca.pt/proc_comp_port.html (retrieved February, 2015)
6. Martínez, H., Vivaldi, J., Villegas, M.: Text handling as a web service for the IULA processing pipeline. In: Proceedings of the Workshop on Web Services and Processing Pipelines in HLT: Tool Evaluation, LR Production and Validation (WSPP 2010) at the Language Resources and Evaluation Conference (LREC 2010), pp. 22–29 (2010)
7. Ogrodniczuk, M., Przepiórkowski, A.: Linguistic processing chains as web services: initial linguistic considerations. In: Proceedings of the Workshop on Web Services and Processing Pipelines in HLT: Tool Evaluation, LR Production and Validation (WSPP 2010) at the Language Resources and Evaluation Conference (LREC 2010), pp. 1–7 (2010)
8. Prokopidis, P., Georgantopoulos, B., Papageorgiou, H.: A suite of NLP tools for Greek. In: The 10th International Conference of Greek Linguistics, Komotini, Greece (2011)
9. Souza, C., Magalhães, J., Costa, E., Fechine, J., Reis, R.: Enhancing the status message question asking process on facebook. In: Murgante, B., et al. (eds.) ICCSA 2014, Part IV. LNCS, vol. 8582, pp. 682–695. Springer, Heidelberg (2014)

Latency Optimization for Resource Allocation in Cloud Computing System

Masoud Nosrati[1(✉)], Abdolah Chalechale[2], and Ronak Karimi[1]

[1] Kermanshah Branch, Islamic Azad University, Kermanshah, Iran
minibigs_m@yahoo.co.uk
[2] Department of Computer Engineering, Razi University, Kermanshah, Iran

Abstract. Recent studies in different fields of science caused emergence of needs for high performance computing systems like Cloud. A critical issue in design and implementation of such systems is resource allocation which is directly affected by internal and external factors like the number of nodes, geographical distance and communication latencies. Many optimizations took place in resource allocation methods in order to achieve better performance by concentrating on computing, network and energy resources. Communication latencies as a limitation of network resources have always been playing an important role in parallel processing (especially in fine-grained programs). In this paper, we are going to have a survey on the resource allocation issue in Cloud and then do an optimization on common resource allocation method based on the latencies of communications. Due to it, we added a table to Resource Agent (entity that allocates resources to the applicants) to hold the history of previous allocations. Then, a probability matrix was constructed for allocation of resources partially based on the history of latencies. Response time was considered as a metric for evaluation of proposed method. Results indicated the better response time, especially by increasing the number of tasks. Besides, the proposed method is inherently capable for detecting the unavailable resources through measuring the communication latencies. It assists other issues in cloud systems like migration, resource replication and fault –tolerance.

Keywords: Distributed systems · Resource allocation · Resource agent · Optimization in resource allocation · Latency of communication

1 Introduction

The distributed computers emerged to tie together the power of large number of resources distributed across a network [1]. The requirements of each user are shared on the network through a proper communication channel. It helps to utilize the capabilities of the whole of system for all users. Generally, High Performance Computing Systems (HPC) are defined as a collection of single coherent systems that are interconnected through a high speed network, to provide facility of high performance computing [2]. Many applications of High Performance Computing systems such as industrial [3],

© Springer International Publishing Switzerland 2015
O. Gervasi et al. (Eds.): ICCSA 2015, Part I, LNCS 9155, pp. 355–366, 2015.
DOI: 10.1007/978-3-319-21404-7_26

educational [4], medical [5] and commercial [6] came to existence after provision of hardware infrastructures. HPCs are preferred for the following reasons [7]:

- The nature of distributed applications is based on the network connections.
- Parallelism is provided in HPC by executing parallel grains on different machines.
- Higher reliability rather than single systems.

Previous studies in this area, categorizes the distributed systems to 3 types: Cluster, Grid and Cloud. In classic texts, Cluster is known as a distributed system with homogeneous nodes and Grid with heterogeneous ones [8][9]. But, in recent researches, there are Clusters with heterogeneous nodes implemented. It shows that homogeneity can't be a good metric for classification. Due to it, [7] categorizes the distributed systems by their resource allocation features as the fig.1. In this study, we will get into the resource allocation in Cloud systems.

Modeling of a Cloud system is not an easy errand to run. It is complicated because of wide range of different factors influencing the systems, such as: number of machines, types of applications, processing load and other important factors which can affect the system. Type of services is also a critical point. It can be Software as a Service (SaaS), Platform as a Service (PaaS) and Infrastructure as a Service (IaaS). Fig.2 shows some instances of these services. Another important notion is the issue of migration. Migration lets the system to achieve better performance, fault-tolerance, system management and load distribution. So, it is always considered in design and modeling of all Clouds. Quality of Service is other challenge in Cloud systems. QoS can be set in run time. This feature is called "Dynamic QoS Negotiation" [10]. DQoSN is implemented by a special entity or through self-algorithms [10].

The challenges that were talked can clearly indicate the delicacy and toughness of resource allocation in Cloud systems. Accordingly, [11] classifiés the recent studies in Cloud resource allocation in 3 types:

- Researches focusing on processing resources like [12] and [13].
- Researches focusing on network resources like [14] and [15].
- Researches focusing on power and energy resources like [16] and [17].

Also, [11] states that challenges are external or internal. External challenges include regulative and geographical challenges (it is about the geographical location, and regulative and security issues of data caused by distribution) and charging model issues (it is about the charges for customers to utilize the Cloud). In other hand, internal challenges include the data locality: combining compute and data management; reliability of network resources inside a data center; and Software Defined Networking design challenges inside the data centers. SDN is a networking paradigm that separates the forwarding plane from software control. Details are mentioned in [11].

One of the important issues in real world implementation of distributed systems is about the power and energy resources optimization. Different strategies are introduced in order to decrease the energy consumption. Most of them try to aggregate the resources on smaller number of servers, in order to shut down the non-busy ones. Resource aggregation has a trade-off with the performance of the system. Also, it will cause to have a bottle neck on the I/O of the running servers. Also, it may affect the Quality of Service of system. Average response time will be increased, respectively.

As the conclusion of this section, it should be restated that in every resource allocation method, all the aspects of processing capabilities, network resources and energy consumption must be considered. Regarding the trade-off among these factors, the best configuration should be found and set. Besides, other issues like the portability and fault-tolerance should be paid attention, respectively. Recent studies go through these issues separately and many approaches for optimization of recourse management are offered by them.

In the rest of this paper, we will have a brief look at the resource allocation and the strategies that Resource Agent (the entity that allocates the resources to applicants) utilizes to choose the best resource for best applicant. After pointing out the standard method of resource allocation, we offer our contribution based on the optimization of latencies between the resource and applicant. Results of simulation of proposed method shows the better performance of this method rather than the standard resource allocation approach. Finally it the end of paper, we will have a discussion and conclusion on the features of proposed method.

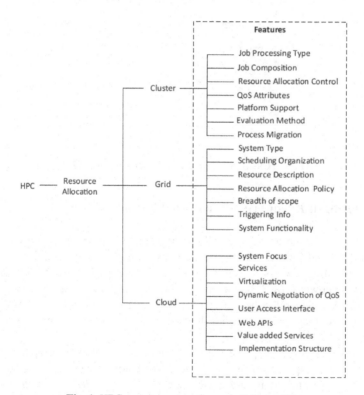

Fig. 1. HPC systems categories and attributes [7]

Fig. 2. Instances of Cloud services
Source: http://ohsweb.ohiohistory.org/ohioerc/?page_id=187

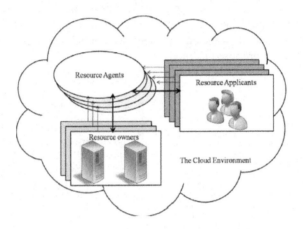

Fig. 3. Cloud resources allocation environment [21]

2   Common Resource Allocation Method

In this section, a common resource allocation method is talked that many researches like [18], [19] and [20] utilized it for their resource allocation optimization. In this method, a resource agent is considered as the entity that performs the resource allocation. As it is shown in fig.3, both resource owners and resource applicants send their costs to ResAg. General policy of ResAg is to sort the requests and resources and totally allocate the applicants with highest budget to the resource with lowest price [21].

Let $U = \{u_1, u_2, u_3, \dots, u_m\}$ be the set composed of m Resource applicants, each task of resource applicant u_i is t_i, so the task set of U can be described as $T = \{t_1, t_2, t_3, \dots, t_m\}$. And t_i has four attributes $t_i = \{tid_i, l_i, b_i, d_i\}$, where tid_i is the ith task's identify, l_i is the ith task's length, b_i is the ith task's budget, and d_i is the deadline of the task.

Let $O = \{o_1, o_2, o_3, \ldots, o_n\}$ be the set composed of n resource owners, each resource of resource owner o_j is r_j, so the resource set of O can be described as $R = \{r_1, r_2, r_3, \ldots, r_n\}$. And r_j has five attributes $r_j = \{rid_j, cpu_j, st_j, lp_j, hp_j\}$, where rid_j is the jth media resource's identify, cpu_j is the jth media resource's computing ability of solving the task, st_j is the start time to deal with a new task (i.e. the current workload of resource r_j), lp_j is the jth media resource's lowest price, and hp_j is the jth media resource's highest price.

The media resources allocation probability matrix is shown as P which each p_{ij} is the probability of resource j to be allocated to applicant i:

$$P = \begin{bmatrix} p_{11} & p_{12} & \cdots & p_{1n} \\ p_{21} & p_{22} & \cdots & p_{2n} \\ \vdots & \vdots & \vdots & \vdots \\ p_{m1} & p_{m2} & \cdots & p_{mn} \end{bmatrix} \quad \text{s.t.} \quad 0 \le p_{ij} \le 1 \;,\; \sum_{i=1}^{m} p_{ij} = 1 \;,\; \sum_{j=1}^{n} p_{ij} = 1$$

Before any decision about resource allocation, the budget of applicant and the price of resource should be calculated and submitted to ResAg. It is important to consider the following point all the time:

Resource must be capable enough to process the request of applicant in the deadline: $d_i - st_j - \frac{l_i}{cpu_j} \ge 0$

The price of resource must be less than or equal to the applicant: $b_i/l_i \ge lp_j$

Budget of applicant must be at least equal to the average of the price of remaining resources. $(\overline{lp} = \left(\frac{1}{n}\right) \sum_{j=1}^{n} lp_j)$

Budget of applicant is calculated from (1), where $bid_i^{resource}$ has inverse relationship with the number of remaining resources. It means, when the number of unallocated resources is decreasing, proposed budget of applicant for the resource will be increased, and vice versa. Other impressing factor that is Average Remaining Time, that [22] calculated it as (2). Total budget based on the (3) is the sum of both (1) and (2) with the weights of α' and β'. Weights might be changed based on the policies of the system.

$$bid_i^{resource}(t) = \overline{lp} + \left(\frac{b_i}{l_i} - \overline{lp}\right)\left(1 - \frac{n_i^t}{n_i^{max}}\right)^{\frac{1}{\alpha}} \tag{1}$$

$$\overline{rt}_i(t) = \sum_{j=1}^{n} \frac{(rt_{ij}(t)\omega_{ij})}{n_i^{max}} \;,\; \omega_{ij} = \begin{cases} 1 & if \; rt_{ij}(t) \ge 0 \\ 0 & otherwise \end{cases}$$

$$bid_i^{time}(t) = \overline{lp} + \left(\frac{b_i}{l_i} - \overline{lp}\right)\left(1 - \frac{\overline{rt}_i(t)}{rt_i^{max}}\right)^{\frac{1}{\beta}} \tag{2}$$

$$bid_i(t) = \alpha' bid_i^{resource}(t) + \beta' bid_i^{time}(t) \tag{3}$$

$0 \le \alpha', \beta' \le 1$

In these equations, n_i^t is the number of remaining resources in the time t, which can be applied by applicant t_i, and n_i^{max} is the maximum number of resources that might be applied by t_i. Remaining time of t_i who is utilizing r_j is calculated as $rt_{ij}(t) = d_i - st_j - l_i/cpu_j$. Let rt_i^{max} be the maximum time of waiting for t_i. Different applicant's budget curve can be adjusted by changing α and β. In fig.4 different values of α is shown. It has the same shape for β and σ (that will be introduced in next parts of paper).

After calculation of the budget of applicant, it is time to calculate the price of the resource. General policy of the resource owner is to service the applicant with the highest budget, in order to increase the utilizing the resource. In other words:

$$rp_j(t) = lp_j + (hp_j - lp_j)\left(\frac{st_j(t)}{wl_j(t)}\right)^{\frac{1}{\sigma}} \qquad (4)$$

Where, $rp_j(t)$ is the price of resource at time t and $st_j(t)$ is the current workload or the workload at the start time of the task at time t. $wl_j(t)$ is the workload of r_j after the last allocation. In this equation, general trend is to decrease the price of the resource when the allocated resource is going to finish the task, in order to let other applicants to apply for it easier.

After the calculation of both budget and price, they are submitted to ResAg. It sorts the applicants' budgets in descent order and the resource prices in ascent order. Then, according to (5) final price is calculated as the average of the richest applicant (bid_i^{max}) and cheapest resource (rp_j^{min}):

$$fp(t) = \frac{1}{2}\left(bid_i^{max}(t) + rp_j^{min}(t)\right) \qquad (5)$$

Matrix P is then constructed according to the final prices. Resource allocation methods utilize this P for binding the resource to the applicant. But, an important issue is optimizing the values of matrix P to achieve more valuable goals like green computing.

This section was mostly adopted from [21]; and readers can refer to it for further details about the construction P and strategies of applicant and resource owners.

3   Optimization of Matrix P Based on the History of Latencies

3.1   Taking the Communication Latencies into Account

Definitely, there might be a geographical distance between the resource and applicant that causes latencies in communication. These latencies affect the whole performance of the system and increase the average response times. These negative results will be emerged while execution of fine-grained parallel programs. Trade-off between the latency factors that might be forked from the faults in the system of from the geographical distance and performance of system, encourages us for optimization of matrix P to achieve better performance. The main contribution is to maintain a history of latencies. The history of latencies can be taken into the account for constructing P. It will help to have better performance of system when there are similar resources. Then, ResAg can consider the latencies as a part of the weight of fp.

Construction of the table of latencies is done gradually. After each resource allocation, a record is added to the table of latencies that shows the latency between t_i and r_j;
or modifies the previous records between them. Let resource r_j is allocated to applicant t_i for the first time. Latencies of messages communication can be measured
through sending acknowledge packets. Acknowledge packets might be more than
once submitted from applicant to resource (and vice versa) to collect the average of
the latencies:

$$\overline{LC}_{ij} = \sum_{q=1}^{p} PL_q$$

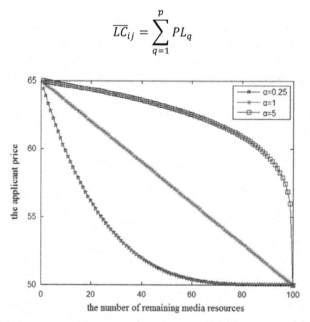

Fig. 4. Resource applicants' price considering remaining resources with different values of α
[21]

Where $\overline{LC}_{ij}$ is the average latency of communication between applicant t_i and resource r_j and PL_q is the latency of qth acknowledge message. Note that, there are no
lower and upper bounds for $\overline{LC}_{ij}$ ($0 \le \overline{LC}_{ij} \le \infty$). Value 0 for the $\overline{LC}_{ij}$ is when the
resource and applicant are located at the same node and there is no communication
latency; and value ∞ is when the resource is faced with failure and it sends no response to the acknowledge message. So, it can't be utilized directly in the P. Here we
need to change it to the scale of 0 to 1 in order to affect the P with the value of $\overline{LC}_{ij}$.
Due to it, the average of all $\overline{LC}_{ij}$ that are stored in the specified table should be calculated as (6):

$$ALC(t) = \frac{1}{m \times n} \times \sum_{i=1}^{n} \sum_{j=1}^{m} \overline{LC}_{ij}$$

if there is a record in table of latencies for t_i and r_j   (6)

Where $ALC(t)$ is the average of latencies of communications between nodes t_i and r_j where they were allocated previously and there is a record in the table of latencies for them.

Equation (7) generates a number between 0 and 1, for modification of matrix P:

$$TLC_{ij}(t) = 1 - \left(\frac{\overline{LC}_{ij}}{\overline{LC}_{ij}+ALC(t)}\right) \tag{7}$$

Where $TLC_{ij}(t)$ is the impact of total latency between applicant t_i and resource r_j at the time t. $TLC_{ij}(t) = 0$ means that the resource is unavailable and it gave no response to the acknowledge packet ($\overline{LC}_{ij} = \infty$); accordingly, $TLC_{ij}(t) = 1$ means that there is no latency and resource and applicant are located at the same node ($\overline{LC}_{ij} = 0$).

Now, matrix LC can be constructed as (8):

$$LC = \begin{bmatrix} l_{11} & l_{12} & \cdots & l_{1n} \\ l_{21} & l_{22} & \cdots & l_{2n} \\ \vdots & \vdots & \vdots & \vdots \\ l_{m1} & l_{m2} & \cdots & l_{mn} \end{bmatrix} \ s.t. \ 0 \leq l_{ij} \leq 1 \ , \ \sum_{i=1}^{m} l_{ij} = 1 \ , \ \sum_{j=1}^{n} l_{ij} = 1 \tag{8}$$

Where l_{ij} is the impact of latency between applicant t_i and resource r_j which is obtained from $TLC_{ij}(t)$.

3.2   Modification of Matrix P with LC

Now, it is time to have a consequent matrix at ResAg to do the resource allocation efficiently. It should be pointed out that in all the systems, policies regulate everything. So, some facilities to implement the policies should be considered. So, we will put a weight on the P and LC to be able to control them. The consequent is matrix FP as in (9):

$$FP(t) = \frac{1}{\theta+\lambda} \times (\theta P + \lambda LC) =$$
$$\frac{1}{\theta+\lambda} \times \begin{bmatrix} \theta p_{11} + \lambda l_{11} & \theta p_{12} + \lambda l_{12} & \cdots & \theta p_{1n}+\lambda l_{1n} \\ \theta p_{21} + \lambda l_{21} & \theta p_{22} + \lambda l_{22} & \cdots & \theta p_{2n} + \lambda l_{2n} \\ \vdots & \vdots & \vdots & \vdots \\ \theta p_{m1} + \lambda l_{m1} & \theta p_{m2} + \lambda l_{m2} & \cdots & \theta p_{mn} + \lambda l_{mn} \end{bmatrix} \tag{9}$$

Where, θ is the weight of P and λ is the weight of LC to implement the policies of system.

3.3   Side Issues

Inherently, the proposed method has some features to detect the unavailable resources. In fact, $\overline{LC}_{ij}$ can indicate the availability of resource r_j. It can be a good feature for handling the faults of system by isolating the crashed resources. This issue will also assist the migration strategies to have a better performance. Replication is not out of the circle, too.

The way of calculation of $\overline{LC}_{ij}$ for unavailable resources has a major drawback. When a node is unavailable, $\overline{LC}_{ij} \to \infty$; so, FP_{ij} will be take a lower number, so that the resource r_j never be allocated to any applicant task even after becoming available. This trap will practically omit the resource from the overlay network of resource-applicant graph. After repairing the failed resource, it won't be able to come back to the network. Due to solve this problem, some solutions must be taken into the account. For example, ResAg can set a timestamp for the resources that $\overline{LC}_{ij} = \infty$ to check their availability every time to time.

4   Simulation and Results

Implementation of proposed method and analyzing the results lead to better understanding about the efficiency of this method. In order to evaluate the performance of the proposed algorithm, we implement it by the CloudSim toolkit [23]. Each task is submitted according to Poisson distribution after its previous tasks, the length of each task is considered as a random number within [100000,200000], the number of tasks are considered between [100,1000], while the number of resources is between [30,50], the deadline d_i of task t_i is set according to (10), and the budget bi of task t_i is set according to(11) [21].

$$d_i = st_j + random\left(\frac{l_i}{1.1 \times cpu_j}, \frac{l_i}{0.9 \times cpu_j}\right) \tag{10}$$

$$b_i = l_i \times random(0.9\overline{lp}, 1.1\overline{hp}) \tag{11}$$

Where $\overline{lp}$ and $\overline{hp}$ are the average values of the media resources' lp and hp [21].

Fig.5 shows a comparison between the response times of common method (which was talked in section 2) and the proposed method with optimization based on the communication latencies between the nodes. This indicates the efficiency of the LO method especially by increasing the number of tasks and passing more times.

In standard resource allocation, the common method is implemented, which the tasks come into the system and execute normally. In this way, latencies between the nodes are waivered. For example, when ResAg is allocating a resource to applicant, it does not consider the factor of latencies. So, it might choose a resource with highest latencies to be allocated to applicant. It will cause longer response times. So, the basic strategy is to let the ResAg to know about the latencies among the nodes of resources and applicants. Then, it can include the factor of latencies to make decisions about choosing the resources for allocation. This advantage improves the response times especially in the case of increasing the tasks. On the other hands, increasing the resources from similar types can also improve the performance of system in comparison with the common resource allocation. Because, it provides more choices for ResAg for selecting the best latencies.

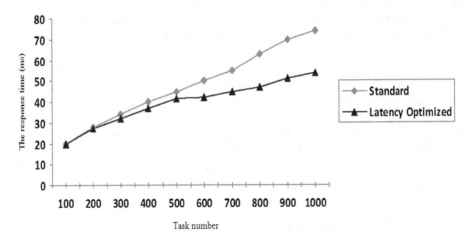

Fig. 5. Comparison of response time between the common and latency optimized methods

5   Discussion and Conclusion

The impact of communication latencies on total performance of Cloud systems en-couraged us to optimize one of the most common resource allocation methods. Al-most in all distributed systems, resource allocation is counted as a duty of Resource Agent. Both resources and applicants calculate their prices and budgets and send it to the ResAg. ResAg then makes decision to allocate the most appropriate resource to the best applicant. Due to it, ResAg constructs the matrix P (as in section 2) based on the prices and budgets. Optimization of P can lead to better performance. In our method, we considered a table in ResAg that holds the history of the resource alloca-tion bindings with their average latencies. At first, this table has no record and after each allocation a record is added or updated. For next allocations, the values of laten-cy impact will be taken into account for making decision. Accordingly, Matrix FP is constructed where $FP_{ij}(t)$ is the possibility of resource r_j to be allocated to applicant t_i. This value is partially obtained from the average latencies of previous allocations. Response time is considered as a metric for evaluation of current method. Results indicate the better response time rather than standard method, especially by increasing the number of tasks and passing time. Increasing the number of tasks and especially more resources from similar types, let the ResAg to have a more options for alloca-tion. In this war, ResAg can select the resources with the best latencies as a part of their decisions. Besides, this method can be utilized to detect the failure of resources by measuring the latency of communications; so that the nodes with very high laten-cies are considered to be disconnected from the network. It is an important point for the other issues like migration, resource replication and fault-tolerance.

References

1. Nezarat, A., Raja, M., Dastghaibifard, G.: A New High Performance GPU-based Approach to Prime Numbers Generation. World Applied Programming **5**(1), 1–7 (2015)
2. Coulouris, G., Dollimore, J., Kindberg, T., Blair, G.: Distributed Systems: Concepts and Design, 5th edn. Addison-Wesley, Boston (2011) ISBN 0-132-14301
3. Yi-wei, F.: Limitation on Stability and Performance of Control System over a Communication Channel. International Journal of Engineering Sciences **4**(3), 19–27 (2015)
4. Sharma, G., Kharel, P.: E-Participation Concept and Web 2.0 in E-government. General Scientific Researches **3**(1), 1–4, (2015)
5. Edessy, M., EL-Darwish, A.G., Nasr, A.A., Ali, A.A., El-Katatny, H., Tammam, M.: Different Modalities in First Stage Enhancement of Labor. General Health and Medical Sciences **2**(1), 1–4 (2015)
6. Malekakhlagh, E., Meysamifard, S.: Industry Pathology to Develop Global Market Entry Strategies: Emphasizing on Small and Medium-Sized Enterprises. International Journal of Economy, Management and Social Sciences **4**(2), 188–193 (2015)
7. Hussain, H., et al.: A survey on resource allocation in high performance distributed computing systems. Parallel Computing **39**, 709–736 2013. http://dx.doi.org/10.1016/j.parco.2013.09.009
8. Tanenbaum, A.S., van Steen, M.: Distributed systems: principles and paradigms. Pearson Prentice Hall, Upper Saddle River (2007). ISBN 0-13-239227-5
9. Shorbi, M., Wan Hussin, W.: The use of Spatial Data in Disaster Management. World Applied Programming **5**(4), 73–78 (2015)
10. Pinel, F., Pecero, J., Bouvry, P., Khan, S.: A two-phase heuristic for the scheduling of independent tasks on computational grids. In: ACM/IEEE/IFIP International Conference on High Performance Computing and Simulation (HPCS), pp. 471–477, July 2011
11. Sharkh, M.A., Jammal, M., Shami, A., Ouda, A.: Resource Allocation in a Network-Based Cloud Computing Environment: Design Challenges. IEEE Communications Magazine (November 2013)
12. Maguluri, S., Srikant, R., Ying, L.: Stochastic Models of Load Balancing and Scheduling in Cloud Computing Clusters. In: Proc. IEEE INFOCOM 2012, March 25–30, pp. 702–10 (2012)
13. Alicherry, M., Lakshman, T.V.: Network Aware Resource Allocation in Distributed Clouds. In: Proc. IEEE INFOCOM 2012, March 25–30, pp. 963–71 (2012)
14. Sun, G., et al.: Optimal Provisioning for Elastic Service Oriented Virtual Network Request in Cloud Computing. IEEE GLOBECOM **2012**, 2541–2546 (2012)
15. Kantarci, B., Mouftah, H.T.: Scheduling Advance Reservation Requests for Wavelength Division Multiplexed Networks with Static Traffic Demands. In: IEEE Symp. Computers and Commun., July 1–4, pp. 806–11 (2012)
16. Srikantaiah, S., Kansal, A., Zhao, F.: Energy Aware Consolidation for Cloud Computing. Cluster Computing **12**, 1–15 (2009)
17. Chase, J.S., et al.: Managing Energy and Server Resources in Hosting Centers. In: 18th ACM Symp. Op. Sys. Principles, October 21, 2001
18. Zhang, B, Zhao, Y, Wang, R.: A resource allocation algorithm based on media task QoS in cloud computing. In: Proceedings of the 4th IEEE International Conference on Software Engineering and Service Science (ICSESS), Beijing, pp. 841–844 (2013)
19. Radu, V.: Application. In: Radu, V. (ed.) Stochastic Modeling of Thermal Fatigue Crack Growth. ACM, vol. 1, pp. 63–70. Springer, Heidelberg (2015)

20. Zhang, M., Zhu, Y.: An enhanced greedy resource allocation algorithm for localized SC-FDMA systems. IEEE Commun. Lett. **17**(7), 1479–82 (2013)
21. Tang, R., et al.: Credibility-based cloud media resource allocation algorithm. Journal of Network and Computer Applications (2014). doi:10.1016/j.jnca.2014.07.018i
22. Anthony, P., Jennings, N.R.: Developing a bidding agent for multiple heterogeneous auctions. ACM Trans. Internet Technol. **3**(3), 185–217 (2003)
23. Calheiros, R.N., Ranjan, R., Beloglazov, A., et al.: CloudSim: a toolkit for modeling and simulation of cloud computing environments and evaluation of resource provisioning algorithms. Soft. w: Pract. Exp. **41**(1), 23–50 (2011)

Optimized Elastic Query Mesh for Cloud Data Streams

Fatma Mohamed$^{(\boxtimes)}$, Rasha M. Ismail, Nagwa L. Badr, and M.F. Tolba

Faculty of Computer and Information Sciences, Ain Shams University, Cairo, Egypt
{fatma_najib1991,rashaismail}@yahoo.com,
nagwabadr@cis.asu.edu.eg, fahmytolba@gmail.com

Abstract. Many recent applications in several domains such as sensor networks, financial applications, network monitoring and click-streams generate
continuous, unbounded, rapid, time varying datasets which are called data
streams. In this paper we propose the optimized and elastic query mesh (OEQM)
framework for data streams processing based on cloud computing to suit the
changeable nature of data streams. OEQM processes the streams tuples over
multiple query plans, each plan is suitable for a sub-set of data with the nearest
properties and it provides elastic processing of data streams on the cloud environment. We also propose the Auto Scaling Cloud Query Mesh (AS-CQM)
algorithm that supports streams processing with multiple plans and provides
elastic scaling of the processing resources on demand. Our experimental results
show that, the proposed solution OEQM reduces the cost for data streams processing on the cloud environment and efficiently exploits cloud resources.

Keywords: Data streams · Cloud computing · Elastic processing · Continuous
query optimization · Query mesh

1 Introduction

Recently a new class of applications such as sensor networks, traffic analysis, web
logs and transaction analysis generate continuous, dynamic, unbounded, time varying
and rapid data elements which are called data streams. New challenges were presented by these applications that were not dealt with by traditional techniques. Continuous
queries are considered to process these data streams that have a continuous nature and
these queries are evaluated continuously with the continuous arrival of data streams
over time [1,2,3,4,5,6].

Traditional database systems use single query optimizers to process data streams,
which select the best query plan to process all data based on the average statistics of
that data. However this single plan isn't suitable with data streams which have continuous variations over time. For example, consider a continuous query which has three
operators (operator1, operator 2 and operator 3) and the result of this query is provided
for the latest incoming data streams within the last 3 minutes. Also consider that the
query result is updated every 1 minute. It is not efficient to process all incoming data
streams over all runs (each 1 minute) or even in the same run with the same query plan,
because data streams have continuous change in statistics over time. Consider that the

© Springer International Publishing Switzerland 2015
O. Gervasi et al. (Eds.): ICCSA 2015, Part I, LNCS 9155, pp. 367–381, 2015.
DOI: 10.1007/978-3-319-21404-7_27

best query plan is to process the incoming streams tuples in the first 10 seconds is (operator1 - operator 2 - operator1) but the best plan for the incoming tuples in the next 20 seconds is (operator 3 operator 2- operator1) and so on. So it is not efficient to process all incoming streams tuples with the same query plan so to avoid the disadvantages of using a single plan, it's a good solution to use multiple plans, each one is the best plan to process a sub-set of data with the same characteristics [2].

Cloud computing is a flexible environment for resource management for elastic application deployments. The Cloud resource management system continuously monitors resource utilization, manages and adjusts these resources in real time to meet the predefined customers' needs [1], [7]. In this paper we propose the optimized and elastic query mesh for data streams processing based on a cloud computing (OEQM) framework which is based on the query mesh (QM) solution. Using the QM solution, data streams are processed over multiple query plans, each plan is suitable for a sub-set of data with the same properties and the OEQM also provides elastic processing of data streams that efficiently scale the number of the processing resources on demand. We also propose the Auto Scaling Cloud Query Mesh (AS-CQM) algorithm for data streams processing over the elastic cloud environment.

The rest of the paper is organized as follows; Section 2 presents the related work to the proposed solution. Section 3 explains the proposed OEQM framework. Section 4 is the proposed AS-CQM algorithm. Section 5 shows the experimental evaluation. Section 6 is the conclusion.

2   Related Works

Different algorithms were proposed in [8,9,10,11,12] for streams processing based on multiple query plans, where the best query plan is selected in runtime for each incoming tuple. In [8] the Query Mesh (QM) framework was proposed to generate multiple execution plans, each one optimized for a sub-set of data with distinct properties. Then a classifier model is used to determine the best execution plans for incoming tuples based on these multiple plans. So each data tuple will be processed over its optimized plan. They proposed the II-QM algorithm to generate the multiple execution plans. It formed a lattice of all possible partitions of the input training tuples and uses the Iterative Improvement search strategy to search for the least costly solution in this lattice that appears until meeting the stop condition. Each partition in this solution includes tuples with distinct properties. The II-QM generates a query plan to each partition tuples and it processes the incoming tuples based on the query plans of these partitions. But despite the advantages of the QM, it increases the memory usage to apply its optimization.

In [9] the early terminated Weight Robust Partitioning algorithm was proposed to produce robust logical plans, each logical plan is designed for a particular sub-region of the parameter space. In [10] the Semantic Query Optimization (SQO) approach was proposed to determine query optimization opportunities. SQO uses dynamic sub-stream metadata at runtime to identify these optimization opportunities but the SQO is not suitable with the semantics of sensor networks or with uncertain data streams. In

[11] the cyclops platform was proposed to process streams tuples over a combination selected from various execution plans and execution engine choices. However cyclops is not a general optimization platform because it is only based on three execution engines (Esper, Storm and Hadoop). In [12] the adaptive multi-route query processing (AMR) was proposed to route the incoming tuples in runtime based on the most recent system statistics.

Different algorithms were proposed in [13,14,15] to process sensor data streams. In [13] the checkpointing and reconstruction algorithms were proposed to recover failure in sensor devices. The FSKY filtering algorithm in [14] and the probabilistic filters protocol in [15] were proposed to reduce the communication costs and save energy of sensor devices when answering continuous queries in sensor networks. Although the advantages of them, the FSKY filtering algorithm didn't deal with multiple data streams' processing and the complex queries didn't get addressed in the probabilistic filters protocol. Also different algorithms were proposed in [16,17,18,19] to deal with the uncertain data streams but none of them addressed streams processing over multiple query plans.

In [7], [20], [21], [22], [23] different algorithms were proposed to dynamically manage, monitor resource utilization and scale up or down the number of the cloud processing resources (VMs) based on the input rate changes of the incoming stream tuples. All of them can adjust the number of the processing virtual machines to the optimal number in runtime based on user specified rules. In [24] the Fault-Tolerant Scale Out approach was proposed to scale out the number of cloud processing VMs on demand, it can also recover cloud resources failures.

In this paper the OEQM framework is proposed to solve previously mentioned problems; which are streams processing over a single plan, specific engines and static environments, also increasing the memory usage and not handling sensor applications and uncertain data streams. So our objective is to decrease the memory usage for applying optimization and also proving that it is suitable to use with sensor networks applications and uncertain data streams processing. In addition, the proposed OEQM is a general optimization solution because it is not suitable for specific engines for generating optimized query plans and, it efficiently processes data streams over multiple plans. Also the Auto Scaling Cloud Query Mesh (AS-CQM) algorithm is proposed to provide optimized and elastic processing of data streams over multiple query plans on the cloud environment. The results prove the efficiency of the proposed solution where it exploits cloud processing resources. Also it reduces the cost for data streams processing on the cloud environment.

3   The Proposed Framework

The optimized and elastic query mesh (OEQM) framework for data streams processing based on a cloud computing in fig.1consists of four sub-systems, the first two sub-systems are the offline phase and the last two sub-systems are the online phase. In the offline phase, multiple query plans are generated. These plans will be used for processing the incoming streams tuples and in the online phase, an elastic cloud environment is provided to process the streams tuples based on the generated multiple plans in the offline phase.

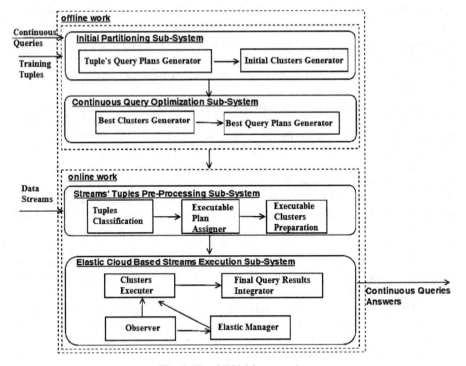

Fig. 1. The OEQM framework

3.1   The Initial Partitioning Sub-system

It is the first sub-system in the offline phase. Given a continuous query and training dataset, this sub-system generates the initial partitions of the training tuples and consists of two main blocks, tuple's query plans generator and initial clusters generator.

- **Tuple's Query Plans Generator.** Is responsible for generating the query plan for each training tuple separately.

- **Initial Clusters Generator.** Is responsible for grouping the tuples based on their query plans. Thus each cluster includes the tuples which have the same query plan.

3.2   The Continuous Query Optimization Sub-system

It is the second sub-system in the offline phase. Using the generated initial clusters, this block generates the best clustering of the training tuples and consists of two main blocks, best clusters generator and best query plans generator.

- **Best Clusters Generator.** Is responsible for generating the best clusters of the training tuples and each cluster includes the tuples which have nearest properties.
- **Best Query Plans Generator.** It generates the best plan for each cluster in the best clusters. Each plan is the most suitable plan for each tuple in the cluster.

3.3   The Streams' Tuples Pre-processing Sub-system

This sub-system is the first sub-system in the online phase. It assigns the best executable plan to each incoming tuple. Then it groups these tuples based on their best executable plan. This sub-system consists of three blocks; tuples classification, executable plan assigner and executable clusters preparation.

- **Tuples Classification.** It classifies each incoming tuple to determine its nearest cluster of the best clusters which are generated in the offline phase.

- **Executable Plan Assigner.** Based on the tuples classification results, it assigns the best executable plan for each incoming tuple.

- **Executable Clusters Preparation.** Is responsible for clustering the incoming tuples based on their assigned executable plans.

3.4   The Elastic Cloud Based Streams Execution Sub-system

This sub-system is the last sub-system. It separately executes each executable cluster using its assigned executable plan and provides elastic processing for these clusters over the cloud environment and then it integrates the results of all clusters to generate the final query results. This sub-system consists of four main blocks; clusters executer, observer, elastic manager and final query results integrator.

- **Clusters Executer.** Is responsible for executing each executable cluster based on its executable query plan separately.

- **Observer.** It monitors the utilization of the processing resources and notifys the elastic manager about the need to scale (up / down) the processing resources.

- **Elastic Manager.** Is responsible for scaling (up / down) the processing resources after a notification from the observer.

- **Final Query Results Integrator.** It integrates the sub-results of each executable cluster to generate the final query results.

4   The Proposed Auto Scaling Cloud Query Mesh (AS-CQM) Algorithm

In this paper, we propose the Auto Scaling Cloud Query Mesh (AS-CQM) algorithm in fig. 2. It's based on data streams processing using multiple query plans and each plan is the best plan for a sub-set of data with the same statistics, it also provides elastic streams processing over the cloud environment.

In the offline phase of the AS-CQM algorithm, it generates the best clusters of the training tuples and the best plan for each cluster. To generate the best clusters, the AS-CQM uses the CluStream methodology which applies the k-means clustering [25,26]. Given the initial clusters of the training tuples, the AS-CQM generates the least costly solution (the best clusters) which is generated until meeting the stop condition. In each iteration in the offline phase, it computes the centroids of the current clusters to generate the next QM solution (next clusters) based on those centroids. In each iteration, the value of the current k (the number of the current clusters) is changed to the number of the non-empty clusters in the current solution. Initially the value of k is the number of the initial clusters of the training tuples [fig.2].

- **Stop Condition.** There are many different cases that the AS-CQM algorithm may stop in. For example; after predefined iterations or when the results didn't improve in the last consecutive iterations or after a specific time are just a few of the many other cases.

- **Cost Model.** The AS-CQM algorithm considers the query plans costs, the online classification costs and the multiple plans overheads for its best solution.

In the online phase of the AS-CQM algorithm, it firstly prepares the incoming tuples for execution, where it then classifies each incoming tuple to determine its nearest cluster from the generated best clusters in the offline phase, it then assigns its most suitable plan from the generated multiple plans. The assigned plan to each tuple is the best plan to its nearest cluster and finally it clusters the tuples based on their assigned best plans. Also the AS-CQM algorithm provides elastic streams processing where it continuously monitors the utilization of the processing resources and scales (up / down) the needed processing resources based on predefined upper and lower thresholds [fig.2]. For example, if there is notification that the average CPU utilization is greater than or equal to 70% then the AS-CQM scales up the number of VMs and if there is notification that the average CPU utilization is lower than or equal to 30% it then scales down the number of VMs. The proposed elastic configurations will be presented in Section 5.

Running Example

Consider a query with three operators (op1,op2,op3), training data of 7 tuples, the initial clusters of the training tuples is {{12},{356},{47}} or for simplicity 12|356|47. So there are three clusters, the first one includes tuples number 1,2 .And the second cluster includes tuples number 3,5,6. And the last cluster includes the tuples number 4,7. Thus the initial number of clusters (k) equals to 3. If the AS-CQM algorithm generates the three centroids c1,c2,c3 for these three clusters, then the training tuples are clustered according to c1,c2,c3. For example, the resulted clusters are 1256|347|empty_cluster. Thus the value of k will be changed to the non-empty clusters which are 2 clusters.   The AS-CQM algorithm completes working until the stop condition is reached. For example, the AS-CQM generates 245|1367 as the best clusters, the query plan p1(op1-op2-op3) for the first cluster in the best clusters, and the query plan p2(op3-op1-op2) for the second cluster. Each incoming tuples are assigned its best plan from the generated plans p1,p2 then the tuples are clustered based on their assigned plans. Initially the number of the processing instances is one, if there is an overload condition then the number of the instances is increased. For example each cluster of tuples is processed on an instance, and so on the number of the instances can be scaled up/down based on the predefined upper/lower thresholds.

```
Algorithm AS-CQM
Offline Phase
Input: initial clusters, best_solution, k, training_tuples
Output: best_solution, best_plans
    1. bestQMCost ← best_solution .cost
    2. QM ← initial clusters
    3. while (not stop condition) do
    4. QM_Centroids ← get_centoids( QM)
    5. QM ← k mean-clustering( training_tuples, QM_Centroids, k)
    6. k ← Non_Empty_Clusters_Count
    7. end while
    8. if (QM.cost < cost(best_solution)) then
    9. best_solution ← QM
    10. end if
    11. end while
    12. for each cluster in best_solution
    13. best_plan ← generate_best_plan ( cluster)
    14. best_plans. add ( best_plan)
    15. end for
    16. return best_solution, best_plans;
Online Phase
Pre-processing
Input: incoming streams tuples, best_solution, best_plans
Output: executable_clusters
    1. for each tuple in  incoming streams tuples
    2. best_cluster ← match_best_cluster (tuple, best_solution)
    3. best_executable_plan ← assign_best_plan(best_cluster, best_plans)
    4. tuple_ plan_pairs. add_pair(tuple, best_executable_plan)
    5. end for
    6. executable_clusters ← generate_ executable_clusters(tuple_plan_pairs)
    7. return executable_clusters;
Elastic Execution
Input: upper_threshold, lower_threshold, executable_clusters
Output: scaling_alert
    1. if (average_CPU_utilization < =lower_threshold) then
    2. scaling_alert ← generate_scale_down_alert()
    3. else if (average_CPU_utilization >= upper_threshold) then
    4. scaling_alert ← generate_scale_up_alert()
    5. end if
    6. return scaling_alert;
```

Fig. 2. The proposed AS-CQM algorithm

5   Experimental Evaluation

All the experiments were run on an Amazon real cloud environment and based on Amazon auto scaling that automatically scales the number of Amazon instances up or down according to configurations conditions we were set, and elastic load balancing to distribute traffic within the auto scaling group instances. Table 1 shows the specifications of Amazon instances that were used in the experiments. The proposed Auto Scaling Cloud Query Mesh (AS-CQM) algorithm was run over five different Amazon auto scaling configurations with different conditions (scaling down thresholds (lower thresholds) and scaling up thresholds (upper thresholds)). We compared the performance of the proposed elastic configurations provided by the proposed AS-CQM algorithm with the static configuration provided by the model view approach in [27]. The static configuration which is used in the experiments is five static processing virtual machines. Table 2 shows the specifications of the five different Amazon auto scaling configurations that were used in the experiments and a symbol is given for each configuration from T1 to T5. The lower and the upper thresholds mean in which conditions the AS-CQM scale down / up the number of processing VMs. The starting processing VMs number means the number of VMs to start processing and the start VMs processing number is one VM in all configurations. The maximum possible processing VMs number means the maximum number of VMs that scale to. The maximum number in all configurations is 5 to be equal to the number of static processing VMs used in the static configuration. The number of VMs to scale up / down at a time is how many VMs can be scaled up / down in each scaling. In all configurations we scale up / down 1 VM in each time. For example, the first elastic configuration T1 starts processing with 1 VM, if the average CPU utilization is lower than or equal to 20% then the number of VMs is decreased by one and if the average CPU utilization is greater than or equal to 70% the number of VMs is increased by 1, then the number of VMs will be 2 and so on but the maximum number is 5.

The experiments include several measurements: 1) The average overall execution time for the static configuration and our different elastic configurations. 2) The average CPU utilization percentages over the static configuration and the elastic configurations. 3) The throughput using the static configuration and the elastic configurations. 4) The cost saving percentage for the elastic configurations over the static configuration.

Table 1. Amazon instances Specifications

Amazon instance	CPU cores	Memory
T2.micro	1	1 GiB = 1.073 GB

Table 2. Amazon Specifications over five different auto scaling configurations

Symbol	Lower threshold	Upper threshold	Starting processing VMs number	Max possible processing VMs number	Number of VMs to scale (up / down) at a time
T1	<= 20	>= 70	1	5	1
T2	<= 30	>= 70	1	5	1
T3	<= 40	>= 70	1	5	1
T4	<= 20	>= 80	1	5	1
T5	<= 30	>= 80	1	5	1

The incoming tuples were processed using a tumbling sliding window. It divides a stream into non-overlapping consecutive windows, so each tuple is classified only one time. Our sliding window size is 200 streams tuples. The training datasets size is equal to 10 tuples to reduce the optimization time and the AS-CQM algorithm stops after ten iterations in its offline phase to get more accurate clusters of the training datasets.

5.1 Data Sets and Queries

We used a real sensor dataset from Intel Research, Berkeley Lab [28]. It contains readings from 54 sensors which were taken between February and April of 2004. The schema of this dataset is (date, time, epoch, mote ID, temperature, humidity, light and voltage), the data from these sensors were divided into ten groups (streams) based on thier sensors locations. The XLSTAT sampling tool was used to sample our sub-sets of data from this dataset [29]. Different join queries were run on different sampled sub-sets with the same size from this dataset. Each sub-set includes data from the ten sensors data groups (streams). We choose the join queries because these queries are the most expensive queries in a database system [16], [30]. Example for these queries: Select * from S1, S2, .. ,Sn Where S1.ts = S2.ts = ... = Sn.ts; where ts is the timestamp of a stream tuple, n is the total number of streams.

5.2 Results and Analysis

Three different join queries were run. Each query is run three times on three different sub-sets from the dataset with a size equal to 200 tuples, the results are averaged over all runs. In the experiment results, the maximum number used of virtual machines (maximum actual number of VMs after scaling)for processing streams tuples over the elastic configurations A (T1, T2, T3) is three VMs and over the elastic configurations B (T4, T5) two VMs (Table 2).

The Average Overall Execution Time. It is measured by the average time which is taken to execute the queries over all runs. In fig.3 the average execution time for the static configuration [27] is compared with the elastic configurations. Fig.3 (a) presents comparisons between the average execution time for the static configuration and the elastic configurations A with maximum actual number of processing VMs after scaling=3 (T1, T2, T3) and in Fig.3 (b) we compare the average execution time for the static configuration and the elastic configurations B with maximum actual number of processing VMs after scaling=2 (T4, T5). The average execution time using the static configuration is improved by 37.5% over using the elastic configurations A and improved by 45.4% over using the elastic configurations B.

The Average CPU Utilization Percentages. In fig.4 the average CPU utilization percentages for the static configuration is compared with the elastic configurations. In Fig.4 (a) the comparison between the static configuration and the elastic configurations A with maximum actual number of processing VMs after scaling=3 (T1, T2, T3) were presented and in Fig.4 (b) we compared the static configuration with the elastic configurations B with maximum actual number of processing VMs after scaling=2 (T4, T5). The CPU utilization percentage in the elastic configuration A is better than using static configurations by 28.2% and the elastic configuration B is improved by 39.5% over the static configurations. Thus the elastic configuration outperforms the static configurations in exploiting the processing resources where the elastic configurations (A,B) used only from two to three VMs and their average CPU utilization over these processing resources is 53.6% . However the static configuration used five VMs and its average CPU utilization over these processing resources is 20.8%.

(a)

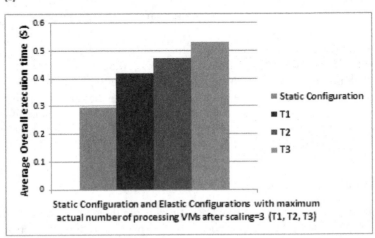

(b)

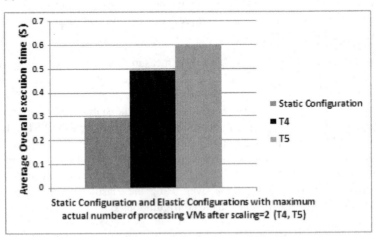

Fig. 3. The Average Overall Execution Time

The Average Throughput. It is measured by the average number of the processed tuples per millisecond over all runs. Fig.5 shows the average throughput over the static configuration and the elastic configurations. Fig.5 (a) presents the comparison between the average throughput for the static configuration and the elastic configurations A with maximum actual number of processing VMs after scaling= 3 (T1, T2, T3) and Fig.5 (b) presents the comparisons between the average execution time for the static configuration and the elastic configuration B with maximum actual number of processing VMs after scaling=2 (T4, T5). The average throughput using the static configuration result improves by 31.2% over the elastic configuration 1 and improves by 38.3% over the elastic configuration 2.

(a)

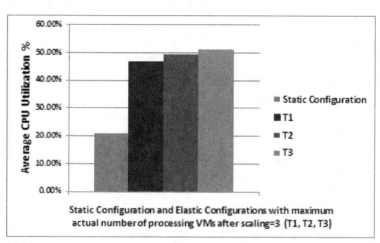

(b)

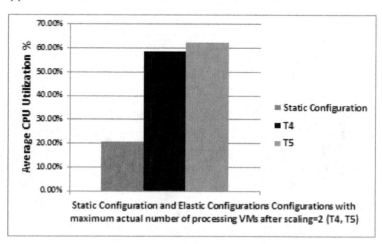

Fig. 4. The average CPU utilization percentages

The Cost Saving Percentage. It is measured by the percentage of cost saving of the elastic configurations over the static configurations. It based on the maximum used number of virtual machines for processing streams tuples (maximum actual number of VMs after scaling). The maximum actual number for processing streams tuples over the elastic configurations A (T1, T2, T3) is three VMs and over the elastic configurations B (T4, T5) is two VMs but the static configuration has five static processing VMs. Thus the elastic configurations A and B reduce the costs more than the static configuration. Fig.6 presents the cost saving percentage of the elastic configurations over the static configurations. The cost saving percentage using the elastic configuration A are better than the static configurations by 40% and the elastic configurations B are better than the static configurations by 60%. Thus the static configuration is too expensive especially with the large datasets of most applications, and the elastic configurations is a good alternative because it uses a smaller number of VMs and improves the cost by 50% over the static configuration.

(a)

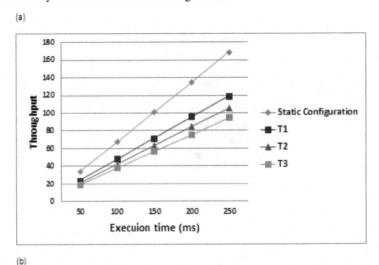

(b)

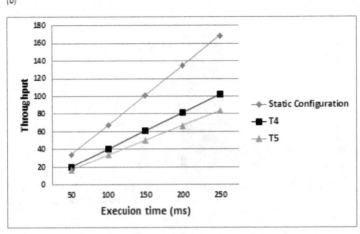

Fig. 5. The average throughput

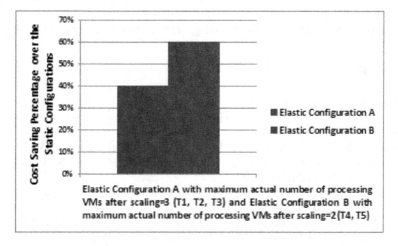

Fig. 6. The cost saving percentage

Based on the results, the elastic configuration improves the CPU utilization by 33.8% and the cost by 50% over the static configuration with acceptable execution time and throughput. Thus we conclude that there are two main measurements for streams processing on the cloud environment to take into account. The First measurement is the need to process streams with the least possible execution time and the most possible throughput, regardless of the number of the processing resources and the processing costs (the static processing) but this is non-preferable and too expensive with the large datasets of most applications. The second measurement is stream processing with the lowest possible cost and the most possible utilization of the processing resources over the minimum number of processing resources and with both acceptable execution times and throughput (the elastic processing). Thus you should order your priorities and choose the most suitable measurement to your application but in general the second measurement is the most suitable for the majority of applications with large data streams.

6   Conclusion

In this paper, the optimized and elastic query mesh (OEQM) framework for data streams processing based on a cloud computing is proposed. The proposed solution provides elastic data streams processing on the cloud environment and also it processes data streams using multiple query plans instead of using a single plan to process all data. Each plan is the most suitable plan for a sub-set of data with the nearest properties. OEQM dynamically assigns the best plan to each incoming stream tuple. Also the Auto Scaling Cloud Query Mesh (AS-CQM) algorithm is proposed. It provides optimized data streams processing using multiple query plans where it assign the most suitable plan to each incoming tuple at runtime. Also it dynamically scales (up /down) the cloud processing VMs on demand. The experimental results prove the efficiency of the proposed solution in terms of utilizing the cloud processing resources (VMs) and reducing the costs for data streams processing on the cloud environment.

References

1. Heinze, T., Pappalardo, V., Jerzak, Z., Fetzer, C.: Auto-Scaling Techniques for Elastic Data Stream Processing. In: 30th International Conference on Data Engineering Workshops, pp. 318–321. IEEE, Chicago (2014)
2. Mohamed, F., Ismail, R., Badr, N., Tolba, M.F.: Efficient Optimized Query Mesh for Data Streams. In: 9th International Conference on Computer Engineering & Systems (ICCES), pp. 157–163. IEEE, Egypt (2014)
3. Chen, H.: Mining Top-K Frequent Patterns over Data Streams Sliding Window. Intelligent Information Systems 42, 111–131 (2014)
4. Ajwani, D., Ali, S., Katrinis, K., Li, C.H., Park, A.J., Morrison, J.P., Schenfeld, E.: Generating Synthetic Task Graphs for Simulating Stream Computing Systems. Parallel and Distributed Computing 73, 1362–1374 (2013)
5. Anceaume, E., Busnel, Y.: A Distributed Information Divergence Estimation over Data Streams. IEEE Trans. on Parallel and Distributed Systems 25, 478–487 (2014)
6. Cao, J., Zhang, W., Tan, W.: Dynamic Control of Data Streaming and Processing In: A Virtualized Environment. IEEE Trans. on Automation Science and Engineering 9, 365–376 (2012)
7. Saleh, O., Gropengießer, F., Betz, H., Mandarawi, W., Sattler, K.U.: Monitoring and Autoscaling IaaS Clouds: A Case for Complex Event Processing on Data Streams. In: 6th International Conference on Utility and Cloud Computing, pp. 387–392. IEEE Computer Society, Dresden (2013)
8. Nehme, R.V., Works, K., Lei, C., Rundensteiner, E.A., Bertino, E.: Multi-Route Query Processing and Optimization. Journal of Computer and System Sciences 79, 312–329 (2013)
9. Lei, C., Rundensteiner, E.A., Guttman, J.D.: Robust Distributed Stream Processing. In: 29th International Conference on Data Engineering, pp. 817–828. IEEE, Washington (2013)
10. Ding, L., Works, K., Rundensteiner, E.A.: Semantic Stream Query Optimization Exploiting Dynamic Metadata. In: 27th International Conference on Data Engineering, pp. 111–122. IEEE, Hannover (2011)
11. Lim, H., Babu, S.: Execution and Optimization of Continuous Queries with Cyclops. In: The 2013 SIGMOD International Conference on Management of Data, pp. 1069–1072. ACM, New York (2013)
12. Works, K., Rundensteiner, E.A., Agu, E.: Optimizing Adaptive Multi-Route Query Processing via time-partitioned indices. Journal of Computer and System Sciences 79, 330–348 (2013)
13. Dou, A., Lin, S., Kalogeraki, V., Gunopulos, D.: Supporting Historic Queries in Sensor Networks with Flash Storage. Information Systems 39, 217–232 (2014)
14. Yin, B., Lin, Y., Yu, J., Luo, Q.: Energy-Efficient Filtering for Skyline Queries in Cluster-Based Sensor Networks. Computers and Electrical Engineering 40, 350–366 (2014)
15. Zhang, Y., Cheng, R.: Probabilistic Filters: A Stream Protocol for Continuous Probabilistic Queries. Information Systems 38, 132–154 (2013)

16. Qian, J., Li, Y., Wang, Y., Chen, H., Dong, Y.: An Embedded Co-processor for Accelerating Window Joins over Uncertain Data Streams. Microprocessors and Microsystems **36**(6), 489–504 (2012)
17. Ding, X., Lian, X., Chen, L., Jin, H.: Continuous Monitoring of Skylines over Uncertain Data Streams. Information Sciences **184**, 196–214 (2012)
18. Liu, Z., Wang, C., Wang, J.: Aggregate Nearest Neighbor Queries in Uncertain Graphs. World Wide Web **17**, 161–188 (2014)
19. Fangzhou, Z., Guohui, L., Li, L., Xiaosong, Z., Cong, Z.: Probabilistic Nearest Neighbor Queries of Uncertain Data via Wireless Data Broadcast. Peer-to-Peer Networking and Applications **6**, 363–379 (2013)
20. Gulisano, V., Jimenez-Peris, R., Patino-Martinez, M., Soriente, C.: StreamCloud:An Elastic and Scalable Data Streaming System. IEEE Trans. on Parallel and Distributed Systems **23**(12) (2012)
21. Cervino, J., Kalyvianaki, E., Salvachua, J., Pietzuch, P.: Adaptive Provisioning of Stream Processing Systems in the Cloud. In: 28th International Conference on Data Engineering Workshops, pp. 295–301. IEEE, Washington (2012)
22. Hu, R., Jiang, J., Liu, G., Wang, L.: Efficient Resources Provisioning Based on Load Forecasting in Cloud. The Scientific World Journal 2014, Article ID 10152, 14 pages (2014)
23. Kailasam, S., Gnanasambandam, N., Dharanipragada, J., Sharma, N.: Optimizing Ordered Throughput Using Autonomic Cloud Bursting Schedulers. IEEE Trans. on Software Engineering **39**(11), 1564–1581 (2013)
24. Castro Fernandez, R., Migliavacca, M., Kalyvianaki, E., Pietzuch, P.: Integrating Scale Out and Fault Tolerance in Stream Processing using Operator State Management. In: SIGMOD International Conference on Management of Data, pp. 725–736. ACM, New York (2013)
25. Yogita, Y., Toshniwal, D.: Clustering Techniques for Streaming Data-a Survey. In: 3rd International in Advance Computing Conference, pp. 951–956. IEEE (2013)
26. Aggarwal, C.C.: A Survey of Stream Clustering Algorithms. In: Data Clustering: Algorithms and Applications, pp. 231–258 (2013)
27. Guo, T., Papaioannou, T.G., Aberer, K.: Efficient Indexing and Query Processing of Model-View Sensor Data in the Cloud. Big Data Research **1**, 52–65 (2014)
28. Intel Lab Data. http://db.csail.mit.edu/labdata/labdata.html
29. XLSTAT. http://www.xlstat.com/en/products-solutions/feature/data-sampling.html
30. Kim, H.G.: A Structure for Sliding Window Equijoins in Data Stream Processing. In: 16th International Conference on Computational Science and Engineering, pp. 100–103. IEEE, Sydney (2013)

An Efficient Hybrid Usage-Based Ranking Algorithm for Arabic Search Engines

Safaa I. Hajeer[✉], Rasha M. Ismail, Nagwa L. Badr, and M.F. Tolba

Faculty of Computer & Information Sciences, Ain Shams University, Cairo, Egypt
safaahajeer@yahoo.com, rashaismail@fcis.asu.edu.eg,
nagwabadr@cis.asu.edu.eg, fahmytolba@gmail.com

Abstract. There are billions of web pages available on the Internet. Search Engines always have a challenge to find the best ranked list to the user's query from those huge numbers of pages. A lot of search results that correspond to a user's query are not relevant to the user's needs. Most of the page ranking algorithms use Link-based ranking (web structure) or Content-based ranking to calculate the relevancy of the information to the user's need, but those ranking algorithms might be not enough to provide a good ranked list for the Arabic search. So, in this paper we proposed an efficient Arabic information retrieval system using a new hybrid usage-based ranking algorithm called EHURA. The objective of this algorithm is to overcome the drawbacks of the ranking algorithms and improve the efficiency of web searching. EHURA was applied to 242 Arabic Corpus to measure its performance. The result shows our proposed EHURA algorithm improves the precision over the Content-Based ranking algorithm representation, as well as the recall is affected too in this improvement.

Keywords: Information retrieval (IR) · Tokenization · Usage-based ranking · Content-based ranking · Link-based ranking · Pagerank · Weighted pagerank · Implicit judgment and explicit judgment

1 Introduction

The amount of information in the world is increasing exponentially through the years. Searching within this huge amount of information becomes a critical behavior of our life. Millions of users interact with search engines daily around the globe; more than 360 of them are Arab ones [1, 2].

Recently, due to the growing number of internet users around the world, information retrieval (IR) has become of great importance as an essential tool for all tasks of searching on the web. The number of Arab Internet users has increased recursively over the years because of the changes in the requirements of the life. Relatively fewer Arabic search engines are currently available despite the enormous efforts to satisfy the needs of the growing number of Arabic internet users. Moreover, Arabic is a highly inflected language and has a complex morphological structure, which makes information retrieval on Arabic texts a challenge [3].

Most of existing web search engines often calculate the relevancy of web pages for a given query by counting the search keywords contained in the web pages, this

© Springer International Publishing Switzerland 2015
O. Gervasi et al. (Eds.): ICCSA 2015, Part I, LNCS 9155, pp. 382–391, 2015.
DOI: 10.1007/978-3-319-21404-7_28

approach is called Content-based ranking algorithms that use the words in each document to determine its ranking. This approach works well when users' queries are clear and specific. However, in the real world, web search queries are often short (less than 3 words) and ambiguous [4] and web pages contain a lot of diverse and noisy information. These will very likely lead to the deterioration   in the performance of web search engines, due to the gap between query space and document space. Another approach, Link-based ranking algorithms assign scores to web pages based on the number and quality of hyperlinks between pages. Links that point to a particular page or endorse a page can help to improve the link-based rankings. Finally, Usage-based ranking algorithms score documents by how often they are viewed by Internet users. For Usage-based ranking, there is limited work to utilize the usage data in the web information retrieval systems, especially in the ranking algorithm. For some systems [5] and [6] that do use the usage data in ranking, they determine the relevance of a web page by its selection frequency. This measurement is not that accurate to indicate the real relevance. The time spent on reading the page, the operation of saving, printing the page or adding the page to the bookmark and the action of following the links in the page, are all good indicators, perhaps better than the simple selection frequency,   so it is worth further exploration on how to apply this kind of actual user behavior to the ranking mechanism. The objective of the paper is to provide a hybrid ranking algorithm to utilize the usage data called EHURA (Efficient Hybrid Usage-based Ranking Algorithm). This ranking algorithm is proposed to improve the ranked list provided from search engines that are based only on content base rankings. This improvement will have a direct effect on the effectiveness and the performance of Information retrieval systems and web search engines.

The rest of the paper is organized as follows. Section 2 presents related Work. Section 3 discusses The System Architecture and explains the EHURA algorithm; Section 4 presents the evaluation of our Information Retrieval System and its experimental results. Finally, the conclusion of the paper appears in Section 5.

2   Related Work

In the context of Information Retrieval, Ranking algorithms have become the most researched area of information retrieval. Many researchers have developed algorithms for improving search engine results. Each research proposed a methodology and measurements to test the performance and compute the accuracy of its algorithm.

Kritikopoulos et al. was studied method in [7] for evaluating the quality of ranking algorithms. Success Index takes into account a user's click-through data and the result shows their method is better than explicit judgment.

A comparison study was proposed on [8] between three methods of ranking in the usage field. Those methods are PageRank, Weighted PageRank and HITS. All of those methods focus on the structure of the page. The result of this comparison shows that HITS is the best.

In [9], this research presented a method based on a combination of click-through of pages by the users (event) and the summarization of documents. They used the advantage of implicit modelling is effectively improving the user model without any extra effort of the user, as result implicit feedback information improves the user modelling process.

Another study was presented by Rekha et al.. This study provided a new model to find a user's preferences from click-through behavior and used the exposed preferences to adapt the search engine's ranking function for improving the search service. In this proposed model, the combination of viewed and stored document summaries is used. The results show that this combining improved the reliability of the ranked list than ever was [10].

Mukherjee et al., presented a method to discover web knowledge for presenting web users with more personalized web content. Their method collected usage data from different users and then found the similarities between all pairs of users. Experimental results generate correct suggestions that retrieve relevant documents to the user [11].

Tuteja's study in [12] was based on user behaviors in order to enhance the weighted PageRank Algorithm by considering a term Visits of Links (VOL) done by the end of 2013. This research idea was presented as modifying the standard Weighted PageRank algorithm by incorporating Visits of Links. The result shows that adding the number of visits of links (VOL) to calculate the values of page rank proves that relevant results are retrieved first. In this way, it may help users to get the relevant information much quicker.

From previous, few researches considered the usage-based ranking in English and it is rare of them are concerned in Arabic usage-based ranking algorithms. On the other hand most researches are based on the pages selection frequency. This might be an incorrect indicator; the reasons might be inadvertent human mistakes, misleading titles of web pages or the returned summaries not representing the real content. As a conclusion, ranking algorithms still have some drawbacks in the ranked list provided by some search engines. So, we decide to develop a hybrid ranking algorithm to utilize the usage data and apply it to an Arabic search engine, This hybrid ranking algorithm is based on Content-Based Ranking which is the more accurate indicator instead of the Link-Based ranking. The algorithm is based on the content of the pages ranked list, in addition to other usage factors which are:

- Frequency of visit that determine the relevance of a web page by its selection frequency
- Time Spent shows how long users spend on a page after removing the download time of the page.

3   The Proposed System Architecture

This section discusses the proposed Arabic search engine system; the basic idea of the system is based on the Efficient Hybrid Usage-based Ranking Algorithm (EHURA). This hybrid ranking algorithm is to improve the ranked list provided from search engines that are based only on content base ranking. The system architecture is shown in Figure 1

The proposed system consists of 5 main modules:

- **Module 1:** Tokenization: This module is parsing the content of text documents and breaking a stream of text of them into words, then keeping the words in a list called a Word's List.
- **Module 2:** Data Cleaning: The Data Cleaning Module removes useless words from the Word's List; these useless words are stored in a stop words database as appears in the figure. The database has 1459 stop words with a size 10KB.
- **Module 3:** Stemming: The hybrid affix removal algorithm is applied in this module, this hybrid affix removal algorithm is explained in detail in section 3.1.
- **Module 4:** Indexing & Ranking: Indexing is a process for describing or classifying a document by index terms; index terms are the keywords that have their own meaning (i.e. which usually has the semantics of the noun). These index terms are grouped in an indexer and a stemmer services this stage by improving the group of these keywords in the indexer. Then the user's query is matched with the index terms to get the relevant documents to the query. Documents are then ranked using the Content-based ranking algorithm which simply tries to find the similarity between the content of the documents and the query. We applied here the cosine similarity measure, this selection is based on studies represented in [14, 15], which proves that the cosine measure is the most efficient one in comparison to other statistical measurements, in order to rank the documents according to the most relevant to the user's query.
- **Module 5:** Usage-based re-ranking: This module is a re-ranking process based on several parameters like: frequency of pages visited by the user and the time spent on those pages, the information of these parameters is taken from log files after analysis as presented in section 3.2; also these log files are kept by web servers. The parameters include time spent on the page, frequency of visits and for more explanation see section 3.3; In addition, considering also the ranking list of pages that comes from applying the Content-based algorithm. This new algorithm is a hybrid usage-based ranking algorithm (EHURA), EHURA's idea is based on the combination of usage-based parameters and the pervious ranking list, their combination provided a new weight for pages in order to re-ranking them as a new ranked list that appears to the user.

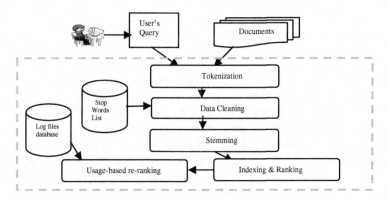

Fig. 1. The Proposed System Architecture

3.1   Hybrid Affix Removal Algorithm

Arafat et al. [13] presented a hybrid Arabic stemmer, this stemmer algorithm is between Root-Based and light stemmers. It removes the suffixes and prefixes of Arabic words and in addition it returns some words to their basic roots.

The algorithm of the hybrid affix removal algorithm [13] is as the following:

- Suffix Removal:

— For Words ending with a suffix & of at least 3 characters without the suffix,
— Replace the suffix with "ه" only if it is "ات" & the word isn't starting with "ال" or its length>5 letters, otherwise Remove "ات". If so, stop searching for more suffixes.
— Replace the suffix by "ي" if it is "ها".
— Stop Searching for Suffixes if the word starts with "ال" and is having any suffix except "ه" and "ي" or its length without the Suffix is less than or equal to 5 characters.
— Otherwise, remove the suffix and stop searching for suffixes.

- Prefix Removal

— I. Phase 1:

❖ For words starting with " ا ا " remove "ا".
❖ For words of at least 5 characters starting with "وال", remove "وال" & stop searching.
❖ For words of more than 4 characters, starting with "لل", remove "لل" & stop searching.
❖ For words of at least 4 characters, starting with "ال", then remove "ال", mark it as a Non Stop Word & stop searching for prefixes.

—   II. Phase 2:
   ❖ For words of at least 3 characters, starting with a prefix, remove the prefix.

Re-check for stop words and remove them.

3.2   Log Files Analysis

Web servers collect large volumes of web usage data. This data is stored in web log files. This usage data is important in the usage-based parameters which are taken from log files after analysis. This section explains the log file analysis. Really, The Log file contains lot of irrelevant entries which need to be removed. To enhance the efficiency of Usage-Based retrieval, any noise should be removed (such as page moved permanently), File does not exist, Server internal error, Service Temporarily Unavailable … etc.) before retrieving the usage data. Log file Analysis consist a series of processes such as data cleaning, user identification, session identification as in the following:

- Data Cleaning is the process of removing unnecessary records like graphics, video and formatted information like css. In addition, this process removes the records of failed HTTP status codes.
- User Identification is the process of identifying users and user agent fields of log entries, its considered as the following:

    1. Different IP addresses refer to different users.
    2. The same IP with different operating systems or different browsers should be considered as different users.
    3. While the IP operating system and browsers are all the same, new users can be determined as to whether the requesting page can be reached by accessed pages before according to the topology of the site.

- A user session is considered to be all of the page accesses that occur during a single visit to a web site. Session Identification is the process of defining users that may access the site more than once.

3.3   Usage-Based Parameters

In this stage the system calculates two Usage-Based Parameters as in the following:
- Frequency of visit: Determines the relevance of a web page by its selection frequency in order to find the frequency weight, which is the admittance frequency of a page u, is the number of times the page is visited and the page rank which appears in the ranked list from the previous stage. The frequency weight formula is:

$$FW = \frac{Number\ of\ Visit\ on\ a\ Page(u)}{Total\ Number\ of\ visit\ on\ all\ Page} \times PR(u) \tag{1}$$

Where:
FW: Frequency Weight.
PR(u): The Page rank of a page u.

- Time Spent: Shows how long the users spend on a page after removing the download time of the page. Because a user generally spends more time on the useful pages and does not waste more time on screening the page and rapidly skipping to another page. So, it's an important parameter to indicate the usefulness of the pages, this parameter is considered to calculate the real time spent on a page by taking the value of time spent on the page from the log file, subtracting from the download time in order to find the Time Spent Weight. It is calculated as follows:

$$TW = \frac{Time\ spent\ on\ a\ Page(u) - Download\ Time(u)}{\max(Time\ spent\ on\ a\ Page(u) - Download\ Time(u))} \tag{2}$$

Where:
TW: Time Spent Weight.

$$Download\ Time\,(u) = \frac{Sizeof\ a\ Page(u)}{Transfer\ Rate\ for\ Page(u)} \tag{3}$$

The combination (i.e. the summation) of the above parameters (frequency of visit and time spent) with the Content-based Ranking results, provide our EHURA algorithm, which is trying to improve the ranking result from Arabic Web search Engines.

4   Performance Studies

In order to study the performance of the proposed system, we used different evaluation measures. These measures are discussed in section 4.1. Then, the data sets used and the experimental results are shown in section 4.2.

4.1   Evaluation IR System

In order to measure the performance of our IR system, we evaluate our proposed EHURA algorithm by measuring the performance of it, then comparing its result with the content-based ranking algorithm. The performance measured by the recall and precision measurements and other measures, are represented in the following formulas:

$$Precision = \frac{|\{relevant\ documents\} \cap \{retrieved\ documents\}|}{|\{retrieved\ documents\}|} \tag{5}$$

$$Recall = \frac{|\{relevant\ documents\} \cap \{retrieved\ documents\}|}{|\{relevant\ documents\}|} \tag{6}$$

Fall-out is the proportion of non-relevant documents that are retrieved, out of all non-relevant documents available.

$$Fall - out = \frac{|\{non-relevant\ documents\} \cap \{retrieved\ documents\}|}{|\{non-relevant\ documents\}|} \tag{7}$$

$$F - Measure = \frac{|\,2*Precision*Recall\,|}{|\,Presision+Recall\,|} \tag{8}$$

Where: F-Measure is the weighted harmonic mean of precision and recall.

$$AveP = \frac{\sum_{i=1}^{N_0} P_i\,(r)}{N_q} \tag{9}$$

Where: AveP: Average precision at recall level r. $P_i(r)$: The precision at recall level r for the ith query. N_q: The number of queries used.

4.2   Experimental Results

For testing the proposed system, it was applied on Ain Shams Arabic corpus. This corpus belongs to the Modern Standard Arabic type; it contains 242 documents with different sizes, and tested the system with 20 queries in order to evaluate the IR system performance.

The system was tested using IR evaluation measurements, which was mentioned in the evaluation section, and it was compared with other information retrieval system called Content-based Information retrieval system (CBIS) in order to prove its effectiveness; the CBIS is based on the Content-Based Ranking Algorithm (CBRA). Figure 2 shows the precision and recall results for each query for the content-based ranking algorithm (CBRA) of CIBS in comparison with the hybrid algorithm (EHURA) in the proposed system. It's clear that EHURA reach a better result than the content-based one. The average precision of our new approach (EHURA) reached 97% while the precision of the Content-Based ranking algorithm is 86%, the results are shown in Table 1. So, our proposed EHURA algorithm improves the precision over the Content-Based ranking algorithm by about 10% while it also improves the recall percentage by 7%.

The proportion of non-relevant documents retrieved (Fall-out) from the system using the Content-based algorithm reached 29%, while our proposed EHURA algorithm reached 22%.

Figure 3 shows the F-Measure for using the content based algorithm and the EHURA and it's clear from the figure that the EHURA algorithm improved the F-Measure over the Content-based algorithm by 9%.

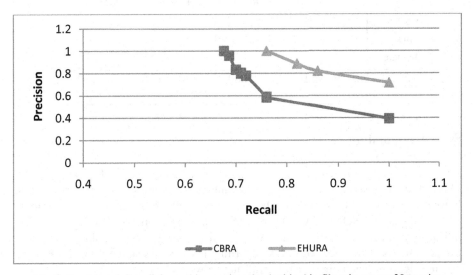

Fig. 2. Precision & Recall for ranking against the Arabic Ain Sham's corpus 20 queries

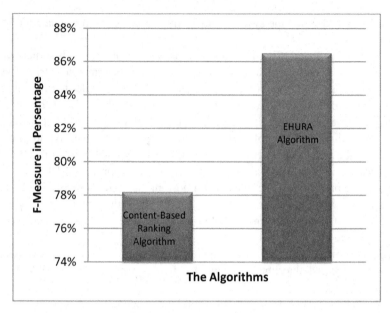

Fig. 3. F-Measure for ranking against the Arabic Ain Sham's corpus

Table 1. Evaluation Summary for the proposed System

	Precision	Recall	Fall-out	F-Measure
Content-Based Ranking	0.8662	0.7125	0.2875	0.7819
EHURA	0.9711	0.7800	0.22	0.8651

5   Conclusion

Arabic is a highly inflected language that has a complex morphological structure. Many researches were done to improve the ranking algorithm for web search engines; however, there are still several drawbacks. Rarely of these researches are done upon Arabic language search engines. Thus, in this paper we proposed an efficient Arabic information retrieval system using a new hybrid Usage-based ranking algorithm called EHURA. The objective of this algorithm is to overcome the drawbacks of the ranking algorithms and improve the efficiency of web searching.

The system was applied to Ain Shams Arabic corpus for testing and evaluation. The results show that the EHURA algorithm improves the performance of the information retrieval system in respect to the recall and precision measures. It also improves the precision over the Content-Based ranking algorithm by about 10% while improving the recall percentage by 7%.

References

1. Internet World Stats: Usage and Population Statistics (2015). http://www.internetworldstats.com/stats.htm
2. Dilekh, T., Behloul, A.: Implementation of New Hybrid Method for Stemming of Arabic Text. International Journal of Computer Applications **46**(8), 14–19 (2012)
3. Dilekh T., Behloul A.: Implementation of a New Hybrid Method for Stemming of Arabic Text. International Journal of Computer Applications **46**(8) (2012)
4. Jiang, X.-M., Song, W.-G., Zeng, H.-J.: Applying associative relationship on the click-through data to improve web search. In: Losada, D.E., Fernández-Luna, J.M. (eds.) ECIR 2005. LNCS, vol. 3408, pp. 475–486. Springer, Heidelberg (2005)
5. Ding, C., Chi, C.-H., Luo, T.: An improved usage-based ranking. In: Meng, X., Su, J., Wang, Y. (eds.) WAIM 2002. LNCS, vol. 2419, pp. 346–353. Springer, Heidelberg (2002)
6. Rodríguez-Mulà, G., Garcia-Molina, H., Paepcke, A.: Collaborative Value Filtering on the Web. Computer Networks **30**(8), 736–738 (1998)
7. Kritikopoulos A., Sideri M., Varlamis I.: Success index: measuring the efficiency of search engines using implicit user feedback. In: Proceedings of the 11th Pan-Hellenic Conference on Informatics. Special Session on Web Search and Mining (2007)
8. Liu, Y., Liu, T., Gao, B., Ma, Z., Li, H.: A framework to compute page importance based on user behaviors. Information Retrieval **13**, 22–45 (2010)
9. Jain, R., Purohit Dr., G.N.: Page Ranking Algorithms for Web Mining. International Journal of Computer Applications **13**(5), 22–25 (2011)
10. Rekha, C., Usharani, J., Iyakutti, K.: Improving the Information Retrieval System through Effective Evaluation of Web Page in Client Side Analysis. International Journal of Computer Applications **15**(6), 35–39 (2011)
11. Mukherjee I., Bhattacharya V., Banerjee S., Gupta P., Mahanti, P.: Efficient Web Information Retrieval based on Usage Mining. IEEE (2012)
12. Tuteja, S.: Enhancement in Weighted PageRank Algorithm Using VOL. Journal of Computer Engineering **14**(5), 135–141 (2013)
13. Arafat, S., Saad, S.: An Affix removal stemming algorithm for Arabic Language. International Journal of Inelligent Computing and Information Science **8**(2), 141–153 (2008)
14. Hajeer S.: Comparison on the Effectiveness of Different Statistical Similarity Measures. International Journal of Computer Applications **53**(8) (2012)
15. Hajeer, S.: Vector Space Model: Comparison between Euclidean Distance & Cosine Measure On Arabic Documents. International Journal Engineering Research and Applications **2**(4), 2085–2090 (2012)

Efficient BSP/CGM Algorithms
for the Maximum Subarray Sum
and Related Problems

Anderson C. Lima$^{(\boxtimes)}$, Rodrigo G. Branco, and Edson N. Cáceres

Faculdade de Computação, Universidade Federal de Mato Grosso do Sul,
Campo Grande, Brazil
{anderson.correa.lima,rodrigo.g.branco}@gmail.com, edson@facom.ufms.br
http://www.facom.ufms.br

Abstract. Given an $n \times n$ array A of integers, with at least one positive value, the maximum subarray sum problem consists in finding the maximum sum among the sums of all rectangular subarrays of A. The maximum subarray problem appears in several scientific applications, particularly in Computer Vision. The algorithms that solve this problem have been used to help the identification of the brightest regions of the images used in astronomy and medical diagnosis. The best known sequential algorithm that solves this problem has $O(n^3)$ time complexity. In this work we revisit the BSP/CGM parallel algorithm that solves this problem and we present BSP/CGM algorithms for the following related problems: the maximum largest subarray sum, the maximum smallest subarray sum, the number of subarrays of maximum sum, the selection of the subarray with k- maximum sum and the location of the subarray with the maximum relative density sum. To the best of our knowledge there are no parallel BSP/CGM algorithms for these related problems. Our algorithms use p processors and require $O(n^3/p)$ parallel time with a constant number of communication rounds. In order to show the applicability of our algorithms, we have implemented them on a cluster of computers using MPI and on a machine with GPGPU using CUDA and OpenMP. We have obtained good speedup results in both environments. We also tested the maximum relative density sum algorithm with a image of the cancer imaging archive.

Keywords: Parallel algorithms · BSP/CGM · Computer vision · GPGPU · Maximum subarray sum

1 Introduction

The maximum subarray sum problem consists in finding the maximum sum among the sums of all rectangular subarrays of an $n \times n$ array A of integers, with at least one positive number [3]. This problem arises in many areas of Science. One such area is Computer Vision, where many applications require the solution of the maximum subarray sum problem. Among these, the identification

O. Gervasi et al. (Eds.): ICCSA 2015, Part I, LNCS 9155, pp. 392–407, 2015.
DOI: 10.1007/978-3-319-21404-7_29

of the brightest region of an image is an important application. This task can aid in medical diagnosis based on images. The best known sequential algorithm for the maximum subarray sum problem has $O(n^3)$ time complexity [3]. Previous works have reported good parallel solutions for the maximum subarray sum problem. Qiu and Akl presented an algorithm that works on interconnection networks (hypercube and star) of length p, using $O(\log n)$ parallel time with $O(n^3/\log n)$ processors [10]. Zhaofang Wen presented a PRAM parallel algorithm using $O(\log n)$ parallel time with $O(n^3/\log n)$ processors [13]. Perumalla and Deo also presented a PRAM algorithm with the same time complexity and number of processors [9]. A BSP/CGM algorithm for the problem was presented by Alves et al. using $O(n^3/p)$ parallel time with p processors and a constant number of communication rounds [1]. Bae [2] presented parallel solutions for the maximum subarray problem on a mesh. Most of these algorithms do not explore the characteristics of their solutions and the output returns only the value of the maximum subarray sum.

In this work, we revisited the maximum subarray sum problem and we propose solutions to five related problems: the maximum largest subarray sum, the maximum smallest subarray sum, the number of subarrays of maximum sum, the selection of the subarray with k-maximum sum and the location of the subarray with the maximum relative density sum. The latter is specially useful when the concentration of elements is more important than the global maximum sum. To the best of our knowledge there are no parallel BSP/CGM algorithms for these problems. Our algorithms use p processors with $O(n^3/p)$ parallel time and a constant number of communication rounds. In order to show the efficiency not only in theory but also in practice, we implemented the algorithms using distributed and shared memory environments. In the distributed memory environment we have implemented the algorithms on a cluster of computers using MPI and in the shared memory we have implemented the algorithms on GPGPU using CUDA and OpenMP. We have got good speedup results in both environments.

This remaining of this paper is organized as follows: Section 2 presents the background and the related works to this study. Section 3 presents our proposed BSP/CGM algorithms. The implementations and results are presented in Section 4. Finally, Section 5 presents the conclusions and future work.

2   Notation and Computational Model

Let A be a $n \times n$ array of integers, with at least one positive value and (g, h) a pair of integers, where $1 \leq g, h \leq n$. We denote $R^{g,h}$ as the set that represents all subarrays $A[i_1 \ldots i_2, g \ldots h]$, where $1 \leq i_1 \leq i_2 \leq n$. Similarly, we also denote the sequence C_j^{gh} of size n as the column resulting from the sum of all elements of each row of $A[1 \ldots n, g \ldots h]$ that are between g e h columns, including g and h, i.e: $C_j^{gh} = \sum_{k=g}^{h} a_{ik}$.

We define the **maximum subarray sum problem** as the task of obtaining the subarray with the maximum sum among all the subarrays $R^{1,n}$ of A [3]. In the array represented by Figure 1 there are three subarrays of maximum sum. All have sum equal to 16.

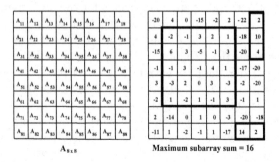

Fig. 1. Array $A_{8 \times 8}$ with $A[2, 6, 2, 6]$, $A[1, 3, 8, 8]$ and, $A[8, 8, 7, 8]$, three subarrays of maximum sum

2.1 Related Problems

Given a $n \times n$ array A of integers, as illustrated in Figure 1, there might be more than one maximum subarray sum. In this array there are three subarrays of maximum sum. All have sum equal to 16, but with different sizes (number of elements): **3**, **25** and **2**, respectively. In this context, at least three new problems arise: the maximum largest subarray sum, the maximum smallest subarray sum and the number of subsequences of maximum sum. In the first two problems, the size is related to the number of elements that make up the subarrays. Besides these three, we expanded our interest including two other problems related to the general problem: the selection of the subarray with k-maximum sum and the location of the subarray with the maximum relative density sum.

2.2 Computational Model

In this paper we use the BSP/CGM parallel computation model [5]. It consists of a set of p processors, each having a local memory of size $O(n/p)$, where n is the size of the problem. An algorithm in this model executes supersteps, that are a series of rounds, alternating well-defined local computation and global communication phases, separated by a barrier synchronization. The cost of communication considers the number of rounds required.

The implementation of a BSP/CGM algorithm generally presents good results with similar performance to those predicted in their theoretical analysis [4]. Since the MPI library is designed for distributed memory environments, a BSP/CGM algorithm can be mapped into a MPI implementation using the message's resources of this library. On the other hand, the implementation of BSP/CGM algorithms on a shared memory environment, like GPGPUs (General Purpose Graphics Processing Units) using CUDA, some abstractions are necessary. In this context, the supersteps of the BSP/CGM model are represented by

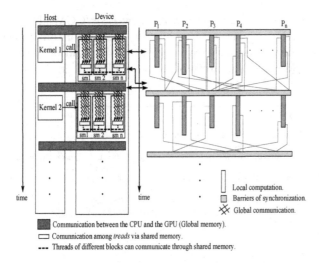

Fig. 2. Abstraction between the BSP/CGM model [7] and GPGPU

sequential invocations of each CUDA kernel. Furthermore, we related the set of processors of the BSP/CGM model with the set of blocks of CUDA. Figure 2 illustrates our suggestion for this process, where the BSP/CGM model [7] is represented on the right.

3   Maximum Subarray Sum

The BSP/CGM algorithm proposed by Alves et al. [1] is well suited for computing the maximum subarray sum in distributed memory environments, since it compress the subarrays in order to decrease the size of the messages between the processors. It is not clear if this compression gives any advantage when using it in a shared memory environment, like a GPGPU. Besides that, the compression makes difficult the computation of the size and localization of the subarrays with maximum sum. On the other hand, the PRAM algorithm for this problem, proposed by Perumalla and Deo [9], does not compress the subarrays. Since we want to explore the informations of the size and localization of the subarrays that have maximum sum, based on the ideas of the later algorithm, we devised a new BSP/CGM parallel algorithm for the maximum subarray sum. This new algorithm can be used in shared or distributed memory environments. In addition, we expanded the algorithm to compute also the five related problems described in Section 2 and implemented the solutions using MPI, OpenMP and GPGPU.

3.1   The BSP/CGM Algorithms for the General and Related Problems

Initially we designed a new BSP/CGM algorithm that solves the general problem of maximum subarray sum (MSS). Since we want to compute the size and localization of the maximum subarray sum and test the algorithm in distributed and shared memory environments, we based the new BSP/CGM algorithm on the results proposed by Perumalla and Deo [9]. The MSS algorithm transforms the computation of the maximum subarray sum in the computation of the maximum subsequence sum (MSqS) in an array of sequence of numbers.

Algorithm 1. Maximum Subarray Sum (MSS)

Input: (1) A set of P processors; (2) The identification of each processor p_i, where $1 \leq i \leq P$; (3) Array $A[1 \ldots n][1 \ldots n]$ of integers.
Output: (1) Array M of maximum values of all subarrays from A. (2) max the maximum value in M.
1. Each processor p_i in **parallel do** $PS[1 \ldots n][1 \ldots n] \leftarrow$ the prefix sums of the rows from array A;.
2. Each processor p_i in **parallel do** $Temp_j[1 \ldots n] \leftarrow C^{(g,h)}(PS)$;
3. Each processor p_i in **parallel do** $max\_local_j \leftarrow MSqS(Temp_j[1 \ldots n])$
4. Each processor $p_i \neq 1$ in **parallel do** $send(max\_local_j, p_i = 1)$;
5. **if** $p_i = 1$ **then**
6.   **for** $k = 2$ **to** P **do**
7. $receive(max\_local_j, p_k)$
8.   **end for**
9.   **for** $j = 1$ **to** $n(n+1)/2$ **do**
10. $M[j] \leftarrow max\_local_j$
11.   **end for**
12. **end if**
13. Each processor p_i in **parallel do** $max \leftarrow M[1, \ldots, n(n+1)/2]$.
14. **return** M, max

The input of the our main algorithm (Algorithm 1) is a $n \times n$ array A of integers, such as the example shown in Figure 1. The first step of the algorithm 1 is the computation of the array PS. Each row of the array PS is given by the prefix sum of the respective row of the array A, where $PS[i, j] = \sum_{k=1}^{j} A[i, k]$. Figure 3 illustrates the computation of the array PS.

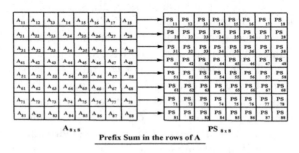

$$A_{8 \times 8} \qquad PS_{8 \times 8}$$

Prefix Sum in the rows of A

Fig. 3. Computation of the array PS

After that we apply the Algorithm 2 on array PS to compute a set of unidimensional subarrays $C^{(gh)'s}$, where $1 \leq g \leq h \leq n$ [9]. Since each line of PS is the prefix sum of the respective line in A, the subarrays $C^{(gh)'s}$, can be easily computed in constant time with $C_j^{gh}[i] = PS[i][h] - A[i][g-1]$. The computation

Algorithm 2. $C^{(g,h)}$

Input: (1) A set of P processors; (2) The identification of each processor p_i, where $1 \leq i \leq P$; (3) Array $PS[1\ldots n][1\ldots n]$ of integers;

Output: Set of one-dimensional arrays $Temp_j[1\ldots n]$.

1. **if** $p_i = 1$ **then**
2. $Temp_j[1\ldots n] \leftarrow PS[i][h]$.
3. **end if**
4. Each processor $p_i \neq 1$ **in parallel do:**
      $Temp_j[1\ldots n] \leftarrow PS[i][h] - PS[i][g-1]$.
5. **return** $Temp_j[1\ldots n]$

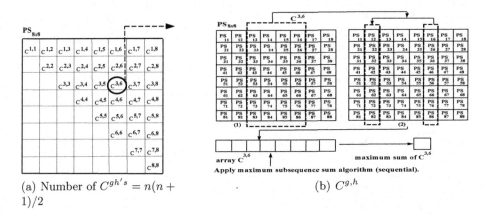

(a) Number of $C^{gh's} = n(n+1)/2$ (b) $C^{g,h}$

Fig. 4. The steps for computing the subarrays $C^{g,h}$

and the number of the subarrays $C^{(gh)'s}$ derived from the array A can be seen in Figure 4.

In the next step, for each subarray C_j^{gh} (compressed in a sequence) the algorithm finds the maximum subsequence sum and the respective maximum sum. In the last step, the algorithm finds the subarray C_j^{gh} with the maximum sum, using the best sequential algorithm for maximum subsequence sum, whose time complexity is $O(n)$ (Algorithm 3). These tasks are illustrate in Figure 4 (b).

Algorithm 3. Maximum Subsequence Sum ($MSqS$)

Input: Integer sequence **S**.

Output: The value of the maximum subsequence sum of **S**.

1. MaxSoFar $\leftarrow 0$.
2. MaxEndingHere $\leftarrow 0$.
3. **for** $i=1$ to n **do**
4.   MaxEndingHere $\leftarrow$ **max**(MaxEndingHere+**S**[i], 0).
5.   MaxSoFar $\leftarrow$ **max**(MaxSoFar, MaxEndingHere).
6. **end for**

We adapt the MSS algorithm (Algorithm 1) in order to have an output with more information than the previous BSP/CGM algorithms. With this we can work on our second goal and we propose solutions for five related problems. They are: 1) the maximum largest subarray sum; 2) the maximum smallest subarray sum; 3) the number of subarrays of maximum sum; 4) the selection of the subarray with k-maximum sum and 5) the location of the subarray with the maximum relative density. The strategies to solve these five problems are described by the Algorithms 4 and 5.

Algorithm 4. Arrays of Maximum Subarray Sum

Input: (1) A set of P processors; (2) The identification of each processor p_i, where $1 \leq i \leq P$; (3) Array $A[1 \ldots n][1 \ldots n]$ of integers.
Output: (1) Array **M** of maximum values of all subarrays from A. (2) Array **E** of number of elements of all maximum subarrays from A.

1. Each processor p_i **in parallel do**
    $PS[1 \ldots n][1 \ldots n] \leftarrow$ the prefix sums of the rows from array $A[1 \ldots n][1 \ldots n]$;
2. $Cgh\_quantity \leftarrow n(n+1)/2$;
3. $k \leftarrow Cgh\_quantity/P$;
4. **for** $j = 1$ to k **in parallel do**
5. $Temp_j[1 \ldots n] \leftarrow C^{(g,h)}(PS)$;
6. $LocalM[1 \ldots j], E[1 \ldots j] \leftarrow MSqS(Temp_j[1 \ldots n])$;
7. **end for**
8. Each processor p_i **in parallel do** $send(LocalM[1 \ldots k], p_i = 1)$;
9. Each processor p_i **in parallel do** $send(LocalE[1 \ldots k], p_i = 1)$;
10. $j \leftarrow 1$;
11. **if** $p_i = 1$ **then**
12.   **for** $i = 2$ to P **do**
13. $receive(LocalMi[1 \ldots k], p_i)$;
14. $receive(LocalEi[1 \ldots k], p_i)$;
15.   **end for**
16.   Computes the arrays $M[1 \ldots n(n+1)/2]$ and $E[1 \ldots n(n+1)/2]$, where:
        $$M[1 \ldots n(n+1)/2]] = [LocalM_{p1_1}, \ldots, LocalM_{p1_k}, \ldots, LocalM_{pp_1}, \ldots, LocalM_{pp_k}] \, .$$
        $$E[1 \ldots n(n+1)/2]] = [LocalE_{p1_1}, \ldots, LocalE_{p1_k}, \ldots, LocalE_{pp_1}, \ldots, LocalE_{pp_k}] \, .$$
17. **end if**

Algorithm 5. Maximum Subarray Sum and Related Problems

Input: (1) A set of P processors; (2) The identification of each processor p_i, where $1 \leq i \leq P$; (3) Arrays $M[1 \ldots n(n+1)/2]$ and $E[1 \ldots n(n+1)/2]$ from Algorithm 4.
Output: (1) Solutions for the maximum subarray sum (general problem) and five related problems.

1. $(M[1 \ldots n(n+1)/2], E[1 \ldots n(n+1)/2]) \leftarrow$ Parallel Sort by Key-value(M,E);
2. **maximum_subarray_sum** $\leftarrow M[n(n+1)/2]$; {general problem}
3. $(NewM[1 \ldots (n(n+1)/2) - i], NewE[1 \ldots (n(n+1)/2) - i]) \leftarrow$ Discard all i elements such as $M[i]$ ¡ **Maximum_subarray_sum** ;
4. **maximum_smallest_subarray_sum** $\leftarrow NewE[0]$; {related problem 1}
5. **maximum_largest_subarray_sum** $\leftarrow NewE[\text{length}(newE)]$; {related problem 2}
6. k-**maximum_subarray_sum** $\leftarrow NewE[k]$; {related problem 3}
7. **number_of_subarrays_of_maximum_sum** $\leftarrow$ size(NewE); {related problem 4}
8. **maximum_relative_density** $\leftarrow$ Parallel Reduce to Max(M,E), using $d_i = M[i]/E[i]$; {related problem 5}

The Algorithms 1 and 4 have similar structures. However, in the Algorithm 4 a new array was created. This array, called $E[1 \ldots n(n+1)/2]]$, stores the number of elements of each position of the array $M[1 \ldots n(n+1)/2]]$. The two arrays generated as output from the Algorithm 4 are used by the Algorithm 5 to solve the general problem of maximum subarray sum and the related problems. For the solutions, the Algorithm 5 uses a parallel sorting algorithm (by key-value) on the arrays $M[1 \ldots n(n+1)/2]]$ and $E[1 \ldots n(n+1)/2]]$. In the last step, to solve the related problem of the location of the subarray with the maximum relative density, we established divisions between valid values of $M[1 \ldots n(n+1)/2]]$ and $E[1 \ldots n(n+1)/2]]$. The Figure 5 illustrates the solution process for all these problems. Particularly in the GPGPU environment, after the sorting, all the steps (except 8) that solve the related problems can run in constant parallel time.

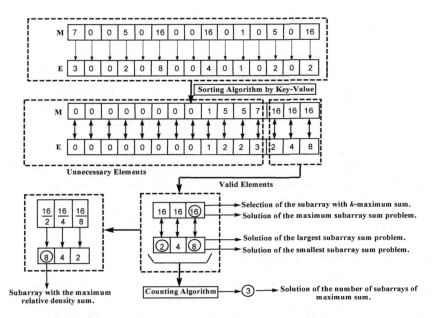

Fig. 5. From the array of valid elements, in constant time, we located the solutions for five problems: The **maximum subarray sum** has sum equal to 16. The **maximum largest subarray sum** has sum equal to 8. The **maximum smallest subarray sum** has sum equal to 2. The **number of subarrays of maximum sum** is 3 (3). The subarray with the **maximum relative density** has value equal to 8.

3.2 Complexity

The complexity of our algorithms is obtained through of an analysis of all steps. Initially, in the Algorithm 1, the prefix sum can be computed by a BSP/CGM parallel prefix algorithm using p processors in $O(n/p)$ time and a constant number of communication rounds. Since there are n columns, the time complexity of this step is $O(n^2/p)$. Then, in the step 2, a set of n^2 one-dimensional arrays is obtained by the processors. Subsequently, the Algorithm 3 $(O(n))$ runs in each array, thus the time complexity of each of these steps is $O(n^3/p)$. The step 4 has only communication. The final steps can be computed by a BSP/CGM parallel algorithm of reduction to maximum, using p processors with $O(n/p)$ time and a constant number of communication rounds. Therefore, we conclude that the final time complexity using p processors is $O(n^3/p)$ with a constant number of communication rounds. Particularly, the Algorithm 4 has the same final complexity, since the final step of parallel sorting does not interfere in the general complexity.

Particularly, the algorithms 4 and 5 have the same final complexity, since the final step of parallel sorting and reduce does not interfere in the general complexity.

4   Implementations and Results

In this section, we discuss the main strategies to development of our algorithms. Regarding the maximum subarray sum (Algorithm 1) two implementations were done. The first one using MPI and the second one using OpenMP. After, we made implementations for the related problems (Algorithm 4) using only CUDA. It is important to state that the Algorithm 4 also solves the problem of maximum subarray sum.

In the MPI solution, we have explored the distributed memory environment. With the OpenMP and CUDA we explore the shared memory environment.

In all these solutions, we search to optimize the most of local processing, running independent operations and thus avoiding unnecessary communications. The process of mapping and subsequent data division is illustrated in Figure 6.

4.1   Computational Resources

We have run the Algorithms on two different computing systems (distributed and shared memory). The Algorithms in CUDA and OpenMP were executed on a machine with 8 GB of RAM, Operating System Linux (Ubuntu 14:04), (R) Intel Core (TM) processor i5-2430M @ 2.40GHz CPU (two cores with 4 threads in each) and an NVIDIA GeForce GTX 680 with 1536 processing cores. The MPI Algorithms were executed on a cluster with 64 nodes, each node consisting of 4 processors Dual-Core AMD Opteron 2.2 GHz with 1024 KB of cache and 8 GB of memory. All nodes are interconnected via a Foundry SuperX switch using Gigabit ethernet. This cluster belongs to High Performance Computing Virtual Laboratory (HPCVL) of Carleton University. In this environment, we performed tests with 16, 32 and 64 processors. In order to compute the speedup (distributed and shared memory environments), the sequential algorithm was run on the cluster environment and on the machine with CPU and GPU. Experiments with different sizes of sequences were performed. For each input sequence, the algorithms were run 10 times and the arithmetic mean of the running times was computed.

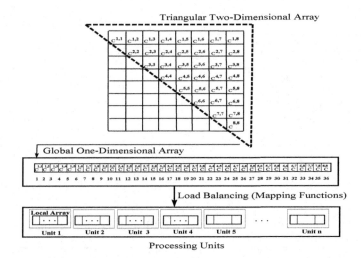

Fig. 6. The process of mapping: each processing unit receives the same number of subarrays $C^{(g,h)'s}$

4.2   Results

Below we present the results achieved by the implemented BSP/CGM algorithms. Three cases of comparison between the algorithms are presented. In these cases we used two-dimensional arrays of integers generated randomly. In all cases we used milliseconds as the time unit. In the speedup calculations, we used the parallel version with the best running time. For equivalent comparisons the pure sequential algorithms were run on different environments, by this they have different times for the same inputs, as presented in the tables.

The First Comparison Case: The first case illustrates a comparison of running times of algorithms for the general problem of maximum subarray sum. We conducted tests with three algorithms: two MPI and one OpenMP. The first MPI algoritm was described in a previous work of Alves et al. [1]. This algorithm involves data compression and a considerable cooperation between the processors. The Figure 7 and the Table 1 illustrate the results of the comparison between this algorithm and the sequential solution. The results prove the efficiency of the solution with increasing speedup values, however an initial overhead seems to elevate the time with 64 processors.

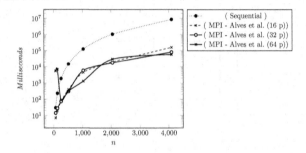

Fig. 7. Sequential × MPI versions (Alves et. al)

Table 1. Running Times: Sequential × MPI versions (16, 32 and 64 p.)

$n \times n$	Sequential Time	MPI Time (16 p.)	MPI Time (32 p)	MPI Time (64 p)	Speedup(Seq./MPI-32p.)
64×64	29.2992	6.8273	14.0064	5422.5232	2.0918
128×128	229.9130	18.6285	27.3484	6877.8257	8.4068
256×256	1849.6825	74.1179	72.2008	87.3313	25.6186
512×512	14879.4916	397.6235	302.5005	316.9507	49.1883
1024×1024	125239.8713	4403.7971	5783.0959	1279.9464	21.6562
2048×2048	1038533.5501	20241.3991	17722.9556	29944.1617	58.5982
4096×4096	8460234.4575	158989.7957	81369.5981	58257.9493	103.9729

The second test represents our MPI solution for the maximum subarray sum problem (Algorithm 1). Differently of the Algorithm of Alves et al. [1], in our solution, each processor is responsible for calculating a set of subarrays $C^{(g,h)'s}$. Furthermore, we do not apply data compression. This is an important factor to extend, more easily, the general solution to the solutions of related problems. The Figure 8 and the Table 2 illustrate the results of the comparison between the Algorithm 1 and the sequential algorithm [3]. In this test we also obtained good results of speeedup, however the values decrease for large input.

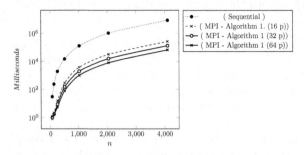

Fig. 8. Sequential × MPI versions (Algorithm 1)

The third test represents our OpenMP implementation for the Algorithm 1. In this test, we worked with 8 threads, because the available machine had a processor with two cores, each with 4 threads. The Figure 9 and the Table 3 illustrate the results of the comparison with the sequential algorithm [3].

Table 2. Running Times (Milliseconds): Seq. × MPI (16, 32 and 64 p.)

$n \times n$	Sequential	MPI Time (16 p.)	MPI Time (32 p.)	MPI Time (64 p.)	Speedup(Seq./MPI-32p.)
64×64	29.2992	0.7725	0.9899	1.1785	29.5981
128×128	229.9130	2.0684	1.6854	1.6009	136.4145
256×256	1849.6825	15.1679	8.5535	5.3154	216.2486
512×512	14879.4916	280.3312	145.2910	77.4639	102.4117
1024×1024	125239.8713	3748.8917	1896.4386	966.3708	66.0395
2048×2048	1038533.5501	31783.7091	15835.1481	8046.7096	65.5841
4096×4096	8460234.4575	261292.5225	130540.967	65800.121	64.8090

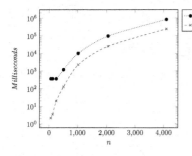

Fig. 9. Sequential × OpenMP (Algorithm 1)

Table 3. Running Times: Seq. × OpenMP

$n \times n$	Sequential	OpenMP	Speedup
64×64	360.9985	2.1495	167.9453
128×128	360.8055	3.4071	105.8981
256×256	360.3314	19.8622	18.1416
512×512	1171.1028	126.3485	9.2688
1024×1024	10099.6082	2221.9307	4.5454
2048×2048	96861.6133	25647.4277	3.7767
4096×4096	835974.2750	242314.7492	3.4500

The Second Comparison Case: The second comparison case illustrates the test based on the CUDA implementation of the Algorithm 4. In this case two versions were implemented, with and without the triangular load balancing described in Section 4. The Figure 10 and the Table 4 illustrate the results of the comparison with the sequential solution. It is possible to note that the version with triangular balancing is a little more efficient. In this implementation, we obtained speedup values that increasing continually.

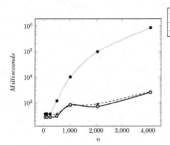

Fig. 10. Sequential × CUDA versions

Table 4. Running Times: Seq. × CUDA

n	Sequential	CUDA	CUDA Triangular	Speedup (Seq./CUDA)
64×64	360.9985	362.1512	271.7961	1.3282
128×128	360.8055	362.0914	272.1506	1.3258
256×256	360.3314	362.9239	271.9573	1.3250
512×512	1171.1028	365.0831	301.2226	3.8878
1024×1024	10099.6082	741.5493	826.9745	12.2127
2048×2048	96861.6133	879.3046	697.1044	138.9485
4096×4096	835974.2750	2699.4783	2525.3061	331.0388

The Third Comparison Case: In the third and final case, we compared the results of speedup of our GPU implementation, with the results of a recent work developed by Cleber et al. [6]. In both cases, GPU algorithms were developed for the problem of the maximum subarray sum. Importantly, the computing resources involved in the two cases were different.

The comparative results are illustrated in Table 5. We observed that our values of speedup are smaller, but the difference tends to decrease. In the last analyzed input ($n = 4096$), our value is better. We believe that this difference occurred because our algorithm has a higher workload, since it presents solutions for the maximum subarray sum and related problems, concomitantly.

Table 5. The best results of speedups and average of speedups

n	Speedup (Cleber et al.)	Speedup (Algorithm 4)
512 × 512	80.2260	3.8878
1024 × 1024	121.9900	12.2127
2048 × 2048	215.2870	138.9485
4096 × 2048	235.9650	331.0388

4.3   Image Application

In medical diagnosis based on X-ray images, the areas of interest are usually the brightest or most luminous region (with more white pixels). In this context, a real application for the maximum subarray sum was presented by Raouf et al. to locate the brightest regions in mammography images, in order to detect macrocalcifications [11].

Particularly, the bitmap images are a good example of input for the maximum subarray sum problem. In this case the RGB pixels are converted to a two-dimensional array of integers or real numbers. This conversion process uses the luminance value. The luminance expresses the color intensity and for the human vision system corresponds the sense of brightness [12]. The luminance of each pixel can be measured by the equation: *Luminance = 0.30R + 0.59G + 0.11B*. The pixels with higher luminance are converted into higher numerical values and lower values (sometimes negatives) are assigned to the pixels with lower value of luminance [2]. The condition of mapping is as follows: If the RGB values of the pixel are within the interval $[0, 1]$, then the luminance value of the pixel will also be within the interval $[0, 1]$. However, in this work, we applied a correction factor equal to -0.5, in order to change the interval of $[0, 1]$ for $[-0.5, +0.5]$ and consequently obtain an array of positive and negative values.

In this work, we also applied our algorithm on real images. The Figure 12 illustrates the result obtained by our algorithm on a real image of X-ray of the human colon. The image belongs to a database of images of various types of cancer [8]. The brightest region is indicated by the dotted rectangle (Figure 12(d)). The Figure 11 illustrates our process of image analysis.

Initially, each pixel from image is converted to an integer using a mapping algorithm. This algorithm uses the formula of luminance and the correction factor proposed. Then, the pixels are placed in a two-dimensional array of same size of the original image. The two-dimensional array is the input to the next stage, which is the location of the area of interest (in this case, the maximum subarray sum). At the end of this step, the output consists of four points, which define

Fig. 11. The steps for finding the brightest region in a X-ray image

the area of interest (rectangle). The points are used as input for the our rendering algorithm, which finally shows the original image, highlighting the region of interest. The four points (x, y) that define the rectangular region are: $(319, 245)$, $(319, 285)$, $(392, 245)$ and $(319, 285)$, as observed in Figure 12 (c). The maximum subarray sum has a value equal to 132382. In our rendering algorithm we used the **devIL** library (Developer's Image Library)[1] and the **GraphicsMagick** package[2], both are open source and are dedicated to image processing.

5 Conclusion and Future Work

In this work we propose efficient BSP/CGM parallel solutions to the maximum subarray sum (Algorithm 1) and five related problems (Algorithm 4): the maximum largest subarray sum, the maximum smallest subarray sum, the number of subarrays of maximum sum, the selection of the subarray with k-maximum sum and the location of the subarray with the maximum relative density sum. To the best of our knowledge, there are no parallel algorithms for these related problems. Our algorithms use p processors and require $O(n^3/p)$ parallel time with a constant number of communication rounds. The good performance of the BSP/CGM parallel algorithms is confirmed by experimental results.

Unlike of the solution presented by Alves et al. [1] in our algorithms, we do not use data compression, and whenever possible, we kept the processors running independently of each other. These strategies were essential to the solutions of the related problems. The results confirmed that, even without data compression, our implementations are effective, with speedups dozens of times better than the sequential solution. It is important to observe that our CUDA implementation (Algorithm 4) solves the general problem and five other problems related to the maximum subarray sum and, even so, obtained a speedup, in the last instance, 331 times better when compared with the sequential solution.

In addition, an experiment of image application is also part of our results. We show that the algorithm can be applied successfully in the localization of the brightest regions in X-ray images.

We also conclude that the BSP/CGM model can be used efficiently to design parallel algorithms in GPGPU/CUDA environments. Besides all the details of these environments, we have mapped our algorithms into implementations with good speedups, where a single GPGPU got better performance than a cluster of workstations. We believe that the abstraction of the Figure 2 can be useful to design BSP/CGM algorithms for GPGPU's environments. Finally, we

---

[1] http://openil.sourceforge.net/

[2] http://graphicsmagick.org/

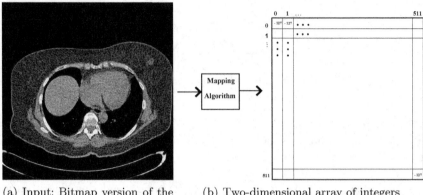

(a) Input: Bitmap version of the X-ray image

(b) Two-dimensional array of integers

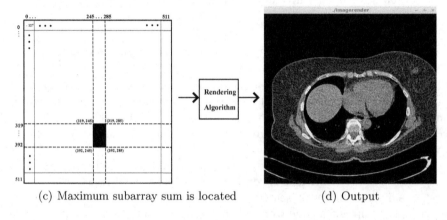

(c) Maximum subarray sum is located

(d) Output

Fig. 12. (a) Bitmap image of a X-ray of colon [8]. (b) Two-dimensional (512 × 512) array generated after the mapping. (c) Location of the maximum subarray sum. (d) The input image with the brightest region highlighted (after the rendering of the brightest region).

also believe that the speedup of our CUDA algorithms can be further improved through the use of multiple GPUs.

5.1 Future Work

A first possibility of future work consists in extending our solutions, in order to address the issue of the tri-dimensional version of the maximum subarray sum problem. To the best of our knowledge, there are no parallel algorithms for this problem. However, we believe it is possible to develop, based on the same principles, real and efficient parallel solutions. With this, practical applications for viewing images can be viewed with even more details.

References

1. Alves, C.E.R., Cáceres, E.N., Song, S.W.: BSP/CGM algorithms for maximum subsequence and maximum subarray. In: Kranzlmüller, D., Kacsuk, P., Dongarra, J. (eds.) EuroPVM/MPI 2004. LNCS, vol. 3241, pp. 139–146. Springer, Heidelberg (2004)
2. Bae, S.E.: Sequential and Parallel Algorithms for the Generalized Maximum Subarray Problem. PhD thesis, University of Canterbury, Christchurch, New Zealand (2007)
3. Bentley, J.: Programming pearls: Algorithm design techniques. Commun. ACM **27**(9), 865–873 (1984)
4. Dehne, F., Ferreira, A., Cáceres, E.N., Song, S.W., Roncato, A.: Efficient para- llel graph algorithms for coarse grained multicomputers and BSP. Algorithmica **33**(2), 183–200 (2002)
5. Dehne, F., Fabri, A., Rau-chaplin, A.: Scalable parallel computational geometry for coarse grained multicomputers. International Journal on Computational Geometry **6**, 298–307 (1994)
6. Ferreira, C.S., Camargo, R.Y., Song, S.W.: A parallel maximum subarray algorithm on GPUs. In: IEEE International Symposium on Computer Architecture and High Performance Computing Workshops, pp. 12–17, October 2014
7. Gotz, S.M.: Communication-Efficient Parallel Algoritms for Minimum Spanning Tree Computation. PhD thesis, University of Paderborn, May 1998
8. National Cancer Institute: Cancer imaging archive (2015) (accessed 11- Jan-2014). https://public.cancerimagingarchive.net
9. Perumalla, K., Deo, N.: Parallel algorithms for maximum subse- quence and maximum subarray. Parallel Processing Letters **5**, 367–373 (1995)
10. Qiu, K., Akl, S.G.: Parallel maximum sum algorithms on interconnection networks. Technical report, Queens University Dept. of Com., Ontario, Canada (1999)
11. Saleh, S., Abdellah, M., Raouf, A.A.A., Kadah, Y.M.: High performance cuda-based implementation for the 2d version of the maximum sub- array problem (msp). In: Cairo International Biomedical Engineering Conference (CIBEC) (2012)
12. ITU International Telecommunication Union. Recommendation ITU-R BT. 470–6,7. Conventional Analog Television Systems (1998) (accessed 03-January- 2015). http://www.itu.int/rec/R-REC-BT.470-6-199811-S/en
13. Wen, Z.: Fast parallel algorithms for the maximum sum problem. Parallel Computing **21**(3), 461–466 (1995)

Set Similarity Measures for Images
Based on Collective Knowledge

Valentina Franzoni[1,2(✉)], Clement H.C. Leung[3], Yuanxi Li[3],
Paolo Mengoni[1], and Alfredo Milani[1,3]

[1] Department of Mathematics and Computer Science, University of Perugia, Perugia, Italy
`valentina.franzoni@dmi.unipg.it, paolo@netdirect.it`
[2] Department of Computer, Control and Mgmt. Engineering,
University of Rome La Sapienza, Rome, Italy
[3] Department of Computer Science, Hong Kong Baptist University,
Kowloon Tong, Hong Kong
`{clement,csyxli}@comp.hkbu.edu.hk, milani@dmi.unipg.it`

Abstract. This work introduces a new class of group similarity where different measures are parameterized with respect to a basic similarity defined on the elements of the sets. Group similarity measures are of great interest for many application domains, since they can be used to evaluate similarity of objects in term of the similarity of the associated sets, for example in multimedia collaborative repositories where images, videos and other multimedia are annotated with meaningful tags whose semantics reflects the collective knowledge of a community of users. The group similarity classes are formally defined and their properties are described and discussed. Experimental results, obtained in the domain of images semantic similarity by using search engine based tag similarity, show the adequacy of the proposed approach in order to reflect the collective notion of semantic similarity.

Keywords: Group similarity · Semantic distance · Image retrieval · Data mining · Collective knowledge · Knowledge discovery

1   Introduction

Search engines, continually exploring the Web using its semantic meaning, are a natural source of information on which to base a modern approach to semantic annotation. The similarity measurement between documents and text has been extensively studied in information retrieval. [15][17]

Content Based Image retrieval (CBIR) [3][18] enables satisfactory similarity measurements of low level features. However, the semantic similarity of deep relationships among objects is not explored by CBIR or other state-of-the-art techniques in Concept Based Image Retrieval [9][10]. The example in Fig. 1 shows the different similarity recognition of humans and computers.

A promising idea is that it is possible to generalize the semantic similarity, under the assumption that semantically similar terms behave similarly [12][18-20]. In this

© Springer International Publishing Switzerland 2015
O. Gervasi et al. (Eds.): ICCSA 2015, Part I, LNCS 9155, pp. 408–417, 2015.
DOI: 10.1007/978-3-319-21404-7_30

work the features of the main semantic proximity measures used in literature are ana-lyzed and used in a new group similarity measure, as a basis to extract semantic con-tent, reflecting the collaborative change made on the web resources. We propose a Context-based Group Similarity to measure deep semantic similarity of images. The proposed algorithm measures the similarity between a pair of images with clouds in-cluding the tags provided by the user i.e., image author/owner.

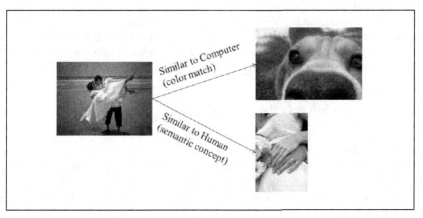

Fig. 1. Image similarity discovery comparison between computer and human

2   Related Work

2.1   WordNet Similarity

WordNet [1], is one of the applications of semantic lexicon propose for the English language and is a general knowledge base and common sense reasoning engine [3]. Recent researches [2] on the topic in computational linguistics has emphasized the perspective of semantic relatedness of two lexemes in a lexical resource, or its inverse, semantic distance. The work in [3] brings together ontology and corpus, defining the similarity between two concepts *c1* and *c2* lexicalized in WordNet, named *WordNet Distance (WD)*, by the information content of the concepts that subsume them in the taxonomy. Then, [4] proposes a similarity measure in WordNet between arbitrary objects:

$$d(c_1, c_2) = \frac{2 \times \log p(lso(c_1, c_2))}{\log p(c_1) + \log p(c_2)} \tag{1}$$

The advantage of a WordNet similarity is that it is a very mature and comprehen-sive lexical database, which provides measures of similarity and relatedness: it re-flects universal knowledge because it is built by human experts, however WordNet Distance is only for nouns and verbs in WordNet, which is not dynamically updated.

2.2   Wikipedia Similarity

WikiRelate [5] was the first research to compute measures of semantic relatedness using Wikipedia. This approach took familiar techniques that had previously been applied to WordNet and modified them to suit Wikipedia. The implementation of WikiRelate follows the hierarchical category structure of Wikipedia.

The *Wikipedia Link Vector Model (WLVM)* [6] uses Wikipedia to provide structured world knowledge about terms of interest.

These approaches use the hyperlink structure of Wikipedia rather than its category hierarchy or textual content [7][11]. The probability of WLVM is defined by the total number of links to the target article over the total number of articles. Thus if t is the total number of articles within Wikipedia, the weighted value w for the link $a \rightarrow b$ is:

$$w(a \rightarrow b) = |a \rightarrow b| \times \log(\sum_{x=1}^{t} \frac{t}{|x \rightarrow b|}) \tag{2}$$

where a and b denote the search terms.

Wikipedia similarity reflects relationships as seen by the user community [18][19], which is dynamically changing, as links and nodes are changed by the users' collaborative effort. However, it only can apply to knowledge base organized as networks of concepts.

2.3   Flickr Similarity

Flickr distance (FD) [8] is another model for measuring the relationship between semantic concepts in visual domain. For each concept, a collection of images is obtained from Flickr, based on which the improved latent topic-based visual language model is built to capture the visual characteristics of the concept. The Flickr distance between concepts *c1* and *c2* can be measured by the *square root of Jensen-Shannon divergence* [10, 11] between the corresponding visual language models as follows:

$$D(C\_1, C\_2) = \sqrt{\frac{\sum_{i=1}^{K} \sum_{j=1}^{K} D_{JS}(P_{Zi}C_1 | P_{Zj}C_2)}{K^2}} \tag{3}$$

where

$$D_{JS} = (P_{Zi}C_1 | P_{Zj}C_2) = \frac{1}{2} D_{KL}(P_{Zi}C_1 | M) + \frac{1}{2} D_{KL}(P_{Zj}C_2 | M) \tag{4}$$

K is the total number of latent topics, which is determined by experiment. $P_{Zi} C_1$ and $P_{Zj} C_2$ are the trigram distributions under latent topic *zic1* and *zjc2* respectively, with M representing the mean of $P_{Zi} C_1$ and $P_{Zj} C_2$. The FD is based on visual language models (VLM), which is a different concept relationship respect to WordNet Similarity and Wikipedia Similarity.

3   Context-Based Semantic Group Distance

3.1   Web-Based Proximity Measures

We use Web-based proximity Measurements [18] as compositor of Group Distance because it can be applied to any retrieval engine. It makes good use of the occurrence/co-occurrence of words, terms, tags, users, objects etc. It reflects the user community current state of believes and dynamically changes as results change [16][11].

3.1.1   Normalized Google Distance (NGD)
Normalized Google distance (NGD) [9] quantifies the extent of the relationship between two concepts by their correlation in the search results from Google search engine when querying both concepts, with

$$NGD(x,y) = \frac{\max\{\log f(x), \log(y)\} - \log[f(x,y)]}{\log N - \min\{\log f(x), \log[f(y)]\}} \tag{5}$$

Where *f(x)* and *f(y)* are the numbers of the web pages and documents returned by Google search engine when typing *x* and *y* as the search term respectively, with *f(x,y)* denoting the number of pages containing both *x* and *y*. *N* denotes the total number of documents indexed by Google search, which is deductible or approximable by any number greater than every page count of the results.

3.1.2   Pointwise Mutual Information (PMI)
Mutual Information (MI) is a measure of the information overlap between two random variables.

Pointwise Mutual Information (PMI) [13][14] is a point-to-point measure of association, which represents how much the actual probability of a particular co-occurrence of events differs from what we would expect it to be on the basis of the probabilities of the individual events and the assumption of their independence. Even though PMI may be negative or positive, its expected outcome over all joint events (i.e., PMI) is positive.

PMI is used both in statistics and in information theory.

PMI between two particular events w_1 and w_2, in this case the occurrence of particular words in Web-based text pages, is defined as follows:

$$PMI(w_1, w_2) = log_2 \frac{P(w_1, w_2)}{P(w_1)P(w_2)} \tag{6}$$

This quantity is zero if w_1 and w_2 are independent, positive if they are positively correlated, and negative if they are negatively correlated.

On particularly low frequency data, PMI does not provide reliable results [19]. Since PMI is a ratio of the probability of w_1, w_2 together and w_1, w_2 separately, in the case of perfect dependence, PMI will be *0*.

PMI is a bad measure of dependence, since the dependency score is related to the frequency of individual words. PMI could not always be suitable when the aim is to compare information on different pairs of words.

3.2   Confidence and Average Confidence (CM)

Given a rule $X \rightarrow Y$, *Confidence* [4] is a statistical measure that, given the number of transactions that contain X, indicates the percentage of transactions that contain also Y.

$$confidence(x \rightarrow y) = \frac{P(x,y)}{P(x)} = \frac{f(x,y)}{f(x)} \tag{7}$$

From a probabilistic point of view, confidence approximates the conditional probability:

$$confidence(x \rightarrow y) = \frac{P(x \wedge y)}{P(x)} = P(y|x) \tag{8}$$

Since Confidence is not a symmetric measure, the Average Confidence (CM) [19] can be defined a

$$CM = \frac{confidence(x \rightarrow y) + confidence(y \rightarrow x)}{2} \tag{9}$$

3.3   Context-Based Group Similarity

We propose a Context-based Group Similarity to measure deep semantic similarity of images. The proposed algorithm measures the image similarities with semantic proximity based one user provided concept clouds. Image semantic concept clouds include any semantic concept associated to or extracted from images. Typical sources for semantic concepts are tags, comments, descriptors, categories, or text surrounding the image. As shown in Fig. 2, Image Ii and Image Ij are a pair of images to be compared. $Ti1$, $Ti2$,..., Tim are original user provided tags of image Ii, while $Tj1$; $Tj2$,..., Tjn are original user provided tags of image IJ .

Given $DIij$ as the distance (or equivalently, the similarity) of image Ii and image Ij, we define the **Group Distance (GD)**:

$$DI_{ij} = AVG2\{AVG1[SEL(dT_{im \rightarrow jn})], AVG1[SEL(dT_{jn \rightarrow im})]\} \tag{10}$$

where *SEL* could be the maximum *MAX*, the average *AVG* or the minimum *MIN* of d, the similarity calculated by algorithm (*Confidence* or *NGD* or *PMI*), as in equations (11-14).

$$dT_{im} \rightarrow dT_{jn} = \begin{pmatrix} dT_{i1 \rightarrow j1}, & dT_{i1 \rightarrow j2}, & dT_{i1 \rightarrow j3}, & \dots & dT_{i1 \rightarrow jn} \\ dT_{i2 \rightarrow j1}, & dT_{i2 \rightarrow j2}, & dT_{i2 \rightarrow j3}, & \dots & dT_{i2 \rightarrow jn} \\ \dots, & \dots, & \dots, & \dots & \dots \\ dT_{in \rightarrow j1}, & dT_{in \rightarrow j2}, & dT_{in \rightarrow j3}, & \dots & dT_{in \rightarrow jn} \end{pmatrix}$$

$$dT_{im} \rightarrow dT_{jn} = \begin{pmatrix} dT_{j1 \rightarrow i1}, & dT_{j1 \rightarrow i2}, & dT_{j1 \rightarrow i3}, & \dots & dT_{j1 \rightarrow im} \\ dT_{j2 \rightarrow i1}, & dT_{j2 \rightarrow i2}, & dT_{j2 \rightarrow i3}, & \dots & dT_{j2 \rightarrow im} \\ \dots, & \dots, & \dots, & \dots & \dots \\ dT_{jn \rightarrow i1}, & dT_{jn \rightarrow i2}, & dT_{jn \rightarrow i3}, & \dots & dT_{jn \rightarrow im} \end{pmatrix}$$

$$(11)$$

$$AVG1[SEL(dT_{im \rightarrow jn})] = avg[SEL(dT_{i1 \rightarrow jn}), SEL(dT_{i2 \rightarrow jn}), , SEL(dT_{im \rightarrow jn})] \qquad (12)$$

$$AVG1[SEL(dT_{jn \rightarrow im})] = avg[SEL(dT_{j1 \rightarrow im}), SEL(dT_{j2 \rightarrow im}), , SEL(dT_{jn \rightarrow im})] \qquad (13)$$

$$AVG2 = AVGAVG1, AVG2 \qquad (14)$$

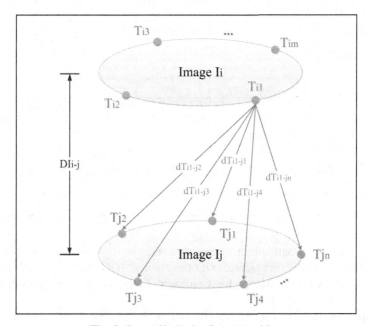

Fig. 2. Group Similarity Core Algorithm

We calculate the similarity of each pair of images, from the users' votes of how much similar of certain pair of images is. Users can score each pair of images with a number from 0 (=very different) to 5 (=very similar), based on their personal opinion.

The quantitative experimental results which compare the users' scores and proposed Context-based Group Distance will be shown in section 4.

The core algorithm of the proposed Context-based Group Distance calculations is shown in Figure 2.

4   Experimental Results

The system is evaluated quantitatively using more than 500 pairs of Web images on Flickr to compare the users' scores and the proposed Context-based Group Distance. We keep the tags as original ones provided by the owner of images in Flickr.

Fig. 3 shows a sample of image pairs similarity experiment results comparing user score and proposed Group Similarity.

Table 1. Experimental Results on Bing for avg-avg-avg GD

image1	image2	User Score	GD-Confidence	GD-NGD	GD-PMI
1	2	0.47826	0.39801	0.744305	5.99773
3	4	0.53043	0.33722	0.721759	6.24246
5	4	0.37391	0.29569	0.705348	6.0838
6	4	0.41739	0.28847	0.680995	5.78023
2	7	0.32174	0.27788	0.732283	5.94255
8	9	0.29565	0.25935	0.595794	4.23411
1	10	0.43478	0.24937	0.658751	5.13127
11	6	0.75652	0.23742	0.576982	5.40974
12	13	0.31304	0.22882	0.711011	6.05964
4	14	0.13043	0.2266	0.674191	5.52523
4	15	0.16522	0.21048	0.734764	6.24088
...	...	...	...	...	...

Table 1 shows a small subset of experimental results. *User Score* represents the average human evaluation score. *GD-Confidence, GD-NGD* and *GD-PMI* are respectively the Group Distance for Confidence, Normalized Google Distance and Pointwise Mutual Information.

Experimental results indicate that this approach can give more accurate and efficient similarity measurements than calculations based on image low level features, in terms of the deep semantic concept similarity.

Image Pair Number	Sample Image Pairs		User Voted Average Score	Similarity Calculated by Group Similarity
19-24	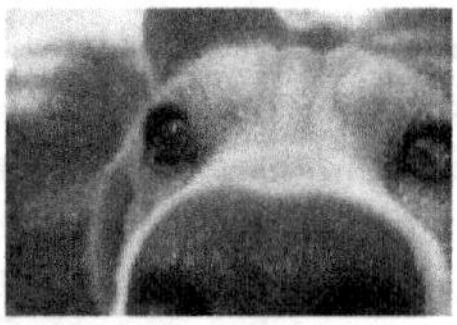		0.47826087	0.398011
Original Tag ID and Tags	19936 beagle 19936 dog 19936 black 19936 white 19936 nose	16054 dog 16054 people 16054 kiss		
19-20		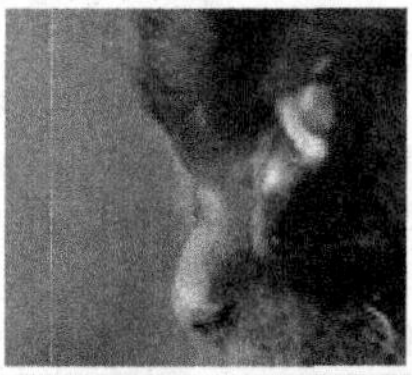	0.434782609	0.249373
Original Tag ID and Tags	19936 beagle 19936 dog 19936 black 19936 white 19936 nose	17405 cat 17405 eye 17405 nose 17405 fur		
19-47			0.07826087	0.168459
Original Tag ID and Tags	19936 beagle 19936 dog 19936 black 19936 white 19936 nose	18239 airplane 18239 aviation 18239 blue 18239 california 18239 commercial		
Searching Engine: Bing Similarity Type: Confidence SEL Option Used: Average (AVG)				

Fig. 3. Sample set of image pairs similarity experiment results – comparing user score and proposed Group Similarity

5   Conclusion

In this work we present our approach to measure the context-based group similarity among concepts and images. Comparing with Content Based Image Retrieval (CBIR), which measures the image content similarities by low level features, the proposed Context-based Image Similarity outperforms CBIR in effectiveness and accuracy of comparing the deep concept among images, since in order to recognize low level features in the CBIR higher computation capabilities are required. The systems are evaluated quantitatively involving the user judgments. Experimental results indicate that our approaches could enable advanced degree of semantic similarity measurements and deep meaning enrichment of images, also delivering highly competent performance, attaining excellent precision and efficiency.

References

1. Miller, G.A.: Wordnet: a lexical database for english. Communications of the ACM **38**(11), 39–41 (1995)
2. Budanitsky, A., Hirst, G.: Semantic distance in wordnet: an experimental, application-oriented evaluation of five measures. In: Proceedings of Workshop on WordNet and Other Lexical Resources, p. 641. North American Chapter of the Association for Computational Linguistics, Pittsburgh (2001)
3. Resnik, P.: Using information content to evaluate semantic similarity in a taxonomy. In: Proceedings of the 14th International Joint Conference on Artificial Intelligence, pp. 448–453 (1995)
4. Lin, D.: An information-theoretic definition of similarity. In: Proceedings of the 15th International Conference on Machine Learning, pp. 296–304. Morgan Kaufmann (1998)
5. Strube, M., Ponzetto, S.P.: WikiRelate! computing semantic relatedness using wikipedia. In: Proceedings of the Twenty-First National Conference on Artificial Intelligence. AAAI Press, July 2006
6. Milne, D., Witten, I.H.: An effective, low-cost measure of semantic relatedness obtained from wikipedia links. In: Proceedings of first AAAI Workshop on Wikipedia and Artificial Intelligence, WIKIAI 2008, Chicago, IL, USA (2008)
7. Völkel, M., Krötzsch, M., Vrandecic, D., Haller, H., Studer, R.: Semantic wikipedia. In: Proceedings of the 15th International Conference on World Wide Web, WWW 2006, pp. 585–594. ACM, New York (2006)
8. Wu, L., Hua, X.-S., Yu, N., Ma, W.-Y., Li, S.: Flickr distance. In: Proceedings of the 16th ACM International Conference on Multimedia, MM 2008, New York, NY, USA, pp. 31–40 (2008)
9. Enser, P.G., Sandom, C.J., Lewis, P.H.: Surveying the reality of semantic image retrieval. In: Bres, S., Laurini, R. (eds.) VISUAL 2005. LNCS, vol. 3736, pp. 177–188. Springer, Heidelberg (2006)
10. Li, X., Chen, L., Zhang, L., Lin, F., Ma, W.: Image annotation by large-scale content-based image retrieval. In: Proceedings of the 14th Annual ACM International Conference on Multimedia, pp. 607–610 (2006)
11. Franzoni, V., Milani, A.: PMING Distance: A collaborative semantic proximity measure. In: 2012 IEEE/WIC/ACM International Conferences on Web Intelligence and Intelligent Agent Technology (IAT), vol. 2, pp. 442–449 (2012)

12. Leung, C.H., Li, Y., Milani, A., Franzoni, V.: Collective evolutionary concept distance based query expansion for effective web document retrieval. In: Murgante, B., Misra, S., Carlini, M., Torre, C.M., Nguyen, H.-Q., Taniar, D., Apduhan, B.O., Gervasi, O. (eds.) ICCSA 2013, Part IV. LNCS, vol. 7974, pp. 657–672. Springer, Heidelberg (2013)
13. Manning, D., Schutze, H.: Foundations of statistical natural language processing. The MIT Press, London (2002)
14. Turney P.: Mining the web for synonyms: PMI versus LSA on TEOFL. In Proc. ECML (2001)
15. Chan, A.W.S., Liu, J., et al.: Intelligent Social Media Indexing and Sharing Using an Adaptive Indexing Search Engine. ACM TIST **3**(3), 47 (2012). doi:10.1145/2168752. 2168761
16. Li, Y.X.: Semantic Image Similarity Based on Deep Knowledge for Effective Image Retrieval. Research Thesis (2014)
17. Cheng, V.C., Liu, J., et al.: Probabilistic Aspect Mining Model for Drug Reviews. IEEE Transactions on Knowledge and Data Engineering **99**, 1 (2014). doi:10.1109/TKDE. 2013.175. vol. 99, no. PrePrints, p. 1
18. Franzoni, V., Milani, A.: Heuristic semantic walk for concept chaining in collaborative networks. International Journal of Web Information Systems **10**(1), 85–103 (2014). doi:10.1108/IJWIS-11-2013-0031
19. Franzoni, V., Mencacci, M., Mengoni, P., Milani, A.: Heuristics for semantic path search in Wikipedia. In: Murgante, B., et al. (eds.) ICCSA 2014, Part VI. LNCS, vol. 8584, pp. 327–340. Springer, Heidelberg (2014)
20. Franzoni, V., Milani, A.: Heuristic Semantic Walk. In: Murgante, B., Misra, S., Carlini, M., Torre, C.M., Nguyen, H.-Q., Taniar, D., Apduhan, B.O., Gervasi, O. (eds.) ICCSA 2013, Part IV. LNCS, vol. 7974, pp. 643–656. Springer, Heidelberg (2013)
21. Franzoni V., Milani A.: Semantic Context Extraction from Collaborative Networks. In: IEEE International Conference on Computer Supported Cooperative Work in Design (CSCWD), Calabria, Italy (2015)

A Classification of Test Purposes
Based on Testable Properties

Simone Hanazumi[(✉)] and Ana C. V. de Melo

Department of Computer Science, University of São Paulo, São Paulo, Brazil
{hanazumi,acvm}@ime.usp.br

Abstract. Test purposes are today key-elements for making the formal testing approach applicable to software. They are abstractions for a set of test cases to be observed in programs and some tools for formal testing concentrate their effort to apply test purposes to programs. Defining test purposes for actual systems, however, are not very straightforward. We must first realize what is the property behind the set of test cases we want to address. To help in defining the test purposes for a system under test, this paper presents a classification for test purposes based on the testable properties patterns. First, we analyze the meaning of properties patterns in the light of test purposes and then check the testability of these patterns. As a result, we provide a classification for testable patterns applicable to test purposes and users can choose which pattern better fits his/her set of test cases to be observed. Moreover, since test purposes are defined as standard properties, a model checker can be used to find executions for which test purposes can be satisfied.

Keywords: Program specification · Linear temporal logic · Software testing · Test purposes · Formal verification

1 Introduction

Model based testing is a promising approach to automate test generation and give a better coverage of the overall system behavior. It relies on formal models to describe systems specifications and equivalence testing relations (based on labelled transition systems) to provide the conformance evaluation of the Implementation Under Test (IUT) in relation to the given specification [17,18]. Following this approach, the IUT must be executed and observed to comply with the given system specification.

One of the advantages of using formal system specification models is the ability to automate the test cases [7]. However, if the whole system specification is taken into account, the number of test cases to be observed becomes infeasible for complex systems, considering the theory of formal testing. Then, *test purposes* were proposed as abstractions of test cases [2,10] in which the goals of testing are defined. The aim of creating test purposes is twofold. First, the tester focuses on some goals to be tested, driving to a separation of concerns at testing level.

© Springer International Publishing Switzerland 2015
O. Gervasi et al. (Eds.): ICCSA 2015, Part I, LNCS 9155, pp. 418–430, 2015.
DOI: 10.1007/978-3-319-21404-7_31

Second, instead of generating test cases for the whole system specification, they are confined to certain goals, making the formal testing more applicable to real problems. As a result, test purposes have been adopted by tools to generate test cases in the formal testing approach.

To realize which test purposes one wants to observe in a system is still a difficult task, since he/she needs to look at the system specification and programs counterparts at the same time. In practice, the events observed at specification level are different from the ones at program level [19], they are written in different abstraction levels. To actually define the test purposes for a given system and programs, one must observe the events inside the specification and somehow map them to programs events. From these elements, the abstract model to represent the set of test cases to be observed are defined. All this makes test purposes as difficult to be defined as properties that are formally verified in programs.

To simplify the task of property specification to be verified in systems, Dwyer et al. [3] proposed a set of property patterns and scopes that can be used to derive several types of system properties. Each pattern is combined with a scope to provide a property formula. These formulas are written in different formal languages, including linear temporal logic (LTL), computational tree logic (CTL) and regular expressions [14]. The set of these combinations of patterns and scopes are called Specification Pattern System (SPS). Although the SPS formulas are useful for property specification, no studies have demonstrated how to use these patterns to make easier the task of defining test purposes.

In fact, from a theoretical point-of-view the patterns can only be directly used for programs testing if a relationship between SPS formulas and the testability concept is established. The testability concept [4,13] uses the relations between property specification traces and the implementation under test (IUT) traces to guarantee that the property will provide either a fail verdict or a weak-pass verdict (i.e. at least one trace of the program will satisfy the property). Assuring that the SPS formulas are testable, we have as result that all properties derived from the SPS are also testable.

In our project, we have studied ways of making test purposes more suitable to end users. This paper presents a classification of test purposes based on testable properties for the Specification Pattern System (SPS). From a practical point-of-view, users can take these patterns and customize them to the system under test. This is a way of shortening the steps to realize which abstract model better represents the set of test cases one wants to address, instead of building from scratch. In addition, since the test purpose patterns have been proved testable, we make sure that executable traces of programs can be generated from them.

In this study, we have chosen the Linear Temporal Logic (LTL) to represent the test purpose patterns because they are more intuitive for end users. From the LTL formulas, the corresponding labelled transition systems are created and users end up with the customized test purposes. Besides that, executable traces of programs that satisfy the given test purposes can be generated using model checkers [8], either for successful and fail verdicts.

The paper is organized as follows: Section 2 describes related work; Section 3 presents the underlying theory; Section 4 presents the testability concept and relates it with the SPS; Section 5 describes the process of using the proposed classification; and, Section 6 presents the conclusion and future work.

2   Related Work

Dwyer et al. [3,14] proposed the specification patterns system (SPS) aiming to simplify the process of properties specification for finite-state verification. To assess the SPS, an empirical study was performed. In this empirical study, a large set of program properties was analyzed to determine the occurrence proportion of each SPS pattern/scope. However, no study relating the SPS to the concept of test purposes and testability was done.

Silva and Machado [16] described in their work the process of using model checking to generate test purposes, thus, relating the concept of test purposes to the concept of program properties. Weiglhofer et al. [20,21] also worked on an automated process to generate test purposes. In this case, they used the coverage data of program specifications to guide the generation of test purposes. But these works do not provide any testable patterns that would be derived using this process.

The works of de Vries and Tretmans [2] and Ledru et al. [10] emphasize the use of test purposes to simplify the use of formal testing in practice. Nevertheless, they do not provide an easier way to write the test purposes formally.

What makes our work different from the others is the connection that we made between test purposes and testability. The testability [13] is an important concept that relates formal properties to program execution traces, making it possible to restrain the universe of program properties to the properties that can provide a success/fail verdict [4]. We do consider test purposes as program properties, and we use the SPS as a starting point to provide a test purposes classification based on the testability concept. This test purposes classification can then be used to simplify the process of specifying a property to be inputted in a model checker.

3   LTL and the SPS

In this section, we describe the underlying theory regarding linear temporal logic (LTL) and the specification pattern system (SPS). The SPS patterns and scopes can be defined using several formulas and representations. In the present work we use LTL to represent them due to its large use in formal verification theory and its simple syntax and semantics. LTL and SPS are briefly described here.

3.1   Linear Temporal Logic (LTL)

Linear Temporal Logic [12,15] extends propositional logic by modalities that allow to refer to the behavior of reactive systems [1]. It considers, besides the inputs and the outputs of a computation, the executions of systems to analyze their correctness. Next we present some of its main concepts and operators.

- **Grammar.** Let φ, φ_1 and φ_2 be LTL formulas. The LTL grammar can be represented as follows.

$$
\begin{aligned}
\varphi ::= \ & True \mid False \mid & \textit{(Boolean Variables)} \\
& P \mid & \textit{(Atomic Proposition)} \\
& \varphi_1 \wedge \varphi_2 \mid \varphi_1 \vee \varphi_2 \mid \\
& \varphi_1 \oplus \varphi_2 \mid \varphi_1 \rightarrow \varphi_2 \mid \\
& \varphi_1 \leftrightarrow \varphi_2 \mid \neg\varphi \mid & \textit{(Boolean Operators)} \\
& \bigcirc\varphi \mid \Box\varphi \mid \Diamond\varphi \mid \\
& \varphi_1 \ U \ \varphi_2 \mid \varphi_1 \ \mathcal{W} \ \varphi_2 & \textit{(Temporal Operators)}
\end{aligned}
$$

- **Atomic Proposition.** An **atomic proposition** is used to express simple known facts about the system under consideration (e.g. system events). They can be denoted by letters.
- **Boolean Operators.** The unary logic operator for negation ($\neg$), and the binary logic operators for conjunction ($\wedge$), disjunction ($\vee$), implication ($\rightarrow$), equivalence ($\leftrightarrow$) and exclusive or ($\oplus$), can be applied to LTL formulas. Their usual meanings in propositional logic are preserved in LTL.
- **Temporal Operators.** The most used LTL temporal operators are the binary operators "until" ($\mathcal{U}$) and "weak until" ($\mathcal{W}$), and the unary operators "next" ($\bigcirc$), "eventually" ($\Diamond$) and "always" ($\Box$). The intuitive semantics of these temporal modalities are given below.
  - *"Eventually"* - $\Diamond$. Let φ be a LTL expression. The formula "$\Diamond\varphi$" holds at the current moment if now or eventually in the future φ holds. The "eventually" operator can also be defined as: "$\Diamond\varphi = true \ \mathcal{U} \ \varphi$".
  - *"Always"* - $\Box$. Let φ be a LTL expression. The formula "$\Box\varphi$" holds at the current moment if now and forever in the future φ holds. This temporal modality can also be defined as: "$\Box\varphi = \neg\Diamond\neg\varphi$".
  - *"Next"* - $\bigcirc$. Let φ be a LTL expression. The formula "$\bigcirc\varphi$" holds at the current moment if φ holds in the next step of the execution of the system under consideration.
  - *"Until"* - $\mathcal{U}$. Let φ_1 and φ_2 be LTL formulas. The expression "$\varphi_1 \ \mathcal{U} \ \varphi_2$" holds at the current moment, if some future moment in which φ_2 holds exists, and φ_1 holds at all moments until that future moment.
  - *"Weak Until"* - $\mathcal{W}$. Let φ_1 and φ_2 be LTL formulas. The expression "$\varphi_1 \ \mathcal{W} \ \varphi_2$" holds at the current moment, if either (i) at some future moment in which φ_2 holds exists, φ_1 holds at all moments until that future moment ($\varphi_1\mathcal{U}\varphi_2$) or (ii) now and forever in the future φ_1 holds ($\Box\varphi_1$).

Safety-Progress (SP) Classification. The SP classification or the Borel Hierarchy [11] categorizes the linear temporal properties in six classes that are organized in a hierarchical manner: *safety* (it states that some bad thing *never* happens); *guarantee* (it states that some good thing occurs *at least once*); *obligation* (it is expressed by a disjunction of safety and guarantee formulas); *recurrence* (it states that some good thing occurs infinitely often); *persistence* (it states that some good thing occurs continuously from a certain point on); and, *reactivity* (it is expressed by a disjunction of recurrence and persistence formulas).

Table 1. SP Classification [11]

Property Class	LTL Canonical Formula
Safety	$\Box\varphi$
Guarantee	$\Diamond\varphi$
Obligation	$\wedge_{i=1}^{n}[\Box\varphi_i \vee \Diamond\gamma_i]$
Recurrence	$\Box\Diamond\varphi \equiv \Box(\varphi \rightarrow \Diamond\gamma)$
Persistence	$\Diamond\Box\varphi$
Reactivity	$\wedge_{i=1}^{n}[\Box\Diamond\varphi_i \vee \Diamond\Box\gamma_i]$

3.2   Specification Pattern System (SPS)

Specification patterns [3] are high-level, formalism independent specification abstractions defined for finite state verification. Their purpose is to assist practitioners in mapping descriptions of system behavior into their formalism of choice, improving the transition of these formal methods to practice. To define a property using the SPS, one must define its scope and then, the corresponding pattern. A brief description of scopes and patterns are presented below. For a complete description of the SPS, please refer to [14].

- **Scopes.** A scope is the extent of the program execution over which the pattern must hold, and is determined by specifying a starting and an ending event for the pattern. They are divided in five categories: *global*; *before* the occurrence of R; *after* the occurrence of L; *between* the occurrences of L and R; and, *after-until*, i.e., after the occurrence of L until the occurrence of R.
- **Patterns.** A pattern is a specification abstraction that can be mapped to various formalisms, including LTL. The patterns are organized in a hierarchy based on their semantics. Here, we are going to describe the patterns that appear in our work: *absence* (an event does not occur within a scope); *existence* (an event must occur within a scope); *universality* (an event occurs throughout a scope); *precedence* (an event P must always be preceded by an event T within a scope); and, *response* (an event P must always be followed by an event T).

Table 2 presents the LTL formulas for each SPS pattern that is considered in our work [14].

Table 2. SPS - LTL formulas for Patterns and Scopes [14]

Pattern	Scope	LTL Formula
Absence (P)	Global	$\Box(\neg P)$
	Before R	$\Diamond R \to (\neg P \, \mathcal{U} \, R)$
	After L	$\Box(L \to \Box(\neg P))$
	Between L and R	$\Box((L \wedge \neg R \wedge \Diamond R) \to (\neg P \, \mathcal{U} \, R))$
	After L-Until R	$\Box(L \wedge \neg R \to (\neg P \mathcal{W} R))$
Existence (P)	Global	$\Diamond(P)$
	Before R	$\neg R \mathcal{W}(P \wedge \neg R)$
	After L	$\Box \neg L \vee \Diamond(L \wedge \Diamond P)$
	Between L and R	$\Box(L \wedge \neg R \to (\neg R \mathcal{W}(P \wedge \neg R)))$
	After L-Until R	$\Box(L \wedge \neg R \to (\neg R \, \mathcal{U} \, (P \wedge \neg R)))$
Universality (P)	Global	$\Box(P)$
	Before R	$\Diamond R \to (P \, \mathcal{U} \, R)$
	After L	$\Box(L \to \Box(P))$
	Between L and R	$\Box((L \wedge \neg R \wedge \Diamond R) \to (P \, \mathcal{U} \, R))$
	After L-Until R	$\Box(L \wedge \neg R \to (P \mathcal{W} R))$
Precedence (T, P)	Global	$\neg P \mathcal{W} T$
	Before R	$\Diamond R \to (\neg P \, \mathcal{U} \, (T \vee R))$
	After L	$\Box \neg L \vee \Diamond(L \wedge (\neg P \mathcal{W} T))$
	Between L and R	$\Box((L \wedge \neg R \wedge \Diamond R) \to (\neg P \, \mathcal{U} \, (T \vee R)))$
	After L-Until R	$\Box(L \wedge \neg R \to (\neg P \mathcal{W}(T \vee R)))$
Response (T, P)	Global	$\Box(P \to \Diamond T)$
	Before R	$\Diamond R \to (P \to$ $(\neg R \, \mathcal{U} \, (T \wedge \neg R))) \, \mathcal{U} \, R$
	After L	$\Box(L \to \Box(P \to \Diamond T))$
	Between L and R	$\Box((L \wedge \neg R \wedge \Diamond R) \to$ $(P \to (\neg R \, \mathcal{U} \, (T \wedge \neg R))) \, \mathcal{U} \, R)$
	After-Until	$\Box(L \wedge \neg R \to ((P \to (\neg R \, \mathcal{U} \, (T \wedge \neg R))))\mathcal{W} R)$

4   Test Purpose Classification

The Specification Pattern System (SPS) was developed to classify properties to be formally verified. In fact, it summarizes the classes of properties that are actually used in formal verification of systems, considering a variety of verification tools. Here, we are interested in giving test purpose patterns to be customized by test users in the same way SPS serves to formal verification users. The patterns here are to represent test purposes and, as such, they must be testable and intuitive to be picked up by users in the testing community. This section first presents the concepts of test purposes and properties testability to then show a classification of test purpose patterns based on testable SPS.

4.1   Test Purposes

A *test purpose* [2,9] is a description of what is being tested concerning to a particular system specification requirement. It presents the desirable system behavior that we want to observe during its execution, describing the execution sequences that lead to an *accept* state and/or to a *refuse* state. The concept of test purposes can be used and described in several ways (e.g. labelled transition systems [2,9], automata [16]).

An important remark is about the relation between *properties* and *test purposes*. As we have stated previously, a test purpose is a description of a behavior that we must observe on an IUT. A property can be represented by a test purpose if it describes a program behavior either. Since our focus is on test purposes, we consider only properties that can be represented by test purposes. We must also consider that test purposes can be represented by temporal formulas, since we are working with the LTL specification language (Section 3.1). As a result, if we have a property P represented by a test purpose E and a temporal formula φ, we assume that they are semantically equal, i.e., $P =_{sem.} E =_{sem.} \varphi$.

4.2   Testable Properties

Considering a finite test execution σ and a property P that is represented by a temporal formula φ, we can conclude that φ is *testable* if at least one of the four relations between the set of executions satisfying φ and the set of (finite or infinite) executions that could be produced by continuations of σ holds [4,13]. Naming the set of φ execution sequences as $Tr(\varphi)$ and the set of the implementation under test (IUT) execution sequences as $Tr(IUT)$, Nahm et al. [13] originally described these four relations as:

- **R1:** $Tr(IUT) \subseteq Tr(\varphi)$;
- **R2:** $Tr(\varphi) \subseteq Tr(IUT)$;
- **R3:** $Tr(\varphi) = Tr(IUT)$;
- **R4:** $Tr(\varphi) \cap Tr(IUT) = \emptyset$.

In addition, Nahm et al. [13] studied whether a connection between these testability relations and the SP classification properties [11] could be established. As a result, they concluded that relations between the execution sequences of a property φ ($Tr(\varphi)$) and the execution sequences of an IUT ($Tr(IUT)$) can be establised in *safety* and in *guarantee* properties, thus, safety and guarantee properties are testable. Falcone et al. [4] revisited these results by analyzing the testability relation $Tr(IUT) \subseteq Tr(\varphi)$ and reached to the conclusions that, under certain conditions, obligation, persistence and recurrence properties are also testable.

Although they stated the conditions for a *fail* verdict, the conditions for a *pass* verdict were not defined. According to Falcone et al., this occurs because the tester may not know the whole set of $Tr(IUT)$. However, concerning to a single execution sequence of the IUT, a *weak-pass* verdict can be established. The weak-pass verdict occurs when an execution sequence of the program positively determines the property φ, i.e., it satisfies φ and every finite or infinite continuation of this sequence also satisfy φ. Thus, the weak-pass verdict guarantees that the IUT exhibits behaviors that will satisfy the property φ. The side-conditions for finding verdicts fail or weak-pass are all related to infinite executions. They are not presented here because we are constrained to finite execution, as with SPS. Readers interested in these conditions are referred to [4].

4.3   Test Purpose Patterns

In our work, we want to relate the testable properties to the SPS so that we can define testable properties using abstract patterns. For an initial analysis, we consider the properties that are satisfied in the whole program execution, i.e., properties that are satisfied under the *SPS global scope*.

The work by Nahm et al. [13] establishes that *safety* and *guarantee* properties are testable. Hence, considering the linear temporal representation of these classes (Section 3.1) [11] and the SPS, we can conclude that the safety properties can be represented by the patterns: universality, absence and precedence (absence and precedence are variations of the canonical safety formula which is represented by the universality pattern [11,12]; and the guarantee properties can be represented by the existence pattern).

In addition, considering the results by Falcone et. al. that some obligation, recurrence and persistence properties are also testable under certain conditions, we have studied how these properties classes can be represented in SPS. The obligation class and the persistence class do not have a direct representation in the SPS, hence they are not discussed here anymore. The recurrence class can be represented by the SPS response pattern. Considering that in model checking we know (or at least we can establish) the set of execution sequences that are verified against a property, we can check for each execution sequence if it produces a weak-pass verdict or a fail verdict. If one fail verdict is obtained during the verification, we can conclude that the property was violated.

Since we are dealing with finite execution programs, we can observe whether the IUT finite execution sequences satisfy (violate) the property within the SPS scope. The SPS scope restricts the IUT execution that must be observed to decide whether a property is satisfied by the system or not, then, we only consider the observable execution sequences within the scope to reach a property verdict. Combining the fact that we are considering a set of known execution sequences, and that these execution sequences are finite and can be checked within the SPS scopes, we conclude that weak-pass and fail verdicts are sufficient conditions to show that the IUT satisfies or not a recurrence property. For this reason, we conclude that *the recurrence property represented by the SPS response pattern is testable* [5].

The complete set of testable properties that can be represented using the SPS patterns in the global scope is presented in Table 3, where P and T represent the program events that can be expressed using temporal formulas. These results connect the SP classification canonical formulas (Table 1 with the SPS formulas (Table 2).

Table 3. Testable Properties Classes x SPS (Global Scope)

Property Class	SPS Pattern	LTL Formula
Safety	Universality (P)	$\Box P$
	Absence (P)	$\Box(\neg P)$
	Precedence (T,P)	$\neg P \, W \, T$
Guarantee	Existence(P)	$\Diamond P$
Recurrence	Response (T,P)	$\Box(P \rightarrow \Diamond T)$

Taking the results from Table 3 as a starting point, we analyzed the whole set of SPS formulas in other scopes (before, after, between and after-until) and concluded that any test purpose that is specified using the formulas for the universality, absence, precedence, existence and response patterns is testable. A full description of this study is depicted in [5]. This means that when this test purpose is checked by a model checker, we can assure that a fail verdict will be given if at least one program execution trace does not satisfy the property and a success verdict will be reached otherwise. Therefore, we can use the formulas from Table 3 to specify the test purposes for a given program by just replacing P and/or T for the program events, and we can certify that when a model checker verifies these properties a success/fail verdict will be reached. A guideline to specify test purposes using these formulas is presented in Section 5.

5   Using Test Purpose Patterns

To show how to use our proposed approach, we present an Java example (Listing 1.1) that performs simple arithmetic operations over an integer x.

Listing 1.1. Arithmetic Example - Partial Code

```
1   public class ArithmeticExample {
2       public static int add1 (int x) {
3           x = x + 1;
4           return x;
5       }
6       public static int reset (int x) {
7           x = 0;
8           return x;
9       }
10      public static boolean isZero (int x) {
11          return x == 0;
12      }
13      public static void main(String[] args) {
14          ...
15      }
16  }
```

This example has three methods:

1. **add1**: this method adds one unit to the number x.
2. **reset**: this method sets the value of x to 0.
3. **isZero**: this method returns *true* if x has value 0, and *false* otherwise.

Considering the above methods as the program events, we can specify test purposes using these methods. For instance, suppose that we want to specify the test purpose: *"the method **add1** must be called at least once during the program execution"*. It states that there is at least one call to the method **add1** for every program execution trace. This is a guarantee property, which is testable (Section 4.3), and can be specified using the existence pattern (Table 3): $\Diamond P$. Replacing P by $add1$ in the formula, we have the specification of the test purpose: $\Diamond add1$. An automaton representation for this test purpose specification in LTL is presented in Fig. 1. The label *!add1* identifies all the events that do not correspond to the **add1** method. The label *true* is used to state that any event that belongs to the set of program events under consideration is accepted. The accepting state is identified by a circle with a thick line border, while the rejecting state is identified by a circle with a dashed line border.

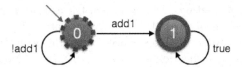

Fig. 1. Automaton Representation for the Test Purpose "$\Diamond$ **add1**"

Similarly, consider that we want to specify the test purpose: *"the method isZero is never called during the program execution"*. It states that there is no call to the method **isZero** in every program execution trace. This is a safety property that is testable (Section 4.3), and can be specified using the absence pattern (Table 3): $\Box \neg P$. To write the test purpose specification, we substitute P by $isZero$ in the formula: $\Box \neg isZero$. Fig. 2 presents the automaton that represents this test purpose specification.

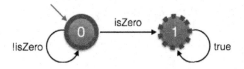

Fig. 2. Automaton Representation for the Test Purpose "$\Box; \neg$**isZero**"

Concerning to test purposes with more than one event, we can specify them using the precedence or the response pattern depending on the program behavior that we want to observe (Section 4.3). For instance, if we want to check that *"when the method* reset *is called, the method* isZero *is called in the future, i.e., a* reset *method call is followed by a call to* isZero*"*, we use the response pattern (Table 3): $\Box(P \rightarrow \Diamond T)$ and replace P by *reset* and T by *isZero*, obtaining: $\Box(reset \rightarrow \Diamond isZero)$ (Fig. 3).

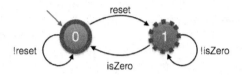

Fig. 3. Automaton Representation for the Test Purpose "$\Box$(**reset** → $\Diamond$**isZero**)"

A tool support for helping the specification of test purposes using this approach was developed. At this moment, it provides a user-friendly interface where the user can choose the program events that should replace the default events (e.g. P, T) in the formulas (Fig. 4), generate the test purpose specification in LTL and submit it to model checking. The model checking can provide the success or fail verdict and the percentage of the program traces that were analyzed during test purpose verification process. These coverage data can guide the process of generating complementary test cases, since we can focus on the traces that were not covered during the verification of the test purpose. A detailed description of the tool is provided in [6].

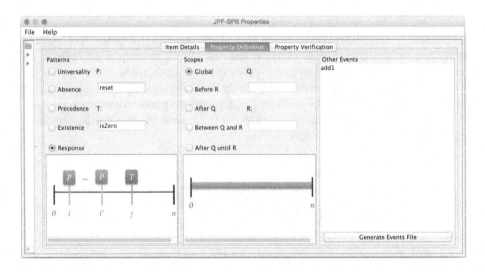

Fig. 4. Tool Screenshot: Defining the property "$\Box$(**reset** → $\Diamond$**isZero**)"

6 Concluding Remarks

The present work describes a classification for test purposes derived from the concept of testable patterns. The analyzed patterns were provided by the SPS, a specification system that was developed to simplify the task of writing a property formally. In our study, we related the SPS patterns to the concept of testability, which establishes a connection between program traces and test purpose verification traces to determine the test purposes that can give a success/fail verdict. By connecting the testability concept to the SPS, we could establish a set of patterns that can be used to specify testable properties (test purposes) (Table 3). The specified test purposes can then be verified with a model checker. The verification result would give not only the success/fail verdict for the property, but also traces coverage information that can be used to analyze the behavior of the program during the verification, and also to guide the generation of a complementary test case.

The current results are restricted for programs with finite execution, and properties concerning to program events (methods). A tool support to automate the process of specifying test purposes, submitting it to a model checker, and analyzing the traces coverage was developed [6]. Future work would include the extension of these results for programs with infinite execution, the specification of properties regarding program states (e.g. variable values), and the addition of these features in the tool support.

Acknowledgments. This project has been co-funded by the State of São Paulo Research Foundation (FAPESP) - Processes: 2011/01928-1, 2012/23767-2, 2013/22317-6, and the Ministry of Education Research Agency (CAPES- Brazil). The authors also would like to thank the NASA Ames Research Center and the Carnegie Mellon University - Silicon Valley for providing a rich environment to develop research activities during their visit.

References

1. Baier, C., Katoen, J.P.: Principles of Model Checking (Representation and Mind Series). The MIT Press (2008)
2. de Vries, R.G., Tretmans, J.: Towards formal test purposes. In: Tretmans, J., Brinksma, H. (eds.) Formal Approaches to Testing of Software 2001 (FATES 2001). BRICS Notes Series, vol. NS-01-4, Aarhus, Denkmark, pp. 61–76, August 2001. http://doc.utwente.nl/66272/
3. Dwyer, M.B., Avrunin, G.S., Corbett, J.C.: Patterns in property specifications for finite-state verification. In: Proceedings of the 21st International Conference on Software Engineering, ICSE 1999, pp. 411–420. ACM, New York (1999)
4. Falcone, Y., Fernandez, J.-C., Jéron, T., Marchand, H., Mounier, L.: More testable properties. In: Petrenko, A., Simão, A., Maldonado, J.C. (eds.) ICTSS 2010. LNCS, vol. 6435, pp. 30–46. Springer, Heidelberg (2010)
5. Hanazumi, S., de Melo, A.C.V.: On the Testability of Properties Patterns, (Manuscript submitted for publication)

6. Hanazumi, S., de Melo, A.C.V., Păsăreanu, C.S.: From testing purposes to formal JPF properties. In: Java PathFinder Workshop. ACM (2014)
7. Hierons, R.M., Bogdanov, K., Bowen, J.P., Cleaveland, R., Derrick, J., Dick, J., Gheorghe, M., Harman, M., Kapoor, K., Krause, P., Lüttgen, G., Simons, A.J.H., Vilkomir, S., Woodward, M.R., Zedan, H.: Using formal specifications to support testing. ACM Comput. Surv. **41**(2), 9:1–9:76 (2009)
8. Holzmann, G.J.: The Model Checker SPIN. IEEE Trans. Softw. Eng. **23**(5), 279–295 (1997)
9. Jard, C., Jéron, T.: TGV: theory, principles and algorithms: A tool for the automatic synthesis of conformance test cases for non-deterministic reactive systems. Int. J. Softw. Tools Technol. Transf. **7**(4), 297–315 (2005)
10. Ledru, Y., du Bousquet, L., Bontron, P., Maury, O., Oriat, C., Potet, M.L.: Test purposes: adapting the notion of specification to testing. In: Proceedings 16th Annual International Conference on Automated Software Engineering (ASE 2001), pp. 127–134. IEEE Comput. Soc. (2001)
11. Manna, Z., Pnueli, A.: A hierarchy of temporal properties (invited paper, 1989). In: Proceedings of the Ninth Annual ACM Symposium on Principles of Distributed Computing, PODC 1090, pp. 377–410. ACM, New York (1990). http://doi.acm. org/10.1145/93385.93442
12. Manna, Z., Pnueli, A.: The Temporal Logic of Reactive and Concurrent Systems - Specification. Springer-Verlag New York Inc., New York (1992)
13. Nahm, R., Grabowski, J., Hogrefe, D.: Test Case Generation for Temporal Properties.Tech. rep., Bern University (1993)
14. Patterns, S.: http://patterns.projects.cis.ksu.edu/ (last access: January 2015)
15. Pnueli, A.: The temporal logic of programs. In: SFCS 1977: Proceedings of the 18th Annual Symposium on Foundations of Computer Science, pp. 46–57. IEEE Computer Society, Washington (1977)
16. da Silva, D.A., Machado, P.D.L.: Towards Test Purpose Generation from CTL Properties for Reactive Systems. Electronic Notes in Theoretical Computer Science **164**(4), 29–40 (2006)
17. Tretmans, J.: Conformance testing with labelled transition systems: Implementation relations and test generation. Computer Networks and ISDN Systems **29**(1), 49–79 (1996)
18. Tretmans, J.: Model based testing with labelled transition systems. In: Hierons, R.M., Bowen, J.P., Harman, M. (eds.) FORTEST. LNCS, vol. 4949, pp. 1–38. Springer, Heidelberg (2008)
19. Tretmans, J.: Model-based testing and some steps towards test-based modelling. In: Bernardo, M., Issarny, V. (eds.) SFM 2011. LNCS, vol. 6659, pp. 297–326. Springer, Heidelberg (2011)
20. Weiglhofer, M., Fraser, G., Wotawa, F.: Using coverage to automate and improve test purpose based testing. Inf. Softw. Technol. **51**(11), 1601–1617 (2009)
21. Weiglhofer, M., Wotawa, F.: Improving coverage based test purposes. In: International Conference on Quality Software, pp. 219–228 (2009)

Dynamical Discrete-Time Rössler Map with Variable Delay

Madalin Frunzete[1,2]([⊠]), Anca Andreea Popescu[1], and Jean-Pierre Barbot[2,3]

[1] Faculty of Electronics, Telecommunications and Information Technology,
POLITEHNICA University of Bucharest, 1-3, Iuliu Maniu Bvd.,
Bucharest 6, Romania
madalin.frunzete@upb.ro
[2] Electronique Et Commande des Systmes Laboratoire, EA 3649
(ECS-Lab/ENSEA), ENSEA, Cergy-pontoise, France
[3] EPI Non-A INRIA, Lyon, France

Abstract. This paper presents an improvement to an existing method used in security data transmission based on discrete time hyperchaotic cryptography. The technique is implemented for a Rössler hyperchaotic generator. The improvement consists in modifying the structure of the existing generator in order to increase the robustness of the new cryptosystem with respect to known plain text attack, particularly the "identification technique".

Keywords: Dynamical-systems · Cryptography · Chaotic map · Identifiability · Chaotic-discrete system with delay

1 Introduction

Secure communications represent an important area of research, therefore there is a constant effort to propose new cryptographic techniques or to improve the existing methods of data enciphering. One approach is based on the usage of hyperchaotic dynamics. [6]

Exchange of information assumes the existence of at least one emitter and one receiver (the number is increasing in multicast or broadcast communications). This paper presents an echiphering method of "Single Input Single Output" (SISO) type - one emitter and one receiver. A very important condition for establishing a communication is the synchronization of both subsystems. Roughly speaking, without this, the receiver could not "understand" the message transmitted by the partner. The possibility of synchronizing two chaotic systems proved in [5] permitted the existence of cryptography based on chaotic behaviour.

Actual chaotic cryptographic techniques with symmetric key are based on the idea that the information scrambled and transported by a pseudo-random signal of chaotic nature, can be restored by using a synchronization process; the control theory of dynamical systems allows for the implementation of the

© Springer International Publishing Switzerland 2015
O. Gervasi et al. (Eds.): ICCSA 2015, Part I, LNCS 9155, pp. 431–446, 2015.
DOI: 10.1007/978-3-319-21404-7_32

existing method, [15]. Cryptographic methods [2] aim at masking the information either by adding it to the chaotic signal in the transition line - "addition method" (AM) - [13], or by including it in the chaotic structure - "inclusion method" (IM) [3]. The usage of dynamical systems in cryptography should take into consideration the noise which can be caused by computational precision [12, 19]

In this paper an I.M. will be treated. An example of how this IM works for enciphering images is given in [9]. The IM proposed in this paper was implemented by using a hyperchaotic system in order to improve transmission security. Such systems have solutions that are more irregular than classical chaos because they have more than one unstable "Lyapunov exponent", [22]. The analyzed method is based on the Rössler map ("Folded-Towel") which presents such behaviour; the proof of the existence of two positive Lyapunov exponents is given in [17].

The most important contribution given in this paper is the idea of adjusting a *variable delay* with exemplification on a cryptosystem already implemented. Our approach lays between the discrete systems with delay and discrete chaotic systems [6]. By adjusting a delay in the structure of a cryptosystem based on chaotic behaviour the distance between the cryptograph and the cryptanalyst increases. It is proven that, if a delay is adjusted in the structure of continuous cryptosystem, the dimension of the system, in the eyes of the cryptanalyst, will become infinite [18]; in discrete case, the delay will only increase this dimension.

In **section 2** the identifiability notion will be presented, see [1], in the context of multi-dimensional systems. This is discussed in order to show why the I.M. proposed in the literature needs some improvements; some parametric cryptanalysis will give a confident answer in the context of identifiability notion.

In **section 3** the improvement and some statistical analysis will be discussed in order to show that the advantages of hyperchaotic behaviour were not lost by through the modification of the system structure. The statistical behaviour of Rössler map is analyzed in [10, 11] and will also be resumed in this paper. The statistical analysis are performed in order to sustain the idea that the eavesdropper cannot find any leads by performing a statistical attack. This is based on the fact that the statistical behaviour of the modified system is the same as the one of the original system. We propose a general statistical analysis to describe the new cryptosystem. At a first glance, it seams that the existing methods cannot be applied in order to brake the new cryptosystem, so it looks to be quite robust.

2   Cryptanalysis on an Existing Cryptosystem

A cryptosystem created by IM is generally characterized through identifiability and some parametric cryptanalysis; hereafter the method is applied to:

$$\begin{cases} x_1^+ = a_1 x_1(1 - x_1) + a_2 x_2 \\ x_2^+ = b_1[(1 - b_2 x_1)(x_2 + b_3) - 1](1 - b_4 x_3) \\ x_3^+ = c_1 x_3(1 - x_3) - c_2(1 - b_2 x_1)(x_2 + b_3) \end{cases} \tag{1}$$

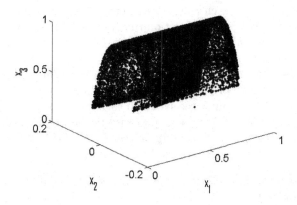

Fig. 1. Rössler map phase portrait, "Folded-Towel", $x(0) = (0.35, 0.05, 0.72)^T$

System (1) is a three-dimensional hyperchaotic system and its behaviour can be observed in Figure 1.

System (1) will be referred with the generic form:

$$x^+ = f(x, p) \tag{2}$$

where $x := (x_1, x_2, x_3)^T \in \Re^3$ represents the state vector evaluated at the step k (i.e. $x(k)$), so $x^+ := x(k + 1)$. The vector $p \in \Re^8$ denotes the parameter vector of system (1), $p = (a_1, a_2, b_1, b_2, b_3, b_4, c_1, c_2)^T$. By definition $f(x, p) := (f_1(x, p), f_2(x, p), f_3(x, p))^T$ such that $f : \Re^3 \times \Re^8 \to \Re^3$ is the analytic hyperchaotic generator of (1).

For the generator (2) a secured communication scheme was proposed in [3]. For the proposed method (as type from the classical cryptography), the secret key can be represented by a part or all of the hyperchaotic system parameters. Also, it was shown that the proposed cryptosystem respects the condition of bidirectional transformations (enciphering/deciphering). For the observability and identifiability analysis the knowledge of *the transmitter* structure is required, [4]. So, the transmitter is constructed as follows:

$$\begin{cases} x^+ = f(x, p) + g(x, p)u \\ y = h(x) \end{cases} \tag{3}$$

where x, p and the vector-field f are defined generically in (2). Vector-field $h : \Re^3 \to \Re$ is such that $h(x) = x_1$ and $g : \Re^3 \times \Re^8 \to \Re^3$ will be specified later. Variable $u \in \Re$ is considered the confidential message to be transmitted; in order not to influence the hyperchaotic behaviour the amplitude of u must be less than 10^{-2}. Hence, system (3) is considered as S.I.S.O. system with u input message and $y = x_1$ as output.

In a communication, a hyperchaotic signal generated by y will be found on the transmission line. Receiving the cryptogram and having the secret key, exchanged through a secure channel, the receiver can understand the original message. For a very precise reconstruction the system (3) has to verify the "*Discrete Observability Matching Condition*" (D.O.M.C.). This one assure, for the receiver, the recovering of the unknown input u.

During the communication, a hyperchaotic signal generated by y is found on the transmission line. The original message can be understood at the receiving end given that the cryptogram is received and the secret key, exchanged through a secure channel, is available. For a very precise reconstruction, the system (3) has to verify the "*Discrete Observability Matching Condition*" (D.O.M.C). This assures that the receiver recovers the unknown input u.

Definition 1. *[3] Considering the n-dimensional system of the form (3) (i.e. $x \in \Re^n$, vector-fields $f, g : \Re^{n \times m} \to \Re^n$ and $h : \Re^n \to \Re$), the D.O.M.C. is given as follows:*

- *Span $\{dh, d(f \circ h), \ldots, d(f_\circ^{n-1} \circ h)\}$ is of rank n almost everywhere in the neighborhood of origin,*
- *$((dh)^T, \ldots, (d(f_\circ^{n-1} \circ h))^T)^T g = (0, \ldots, 0, *)^T$ in the neighborhood of origin,*

where symbol $\circ$ denotes usual composition function, $f_\circ^j$ denotes function f composed j times (i.e. $f_\circ^j = f \circ f_\circ^{j-1}$ for $2 \le j \le n$ with $f_\circ^1 = f$). The symbol $*$ represents a non-null function, almost everywhere of x, containing - if they exists - observability singularities.

Following D.O.M.C. the confidential message u is hidden in the third state variable x_3 of the transmitter, where $g(x, p) = (0, 0, 1)^T$. This ensures the full recovery of the message.

Applying D.O.M.C. and having the cryptosystem structure described above, a singularity manifold S given by (4) is obtained.

$$S = \{x \in \Re^3 \text{ such that } (1 - b_2 x_1)(x_2 + b_3) - 1 = 0\} \tag{4}$$

In the next subsection some definitions and theorems will be given in order to analyze the cryptosystem described above.

2.1   Identifiability. Definitions.

The following definitions are recalled from [16]. These definitions will be useful in order to prove the weakness of the cryptosystem (3).

Definition 2. *An input sequence over a window of iterations $0 \ldots N$, denoted by $\{u\}_0^N$, is called an admissible input on $0 \ldots N$ if the difference equation (3) admits a unique local solution.*

Definition 3. *The system (3) is locally strongly $x(0)$-identifiable at p through the admissible input sequence $\{u\}_0^N$ if there exists an open neighborhood of p, $\nu(p) \subset P$, such that for any $\widehat{p} \in \nu(p)$ and for any $p \in \nu(p)$:*

$$\widehat{p} \neq p \Rightarrow \{y(x(0), u, \widehat{p})\}_0^N \neq \{y(x(0), u, p)\}_0^N \qquad (5)$$

Definition 4. *The system (3) is structurally identifiable if there exists $N > 0$, an open subset $\mathscr{X}(0) \subset \mathscr{X}$ and some dense subsets $\nu(p) \subset P$ and $\mathscr{U}_0^N \subset \mathscr{U}$ such that, for every $x(0) \in \mathscr{X}(0)$, $p \in \nu(p)$ and $\{u\}_0^N \in \mathscr{U}_0^N$ the system (3) is locally strongly $x(0)$-identifiable at p through admissible input sequence $\{u\}_0^N$.*

In order to test the identifiability of system parameters it is necessary to introduce two approaches: the putput equality and input/output relation.

2.2 Outputs Equality Approach

This *outputs equality approach* is directly connected with Definition 4. The following theorem is a sufficient condition for structurally identifiable of the system (3).

Theorem 1. *[1] The system (3) is structurally identifiable if the equations set:*

$$\{y(x(0), u, \widehat{p})\}_0^N = \{y(x(0), u, p)\}_0^N \qquad (6)$$

has a unique solution for $\widehat{p}$, that is $\widehat{p} = p$.

The proof of this theorem can be interpreted as a consequence of Definition 3.

Equation (6) requires explicitly that the initial conditions of the system are known

2.3 Input/Output Relation Approach

The next theorem gives a necessary and sufficient condition for the parametric structural identifiability:

Theorem 2. *[1] The system (3) is structurally identifiable if and only if there exist two integers $K < \infty$ and $K' < \infty$ such that (3) can be rearranged in a linear regression such that, for $i = 1, \ldots, L$ (L the dimension of the parameter vector):*

$$P_i(y, \ldots, y^{K+}, u, \ldots, u^{K'+})p^{(i)} - Q_i(y, \ldots, y^{K+}, u, \ldots, u^{K'+}) = 0 \qquad (7)$$

where P_i and Q_i are functions depending only on y, u and on their iterates.

Following Theorem 2 it is obvious that every parameter $p^{(i)}$ (i.e. $p^{(1)} = a_1$) can be determinate as follows:

$$p^{(i)} = \frac{Q_i(y, \ldots, y^{K+}, u, \ldots, u^{K'+})}{P_i(y, \ldots, y^{K+}, u, \ldots, u^{K'+})} \qquad (8)$$

unless P_i vanishes. The set of y_k and u_k corresponding to $P_i = 0$ is a set of zero measure, in general. The set of zero measure is omitted because only the set of admissible inputs (Definition 2) will be considered.

In order to obtain a relation such as (8) for system (3) it is necessary that variable x and its iterates be in algebraic relation; an example of such systems are those which are used for building cryptosystems based on I.M.. That leads to a relation which is written only with the parameter vector p, the output y, the input u and their iterates, called input/output relation:

$$\mathscr{L}(p, y, \ldots, y^{s+}, u, \ldots, u^{s'+}) = 0 \qquad (9)$$

where s is the observability index [15] of the system (3), and $s' < s$.

This approach constructs a way of retrieving the parameters in the context of known plain text attack. So, input/output relation approach allows concluding on the parameter identifiability but is also an instrument for eavesdropper. In the next subsection an exemplification of attack on cryptosystem (3) will be presented.

Remark 1. Theorem 2 is strongly related to the differential algebra and numeric algebra approach [7] and identifiability approach [14]. The purpose of the paper is not to present a complete algebra identifiability theory and the reader can be referred to both references [7] and [14] for more details.

2.4   Parametric Cryptanalysis

This section presents a connection between parametric cryptanaylsis and identifiability introduced before. Usually in cryptanalysis the eavesdropper knows the structure of the system and the output signal y but lacks information about the secret key (system parameters).

The robustness of a cryptosystem can be tested by analyzing the resistance to brute force attack: by testing exhaustively every possible value of its secret key. If a given input/output relation leads to a unique set of parameters then it can be said that the cryptosystem will easily fail to a known plain text attack. In this kind of analysis Theorem 1 is used and it leads to the idea that in order to perform a plain text attack it is necessary to perform an exhaustive search of the key.

The known plain text attack (where leaded Theorem 2) assume treating the cryptosystem like a *gray box* - the difference from *black box* is that in the content of a gray box there can exist other things than in the content of the black box, like the structure of the cryptosystem. In the case of the black box the cryptanalyst has the possibility to send the message u by using the cryptosystem and he has access only to the cryptogram y.

In this context the security can be measured by verifying the input/output relation approach, in what is called *algebraic attack*, [8]. An example of this kind of attack will be applied in the following. The message u is embedded in the third dynamics x_3, then the extended form of system (3) results:

$$\begin{cases} x_1^+ = a_1 x_1(1 - x_1) + a_2 x_2 \\ x_2^+ = b_1[(1 - b_2 x_1)(x_2 + b_3) - 1](1 - b_4 x_3) \\ x_3^+ = c_1 x_3(1 - x_3) - c_2(1 - b_2 x_1)(x_2 + b_3) + u \\ y = x_1 \end{cases} \tag{10}$$

The structural identifiability of parameter vector $p = (a_1, a_2, b_1, b_2, b_3, b_4, c_1, c_2)^T$ (L=8) is treated with respect to input/output relation approach. The observability index of the system (3) is equal with the system dimension, $s = 3$ (see [3]).

For obtaining the input/output relation $\mathscr{L} = 0$ the system (10) is reiterated three times:

$$\begin{cases} x_1^+ - a_1 x_1(1 - x_1) - a_2 x_2 = 0 \\ x_1^{++} - a_1 x_1^+(1 - x_1^+) - a_2 x_2^+ = 0 \\ x_1^{+++} - a_1 x_1^{++}(1 - x_1^{++}) - a_2 x_2^{++} = 0 \\ x_2^+ - b_1[(1 - b_2 x_1)(x_2 + b_3) - 1](1 - b_4 x_3) = 0 \\ x_2^{++} - b_1[(1 - b_2 x_1^+)(x_2^+ + b_3) - 1](1 - b_4 x_3^+) = 0 \\ x_2^{+++} - b_1[(1 - b_2 x_1^{++})(x_2^{++} + b_3) - 1](1 - b_4 x_3^{++}) = 0 \\ x_3^+ - c_1 x_3(1 - x_3) + c_2(1 - b_2 x_1)(x_2 + b_3) + u = 0 \\ x_3^{++} - c_1 x_3^+(1 - x_3^+) + c_2(1 - b_2 x_1^+)(x_2^+ + b_3^+) + u^+ = 0 \\ x_3^{+++} - c_1 x_3^{++}(1 - x_3^{++}) + c_2(1 - b_2 x_1^{++})(x_2 + b_3^{++}) + u^{++} = 0 \\ y - x_1 = 0 \\ y^+ - x_1^+ = 0 \\ y^{++} - x_1^{++} = 0 \\ y^{+++} - x_1^{+++} = 0 \end{cases} \tag{11}$$

From system (11) an input/output relation can be obtained:

$$\begin{aligned} & y^{+++} - a_2 b_1 \left((1 - b_2 y^+) \left(\frac{y^{++} - a_1 y^+(1 - y^+)}{a_2} + b_3 \right) - 1 \right) \\ & \left[1 - \left(1 - \frac{y^{++} - a_1 y^+(1 - y^+)}{a_2 b_1((1 - b_2 y)((y^{++} - a_1 y^+(1 - y^+))/a_2 + b_3) - 1)} \right) \right. \\ & \left. \left(1 - \frac{1}{b_4} \left(1 - \frac{y^{++} - a_1 y^+(1 - y^+)}{(a_2 b_1(1 - b_2 y)((y^{++} - a_1 y^+(1 - y^+))/a_2 + b_3) - 1)} \right) \right) \right) - \\ & - c_2 (1 - b_2 y) \left(\frac{y^+ - a_1 y(1 - y)}{a_2} + b_3 \right) + u \right] - a_1 y^{++} \left(1 - y^{++} \right) = 0 \end{aligned} \tag{12}$$

The input/output relation given in (12) shows dependence only between the parameter vector p, the output y and the message u; it can be summarized:

$$\mathscr{L}(p, y^{+++}, y^{++}, y^+, u) = 0 \tag{13}$$

Then, the input/output relation (13) is iterated $L - 1 = 7$ times and yields $\mathscr{L}_1, \mathscr{L}_2, \ldots, \mathscr{L}_8$. In this case there are 8 equations for 8 parameters, where:

$$\mathscr{L}_1(a_1, y^{+++}, y^{++}, y^+, u) = 0 \tag{14}$$

So, performing plain text attack and having the parameter identifiability proved the cryptosystem (3) can be easy broken. In the first section the idea of adjusting a variable delay was introduced as an improvement for the cryptosystem robustness and will be presented in the following section.

3   The New Cryptosystem

The proposed improvement refers to increasing the distance between the cryptograph and the cryptanalyst. The structure of the resulting cryptosystem is:

$$\begin{cases} x_1^+ = a_1 x_1 (1 - x_1) + a_2 x_2^{d-} \\ x_2^+ = b_1 [(1 - b_2 x_1)(x_2^{d-} + b_3) - 1](1 - b_4 x_3) \\ x_3^+ = c_1 x_3 (1 - x_3) - c_2 (1 - b_2 x_1)(x_2^{d-} + b_3) \end{cases} \tag{15}$$

where $x^{d-} := x(k-d(k))$ and d is a vector which contains the delay corresponding to the step k.

The system (15) can be referenced with the generic form:

$$x^+ = \tilde{f}(x, p) \tag{16}$$

where $x := (x_1, x_2, x_3)^T \in \Re^3$ represents the state vector evaluated in the step k (i.e. $x(k)$), the vector $p \in \Re^8$ denotes the parameter vector of system (15) and by definition $\tilde{f}(x, p) := (f_1(x_1, x_2^{d-}, x_3, p), f_2(x_1, x_2^{d-}, x_3, p), f_3(x_1, x_2^{d-}, x_3, p))^T$ such that $\tilde{f} : \Re^3 \times \Re^8 \to \Re^3$ is the analytic hyperchotic generator of (15).

By modifying the structure of (1) the Rössler map behaviour was not changed. At a first glance this can be observed by comparing with Figures 3 and 1

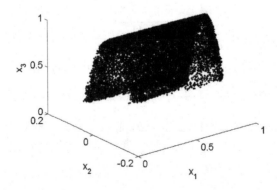

Fig. 2. Rössler map phase portrait (with delay), $x(0) = (0.35, 0.05, 0.72)^T$

In this section we analyse whether the hyperchaotic behavior was affected by introducing the delay in the structure of the system. Smirnov tests were applied in two different ways in order to show that nothing was lost by modifying the system structure, from a statistical point of view. We verify if the transient time (the time elapsed from the initialization of the system until its entrance in the stationarity region) is quite the same. Also the cumulative distribution function of each state variable x_i of system (1) is conserved for the same state variable of system (15). By showing that both systems present the same behaviour, a new advantage is introduced for the cryptograph. The eavesdropper cannot conclude, by performing this kind of statistical analysis, which system is used for ciphering. No information on the delay is obtained by performing this kind of statistical investigation, if the structure of the cryptosystem is known.

The proposed cryptosystem has a similar structure with 3, but the reconstruction of the original message will need a few more steps.

3.1 Transmitter

This new secured transmission scheme is using the generator (16) to build a cryptosystem of I.M. type in the same way as (3), presented in section 2. The secret key is composed by a part or all of the parameters plus $d(0)$ which is the initial condition used for a logistic map with parameter a_1; this logistic map will serve for constructing the delay vector used in the structure of the new cryptosystem. The transmitter is constructed as follows:

$$\begin{cases} x^+ = \tilde{f}(x,p) + g(x,p)u \\ y = h(x) \end{cases} \tag{17}$$

where x, p and the vector-field $\tilde{f}$ are defined in (16). Vector-field $h : \Re^3 \to \Re$ is such that $h(x) = x_1$ and $g : \Re^3 \times \Re^8 \to \Re^3$ is the same as in (3). Variable $u \in \Re$ is considered as the confidential message to be transmitted; the amplitude of u must be less than 10^{-2} in order to not influence the hyperchaotic behaviour.

Following D.O.M.C. (see Definition 1) the confidential message u is hidden in the third state variable x_3 of the transmitter, in this case $g(x,p) = (0,0,1)^T$. This ensures a very good reconstruction of the message.

Applying D.O.M.C. and having the cryptosystem structure described before, a singularity manifold S given by (18) is obtained.

$$S = \{x \in \Re^3 \text{ such that } (1 - b_2 x_1)(x_2^{d-} + b_3) - 1 = 0\} \tag{18}$$

The statistical behaviour of the new system which will be presented next will show that the delay does change the statistical properties of the system without delay.

3.2 Statistical Analysis on System Behaviour

Here the statistical analysis are based on applying the Smirnov test in two ways:

(i) by measuring the transient time for each state variable of the system (15),

(ii) by verifying if the probability law for x_i, with $i = 1 \ldots 3$, from system (15) is the same with the probability law corresponding to x_i from the system (1).

For the Rössler map the first order probability law appropriate to the random process (in the stationarity region) is unknown, an this justifies why the Smirnov test was selected for the measurements of the transient time, [20]. The test is based on two independent experimental data sets, $(x_1, x_2, \ldots, x_n)$ and $(y_1, y_2, \ldots, y_m)$, considered as the observed values of two random variables X and Y, which complies the $i.i.d.$ model. The two hypotheses of the test are:

- H_0 : the two random variables X and Y have the same probability law,
- H_1 : the two random variables X and Y do not have the same probability law.

The test relies on the experimental cumulative distribution function (c.d.f.) $Fe_X(x)$ and $Fe_Y(y)$ for the two random variables X and Y.

So the aim of the test is to establish if the two random variables X and Y obey the same probability law. The way in which X and Y are chosen will assure the conclusions needed for describing the statistical behaviour of the modified system.

Measuring the Transient Time. A similar way of applying the mentioned test as in [10] will be applied for system (15). As was measured in [10], the transient time for each of the three dynamics of system (1) is approximately 25 iterations.

For the measuring of the transient time there were considered $n = m = N = 100000$ (the size of the experimental data sets) and the significance level $\alpha = 0.05$. For generating these data sets are needed $2N = 200000$ different initial conditions of the type $(x_1(0), x_2(0), x_3(0))^T$ which were generated according to the uniform law in the domain $[0.2; 0.8] \times [-0.1; 0.1] \times [0.2; 0.8]$. Using the $2N$ initial conditions for system (15) was obtained the same number of trajectories assigned to the three random processes of the Rössler map. For the two random variables X and Y the experimental data sets are obtained as follows: by sampling one of the random processes at the iteration k_1 result the set $(x_1, x_2, \ldots, x_N)$ and sampling, the same state variable, at iteration k_2 results the set $(y_1, y_2, \ldots, y_N)$. The set $(x_1, x_2, \ldots, x_N)$ is obtained from the first N initial conditions and the other set $(y_1, y_2, \ldots, y_N)$ comes from the other N initial conditions (from the ensemble of $2N$). This way of choosing the two data sets ensures the independence between them. The Smirnov test was applied for different values of k_1 in the range $[10; 50]$ and each time k_2 value was kept the same; this k_2 value was selected in the stationarity region. The Smirnov test is applied for each pair (k_1, k_2) and if the hypothesis H_0 is accepted, k_1 may indicate the entrance in the stationarity region.

For an accurate decision a Monte Carlo analysis is applied, which consists in resuming the Smirnov test 500 times. Thus, for each pair (k_1, k_2) and keeping the same volume of experimental data sets $n = m = N = 100000$ is recorded the proportion of acceptance of H_0 null hypothesis.

The experimental results referring to the three random processes assigned to the modified Rössler map (15) are presented in Table 1. On the first row there are indicated the iterations k_1 from where it was obtained the set corresponding to the random variable X; for Y was considered the iteration $k_2 = 200$, which is supposed to be in the stationarity region. The other three rows contain the results for the state variables of the system (15). How to read this table: considering state variable x_2 and $k_1 = 25$ the proportion of accepting H_0 null hypothesis is 94.1% from the total of 500 Smirnov tests applied for the pair $(k_1, k_2) = (25, 200)$. Analyzing the results, the value $k_1 = 25$ iteration may indicate the beginning of the stationarity region, because for $k_1 \geq 25$ the proportion of accepting the tests stays in the interval $[0.93; 07]$.

Table 1. The proportion of acceptance for H_0 for Smirnov test (transient time)

k_1	10	15	20	25	30	35	40
x_1	21.2	31.5	87.8	**93.4**	**95.1**	**93.7**	**94.9**
x_2	51.2	88.6	81.3	**93.9**	**95**	94.1	**95.7**
x_3	11.1	12.4	91.3	**95.3**	**94.1**	**93.7**	**95.1**

Verifying the Probability Law. By also using Smirnov tests it is verified if the probability law of x_i from system (15) is the same as in the case of system (1). This will confirm that the statistical behaviour of the two systems is the same, so the advantages of hyperchaotic dynamics are still useful for the new cryptosystem projected.

The experimental results are obtained in the same way as before, only the data set $(y_1, y_2, \ldots, y_N)$ was changed by considering N trajectories of system (1) and sampling each of the three random processes assigned to this system at the iteration $k_2 = 100$ (this iteration is considered in the stationarity region). For the second data set $(x_1, x_2, \ldots, x_N)$ the system (15) was initialized by other N initial conditions and was sampled for each state variable at the iteration k_1. The values k_1 were chosen in the stationarity region, so more than 25. For each pair (k_1, k_2) the Smirnov test was resumed 500 times for a more accurate decision. The results are presented in Table 2 and determine how the decision upon the probability law of the new system will be interpreted. This table is read in the same way as Table 1, so: taking in consideration state variable x_3 and iteration $k_1 = 40$ in the table the result found indicates that in 96.4% of times from the ensemble of 500 the Smirnov test indicates that the random variable X obtained from the third state variable of system (15) at $k_1 = 40$ has the same probability law as the random variable Y obtained from the system (1) at the iteration $k_2 = 100$ also from the third dynamics.

Table 2. The proportion of acceptance for H_0 for Smirnov test (verifying probability law for each state variable of systems (1) and (15))

k_1	30	40	50	60
x_1	96.6	94.7	95	94.1
x_2	95.7	97.4	96.2	95.9
x_3	95.2	96.4	95.7	94.8

Remark 2. The two tables with the experimental results confirm that the statistical behaviour of the systems (1) and (15) is quite the same, so the first visual observation concerning Figure 1 and Figure 3 is proven. This seams to be sustained by the idea that both systems are described by the same *strange attractor*. All the advantages of hyperchaotic behaviour could be taken in consideration in the context of the new cryptosystem.

3.3   Consecutive Condition

The idea of modifying discrete dynamical systems by adding a delay in its structure was equivalent with the idea of increasing the dimension of these systems. Because the aim of this modification was to build a more powerful cryptosystem the decision was taken to adjust a variable delay in the structure of the second state variable of system (1). For simplicity reasons, to reconstruct the original message some conditions are needed:

- the delay has to be added in the structure of the second dynamics x_2,
- *consecutive condition*, which means that during the reconstruction process all the samples of x_2 has to be reconstructed.

The first condition will be explained in the description of the *receiver* and the second one will be discussed here.

Definition 5. *The inversion sufficient condition (consecutive condition) is given as follows: the difference vector $w = k - d(k)$, with $k = 1 \ldots n$, must contain all nonzero natural numbers from 1 to $n - max(d)$, where $max(d)$ is the maximum value of the delay.*

This consecutive condition can be interpreted as a *left invertibility* condition. see [21]. As it can be observed from the system (15) and from the construction of the *transmitter* (17) the delay $d(k)$ has a specific value at each step k (see also Table 3); this value is different from iteration to iteration. It was already mentioned that this value will be a different function of the moment were the delay vector is analyzed.

For exemplification the technique was implemented for a maximum value for the delay of 3 and an algorithm of generating this variable delay is proposed:

- a number of $max(d) = 3$ iterations, starting from the initial conditions, the system (17) is working with no delay, so $d(k) = 0$, for $k = 1 \ldots 3$,

- starting from iteration $max(d)+1 = 4$ to $2max(d) = 6$ the value of the delay will be 1, for reasons of initializing the system, so $d(k) = 1$, for $k = 4\dots6$,
- starting from $2max(d) + 1 = 7$ iterations the value of the delay will be generated by using an algorithm based on the logistic map presented in the matlab code.

An exemplification that the proposed algorithm is according to the Definition 5 is presented in Table 3.

Table 3. Exemplification of generating variable delay

k	1 2 3 4 5 6 7 8 9 10 11 12 13 14 15 16 17 18 19 20 21
$d(k)$	0 0 0 1 1 1 3 1 3 1 3 2 1 3 2 1 3 2 1 3 2
$w = k - d(k)$	1 2 3 3 4 5 4 7 6 9 8 10 12 11 13 15 14 16 18 17 19

Looking at the last row of Table 3 it can be observed that the *consecutive condition* is respected and now everything is prepared for building the *receiver*.

3.4   Receiver

Message reconstruction is the last step in the transmission process. The partner has received now the ciphered message and knowing the secret key and the structure of the cryptosystem has to reconstruct the original message. For different reasons some conditions were presented before and their utility will be observed here.

The extended form of system (16) is presented for a better understanding of building the reconstructor:

$$\begin{cases} x_1^+ = a_1 x_1(1 - x_1) + a_2 x_2^{d-} \\ x_2^+ = b_1[(1 - b_2 x_1)(x_2^{d-} + b_3) - 1](1 - b_4 x_3) \\ x_3^+ = c_1 x_3(1 - x_3) - c_2(1 - b_2 x_1)(x_2^{d-} + b_3) + u \\ y = x_1 \end{cases} \tag{19}$$

The receiver knows the secret key which is composed by the parameter vector p and from the initial condition $d(0)$ for the delay vector; also the partner who has received the cryptogram y knows the structure of the cryptosystem and the way of generating the variable delay d. Starting from the form (19) for identifying the original embedded message u three steps are needed:

1. Knowing the transmitter output y the second dynamics of the system (19) can be obtained:

$$\tilde{x}_2^{d-} = \frac{y - a_1 y^-(1 - y^-)}{a_2}$$

2. In this step the third dynamics of the receiver will be obtained. Introducing the notations $q := (1 - b_2 x_1)(x_2^{d-} + b_3) - 1$ and remembering that $x_2^- := x_2(k-1)$ with consecutive condition respected it can be written:

$$\tilde{x}_3^{2-} = \begin{cases} \frac{b_1 q^{2-} - \tilde{x}^-}{b_1 b_4 q^{2-}} & \text{for } \tilde{x}^{2-} \in \Re^3/S \\ \tilde{x}_3^{3-} & \text{for } \tilde{x}^{2-} \in S \end{cases}$$

if a value is obtained from singularity manifold S then $\tilde{x}_3^{2-}$ is forced to take its last buffered value.

3. After reconstructing all the dynamics of the cryptosystem the message can be extracted with a delay, when the reconstructed values of $\tilde{x}_3^{3-}$, $\tilde{x}_3^{2-}$, $\tilde{x}_2^{3-}$ obtained in the previous steps and the value of the output y^{3-}:

$$u^{3-} = \tilde{x}_3^{2-} - c_1 \tilde{x}_3^{3-}(1 - \tilde{x}_3^{3-}) + c_2(1 - b_2 \tilde{x}_1^{3-})(\tilde{x}_2^{3-} + b_3)$$

A way of reconstructing the original message u is presented in the Appendix. In the presented algorithm the *consecutive condition* is verified after the reconstruction of each sample of the second dynamics x_2; the third dynamics x_3 is rebuilt every time after this condition is verified. In this way the reconstruction of the original message is faster and the delay of the transmission is very small.

Synchronization. In case of losing synchronization the receiver and the transmitter can be resynchronized in the same way as in the beginning of the communication. A scheme of a cryptosystem based on the inclusion method provides self synchronization, it is the case of the cipher proposed in this paper. An example to show how this really works is given in the appendix.

3.5   Cryptanalysis

By adjusting the variable delay in the structure of system (1), the identifiability of the system parameters will be very difficult to prove because of the huge dimension of the resulted system, [16]. Also the fact that the delay will be part of the secret key gives another reason to sustain the robustness of the new cryptosystem. So for obtaining an equation of type (13) - input/output approach - reiterating (15) in order to obtain a relation of type (11) the following will show that x_2^+ cannot be reconstructed from the second equation like in (11).

$$\begin{cases} x_1^+ - a_1 x_1(1 - x_1) - a_2 x_2^{d-} = 0 \\ x_1^{++} - a_1 x_1^+(1 - x_1^+) - a_2 x_2^{(d-)+} = 0 \\ \dots \\ x_2^+ - b_1[(1 - b_2 x_1)(x_2^{d-} + b_3) - 1](1 - b_4 x_3) = 0 \\ \dots \end{cases} \tag{20}$$

To solve the system (20) it will be useful to know how many times to reiterate the first equation of system (15) to obtain sufficient samples for x_2 state variable to try to obtain a relation of type (13); when $x_2^{(d-)+} := x_2(k - d(k) + 1)$. Supposing that the cryptanalyst knows the structure of the new cryptosystem, he has to try iteration by iteration all the steps of the reconstruction process to find an input/output relation.

The statistical results on the behaviour of the modified system show that form the statistical point of view nothing was changed, so all the advantages of the hyperchaotic behaviour can be used in the applications of this proposed system.

4   Conclusions

The robustness obtained by adjusting a delay in the structure of an existing cryptosystem can justify the complexity of the computational operations for reconstructing the original message. By using different computational techniques it was proven that the initial cryptosystem could be broken. The idea presented in this paper is just an example how to modify the structure of a cryptosystem in order to make it more robust. The cryptosystem will become more and more robust by increasing the maximum value of the delay.

In regards to the structure of the delay vector a maximum value of 30 can be of interest looking at [11], because the minimum sampling distance which can give the statistical independence was decided as being approximately 30 for the state variable x_2 - if it is the structure of the cryptosystem presented here.

Acknowledgments. The work has been funded by: Sectorial Operational Programme Human Resources Development 2007-2013 of the Romanian Ministry of Labour, Family and Social Protection through the Financial Agreement POSDRU/159/1.5/S/132395; PN-II-PT-PCCA-2011-3.2-1448 project DGI-SAR.

References

1. Anstett, F., Millerioux, G., Bloch, G.: Message-embedded cryptosystems: cryptanalysis and identifiability. In: 44th IEEE Conf. on Proc. and 2005 European Control Conf. Decision and Control CDC-ECC 2005, pp. 2548–2553 (2005)
2. Baptista, M.S.: Cryptography with chaos. Physics Letters A **240**(1–2), 50–54 (1998)
3. Belmouhoub, I., Djemai, M., Barbot, J.-P.: Cryptography by discrete-time hyperchaotic systems. In: Proc. 42nd IEEE Conf. Decision and Control, vol. 2, pp. 1902–1907 (2003)
4. Bhat, K., Koivo, H.: Modal characterizations of controllability and observability in time delay systems **21**(2), 292–293 (1976)
5. Carroll, T.L., Pecora, L.M.: Synchronizing chaotic circuits **38**(4), 453–456 (1991)
6. Cicarella, G., Dalla Mora, M., Germani, A.: A robust observer for discrete time nonlinear systems. Sys. Contr. Lett. **24**(10), 291–300 (1995)
7. Diop, S., Fliess, M.: Nonlinear observability, identifiability, and persistent trajectories. In: Proc. 30th IEEE Conf. Decision and Control, pp. 714–719 (1991)
8. Fliess, M.: Automatique en temps discret et algbre aux diffrences. Mathematicum **2**, 213–232 (1990)

9. Frunzete, M., Florea, B.C., Stefanescu, V., Stoichescu, D.A.: Image enciphering by using rossler map. In: Proceedings of the 2011 IEEE International Conference on Intelligent Computer Communication and Processing, pp. 307–310 (2011)
10. Frunzete, M., Luca, A., Vlad, A.: On the statistical independence in the context of the rössler map. In: 3rd Chaotic Modeling and Simulation International Conference (CHAOS2010), Chania, Greece (2010). http://cmsim.net/sitebuildercontent/sitebuilderfiles/
11. Frunzete, M., Luca, A., Vlad, A., Barbot, J.-P.: Statistical behaviour of discrete-time rössler system with time varying delay. In: Murgante, B., Gervasi, O., Iglesias, A., Taniar, D., Apduhan, B.O. (eds.) ICCSA 2011, Part I. LNCS, vol. 6782, pp. 706–720. Springer, Heidelberg (2011)
12. Frunzete, M., Barbot, J.-P., Letellier, C.: Influence of the singular manifold of nonobservable states in reconstructing chaotic attractors. Physical Review E **86**(2), 026205 (2012)
13. Larger, L., Goedgebuer, J-P.: Le chaos chiffrant. Pour la science (36) (2002)
14. Ljung, L.: System Identification - Theory for the User, 2nd edn. Prentice-Hall, Upper Saddle River (2002)
15. Nijmeijer, H., van der Schaft, A.: Nonlinear dynamical control systems. Springer-Verlag New York Inc., New York (1990)
16. Nomm, S., Moog, C.H.: Identifiability of discrete-time nonlinear systems. In: Proc. of the 6th IFAC Symposium on Nonlinear Control Systems, pp. 477–489. NOLCOS, Stuttgart (2004)
17. Perruquetti, W., Barbot, J.-P.: Chaos in automatic control. CRC Press, Taylor & Francis Group (2006)
18. Richard, J.-P.: Time-delay systems: an overview of some recent advances and open problems. Automatica **39**(10), 1667–1694 (2003)
19. Stefanescu, V., Stoichescu, D., Frunzete, M., Florea, B.: Influence of computer computation precision in chaos analysis. University Politehnica of Bucharest Scientific Bulletin-Series A-Applied Mathematics and Physics **75**(1), 151–162 (2013)
20. Vlad, A., Luca, A., Frunzete, M.: Computational measurements of the transient time and of the sampling distance that enables statistical independence in the logistic map. In: Gervasi, O., Taniar, D., Murgante, B., Laganà, A., Mun, Y., Gavrilova, M.L. (eds.) ICCSA 2009, Part II. LNCS, vol. 5593, pp. 703–718. Springer, Heidelberg (2009)
21. Vo Tan, P., Millerioux, G., Daafouz, J.: Left invertibility, flatness and identifiability of switched linear dynamical systems: a framework for cryptographic applications. International Journal of Control **83**(1), 145–153 (2010)
22. Wolf, A., Swift, J.B., Swinney, H.L., Vastano, J.A.: Determining lyapunov exponents from a time series. Physica, 285–317 (1985)

#Worldcup2014 on Twitter

Wilson Seron[1], Ezequiel Zorzal[1], Marcos G. Quiles[1(✉)],
Márcio P. Basgalupp[1], and Fabricio A. Breve[2]

[1] Institute of Science and Technology, Federal University of São Paulo (UNIFESP),
São José dos Campos, Brazil
{wilsonseron,ezorzal,quiles,basgalupp}@unifesp.br
[2] São Paulo State University (UNESP), Rio Claro, SP, Brazil
fabricio@rc.unesp.br

Abstract. A microblogging, such as the Twitter, is a Social Networking
Service that allows the publication of short messages. Currently, Twitter
has more than 270 million monthly active users, and it is widely used to
discuss the most variety of topics. Due to the large amount of information
circulating on Twitter, and the facility to publish and read messages
through the web or mobile devices, Twitter has attracted the interest
of the general public, companies, media etc. By analyzing the Twitter's
stream of data, one can identify trends, events, or even the feelings of
its users. Here, we introduce a dataset of tweets about the World Cup
2014, collected from January to August of 2014; present some descriptive
statistics about the data; and, finally, we show a sentiment analysis study
about the Brazilian population regarding to the Brazilian national team.

1 Introduction

The Internet and the social networks sites (SNS) have opened uncountable possibilities in the last decades. The Internet instantly allows users to get informed
about any news without any restriction. Many social network sites with distinct
purposes have been developed, e.g. maintain and reinforce professional links,
such as the LinkedIn; joining students, which was the original Facebook's goal;
to establish loving relationships, such as the Friendster; and also connecting
people with common interests (e.g. music and politics), such as MySpace [7].

Lately, social networks have played an important role in disseminating news
throughout the world since any person can become a disseminator of information
by publishing news, pictures, and videos on social networks. Besides, a simple
message published on an SNS can instantly reach the whole globe, thus, providing
a spread of information much faster than traditional media.

Amidst many SNS, the Twitter microblogging has become very popular due
to its simplicity and publishing facilities. For instance, users can publish any
news directly from their mobiles. Currently, Twitter is used worldwide to discuss specific trends, complaints, and to protest about any issue [12]. Aware of
the reachability of Twitter over many different people around the world, several
companies have also adopted Twitter as a powerful marketing tool to adver-
tise their products and to increase their sales. Moreover, several companies and

O. Gervasi et al. (Eds.): ICCSA 2015, Part I, LNCS 9155, pp. 447–458, 2015.
DOI: 10.1007/978-3-319-21404-7_33

research groups have started to analyze streams of tweets in order to identify specific patterns, opinions, sentiments, etc. For example, a company can track mentions of itself on Twitter to evaluate the feelings of its clients and potential costumers [1].

Sentiment Analysis, also known as opinion mining, is a recent topic in the Natural Language Processing (NLP) field. It aims to extract any opinion, sentiment, subjective information from text [13]. The sentiment analysis can be performed on large texts, sentences, and even short messages like tweets.

Recently, several research groups have applied sentiment analysis on Twitter. Specifically, numerous methods have been developed to categorize the sentiment of tweets as positive/neutral/negative [1,9,12,14,18], or even in more specific sentiments, such as, angry, happy, sad, etc. [6]. These sentiment analysis works have been applied to either specific or general topics [2–4,8,9,17,19].

One interesting outcome of the sentiment analysis on the Twitter's streaming flow is that it can be used to detect events in real-time. For example, harsh news being broadcast worldwide; the feeling about a specific event, such as The Oscar [18]; or even, as presented in [15], a high frequency of negative feelings in a specific geolocation could indicate an earthquake.

In this work, we present a study about the tweets related to the World Cup 2014, which took place in Brazil from June 12 to July 13 of 2014. The main contributions of this paper are the following: 1) we introduce a dataset with more than 8 million tweets from January to August of 2014[1]; 2) we present a descriptive statistic analysis about the collected tweets; and 3) a sentiment analysis about the feelings of Brazilians regarding to the World Cup and their National Team is reported.

This work is organized as follows. Section 2 provides a deeper description of the Twitter and also introduces our WorldCup2014 dataset. Section 3 describes our experiments and results. Finally, Section 4 provides some concluding remarks and points some directions of future work.

FIFAWorldCup @FIFAWorldCup · Jul 13

Germany (#GER) are the FIFA World Cup winners! #GERARG #WorldCup - fifa.to /1riFcRi

Fig. 1. A tweet sample

---

[1] The dataset is publicly available on request.

2   The Twitter and The WorldCup Dataset

The Twitter is an online SNS created in 2006 by Jack Dorsey, Biz Stone, and Evan Williams. Each Twitter's user might have a list of *followers* and *friends*. The followers represent the users that receive information from an active user. Friends, on the other hand, are those that provide information to their followers.

Through the Twitter network, users can publish and read short messages with up to 140 characters, named *tweets*[2]. Figure 1 depicts a tweet sample collected on July 13th of 2014. Moreover, by inserting the "@" sign followed by a username, the publisher can straightly mention or reply to a specific user.

Users can also *retweet* a tweet from a particular user by mention its original publisher. The *retweets* are identified by the string *RT* that appears in bold at the beginning of the tweet. Another important feature to identify trends is the Trending Topics, which represents, in real-time, a list of the most common issues being published worldwide on Twitter. These topics are created by using *hashtags*, which are strings followed by the symbol #. The Twitter's Trending Topics feature is a global search, albeit it can be filtered by specific countries, cities, etc.

2.1   The Dataset

In this work, we collect the tweets by using the Twitter's Streaming API. This API provides low latency access to the global flow of tweets that are circulating on the net in real-time, without restrictions regarding the number of requests. However, this API provides only random samples (up to 1%) of the entire public stream of tweets. Through this API is possible to obtain two kinds of samples: random samples reaching the total of 1% of all tweets; or random samples by filtering the global stream with certain criteria, such as, usernames, keywords, or geographic boundaries, nevertheless also restricted to the same global rate of 1%.

From January 11th of 2014 to August 22nd of 2014 (223 days), we collected an amount of 8, 117, 002 tweets related to the WorldCup 2014. Sixteen keywords were used to filter the tweets from the Streaming API. The complete list of keywords is shown in Table 1.

The list was split into two parts: one with positive or neutral words, such as "vai ter copa[3]", which is positive or "worldcup" that represents a generic or neutral word; the second part is related to negative comments, such as "não vai ter copa[4]". We also consider the following three negative words: "anonbr", "opcup", and "tangodown", which are keywords related to hacker attacks in Brazilian sites, planned to take place during the World Cup[5].

---

[2] https://support.twitter.com/

[3] English translation: there will be world cup.

[4] English translation: there will be no world cup.

[5] http://bits.blogs.nytimes.com/2014/06/20/hackers-take-down-world-cup-site-in-brazil/

Table 1. Keywords used to filter the streaming

Positive	Negative
FinalDraw	NãoVaiTerCopa
Brazil2014	NaoVaiTerCopa
WorldCup2014	Nao vai ter copa
WorldCup	Não vai ter copa
copa2014	anonbr
copadomundo	opcup
VaiTerCopaSim	tangodown

By using the keywords presented in Table 1, we obtained tweets from distinct geographical places and from several languages throughout the world.

The complete dataset is stored in a PostgreSQL database, which contains specific tables for the tweets, profiles, friends and followers, locations, etc. The database was modeled to favor the network analysis and the generation of queries.

To identify the feelings of Brazilians towards the World Cup 2014, only tweets published in Portuguese language were considered. They totalized the amount of 523,707 tweets, which represents about 6.45% of the total collected in the period mentioned above. Each tweet can contain numerous information, such as special characters, hashtags, usernames (*@FIFAWorldCup*), and even URLs (http://t.co/pnaaGOs0Hc), as depicted in Figure 1. This extra information can compromise the sentiment analysis, thus, prior to classifying those messages, a preprocessing phase is mandatory.

Besides the tweets themselves, our dataset also contains the following information: the profiles of the active users, those who had published at least one tweet in the aforesaid period, and their lists of friends and followers.

2.2 Preprocessing the Tweets

The first step consists in preprocessing each message by removing unimportant data. In this work, we considered a quite simple preprocessing phase consisting of four steps:

1. deletion of all accents from the Portuguese language: The Portuguese language has several accents. By eliminating them, we can avoid the misclassification of several words, such as "campeão[6]" and "campeao", or "não[7]" and "nao";
2. conversion all characters into lowercase letters: It consists in a simple conversion, which can avoid redundancies;
3. deletion of all *url* and username mentions: Normally, *url* introduces only a link to any web page related to the subject being published. To the information itself, it rarely adds any useful information;

---

[6] English translation: Champion.
[7] English translation: no.

4. deletion of all stop words: Stop words, such as pronouns, conjunctions, etc. do not add significant information to the message. Thus, by removing them, we help to reduce the noise into the textual data and also to decrease the feature vector size, which can lead to a more accurate classification [16].

It is worth mentioning that the *hashtags* are preserved because they provide important data or even the sentiment of the message, e.g. #IhateWorld-Cup,#love, etc.

3   Experiments and Results

By using our dataset composed of more than 8 million tweets, two analyses were conducted: 1) a descriptive statistics analysis on the global data, and 2) sentiment analysis on the Brazilian tweets (tweets written in Portuguese language).

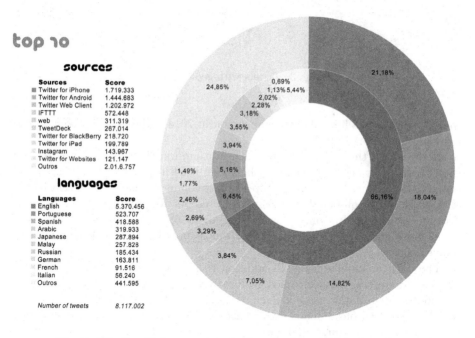

top 10

sources

Sources	Score
Twitter for iPhone	1.719.333
Twitter for Android	1.444.683
Twitter Web Client	1.202.972
IFTTT	572.448
web	311.319
TweetDeck	267.014
Twitter for BlackBerry	218.720
Twitter for iPad	199.789
Instagram	143.967
Twitter for Websites	121.147
Outros	2.01.6.757

languages

Languages	Score
English	5.370.456
Portuguese	523.707
Spanish	418.588
Arabic	319.933
Japanese	287.894
Malay	257.828
Russian	185.434
German	163.811
French	91.516
Italian	56.240
Outros	441.595

Number of tweets	*8.117.002*

Fig. 2. The Top 10 languages and client used to publish the tweets

3.1   Descriptive Statistics about Global Data

The World Cup is considered the largest sporting competition with a single modality in the world; thus, we can expect tweets from all over the world and written in several languages. Even using keywords only in English and Portuguese (see Table 1), our first expectation was to gather most of the tweets in Portuguese and also from Brazilian users. This assumption comes from the fact

Fig. 3. Tweets with geolocation

that Brazil was the host of this competition in 2014 and soccer represented the most popular sport in the country.

However, on the contrary of our expectation, our first surprise was to find that English was the most common language used in the tweets related to the competition. Figure 2 depicts the top 10 most common languages present in our dataset. It can be observed that English represents about 66% of total, whereas Portuguese was used in only a fraction of about 6% of the tweets. Besides the language, Figure 2 also shows the top 10 most popular Twitter clients used to publish the tweets.

The Twitter mobile client, in addition to sending and reading tweets, also offers an interesting feature named geolocation. When enabled, this feature adds the coordinates of the users when publishing the tweets. From the 8 million tweets collected from January to August of 2014, about 2% of them contains the geolocation coordinates. Figure 3 shows the distribution of those tweets throughout the world, demonstrating that the World Cup had a global audience reaching almost every country in the world. Figure 4 depicts the Top 10 countries according to the number of tweets published in this period.

The number of tweets per day on a given subject or geolocation can provide quite useful information regarding any particular event or trends. We have collected tweets during a 223 days window. In this period, the most active day was the eve of the World Cup opening ceremony. Based on this information, we can conjecture that people, in general, were quite anxious about the beginning of the event.

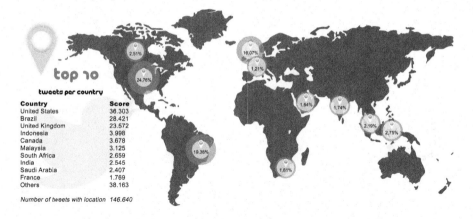

Fig. 4. Top 10 countries

It was also noted that a small fraction of the tweets provides *hashtags*. These words/sentences are considered a very valuable tool to identify specific trends on Twitter. In Figure 5(b)-(c) we show the Top 10 *hashtags* published during the whole window and also the Top 10 used in the final match, respectively. From this analysis, we can see an intersection between the hashtags in the whole period and at the final. However, the following exceptions might be highlighted: *argentina* and *germany*, which represents the two countries in the final match. In Figure 5(a), the Top 10 users are presented.

Fig. 5. The Top 10: Users, Hashtags, and Hashtags in the Final Match.

Another interesting observation is the power-law distribution of the number of tweets per user. As depicted in Figure 6, the Pareto principle, or the 80/20 rule, is observed, in which most of the tweets were published by a small fraction of users. On the other hand, most of the users have published only a few tweets in the considered time window.

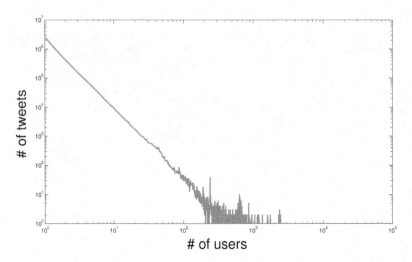

Fig. 6. Distribution of tweets per users in a log-log scale plot

3.2   Sentiment Analysis Results

The sentiment analysis can concentrate on several aspects. Simpler ones, such as classifying the text polarity: positive, negative, or neutral; or even providing a deeper classification in terms of emotional states: sad, angry, happy, etc.

Here, we apply sentiment analysis to evaluate the opinion of the Brazilian people regarding the World Cup Brazil 2014. Specifically, we want to identify whether the Brazilians were against or in favor of the World Cup and their National team.

The sentiment analysis study was divided into two parts: The first one lies in a global sentiment analysis before, during, and after the World Cup; the second part consists of a study restricted to matches in which the Brazilian team takes part. For this second part, the tweets were collected in a 14 hours window (6 hours before and 6 hours after the match plus the match itself).

Before presenting the results themselves, we first explain how the classifier was trained.

Training the Classifier: To categorize the tweets as positive or negative, we built a Naïve Bayes classifier [11]. The Naïve Bayes classifier is a simple probabilistic model, which applies the Bayes' theorem with strong independence assumptions [5]. The Naïve Bayes classifier is a simple model, and it is widely used in text categorization [10]. Roughly speaking, the classifier learns to predict the probability that a tweet belongs to the positive or negative sentiment based on the words it contains.

The classifier was trained with a labeled dataset of 500 tweets. The dataset was equally divided into two classes: 1) positive or neutral tweets, and 2) negative tweets. In order to build this dataset, we intentionally selected a large variety of tweets containing abbreviations, slangs, etc. Each tweet was manually labeled as negative whether it contains any pejorative statement in relation to the World Cup or the Brazilian team or even if it incites any violence, vandalism or protest. Otherwise, the tweet was labeled as positive. In Table 2, we show some samples of positive and negative preprocessed tweets present in our dataset.

Table 2. Samples of positive and negative tweets

Tweet	Sentiment
vaibrasil copa2014	Positive
copa2014 vaitercopa fifa	Positive
perfeita essa abertura worldcup2014	Positive
Tango Down CETEM OpHackingCup	Negative
vemprarua naovaitercopa	Negative
agora podemos protestar naovaitercopa	Negative

In additional to the words of the tweets themselves, we also take a list of words and a list of emoticons into account. The list of words is composed of 500 positive words and 500 negative words. The emoticons list contains 60 emotions equally divided into positive and negative. The main purpose of these two lists is to improve the accuracy of the classifier. Thus, the training set is composed of a set of words (negative or positive) extracted from the tweets combined with these new sets of extra words and emoticons.

The classification accuracy of Naïve Bayes classifier in a 10-fold cross validation was about 73%, which represents a good accuracy for sentiment analysis purposes.

The Sentiment Results: Before the beginning of the event, numerous protests and demonstrations against the World Cup happened in Brazil. This fact can be observed in the amount the negative tweets depicted in Figure 7. If we analyze the percentage of negative tweets, we can see that about 45% of them demonstrate any negative or even repulsive feelings about the World Cup. On the other hand, during the event itself, probably due to the patriotic feeling of watching the National team, we observe an increasing of the tweets with positive sentiment. During the competition, the rate was about 70%/30% of positive tweets against negative ones. This excitement stood until the semi-final when the Germans defeated the Brazilian team. From this point, as illustrated in Figure 7, which the percentage of positive tweets was slightly reduced, albeit still quite positive when comparing to the pre World Cup period.

An interesting observation is that, even after being defeated by the German team in the semifinal and also by the Netherlands, the rate of positive/negative feelings remained higher than before the World Cup. An assumption that sustains this positive sentiment was the defeat of Argentina to Germany, once Brazil has a long history of rivalry with Argentina in soccer.

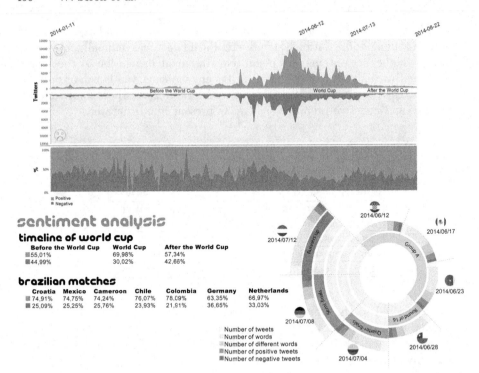

Fig. 7. Sentiment analysis results. On the top: the number of positive (blue) and negative (red) tweets; In the middle: The percentage of positive tweets against the negative ones; On the bottom: summary of results in the three phases taken into account and also the percentage of positive and negative feelings in the Brazilian team matches.

Whether we take the Brazilians matches into account, as depicted at the bottom of Figure 7, we can observe that the positive sentiment during the games is much higher than in general. Before the match against Germany, positive tweets represents about 75% of the total. During the last two matches (against Germany and Netherlands), the percentage of positive tweets was reduced to about 65%, albeit still superior to the period pre World Cup.

4   Conclusions and Future Work

In this work, we have collected and studied *tweets* about the World Cup 2014. By using the Twitter Streaming API, we filtered the Twitter's global stream storing only tweets related to the World Cup. In total, more than 8 million tweets were collected from January to August of 2014.

By using this dataset, two distinct studies had been conducted and presented in this paper. The first was a descriptive statistics analysis to quantify the information about the tweets, such as the most common language, countries, users, hashtags, daily activity on Twitter, etc. The second one has focused only in tweets published in the Portuguese language. In this late scenario, our purpose was to analyze the sentiment of the Brazilian people regarding the World Cup and their national team.

From these studies, the following results could be summarized: 1) Quite interesting observations have been highlighted, e.g. English was the most common language, even the event having occurred in Brazil, where the native language is Portuguese; June 11th, the day before the World Cup opening ceremony, was the most active day on Twitter, maybe due to the anxiety about the matches; nearly all regions around the world contributed with some tweets about this event; the information about the Twitter client has shown that the iPhone was the most used device, the distribution of tweet per user follows a power-law distribution, etc. 2) The sentiment analysis has indicated that the Brazilian population was quite discontent with the World Cup before the beginning of the competition. During that period, several protests and demonstrations have taken place in several states in Brazil, and the tweets reflected these negative feelings of the population. However, during the event, the magic of the sport covered part of the negative feelings observed a priori and a wave of positive tweets emerged in the network.

The positive sentiment has started diminishing only after the defeat of the Brazilian team to the Germans (Germany 7 × 1 Brazil), albeit more than 50% of the tweets remained positive, even after the competition.

As a future work, we will conduct further analyzes of tweets and perform a sentiment analysis taking other languages into account.

Acknowledgments. This research was supported by the São Paulo Research Foundation (FAPESP) and by the Brazilian National Research Council (CNPq).

References

1. Agarwal, A., Xie, B., Vovsha, I., Rambow, O., Passonneau, R.: Sentiment analysis of twitter data. In: Proceedings of the Workshop on Languages in Social Media, pp. 30–38. Association for Computational Linguistics (2011)
2. Aston, N., Liddle, J., Hu, W.: Twitter sentiment in data streams with perceptron. Journal of Computer and Communications (2014)
3. Barbosa, L., Feng, J.: Enhanced sentiment learning using twitter hashtags and smileys. In: Proceedings of the 23rd International Conference on Computational Linguistics: Posters, pp. 241–249. Association for Computational Linguistics (2010)
4. Bifet, A., Frank, E.: Sentiment knowledge discovery in twitter streaming data. In: Pfahringer, B., Holmes, G., Hoffmann, A. (eds.) DS 2010. LNCS, vol. 6332, pp. 1–15. Springer, Heidelberg (2010)
5. Bishop, C.M.: Pattern Recognition and Machine Learning, 1st edn. Springer (2006)
6. Bollen, J., Mao, H., Pepe, A.: Modeling public mood and emotion: twitter sentiment and socio-economic phenomena. In: ICWSM (2011)
7. Ellison, N.B., Steinfield, C., Lampe, C.: The benefits of facebook "friends:" social capital and college students' use of online social network sites. Journal of Computer-Mediated Communication 12(4), 1143–1168 (2007)

8. Gautam, G., Yadav, D.: Sentiment analysis of twitter data using machine learning approaches and semantic analysis. In: 2014 Seventh International Conference on Contemporary Computing (IC3), pp. 437–442. IEEE (2014)
9. Go, A., Huang, L., Bhayani, R.: Twitter sentiment analysis. Entropy **17** (2009)
10. Manning, C.D., Raghavan, P., Schütze, H.: Introduction to information retrieval. Cambridge University Press, Cambridge (2008)
11. Mitchell, T.M.: Machine Learning. McGraw-Hill (1997)
12. Pak, A., Paroubek, P.: Twitter as a corpus for sentiment analysis and opinion mining. In: LREC (2010)
13. Pang, B., Lee, L.: Opinion mining and sentiment analysis. Foundations and trends in information retrieval **2**(1–2), 1–135 (2008)
14. Saif, H., He, Y., Alani, H.: Semantic sentiment analysis of twitter. In: Cudré-Mauroux, P., Heflin, J., Sirin, E., Tudorache, T., Euzenat, J., Hauswirth, M., Parreira, J.X., Hendler, J., Schreiber, G., Bernstein, A., Blomqvist, E. (eds.) ISWC 2012, Part I. LNCS, vol. 7649, pp. 508–524. Springer, Heidelberg (2012)
15. Sakaki, T., Okazaki, M., Matsuo, Y.: Earthquake shakes twitter users: real-time event detection by social sensors. In: Proceedings of the 19th International Conference on World Wide Web, pp. 851–860. ACM (2010)
16. Silva, C., Ribeiro, B.: The importance of stop word removal on recall values in text categorization. In: Proceedings of the International Joint Conference on Neural Networks, 2003, vol. 3, pp. 1661–1666. IEEE (2003)
17. Tang, D., Wei, F., Qin, B., Liu, T., Zhou, M.: Coooolll: A deep learning system for twitter sentiment classification. SemEval 2014, 208 (2014)
18. Thelwall, M., Buckley, K., Paltoglou, G.: Sentiment in twitter events. Journal of the American Society for Information Science and Technology **62**(2), 406–418 (2011)
19. Wang, X., Wei, F., Liu, X., Zhou, M., Zhang, M.: Topic sentiment analysis in twitter: a graph-based hashtag sentiment classification approach. In: Proceedings of the 20th ACM International Conference on Information and Knowledge Management, pp. 1031–1040. ACM (2011)

Adaptive Computational Workload Offloading Method for Web Applications

Inchul Hwang[✉]

Software R&D Center, Samsung Electronics, Samsung-ro 129, Yeongtong-gu,
Suwon-si, Gyunggi-do, South Korea
inc.hwang@samsung.com

Abstract. Nowadays, cloud computing has become a common computing infrastructure. As the computing paradigm has been shifted to cloud computing, devices can utilize computing resource any-where/any-time/any-device. Many research papers in mobile cloud called 'cloud offloading' which migrates a device's workload to a server or to other devices have been proposed. However, previous cloud offloading methods are mainly focusing on the cloud offloading between a device and a server. Furthermore, these proposed methods have rarely commonly used because the proposed methods were very complex - difficulty of partitioning application tasks and maintaining execution status sync between a device and a server in the cloud. In this paper, I proposed the adaptive framework for cloud offloading based on the web application standard - HTML5 specification - for web applications based on flexible resource in servers as well as in devices. In HTML5 specification, there is the method for the parallel execution of the task named 'Web Worker' and the method for the communication between a device and a server named 'Web Socket'. Utilizing the property of this specification, I proposed a seamless method to do the cloud offloading for parallelized tasks of the web applications among devices as well as between a device and a server. Based on proposed method, a device can seamlessly migrate a part of web application workload with the Web Worker to other resource owners - devices and servers - with a little modification of web applications. As a result, I can successfully build the environment where a device which has a HTML5 browser such as a mobile phone and a smart TV can share the workload among devices and servers in various situations – out-of-battery, good network connection, more powerful computing needs.

Keywords: Mobile cloud · Cloud offloading · Web application · Workload balancing · HTML5 · Web worker · Web socket

1 Introduction

In recent 10 years, we have seen that cloud computing has the potential to change the IT industry. The hardware and software resources can be accessible with on-demand style. It means that a device can utilize any computing resource whenever/wherever it needs the external resources. Currently, many services based on flexible resources in a cloud such as a cloud storage service have been widely spread because of the

© Springer International Publishing Switzerland 2015
O. Gervasi et al. (Eds.): ICCSA 2015, Part I, LNCS 9155, pp. 459–471, 2015.
DOI: 10.1007/978-3-319-21404-7_34

convenience of the accessibility and flexibility. However, this commercialized utilization of cloud has been just focusing on the coarse-grained resource sharing between a server and a device as a service. With this environment, many system-level resource sharing method named cloud offloading [1, 2] has been proposed. However, most of the previous research could not be commercialized because of the difficulty of partitioning of application and syncing the status of the application between a server and a device. Additionally, it's mainly focusing on the resource sharing between a device and a server. However, a current smart device's computing power is more powerful than a server in 10 years ago. It means that we can utilize a smart device's computing resource as well as server's resource for the other devices if the device wants to use it in various situations – out-of-battery, heavy workload.

In this paper, I proposed an adaptive cloud offloading method for web applications – maximizing a utilization of resources on devices as well as servers based on web standard. According to the news [10] and survey report [3], over 300M smart devices had been shipped in the worldwide market in the first quarter of 2013 and the web applications are over 36% of the applications which the user executes in mobile devices. So, if I can provide the sophisticated method for cloud offloading method of web applications among devices and servers, devices can provide the more flexible resource of multiple devices with users.

Before HTML5 [4] was released as a new standard, the web applications were mainly focusing on the user presentation and interaction. With HTML5, the web application can execute heavy workload based on Web Worker [5] with paralleled style. Based on the Web standard – HTML5 Web Worker, I can provide more seamless cloud offloading method for web applications and eliminate overhead of maintaining the execution stack between a device and a server or among devices because the HTML5 Web Worker thread is separated from the main UI thread. Previously, Elastic HTML5 [16] and WWF [12] were proposed for cloud offloading between a device and a server based on web standards. In Elastic HTML5, the authors were quietly good at analyzing and suggesting method which can utilize the property of Web Worker. However, in this approach, they didn't show the actual implementation and experimental results. In WWF, my colleague and I designed and implemented WWF in order to do the offloading of the workload in a client to a server based on HTML5 specification. In other words, it's designated to utilize all resources in servers for a client device. However, these days devices have a lot of computing powers rather that a server which was 10 years ago. Therefore, I devised the new framework (A-WWF) for sharing resource a server as well as devices adaptively. With A-WWF, if a device has a browser and wants to share its resource with other devices, it can share the computing resource easily with others. Currently, in HTML5 specification, there is a way to communicate with servers with a little overhead named Web Socket [15]. Based on these specifications, I could make devices share their resource with others via a server. With the experiments, I could see the possibility of flexible resource sharing among devices as well as servers. Finally, I expected that I could see the performance improvement as well as other benefits with cloud offloading based on servers and devices – saving processing time and battery consumption, etc.

The paper organization is that I will describe what kinds of research related to cloud offloading has been done regarding the fine-grained resource sharing with the flexibility among devices and servers in Section 2. In Section 3, I will describe regarding HTML5 Web Worker and Web Socket specification and why these method is efficient to be applied cloud offloading to workload of web applications. In Section 4, I will describe how I built the web workload balancing framework among devices (A-WWF). In Section 5, I will show the experimental results with mobile phones as well as servers. Finally, I make a conclusion and explain the future work in Section 6.

2   Related Works

Recently, to utilize the cloud resource more dynamically from a device, many research papers have been published about how to divide the workload of application between a device and a server.   To utilize a server's resource from a device, if a device has very limited resource – lack of computing power/ network/ battery, etc., and it can't handle large and complex workload in some cases, main parts of workload will be executed in a server and then the results from a server are sent back to a device. Traditionally, to divide the workload between a device and a server, the application programmer's support or systematic pre-analysis has been required to detect candidate tasks in an application. After detecting candidates, a device can easily offload candidates to a server based on its resource status. As a result, a device is just to get execution results of candidates from a server instead of executing candidates on it. Therefore, a device can save its resource including a power consumption or CPU time.

CloneCloud [1] is the research to automatically divide the workload of an application. In CloneCloud, a device and a server synchronously execute an application. It means that the clone of a device execution environment can be maintained in a server and when a device decides to migrate some parts of workload to its clone in a server, the device will send the stack of an application to a server and a server executes offloaded workload and returns the results of the execution to the device. This approach has also some limitations – syncing application execution environment status between a device and a server is a mandatory, so that a lot of overhead will be generated. In the research [8], they enhanced the CloneCloud to divide the workload dynamically between a device and a server. They introduced the methodology of dynamic partitioning of an application between a device and a server. The key is to solve the heterogeneity issues regarding device platform, network and cloud. In this system, both a device and a server have all application and instantiate modules to run at a device and a server dynamically at run time.

In MAUI [2], to enhance the traditional method, the paper introduced the research topic how to divide the workload between a device and a server while utilizing the resource as much as possible. MAUI utilized the programming environment to easily provide application programmers with predefined notations. So, when an application is compiled, the system can easily detect the candidate parts of an application which can be migrated to a server. MAUI supported fine-grained code offloading to maximize energy savings with minimal burden on the programmer's effort. MAUI's goal is to mitigate the energy problem for mobile handhelds, which is one of the mobile industry's foremost challenges.

In HCCF [14], to share the workload among devices, they devised framework named HCCF (Home cloud computing framework). In this case, they used the python and embedded JVM as a computing environment and devised new communication protocol for connecting and managing the workload. This research has the restriction that computing environment should be based on the specialized environment which is not common in a smart device – Python, JVM. In WWF [12], my colleague and I developed WWF (Web Workload balancing Framework) for Web applications. This is to share the resources in a server for a device. Even though there are a lot of devices of a user, the user can't utilize the resource in all devices fully.

In Elastic HTML5 [16], the authors suggested two things – computational offloading and storage offloading based on web technology. They analyzed the specification and suggested the method to apply cloud offloading. However, there was no implementation and no consideration to utilize the resource in other devices.

Most of all approaches have proposed for utilizing any resource by effectively dividing the application or by cloning the execution environment in case of the various device situations. However, these researches are difficult to be adapted to a smart phone because of the complexity and required the specific execution environment. Additionally, if the cloud offloading can be applied, the whole stack should be synchronized between a client and a server or among devices. Because this overhead is so heavy, so the previous research also focused on how to efficiently utilize the cloud offloading method with various parameters –network/battery/part of applications. Therefore, I tried to find out the natural way to divide the application and finally to execute whole workload on multiple devices based on browser which is widely spread over smart devices as well as servers.

3   HTML5 Specification

In this section, I will describe the related HTML5 specification and why this specification is important to make cloud offloading framework on multiple devices.

3.1   Web Worker

In HTML5 Specification, there is the specification regarding Web Worker which web applications can execute the parallelized workload like a thread in Figure 1 [11].

Previously before HTML5 without Web Worker as the left-side in Figure 1, the entire web logic was executed as a sequential way. If there is an input from a user, a web application handles it and responds. With Web Worker specification in HTML5 as the right-side in Figure 1, parts of a web application can be executed concurrently separately from a main UI javascript thread. Therefore, more complex/heavy web workload can be executed in web applications based on Web Worker.

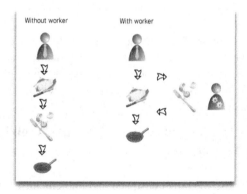

Fig. 1. Web Worker in HTML5

Figure 2 shows how a UI main javascript logic and a web worker thread communicate each other.

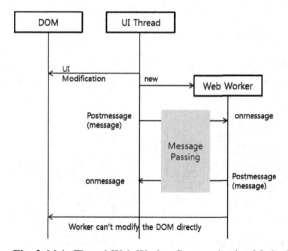

Fig. 2. Main Thread-Web Worker Communication Method

As in Figure 2, the main UI javascript manages the DOM structure in a web application and Web Worker is separated from the main UI javascript and they are communicating each other with the message-passing way (postmessage, onmesssage). This is the reason why I suggest utilizing Web Worker is an effective way to apply cloud offloading for web applications - to execute Web Worker both in a device and a server, a device or a server doesn't need to maintain the whole stack of the main UI javascript stack, and then I can easily eliminate the execution stack maintenance overhead between a client and a server or among devices in previous researches. As a result, if a device wants to execute parts of a web application based on Web Worker in a server or in the other device, a server or the other device can execute Web Worker tasks separately from a device and only needs to communicate with a device for messages between UI main thread and Web Workers.

3.2   Web Socket

In HTML5, there is a method to communication with a server based on browser named Web Socket [15]. With Web Socket, a web application can make a connection with a server like a BSD socket interface. This communication channel is so reliable to use for transmitting all data between a server and a device or among devices.

4   Design and Implementation of Adaptive Web Workload Balancing Framework

In this section, I will describe how I designed and implemented Adaptive Web Workload balancing Framework (A-WWF).

First of all, I selected the browser as an execution environment for devices because a current browser based on HTML5 is widely spread over many devices. Additionally, as my colleague and I devised WWF [12], I utilized the web worker property which is separated from a main UI thread.

4.1   Design of A-WWF

I designed A-WWF based on the WWF. a client side of A-WWF is very similar to WWF, but a server side application on a device has been newly designed as Figure 3.

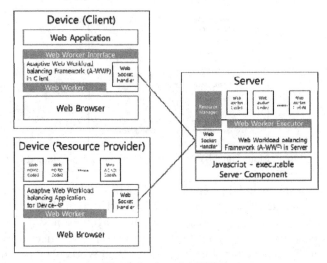

Fig. 3. Design of A-WWF

Initially, when a web browser on a device launches a web application which wants to use flexible resources, A-WWF in a client registers a device as a client to a server. Additionally, if a device can share resource, it will launch a web application named A-WWF Resource Provider (A-WWF-RP). In this case, A-WWF-RP will register itself as a resource provider to a server. As a result, a Resource Manager in a server can manage which devices will use a computing resource as a client and which devices will provide computing resource with other devices as a resource provider. Also, if a server can provide a computing resource with others, it will register itself as a resource provider.

After this registration phase, if a client on a device wants to use a flexible resource, it will send the request to launch Web Worker to a Resource Manager in a server with a Web Worker ID and URL. If a Resource Manager in a server receives a Web Worker launch request from a client in a device, it will determine which Resource Provider will launch a Web Worker and forward a launch request to a server or a Resource Provider in a device. When a Resource Provider receives a launch request from a Resource Manager, it will launch a new Web Worker. Currently, I used a static scheduler which can assign a Resource Provider – if there are any Resource Providers in devices, it will determine them as a Resource Provider, unless it will determine a server as a Resource Provider.

After launching a Web Worker in a Resource Provider, this Web Worker will communicate with a A-WWF via Web Socket connection – if a Resource Provider is in a device, a Web Worker will communicate to a A-WWF in a client via a server.

The detail sequence of A-WWF when there is a Resource Provider in a device is as Figure 4. For communicating exact Web Worker in a Resource Provider and a UI Thread, A-WWF in a client creates and assigns a unique web worker ID when a UI thread wants to create a new Web Worker. With this worker ID, whole system can identify which Web Worker is doing the right task. Additionally, if there is a Resource Provider in a device, a Resource Manager will assign it as a Resource Provider for a new Web Worker.

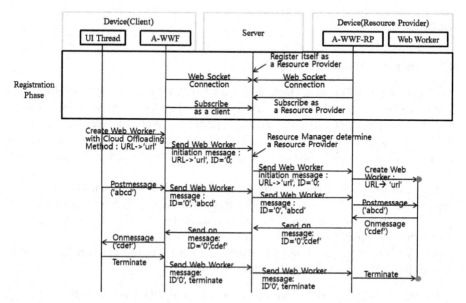

Fig. 4. Sequential Flow for Cloud Offloading in A-WWF when there is a Resource Provider in a Device

Figure 5 describes another case – there is no Resource Provider in a device – in this case, a Resource Manager will assign a server as a Resource Provider – all new Web Worker will be created in a Server.

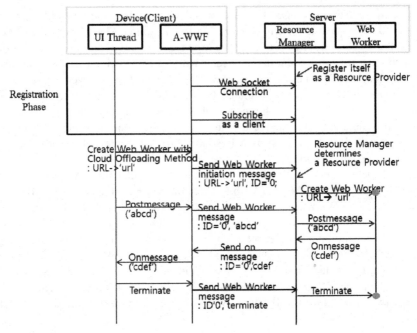

Fig. 5. Sequential Flow for Cloud Offloading in A-WWF when there is no Resource Provider in a Device

4.2   Implementation of A-WWF

To implement A-WWF in a client side, I used the WWF previously developed. A-WWF in a client side is a javascript library which wrapping all the same interfaces of Web Worker for the compatibility and easy porting of Web Applications. To use A-WWF in a client, a web application need to modify each line of code – Web Worker creation code to A-WWF creation code. Additionally, a web application needs to link a A-WWF library.

If a device wants to share its resource with others, it should launch the WWF-RP application based on browser. In WWF-RP application, it will connect with a server based on Web Socket and send a registration request to a server. For this purpose, I implemented a separate application with blank UI– WWF-RF.

For A-WWF in a server side, I implemented a javascript application based on Node.js [6]. I used the package of Node.js as follows:

- http-server: to handle HTTP request from a client, it's the base for websocket
- websocket: to handle Web Socket connection between a server and a client
- webworker-threads: to provide resource in a server based on Web Worker.

5   Experimental Results

I used the Web Worker application named Rayjs [7] which I could easily find on the internet. Figure 6 shows the web application UI which I used.

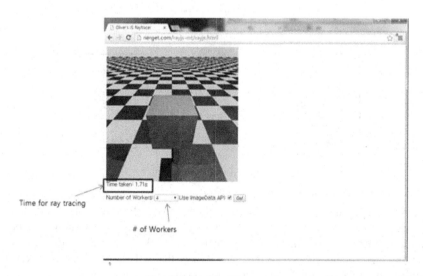

Fig. 6. Rayjs Test Program

With this program, I evaluated the computation time for ray tracing one object with the variance of the Web Worker numbers. The scene is created with the communication about 1.5 MB data between a main UI thread and Web Workers when I executed it on a local browser. In order to apply our framework to this application, I just changed two lines of code of the application (linking A-WWF code into the application-HTML tag, changing Web Worker creation part of the application to A-WWF wrapper creation – javascript code).

I setup the experimental environment as Figure 7:

- Server : Intel core i7 4500U, 8GB Memory, 256GB SSD, Node.js 0.10.31
- Device 1: Galaxy S5 SM-G900T, Qualcomm MSM8974AC Snapdragon 801 2.5GHz, 2GB Memory, 16GB Storage, Chrome version 37.0.2062.117
- Device 2: Intel core i5 3317U, 8GB Memory, 128GB SSD, Chrome version 37.0.2062.124 m
- Device 3: Galaxy Note 2 SHVE250S, Samsung Exynos 4412 1.6GHz, 2GB Memory, 32GB Storage, Chrome version 37.0.2062.117
- Wireless connection : 802.11n

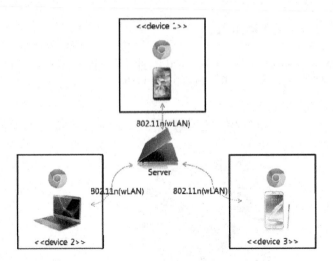

Fig. 7. Test Environment

Before testing the real application, I evaluated the performance of each device and a round-trip latency between a client and a server device. Table 1 shows the performance evaluation with Sunspider 1.0.2 benchmark [9]. As you can see, PC is worse than any other devices. So, I could easily expect that if I use Device 2 as a Resource Provider, the performance is worse than any others.

Table 1. Javascript Benchmark Result

Client	Time(ms)
Device 1	760.1
Device 2	1323.6
Device 3	1025.1

Figure 8 shows the round-trip latency (miliseconds) between a server and a client. X-axis is a packet size and Y-axis is a milli-second latency. I measured round-trip latency10 times trials and average it. As you can see, when a Device 2 is involved, the latency is longer than any other cases. This means that if a client wants to use the resource in another device, the network latency will be impacted on the performance because of the communication- I could expect that if Device 2 will be involved, the performance will be worse than any other cases.

I experimented two cases – first without a client which has a Resource Provider and second with a Resource Provider in a client.

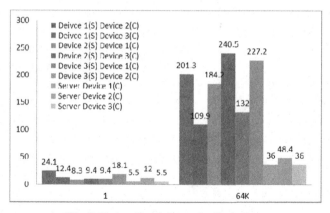

Fig. 8. End-to-End Latency for Each Case

5.1   Experiment without a Resource Manager in a Device

Figure 9 shows the execution time in case of a Resource Manager not in a device with the variation of the number of Web Worker. As the number of Web Worker increases, the performance is saturated on the number of 2~4. Any clients have very similar result from a server because Web Worker will be created in a same server in this case – only difference is coming from network latency.

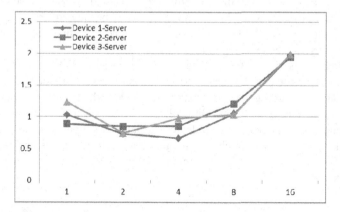

Fig. 9. Execution Time without a Resource Manager in a Device (msec)

5.2   Experiment with a Resource Manager in a Device

Figure 10 shows the execution time in case of a Resource Manager in a device with the variation of the number of Web Worker. As the number of Web Worker increases, I could see the performance improvement. The interesting thing is that when I set up with a PC client, because of the latency between PC and mobile devices, the performance is worse than any other cases.

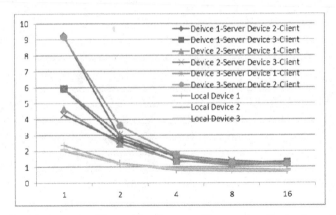

Fig. 10. Execution Time with a Resource Manager in a Device (msec)

6   Conclusion and Future Work

In this paper, I proposed the Adaptive Web Workload balancing Framework (A-WWF) for distributing client web application workload among devices and servers. With A-WWF, the device can migrate adaptively the web application workload based on HTML5 specification (Web Worker and Web Socket) to another device and to a server with very little modification. So, if a device has a browser which supports HTML5 specification, it can utilize flexible resource with A-WWF. Even though the performance improvement of cloud offloading is less because of the communication latency via Web Socket, I believe a device can get benefits with cloud offloading such as saving battery as well as computing resource.

This paper is describing the A-WWF as an extension of WWF. Currently, I used the static schedule for distributing workload among a server and devices. So, I'll investigate more regarding the optimized schedule which can be performance-oriented/power-oriented model. Additionally, I considered to use WebRTC [13] for the communication channel between devices which can support peer-to-peer connection. When I experimented, WebRTC is quite bad performance using Data Channel until now. So, to use WebRTC as a communication channel among devices, I need to optimize the connection settings.

References

1. Chun, B., Ihm, S., Maniatis, P., Naik, M., Patti, A.: Clonecloud: elastic execution between mobile device and cloud. In: Proceedings of the 6th Conference on Computer Systems. ACM (2011)
2. Cuervo, E., Balasubramanian, A., Cho, D., Wolman, A., Saroiu, S., Chandra, R., Bahl, P.: MAUI: making smartphones last longer with code offload. In: Int/ernational Conference on Mobile Systems, Applications, and Services (2010)
3. Nielson research. http://www.randykisch.com/2011/08/use-of-mobile-apps-far-outweighs.html
4. HTML5 Specification. http://www.w3.org/TR/html5/

5. Web Worker Specification. http://www.w3.org/TR/workers/
6. Node.js. http://nodejs.org/
7. Ray Tracing Application. http://nerget.com/rayjs-mt/rayjs.html
8. Chun, B.-G., Maniatis, P.: Dynamically partitioning applications between weak devices and cloud. In: 1st ACM Workshop on Mobile Cloud Computing&Services: Social Networks and Beyond (2010)
9. Sunspider. http://www.webkit.org/perf/sunspider/sunspider.html
10. CNN Smart Device Shipment in Q1 2013. http://tech.fortune.cnn.com/2013/05/09/apple-samsung-android-canalys/
11. Web Worker rise up. http://dev.opera.com/articles/view/web-workers-rise-up/
12. Hwang, I., Ham, J.: WWF: web application workload balancing framework. In: IEEE AINA Joint Workshop – Device Centric Cloud (2014)
13. Web RTC. http://dev.w3.org/2011/webrtc/editor/webrtc.html
14. Lee, J., Choi, K., Kim, Y., Kang, S.: Design and implementation of the lightweight home cloud computing framework. In: IEEE Third International Conference on Consumer Electronics - Berlin (2013)
15. Web Socket. http://www.w3.org/TR/2009/WD-websockets-20091222/
16. Zhang, X., Jeon, W., Gibbs, S., Kunjithapatham, A.: Elastic HTML5: workload offloading using cloud-based web workers and storages for mobile devices. In: International Workshop on Mobile Computing and Clouds (MobiCloud), Santa Clara (2012)

Clustering Retrieved Web Documents
to Speed Up Web Searches

Rani Qumsiyeh and Yiu-Kai Ng[✉]

Computer Science Department, Brigham Young University, Provo, UT 84602, USA
raniq@microsfot.com, ng@compsci.byu.edu

Abstract. Current web search engines, such as Google, Bing, and Yahoo!, rank the set of documents S retrieved in response to a user query and display the URL of each document D in S with a title and a snippet, which serves as an abstract of D. Snippets, however, are not as useful as they are designed for, which is supposed to assist its users to quickly identify results of interest, if they exist. These snippets fail to (i) provide distinct information and (ii) capture the main contents of the corresponding documents. Moreover, when the intended information need specified in a search query is *ambiguous*, it is very difficult, if not impossible, for a search engine to identify precisely the set of documents that satisfy the user's intended request without requiring additional inputs. Furthermore, a document title is not always a good indicator of the content of the corresponding document. All of these design problems can be solved by our proposed query-based cluster and labeler, called *QClus*. *QClus* generates concise clusters of documents covering various subject areas retrieved in response to a user query, which saves the user's time and effort in searching for specific information of interest without having to browse through the documents one by one. Experimental results show that *QClus* is *effective* and *efficient* in generating high-quality clusters of documents on specific topics with informative labels.

Keywords: Clustering · Cluster labels · User queries · Web documents

1 Introduction

With new electronic documents added to the Web each day, it is essential for web search engines to continuously enhance their current design in extracting and ranking relevant results to meet web users' information needs effectively and efficiently. Current web search engines rank retrieved documents based on their likelihood of relevance to a user query Q and display the URL of each document with a title and a short snippet[1]. A snippet S, however, is (i) often very similar to others created for documents retrieved in response to the same query and (ii) generated using sentences/phrases in the corresponding document D in where the keywords in Q appear, which may not capture the main content of D. Consider the top-5 results retrieved by Google (on February 16, 2015) for the query

---

[1] A snippet of a document D is treated as a summary of D.

© Springer International Publishing Switzerland 2015
O. Gervasi et al. (Eds.): ICCSA 2015, Part I, LNCS 9155, pp. 472–488, 2015.
DOI: 10.1007/978-3-319-21404-7_35

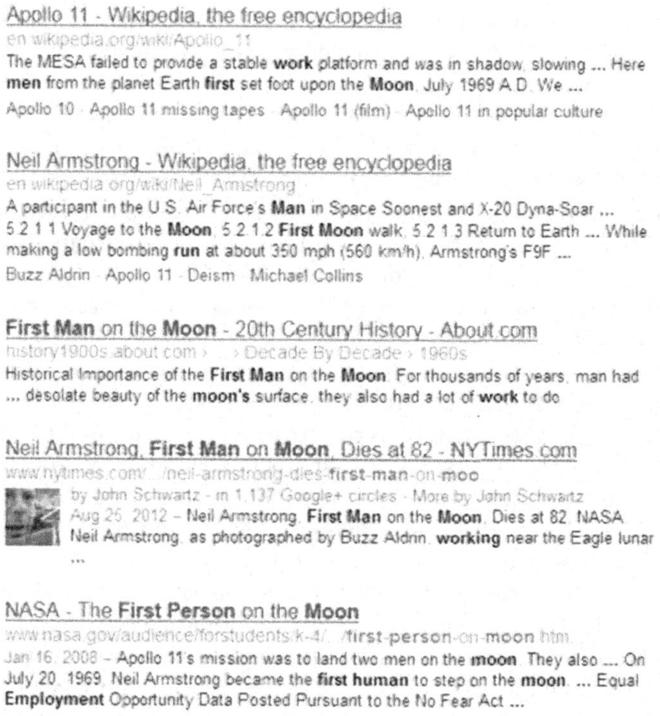

Apollo 11 - Wikipedia, the free encyclopedia
en.wikipedia.org/wiki/Apollo_11
The MESA failed to provide a stable **work** platform and was in shadow, slowing ... Here
men from the planet Earth **first** set foot upon the **Moon**. July 1969 A.D. We ...
Apollo 10 · Apollo 11 missing tapes · Apollo 11 (film) · Apollo 11 in popular culture

Neil Armstrong - Wikipedia, the free encyclopedia
en.wikipedia.org/wiki/Neil_Armstrong
A participant in the U.S. Air Force's **Man** in Space Soonest and X-20 Dyna-Soar ...
5.2.1.1 Voyage to the **Moon**. 5.2.1.2 **First Moon** walk. 5.2.1.3 Return to Earth ... While
making a low bombing **run** at about 350 mph (560 km/h). Armstrong's F9F ...
Buzz Aldrin · Apollo 11 · Deism · Michael Collins

First Man on the Moon - 20th Century History - About.com
history1900s.about.com › ... › Decade By Decade › 1960s
Historical Importance of the **First Man** on the **Moon**. For thousands of years, man had
... desolate beauty of the **moon's** surface, they also had a lot of **work** to do

Neil Armstrong, **First Man** on Moon, Dies at 82 - NYTimes.com
www.nytimes.com/.../neil-armstrong-dies-first-man-on-moo
by John Schwartz - in 1,137 Google+ circles - More by John Schwartz
Aug 25, 2012 - Neil Armstrong, **First Man** on the **Moon**, Dies at 82. NASA.
Neil Armstrong, as photographed by Buzz Aldrin, **working** near the Eagle lunar
...

NASA - The **First Person** on the **Moon**
www.nasa.gov/audience/forstudents/k-4/.../first-person-on-moon.htm
Jan 16, 2008 - Apollo 11's mission was to land two men on the **moon**. They also .... On
July 20, 1969, Neil Armstrong became the **first human** to step on the **moon**. ... Equal
Employment Opportunity Data Posted Pursuant to the No Fear Act ...

Fig. 1. Top-5 results retrieved by Google for the query "First man to walk on the moon" on February 16, 2015

"First man to walk on the moon" as shown in Figure 1. The titles and snippets of the results show the same information, i.e., Neil Armstrong was the first man to walk on the moon. If the user who submitted the query was interested in specific information, such as the shuttle used during the mission, astronauts that accompanied Neil Armstrong, length of the journey, etc., the user must scan through the retrieved documents one by one, since there is no indication in which retrieved documents additional information might be included. A solution to this problem is to cluster retrieved documents based on their *subject areas* and capture the subject area of documents in each cluster using a *label*, which allow the users to quickly draw a conclusion on the cluster that includes materials satisfying their information needs. Labels created for clusters of documents offer ordinary web users, as well as professional information consumers and researchers, a mechanism to quickly familiarize themselves with a very large volume of retrieved information segregated according to their contents into various clusters.

To create clusters of retrieved documents and assign cluster labels, we introduce *QClus*, a *query-based multi-document cluster and labeler*, which enhances web search. *QClus* allows novice, as well as expert, users to post a query Q and quickly locate the desired information, if they exist, captured in a cluster of documents. *QClus* (i) queries three major web search engines, Google, Bing, and Yahoo!, using Q, (ii) generates meaningful cluster labels, and (iii) assigns retrieved documents (on the same topic) to labeled clusters. Clusters of retrieved documents and their labels are useful for quickly identifying and accessing relevant information, especially when Q is *short* and *ambiguous* in meaning. Each cluster label of a set of retrieved documents achieves conciseness in capturing the major content of the documents. For example, if the search query is "tiger," the retrieved documents can be various in terms of their contents, which might discuss the Mac OS, a fish, the golf player Tiger Woods, etc. A cluster label distinguishes the content of the clustered documents from other cluster labels on different subject areas.

We have evaluated the quality of *QClus*-generated clusters using the 20 News Groups and DMOZ (DUC, respectively) datasets and compared *QClus*-generated clusters against (i) those created by existing state-of-the-art query-based multi-document clusters, and (ii) snippets generated by Google in terms of the time required to locate desired information. Experimental results show that *QClus* is *highly effective* and *efficient* in generating cohesive clusters and concise cluster labels of documents retrieved for a query.

QClus is a contribution to the Web and information retrieval community, since it (i) provides the user with an unbiased information source on a particular topic, since the creation of each cluster and label is fully automated, without any subjective human intervention, (ii) enhances web search by helping the user quickly locate desired information, if they exist, and (iii) establishes, as a by-product, a new source of information for answering questions, since a cluster which contains closely-related information in documents is likely to contain the answer to a user's question.

QClus is unique, since it generates cluster labels, each of which identifies the common content of its corresponding collection of retrieved documents, instead of titles created by existing web search engines which may not be good indicators of the contents of the corresponding documents. Furthermore, *QClus* does not require training/learning in clustering documents nor generating labels.

We present our work as follows. In Section 2, we discuss existing web clustering methods. In Section 3, we introduce the clustering and labeling approaches of *QClus*. In Section 4, we present the performance evaluation of *QClus*. In Section 5, we give a conclusion.

2   Related Work

Clustering of web search results was first introduced in the Scatter-Gather system [8]. Hereafter, a variety of web clustering paradigms have been proposed, which include graph-theory-based [19] and concept-lattices-based [2] clustering approaches. Suffix Tree Clustering (STC) [21], which uses recurring phrases to determine the similarity of web documents for clustering purpose, was enhanced by minimizing the size of a suffix tree [3]. Ferragina and Guli [5] apply a strategy similar to frequent phrases which extracts meaningful labels and builds cluster labels using a bottom-up hierarchy construction process. Their web search results, however, contain *non-relevant* terms in their navigational aids. *QClus*, which uses the *titles* and *snippets* of documents retrieved by web search engines in clustering, avoid *non-relevant* terms found in documents.

Xide et al. [19] expand the set of retrieved snippets by including the in- and out-linked pages to improve clustering precision. Since web search engines do not provide cheap access to the web graph, the efficiency of the link-retrieval approach is an issue. Zeng et al. [22] introduce a system that extracts (contiguous) sentences of various lengths via regression for clustering. Regression, however, requires a training phase, which is non-feasible to the heterogeneous Web. *QClus* performs clustering on-the-fly without training.

Kang and Kim [11] attempt to solve the ambiguous short query problem by classifying queries into one of the three different tasks: topic relevance, homepage finding, or a service. Depending on the task, a different information emphasis is presented to the user. Although their method can identify a user query into its task, it fails to solve the multiple interpretation problem of a given query, which is still a main issue when dealing with ambiguous short queries.

Yippy (search.yippy.com) and Carrot2 (search.carrot2.org) are search-result cluster engines spun off from their respective academic projects. Yippy imposes censorship on results retrieved for "inappropriate" search terms. Since search keywords can carry multiple meanings, an effect of filtering can exclude reprehensible material that are relevant to users' queries. Carrot2 employs STC for clustering. A critique on Carrot2 is that its clusters are less scannable/granular than the clusters created by Clusty, the predecessor of Yippy.

3   The Clustering Approach

According to [16], the *web search problem* is defined as finding the set of documents on the Web relevant to a given user query. However, these days the problem is more complicated than simply finding relevant documents due to (i) information overload caused by the increasing information available on the Web, which has become a highly dynamic collection of documents and (ii) the large number of potential documents relevant to a user query. To solve the problem, a web search engine, such as Google, Bing, and Yahoo!, retrieves a ranked list of documents in response to a user query; the *higher* a retrieved document is on the list, the *higher* its relevance to the query is. Many algorithms for computing the degree of relevance of retrieved documents have been proposed [1,17]; however, these algorithms work well only with queries that are *precise* and *narrow* [13]. If the query is too *general* or *broad*, it is difficult, if not impossible, for a search engine to identify precisely the specific set of documents that satisfy the user's information needs. The side effect is that the user is required to sort through a list of documents to locate the relevant ones that (s)he is particularly interested in. Such a search has been identified as a *low precision search* [21]. The low precision search problem is caused by the fact that web search queries are typically *short* and often *ambiguous* [18][2].

Clustering, which is an elegant approach to solve the low precision search problem, first identifies cluster labels and then assigns similar search results to different categories according to their subject areas. Providing clustered search results would be much more appealing and useful to a user than a ranked list of retrieved documents that intermix with results based on different interpretations of a given query.

Upon receiving a user query Q, *QClus* (i) gathers the top 33 documents retrieved by each of the three major web search engines, Google, Bing, and Yahoo! (with *duplicates eliminated*) in response to Q, since a collection of 100 documents is an *ideal* set for generating clusters [4]. *QClus* retrieves the title and snippet of each one of the 100 documents. During the clustering process, *QClus* (i) generates a set of cluster labels, which are non-stopwords in the titles and snippets of retrieved documents (presented in Section 3.1), (ii) ranks a subset of the labels generated in Step (i) using a set of features introduced in Section 3.2, (iii) applies the Vector Space Model (VSM) to calculate the similarity between each label and each retrieved document (as discussed in Section 3.3), and (iv) assigns the top-N ($N \geq 1$) retrieved documents, ranked by VSM using the cosine similarity measure between documents and the labels, to their corresponding cluster identified by a label.

3.1   Generating Cluster Labels

In a typical text clustering algorithm, cluster labeling follows the document clustering step and the entire document text is used for clustering. Web-based clustering is different as it involves different challenges. Web-based clustering must be (i) *fast* and document titles and snippets can be processed faster than using the entire documents in extracting and ranking cluster labels, (ii) *flexible* and *fully automated* as web contents change constantly and user feedback must be avoided, and (iii) *user-oriented*, i.e., eases the process of finding the required information by producing cluster labels that are *meaningful* and *informative* to the user, which is not a concern in document clustering due to different types of users—the users of traditional clustering tools are typically skilled professionals, whereas users of web-based clustering are often ordinary users who are less tolerant to errors.

QClus generates *cluster labels* prior to creating clusters using the titles and snippets of the retrieved documents utilizing the *suffix tree* structure[3]. A *suffix tree* is a data structure that retains all the n-grams in a string S, which is the concatenation of all the sentences in the titles and snippets of the retrieved documents in our case. Words in S can be inserted into the suffix tree incrementally

---

[2] According to a survey on iProspect (iprospect.com), majority of the queries submitted to web search engines are general 1-3 words in length.

[3] Although the titles and snippets are inadequate in capturing the contents of their corresponding documents, they provide sufficient information for document clustering and labeling.

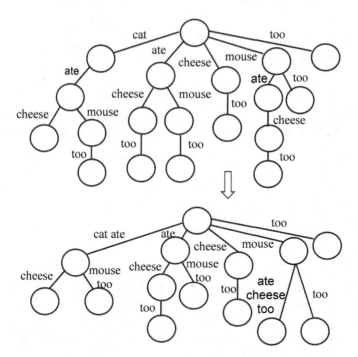

Fig. 2. An initial and the final suffix tree of strings "cat ate cheese", "cat ate mouse too", and "mouse ate cheese too"

in time *linear* to the number of words in S. The root node of the initial suffix tree T of S has a number of child nodes, each of which is labeled by the first word in each suffix F of each sentence in S. Subsequent words in F form individual nodes labeled by the words created in a top-down manner, which yield a path rooted at the corresponding child node. Hereafter, T is modified so that nodes between A and B are *combined* if each node on a path P from A to B (including A) has only a single child and none of the labeled nodes (excluding B) on P is the last word of any sentence in S. In traversing the suffix tree, the nodes are concatenated on a path from the root to an internal node. A node N in T is a *candidate cluster label* if N is a *parent* node, except the dummy root node of T. (See Figure 2 for the initial and final suffix trees of the three different strings as shown in [20].)

As the number of candidate cluster labels can be large and not all of them capture the subject area shared by a non-empty subset of retrieved documents, candidate cluster labels, which are non-numerical keywords, are *selected* if they do not (i) *cross* sentence boundaries in S, since sentence markers indicate a topical shift, (ii) *begin or end* with a stopword, since stripping leading and trailing stopwords from a phrase is likely to increase its readability [21], and (iii) *end* in apostrophe-s. Moreover, a selected label is *complete*, i.e., is not a substring of

Query: Tiger   Candidate Cluster Labels   :
Document 1: Tiger OS reviews. Review center   tiger,
Document 2: Tiger reviews   review,
Document 3: View Panthera tigris images   Panthera tigris,
Document 4: Panthera tigris   tigris

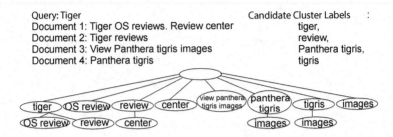

Fig. 3. Candidate cluster labels generated for the query "Tiger"

another label. Figure 3 shows sample candidate cluster labels extracted from a suffix tree.

3.2   Ranking Cluster Labels

Selected cluster labels differ in terms of the number of retrieved documents assigned to the (cluster) labels that capture their topics due to the different degrees of popularity of various topics. *QClus ranks* the selected cluster labels using the measures below, where L is a selected cluster label and C is the set of documents retrieved in response to a query.

1. The *Number of Title Words* (NTW) in L, which is the number of keywords in the *titles* of documents in C that are also in L. The *more frequently* these keywords occur, the *higher* the ranking score of L is, which reflects its higher possibility as a potential *subject area* in C.
2. The *Frequency of Label* (FoL) L, which is the *frequency of occurrence* of L in the *titles* and *snippets* of documents in C, which represents the *content* of the documents to a certain degree.
3. *The label stability* of L, which measures the *mutual information*, denoted MI, of L. The *higher* the MI of L is, the *more dependent* L is as a cluster label. Assume that $L = $ "$c_1 c_2 \ldots c_n$", where c_i ($1 \leq i \leq n$) is a (stop)word in L, the *stability* of L is defined in [23] as

$$MI(L) = \frac{f(L)}{f(L_L) + f(L_R) - f(L)} \qquad (1)$$

where $L_L = $ "$c_1 \ldots c_{n-1}$", $L_R = $ "$c_2 \ldots c_n$", and $f(L_L)$, $f(L_R)$, and $f(L)$ are the frequencies of occurrence of L_L, L_R, and L, respectively in titles and snippets in C.

4. The *Label Significance* (LS) of L, which indicates the *significant factor* of a cluster label. The *longer* a cluster label is, the *more significance* it is, since longer cluster labels are more *informative* of the subject area covered in the corresponding subset of documents than shorter ones. The *significance* of L is defined in [23] as

$$LS(L) = FoL(L) \times g(|L|) \qquad (2)$$

where $|L|$ is the number of (stop)words in L and

$$g(x) = \begin{cases} 0 & \text{if } x = 1 \\ log_2 \ x & \text{if } 2 \le x \le 8 \\ 3 & \text{if } x > 8. \end{cases}$$

$QClus$ computes a *ranking* score for L, denoted *LabWeight* (L), which reflects the *significance* of L in capturing a particular *subject area* of documents in C.

$LabWeight(L) =$

$$\frac{NTW(L) + FoL(L) + MI(L) + LS(L)}{1 - Min\{NTW(L), FoL(L), MI(L), LS(L)\}} \tag{3}$$

which is the Stanford Certainty Factor [14] on NTW, FoL, MI, and LS of L. Since NTW, FoL, MI, and LS are in different scales, $QClus$ normalizes them to the same range using a logarithmic scale before computing *LabWeight*.

Figure 4 shows the top-10 ranked labels created from the retrieved documents in response to the query "Libya".

3.3   Clustering Retrieved Documents

VSM, which compares textual data by using algebraic vectors in a multidimensional space, is adopted by $QClus$ for assigning retrieved documents to their clusters identified by the corresponding labels. The linear algebra operations calculate the similarities among a given set of documents using VSM, which are effective and efficient. $QClus$ considers the retrieved document title, document snippet, and the cluster labels, such that each cluster label is treated as a separate document during the comparison process, and the degree of similarity between each label and the retrieved set of documents is computed. Every unique keyword in the retrieved documents to be analyzed yields a dimension in the VSM, and each retrieved document is represented as a vector spanning all these dimensions. For example, if vector v represents document Doc_j in a k-dimensional space, then component t of vector v, where $t \in 1..k$, denotes the *degree of similarity* between document Doc_j and t in (the document representing) the label. The degree of similarity between retrieved documents and labels can be captured in a $k \times d$ matrix, known as the *term-document matrix*, where k is the number of unique terms in all the documents and d is the number of documents in the collection of 99 retrieved documents, in addition to the chosen cluster labels. Element a_{i_j} of the term-document matrix is the *similarity value* between term i of the label and document j of the set of retrieved documents. VSM calculates the *similarity* between label L and a document D in C using the well-known term (i.e., word) frequency (TF) and inverse document frequency (IDF) measures.

After selecting and ranking cluster labels (as discussed in Sections 3.1 and 3.2), $QClus$ determines and assigns the top-N (≥ 1) retrieved documents in a

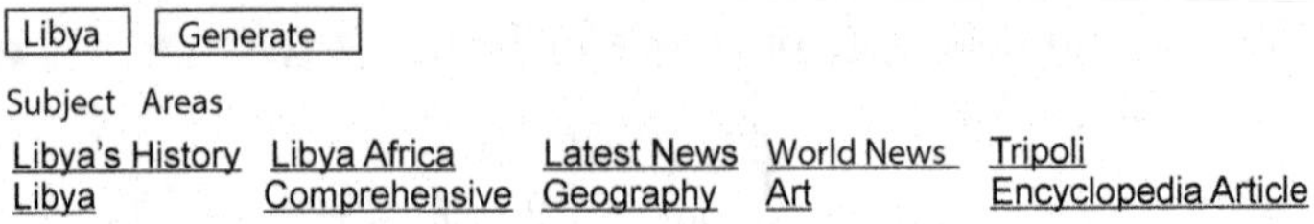

| Libya | | Generate |

Subject Areas

Libya's History   Libya Africa   Latest News   World News   Tripoli
Libya   Comprehensive   Geography   Art   Encyclopedia Article

Fig. 4. The top-10 ranked cluster labels for the query "Libya" ordered from left-to-right

| World News |

Libya | World news | guardian.co.uk
Latest news and comment on Libya from guardian.co.uk

The New York Times - Breaking News, World News & Multimedia
Find breaking news, multimedia, reviews & opinion on Washington, business.
sports, movies, travel, books, jobs, education, real estate, cars & more.

Libya News - Protests and Revolt (2011)
World news about Libya. Breaking news and archival information about its
people, politics and economy from The New York Times.

Libya News From AOL.News
Libya news. See the latest headlines on libya from AOL News. Videos, photos
& libya pictures on all of the latest news from the US & around the world.

Libya - Bing News
News from world, national, and local news sources, organized to give you in-depth
news coverage of sports, entertainment, business, politics, weather, and more.

Libya - Africa - World - News
Libya. Twitter. With Wikio you can create personal pages by selecting the info
that ... Residents of Libya's de facto rebel capital of Benghazi joined in a ...

Libya News - Breaking World Libya News - The New York Times
World news about Libya. Breaking news and archival information about its people,
politics and economy from The New York Times.

More Canadians evacuated from Libya - World - CBC News
Ottawa confirms a military flight was able to pluck more Canadians and other
foreign nationals from Libya on Saturday.

Fig. 5. Documents in the cluster labeled "World News" created and retrieved by $QClus$

collection C to a cluster labeled L. The top-N documents are ranked by VSM
with respect to L such that N is the *ratio* of the *LabWeight* of L to the sum of
the *LabWeights* of all the selected cluster labels rounded to the nearest whole
number as defined below.

$$ROUND \left(\frac{LabWeight(L)}{\sum_{M=1}^{Number\ of\ Selected\ Labels} LabWeight(M)} \times |C| \right) \quad (4)$$

The N documents in C with the *highest similarity scores* with respect to L
are assigned to the cluster labeled L such that the assigned documents address
the same or very similar topic identified by L. The assignment method is effective, since it (i) allows *overlapping* of documents among clusters, which is desirable, since some documents may include contents of multiple subject areas

Table 1. Datasets used for evaluating the effectiveness and efficiency of $QClus$'s clusters

Dataset	CS_1	CS_2	CS_3
Number of Articles/Documents	20,000	20,000	19,997
Number of Topics	20	20	20
Data source	DMOZ	DMOZ	20NG

represented by their corresponding cluster labels of different clusters, and (ii) assigns *highly-ranked* cluster labels *more* documents, since the distribution of different subject areas covered in C is not uniform, and highly-ranked labels capture the corresponding subject areas of more documents than others.

3.4 User Interface

The user interface of $QClus$ is very simple, following the design of the user interface of popular web search engines being used these days. The interface consists of a query text *box* and a *search button*. After a search for a query Q is processed by $QClus$, a list of ranked clusters (represented by their corresponding labels) for Q is displayed. Clicking on a cluster label L shows the *titles* and *snippets* of the documents contained in the cluster of L.

Example 1 Figure 4 shows the interface of $QClus$ with the user query "Libya," along with the top-10 ranked labels created by $QClus$ for the query[4], whereas Figure 5 displays all the documents in the cluster labeled "World News", which are shown after the cluster label in Figure 4 is clicked. □

4 Experimental Results

In this section, we assess the overall performance of $QClus$. We first describe the datasets (in Section 4.1) used for the empirical study and then detail the statistical approach (in Section 4.2) that determines the ideal number of appraisers and queries required for evaluating the quality of $QClus$-generated cluster labels. The performance analysis, which is based on the evaluation measure (introduced in Section 4.2) and the assessment of the appraisers, addresses the ability of $QClus$ to generate useful cluster labels (in Section 4.3), in addition to the quality of generated clusters (in Section 4.3). We have also compared the anticipated time to locate desired information on $QClus$ and Google and measured the query processing time of $QClus$ in clustering retrieved documents (in Section 4.4).

---

[4] The labels in the actual interface are listed vertically, instead of horizontally as arranged in the figure to save space.

4.1   The Datasets

Results generated by a clustering algorithm are often evaluated against an *"ideal"* set of clusters. This *distortion measure* compares the difference between a set of generated clusters (to be examined) and a *ground truth set*, which are created on the same set of documents. We evaluated the clustering results of *QClus* using three datasets, as shown in Table 1, which were constructed from various data sources widely used for document clustering evaluation. The first dataset, called CS_3, is the 20 N(ews)G(roup) (people.csail.mit.edu/jrennie/20Newsgroups/) dataset with articles retrieved from the Usenet newsgroup collection, which are clustered into 20 different categories. DMOZ (dmoz.org), on the other hand, is the largest, most comprehensive human-edited directory of the Web, which organizes web pages into their corresponding categories. We randomly selected 20 categories from DMOZ with a total of 20,000 documents to create the multi-topic dataset CS_1 and another 20 categories from DMOZ with another set of 20,000 documents to obtain another dataset CS_2. Categories and documents in the three datasets are disjoint.

4.2   Appraisers and Test Queries Used for the Controlled Experiments

To verify the usefulness of *QClus*-generated cluster labels, we determine the ideal number of appraisers (in Section 4.2) and test queries (in Section 4.2) used in our study to ensure that the evaluation is reliable and objective.

The Number of Appraisers. In statistics, two types of errors, Types I and II, are defined [10]. Type I errors, also known as α errors or *false positives*, are the *mistakes* of *rejecting* a null hypothesis when it is true, whereas Type II errors, also known as β errors or *false negatives*, are the *mistakes* of *accepting* a null hypothesis when it is false. We apply the formula in [10] below to determine the ideal number of appraisers, n, which is dictated by the probabilities of occurrence of Types I and II errors, to evaluate *QClus*-created cluster labels.

$$n = \frac{(Z_{\frac{\alpha}{2}} + Z_{\beta})^2 \times 2\sigma^2}{\triangle^2} + \frac{(Z_{\frac{\alpha}{2}})^2}{2} \tag{5}$$

where (i) $\triangle$ is the *minimal expected difference* to compare *QClus* with Google, which is set to 1 in our study as we expect *QClus* to perform as good as Google in terms of generating high-quality cluster labels in comparison with document titles and snippets created by Google, respectively; (ii) σ^2 is the *variance*[5] of data, i.e., generated cluster labels, which is 6.00 in our study; (iii) α (β, respectively) denotes the probability of making a Type I (II, respectively) error, which is 0.05 (0.20, respectively) in our study, and 1 - β determines the probability of a

---

[5] *Variance* is widely used in statistics, along with standard deviation (which is the square root of the variance), to measure the average dispersion of the scores in a distribution.

false null hypothesis that is correctly rejected; and (iv) Z is the value assigned to the standard *normal distribution* of cluster labels. Based on the standard normal distribution, when $\alpha = 0.05$, $Z_{\frac{\alpha}{2}} = 1.96$, whereas when $\beta = 0.20$, $Z_\beta = 0.84$.

To determine the values of $\triangle$ and σ^2 for evaluating *QClus*-created cluster labels, we conducted an experiment using a randomly sampled 100 test queries extracted from the *AOL log*[6]. We chose only 100 queries, since the *minimal expected difference* and *variance*, which are computed on a *simple random sample*, do not change with a larger sample set of queries. σ^2 is computed by averaging the sum of the square difference between the mean and the actual number of cluster labels created for each one of the 100 test queries. We obtained 6.00, which is the value of σ^2 for cluster labels.

The values of α and β are set to be 0.05 and 0.20, respectively, which imply that we have 95% *confidence* on the correctness of our analysis and that the *power* (i.e., probability of avoiding false negatives/positives) of our statistical study is 80%. According to [12], 0.05 is the commonly-used value for α, whereas 0.80 is a conventional value for 1 - β, and a test with $\beta = 0.20$ is considered to be statistically powerful.

Based on the values assigned to the variables in Equation 5, the ideal number of appraisers for our study is

$$n(Cluster\ Labels) = \frac{(1.96+0.84)^2 \times 2 \times 6}{1^2} + \frac{1.96^2}{2} \cong 96$$

Note that the values of α, β, σ^2, and $\triangle$ directly influence the size of n. Furthermore, the results collected from the $n = 96$ appraisers in the study are expected to be comparable with the results that are obtained by the actual population [10], i.e., web users who query web search engines.

The Number of Test Queries. In determining the ideal number of test queries to be included in the controlled experiments, we rely on two different variables: (i) the *average attention span* of an adult and (ii) the *average number of search queries* that a person often creates in one session when using a web search engine. As mentioned in [15], the average attention span of an adult is between twenty to thirty minutes. Furthermore, Jansen et al. [9], who have evaluated web users' behavior especially on (i) the amount of time web users spend on a web search engine, (ii) the average size of users' queries, and (iii) the average number of queries submitted by a user, estimate that the average number of queries created by each user on a web search engine in one session is approximately 2.8. Based on these studies, each appraiser was asked to evaluate *QClus* using *three* queries, since evaluating the cluster labels on the retrieved results of each one of the three queries takes approximately *thirty* minutes, which falls in the time span of an adult. We randomly selected *288* (= 96 × 3) queries for evaluating *QClus*-created *cluster labels*.

---

[6] The logs of AOL (gregsadetsky.com/aol-data/) include 50 million queries created by millions of AOL users over a three months period between 03/01/06 and 05/31/06, and the AOL logs are publicly available.

The Quality Measure of *QClus*-generated Labels. We developed various applications on Facebook for its appraisers to evaluate *QClus* on the *usefulness* of *QClus*-created cluster labels. Facebook appraisers were used, since Facebook is a social network with users diverse in genders, ages, and cultures who can provide unbiased evaluations.

For each set of cluster labels created by *QClus* in response to each one of the 288 randomly chosen test queries Q, the designated appraisers are required to examine and indicate the *usefulness* (*useless*, respectively) of the cluster labels for Q. A cluster label is *useful* if it includes *concise* and *meaningful* keywords with respect to Q, whereas a *useless* label either is *ambiguous* or includes *senseless* description.

Given a set of cluster labels CL created by *QClus* on Q, the quality of CL is measured as the *portion* of labels in CL, denoted *Total_Cls*, that the appraisers judged as *Useful*.

$$Qual_{Val} = \frac{Useful}{Total\_Cls}. \tag{6}$$

The *higher* $Qual_{Val}$ is, the *more useful* the corresponding set of *QClus*-created labels are.

4.3   Performance Evaluation of *QClus*

In this section, we present the experimental results that verify the quantify of *QClus*-generated *clusters* and *labels*.

Evaluation of Cluster Labels. We have collected the evaluations provided by the 96 Facebook appraisers who examined and assessed the usefulness of *QClus*-created cluster labels for the 288 test queries (as discussed in Section 4.2). Based on the assessments of the appraisers, *92.2%* of *QClus*-created cluster labels were considered *useful*. The percentages of cluster labels treated as *useful* by the 96 appraisers are almost always in the upper eightieth and lower ninetieth percentile. The few cases where the average percentages are below 80 are due to the *low quality* of retrieved document titles and snippets from where the labels were constructed by *QClus*. Figure 6 shows the average percentages of useful cluster labels determined by 20 out of the 96 Facebook appraisers involved in the evaluation.

Evaluation of *QClus*-Created Clusters. We have also evaluated the effectiveness and efficiency of *QClus*'s clustering approach using the three datasets CS_1, CS_2, and CS_3 defined in Section 4.1. *Efficiency* is determined by the *processing time* required to cluster each dataset, whereas *effectiveness* is measured using the well-known *F-measure*, which is defined as

$$F - measure = \frac{2}{\frac{1}{Precision} + \frac{1}{Recall}} \tag{7}$$

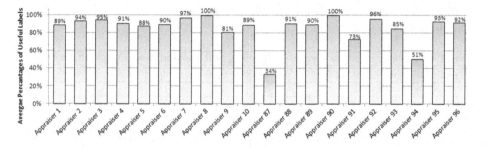

Fig. 6. Average percentages of useful *QClus*-created cluster labels determined by 20 (out of the 96) Facebook appraisers

where *Precision* (*Recall*, respectively) is the *ratio* of the number of common documents in a *generated* cluster C and the corresponding *ideal* cluster I to the total number of documents in C (I, respectively). The *ratio* is *averaged* over generated clusters.

The *F-measure* and *processing time* are compared against the ones achieved by Armil [6] and WhatsOnWeb [7] using the same three datasets. WhatsOn-Web is one of the few web clustering engines based on graph theory, whereas Armil clusters documents retrieved in response to a search query by mapping document snippets into a vector space. We chose the two clustering methods for comparisons, since (i) unlike other clustering approaches, their algorithms are publicly available and (ii) they outperform the Vivisimo clustering engine, which is considered an industrial standard on the evaluation of clustering quality and user satisfaction. The *average F-measure* achieved by each of the three clustering approaches are **0.90** for *QClus*, **0.81** for Armil, and **0.84** for WhatsOnWeb, whereas the *average processing time* for the clustering method of *QClus* (Armil and WhatsOnWeb, respectively) on the three datasets is 10 (14 and 21, respectively) seconds. The results demonstrate that *QClus* outperforms the other two clustering methods.

4.4   *QClus* Versus Google

We created another two Facebook application, *App₁* and *App₂*, which includes a number of performance evaluation questions for another group of Facebook appraisers, other than the 96 appraisers mentioned in Section 4.2. The application was posted under Facebook for the appraisers to provide their feedbacks.

For *App₁*, the application includes two pages in a panel, the *left* page displayed the (traditional) top-10 results generated by Google for a query Q arbitrarily created by an appraiser, whereas the *right* one is the *QClus*-created cluster labels on the documents retrieved by *QClus* for Q.

The purpose of this study is to analyze whether *QClus*-generated cluster labels are really useful to its users for browsing search results and enriching their search experiences. After submitting a query and examine the results displayed

Table 2. Facebook appraisers' responses to different tasks posted as queries under Google and *QClus*

Tasks (Posted as Queries on Google/*QClus*)	No. of Responses	Prefer Google	Prefer *QClus*
Research a Topic	9	3	6
Find News on an Event	11	3	8
Find Answers to Questions	5	3	2
Find Information on an Item	17	6	11
Find Tools/Software	8	7	1
Navigate to a Site	8	8	0

on each (left/right) page, an appraiser responded to each of the following posted questions:

1. "On which system did you spend less time locating the intended information?"

2. "Did the system on the left offer vital information not contained in the system on the right?"

For the first question, the responses are 12% for *Google*, 6% for *QClus*, and 82% for the same, whereas for the 2nd question, 27% said 'Yes' and 73% said 'No.' Based on the responses, we conclude that the appraisers have found *QClus*-generated cluster labels to be *useful* and *informative* compared with the traditional results retrieved by Google. Altogether, there are 288 responses to the study.

For *App₂*, the application requires the involved appraisers to (i) first identify a task that each one often performs on a search engine, (ii) create a query that represents the task, (iii) submit the query to both systems (Google and *QClus*). Hereafter, the appraisers were asked to answer the question,

"Which system helped you perform this task faster?"

The tasks (which were clustered based on their similarity), the number of responses for each type of tasks, and their answers to the question are shown in Table 2. The responses verify that *QClus*-created cluster labels on results of queries for different tasks were highly regarded by Facebook appraisers than the results generated by Google, with the exception of the two tasks, "Find Tools/Software" and "Navigate to a site." The results are anticipated, since *QClus*-created cluster labels include information on products but exclude links to download them, which are provided in the results generated by Google for its users to access, and finding the URL of a website W using the name of W provided by the user is a strength of Google, while *QClus* on W offers no such value. Overall, there are 58 responses to *App₂*.

Although the empirical study of *App₂* shows that *QClus* cannot handle navigation-type web queries, an online report published by Wordtracker (freekeywords.wordtracker .com/top-keywords/long-term.html) shows that out of the top 500 most popular query keywords created by web search engine users, only 51 of them include keywords explicitly specify a website (such as facebook.com, amazon.com, and ebay.com). The report illustrates that the percentage of navigation-typed web queries is not a dominating type of commonly-used web queries.

Query Processing Time of *QClus*. We have measured the *processing time* of the clustering and labeling module of *QClus* using the 288 queries from the AOL query log. The processing time required to generate document *clusters* and their corresponding *labels* for a query is on an average of *0.23 seconds*.

The Implementation of *QClus*. *QClus* is implemented on an Intel Dual Core PC with dual 2.66 GHz processors, 3 GB RAM size, and a hard disk of 300 GB running under the Windows XP operating system.

5 Conclusions

Current web search engines, such as Google, Bing, and Yahoo!, offer users a mean to locate desired information available on the Web. In response to a user query, current web search engines retrieve a list of ranked documents and display each with a title and a snippet to help users quickly identify the document(s) of interest. However, whenever a user query is *ambiguous*, it is very difficult, if not impossible, for a search engine to determine precisely the set of documents that satisfy the user's information need. Moreover, since snippets are created using sentences/phrases in the corresponding retrieved documents in which the keywords in the user query also appear, they (i) may not always capture the main content of the corresponding documents and (ii) are similar in contents and thus are not useful in distinguishing their differences. To enhance web search, we have developed *QClus* which clusters documents retrieved in response to a user query and creates cluster labels to assist its users in identifying results of interest. Experimental results using well-known datasets and Facebook applications show that *QClus* creates high-quality clusters and cluster labels and outperforms existing clustering approaches.

QClus is a significant contribution to the web search community, since it (i) handles the ambiguous problem of a search by creating clusters generated in response to different interpretations of the search, which offer a "road map" for information access, and (ii) assists users to quickly identify information of interest, which is unique by itself.

For future work, we plan to extend *QClus* so that it can process user queries in multiple languages other than English. The extension requires that *QClus* to be equipped with models that recognize natural language encoding schemes and handle internationalization.

References

1. Braschler, M., Schäuble, P.: Multilingual information retrieval based on document alignment techniques. In: Nikolaou, C., Stephanidis, C. (eds.) ECDL 1998. LNCS, vol. 1513, pp. 183–197. Springer, Heidelberg (1998)
2. Chen, L.: Using a New Relational Concept to Improve the Clustering Performance of Search Engines. IPM **47**, 287–299 (2011)
3. Chim, H., Deng, X.: A new suffix tree similarity measure for document clustering. In: WWW, pp. 121–130 (2008)
4. Dunlavy, D., O'Leary, D., Conroy, J., Schlesinger, J.: QCS: A System for Querying, Clustering, and Summarizing Documents. IPM **43**, 1588–1605 (2007)
5. Ferragina, P., Guli, A.: A Personalized Search Engine Based on Web-snippet Hierarchical Clustering. Software-Practice & Experience **38**(2), 189–225 (2008)
6. Geraci, F., Pellegrini, M., Pisati, P., Sebastiani, F.: A scalable algorithm for high quality clustering of web snippets. In: ACM SAC, pp. 1058–1062 (2006)
7. Giacomo, E., Didimo, W., Grilli, L., Liotta, G.: Graph Visualization Techniques for Web Clustering Engines. TVCG **13**(2), 294–304 (2007)
8. Hearst, M., Pedersen, J.: Reexamining the cluster hypothesis: scatter/gather on retrieval results. In: ACM SIGIR, pp. 76–84 (1996)
9. Jansen, B., Spink, A., Saracevic, T.: Real Life, Real Users, and Real Needs: a Study and Analysis of User Queries on the Web. IPM **36**(2), 207–227 (2000)
10. Jones, B., Kenward, M.: Design and Analysis of Cross-Over Trials, 2nd edn. Chapman and Hall (2003)
11. Kang, H., Kim, G.: Query type classification for web document retrieval. In: ACM SIGIR, pp. 64–71 (2003)
12. Kazmier, L.: Schaum's Outline of Business Statistics. McGraw-Hill (2003)
13. Li, H., Sun, C., Wang, K.: Clustering web search results using conceptual grouping. In: ICMLC, pp. 1499–1503 (2009)
14. Luger, G.: Artificial Intelligence: Structures and Strategies for Complex Problem Solving. Addison-Wesley (2008)
15. Rozakis, L.: Test Taking Strategies and Study Skills for the Utterly Confused. McGraw Hill (2002)
16. Selberg, E.: Towards Comprehensive Web Search. PhD thesis, University of Washington (1999)
17. Shekhar, S., Agrawal, R.: An architectural framework of a crawler for retrieving highly relevant web documents by filtering replicated web collections. In: ACE, pp. 29–30 (2010)
18. Shen, D., Pan, R.: Query Enrichment for Web-Query Classification. ACM TOIS **24**(3), 320–352 (2006)
19. Lin, C.X., Yu, Y., Han, J., Liu, B.: Hierarchical web-page clustering via in-page and cross-page link structures. In: Zaki, M.J., Yu, J.X., Ravindran, B., Pudi, V. (eds.) PAKDD 2010. LNCS, vol. 6119, pp. 222–229. Springer, Heidelberg (2010)
20. Zamir, O., Etzioni, O.: Web document clustering: a feasibility demonstration. In: SIGIR, pp. 46–54 (1998)
21. Zamir, O., Etzioni, O.: Grouper: A Dynamic Clustering Interface to Web Search Results. Computer Networks **31**(11–16), 1361–1374 (1999)
22. Zeng, H., He, Q., Chen, Z., Ma, W.: Learning to cluster web search results. In: ACM SIGIR, pp. 210–217 (2004)
23. Zhang, D., Dong, Y.: Semantic, hierarchical, online clustering of web search results. In: Yu, J.X., Lin, X., Lu, H., Zhang, Y. (eds.) APWeb 2004. LNCS, vol. 3007, pp. 69–78. Springer, Heidelberg (2004)

A Comparative Study of Different Color Space Models Using FCM-Based Automatic GrabCut for Image Segmentation

Dina Khattab[1]($\boxtimes$), Hala Mousher Ebied[1],
Ashraf Saad. Hussein[2], and Mohamed Fahmy Tolba[1]

[1] Faculty of Computer and Information Sciences, Ain Shams University, 11566, Cairo, Egypt
{dina.reda.khattab,fahmytolba}@gmail.com,
hala_mousher@hotmail.com
[2] Faculty of Computer Studies, Arab Open University, Headquarters 13033, Kuwait, Kuwait
ashrafh@acm.org

Abstract. GrabCut is one of the powerful color image segmentation techniques. One main disadvantage of GrabCut is the need for initial user interaction to initialize the segmentation process which classifies it as a semi-automatic technique. The paper presents the use of Fuzzy C-means clustering as a replacement of the user interaction for the GrabCut automation. Several researchers concluded that no single color space model can produce the best results of every image segmentation problem. This paper presents a comparative study of different color space models using automatic GrabCut for the problem of color image segmentation. The comparative study includes the test of five color space models; RGB, HSV, XYZ, YUV and CMY. A dataset of different 30 images are used for evaluation. Experimental results show that the YUV color space is the one generating the best segmentation accuracy for the used dataset of images.

Keywords: Color image segmentation · Automatic GrabCut · FCM clustering · Color space models

1   Introduction

The process of segmentation refers to partitioning a digital image into multiple segments. During this process, each pixel in the image is assigned to a label where pixels with the same visual characteristics share the same label [1]. Segmentation aims to arrange the image into regions that are simpler, meaningful and easier to analyze [2]. These regions may correspond to individual surfaces, objects or natural parts of objects [3]. Image segmentation is usually used as a pre-processing step in many applications such as; object recognition, scene analysis, automatic traffic control systems and medical imaging. Usually, the local information that is incorporated in the image, i.e. color information, edges, boundaries or texture information are used to compute the best segmentation [4].

© Springer International Publishing Switzerland 2015
O. Gervasi et al. (Eds.): ICCSA 2015, Part I, LNCS 9155, pp. 489–501, 2015.
DOI: 10.1007/978-3-319-21404-7_36

The image pixels' colors are considered the main feature for the problem of color image segmentation. It is usually assumed that homogeneous colors in the image correspond to separate clusters and hence meaningful objects in the image. In other words, each class of pixels sharing similar color properties can define a separate cluster. Considering that not all color spaces can provide acceptable results for all kinds of images, many trials [5] have been carried out to define which color space is most suitable for their specific color image segmentation problem.

One of the most powerful color image segmentation techniques is the GrabCut technique [6]. It extends the famous graph cut technique [7] for efficient segmentation of color images which allows it to be advantageous for several applications. One main drawback of GrabCut is being an interactive/semi-automatic technique, i.e. being more appropriate for binary-label segmentation. Binary-label segmentation (i.e. foreground segmentation) is a segmentation class where the image can be segmented into the background and foreground regions only [6]. The need to define a region of interest to be segmented out of the image requires initial user interaction. This initial user intervention is responsible for classifying GrabCut as semi-automatic technique and makes it more subject to error.

The authors in [8] presented a modification of the semi-automatic GrabCut into an automatic one using Self-Organizing Feature Map (SOFM) [9,10] as an unsupervised clustering technique. SOFM was selected as a hard clustering technique to replace the initialization phase of GrabCut and eliminate the user interaction. In this work, Fuzzy C-means (FCM) [11,12] is selected as a Soft/Fuzzy clustering for automatic GrabCut initialization. The segmentation of color images is tested using FCM for different classical color spaces; RGB, CMY, XYZ and YUV; to select the best color space for the considered kind of images. Experiments on a dataset of 30 images are carried out to test the modified automatic GrabCut for binary-label segmentation.

The paper is organized as follows; section 2 reviews the related work of image segmentation based on different color space models, in addition to the use of GrabCut and FCM clustering for image segmentation. Section 3 explains the different color space models. The proposed automatic GrabCut using FCM clustering is illustrated in Section 4. Experimental results and discussion are presented in Section 5. Finally the conclusion and future work are presented in section 6.

2   Related Work

Several segmentation problems had utilized GrabCut in different applications such as; human body segmentation [13,14,15], video segmentation [16], semantic segmentation [17] and volume segmentation [18]. Yi Hu [15] developed an iterative technique for automatic extraction of the human body from color images. In their implementation, a scanning face detector was used to initialize a tri-map that is dynamically updated using the iterated GrabCut technique. One defect was having the research constrained to human poses with frontal side faces.

In video sequences, a fully automatic Spatio-Temporal GrabCut human segmentation methodology was developed by Hernández et al. [14]. In their work, face detection and a skin color models were assigned to generate a set of seeds that were used to initialize the GrabCut algorithm. Another application to human segmentation developed by Gulshan et al. [13] utilized the local color model based GrabCut to automatically segment humans from cluttered images. According to them, segmentation masks were learned from sparsely coded local HOG descriptors using trained linear classifiers. Afterward, the GrabCut local color model was used to refine a crude segmentation of the human figure.

Corrigan et al. [16] had applied a more robust segmentation technique in the field of video segmentation. In order to include temporal information in the segmentation optimization process, they extended the Gaussian Mixture Model (GMM) of the GrabCut algorithm so that the color space was complemented with the derivative in time of pixel intensities. GrabCut was integrated into a semantic segmentation framework by Göring et al. [17] by labeling objects in a given image. In the field of 3D segmentation, a fully parallelized scheme using GrabCut had been adapted to run on GPU by Ramírez et al. [18]. Advantages of the scheme for the case of volume meshes included producing efficient segmentation results, in addition to reducing the computational time.

Fuzzy clustering is a natural type of clustering since no exact division is possible in real life due to the presence of noise. It is the process of assigning membership levels and then using these levels to assign data elements to one or more clusters or classes in the image/data set. In Soft/Fuzzy clustering, data elements can belong to more than one cluster with a degree of some membership value [19]. The Fuzzy C-means (FCM) algorithm [11,12] is one of the most popular fuzzy clustering methods widely used in various tasks of pattern recognition, data mining, image processing and gene expression data recognition, etc.. Various authors [20, 21, 22, 23] have used FCM clustering in recently proposed image segmentation techniques in the literature. Beevi and Sathik [23] had developed a robust segmentation technique that exploited a histogram based FCM algorithm for the segmentation of medical images. Their approach converged more quickly than the conventional FCM and attained reliable segmentation accuracy apart from noise levels [19].

Krinidis and Chatzis [20] had developed a Fuzzy Logic Information C-Means Clustering (FLICM) algorithm with a new factor in the objective function of FCM. The algorithm proved to be more robust because of the new factor incorporated in the objective function which was noise insensitive and preserved image details. Kannan et al. [22] proposed a Novel Fuzzy Clustering C-Means Algorithm (NFCM) where a center knowledge method was presented to reduce the running time of the algorithm. The advantage of NFCM was that it can be applied at an early phase of automated data analysis and was found to deal effectively with image intensity inhomogeneities and noise present in the image [19]. Beevi et al. [21] proposed an improved Spatial Fuzzy C-Means algorithm (ISFCM), where spatial neighborhood information was incorporated into the standard FCM by a priori probability. The advantage of ISFCM was that it can overcome the noise sensitiveness of the standard

FCM. The incorporation of spatial information in the clustering process made the algorithm robust to noise and blurred edges.

Because no specific color space can be best for every image segmentation problem, several researchers [24,25] worked out on different color spaces to show which is useful for their works. A comparative study of different color spaces has been carried out by Jurio et al. [26] using two similar clustering algorithms in cluster based image segmentation. In order to identify the best color representation, they tested four color spaces; RGB, HSV, CMY and YUV. The best results were obtained using the CMY color space for most cases.

An automatic method to select a specific color space between classical color spaces was proposed by Busin et al. [5]. An evaluation criterion was used for the selection that was based on a spectral color analysis. The best color space was selected based on the quality of its segmentation, i.e. the one that preserved its own specific properties. Chaves-González et al. [27] presented a study of the ten most common color spaces for skin color detection. Based on their study, HSV was the best color space to detect skin in an image. Du and Sun [28] applied another study for the classification of pizza topping. Again among five different color spaces, they proved that the polynomial SVM classifier combined with HSV color space is the best approach. Another best accuracy was achieved using HSV representation by Ruiz-Ruiz et al. [29] in order to achieve real time processing in real farm fields for crop segmentation. They compared between the RGB and HSV models.

3   Color Space Models

The most widely used color space is the RGB color space, where a color point in the space is characterized by three color components of the corresponding pixel which are Red (R), Green (G) and Blue (B). However, since there exists a lot of color spaces, it is useful to classify them into fewer categories with respect to their definitions and properties. Vandenbroucke [30] proposed the classification of the color spaces into the following categories:

1. The primary spaces

Which are based on the theory that assumes it is possible to match any color by mixing an appropriate amount of the three primary colors. The primary spaces are the real RGB, the subtractive CMY and the imaginary XYZ primary spaces. The conversion from RGB to CMY is:

$$C' = 1 - R \qquad C = \min(1, \max(0, C'-K'))$$

$$M' = 1 - G \qquad M = \min(1, \max(0, M'-K')) \qquad (1)$$

$$Y' = 1 - B \qquad Y = \min(1, \max(0, Y'-K'))$$

$$K' = \min(C', M', Y')$$

And the conversion from RGB to XYZ is:

$$\begin{bmatrix} X \\ Y \\ Z \end{bmatrix} = \begin{bmatrix} 0.412453 & 0.357580 & 0.180423 \\ 0.212671 & 0.715160 & 0.072169 \\ 0.019334 & 0.119193 & 0.950227 \end{bmatrix} \begin{bmatrix} R \\ G \\ B \end{bmatrix} \tag{2}$$

2. The luminance-chrominance spaces

Which are composed of one color component that represents the luminance and two color components that represent the chrominance. The YUV color space is an example of the luminance-chrominance spaces. The conversion from RGB to YUV is:

$$\begin{bmatrix} Y \\ U \\ V \end{bmatrix} = \begin{bmatrix} 0.2989 & 0.5866 & 0.1145 \\ -0.147 & -0.289 & 0.436 \\ 0.615 & -0.515 & -0.100 \end{bmatrix} \begin{bmatrix} R \\ G \\ B \end{bmatrix} \tag{3}$$

3. The perceptual spaces

That try to quantify the subjective human color perception by means of three measures; intensity, hue and saturation. The HSV is an example of the perceptual color space. The conversion from RGB to HSV is:

$$H = \begin{cases} 0, & \text{if Max = Min} \\ \left(60° \times \dfrac{G-B}{\text{Max}-\text{Min}} + 360°\right) \text{mod } 360°, & \text{if Max = R} \\ 60° \times \dfrac{B-R}{\text{Max}-\text{Min}} + 120°, & \text{if Max = G} \\ 60° \times \dfrac{R-G}{\text{Max}-\text{Min}} + 240°, & \text{if Max = B} \end{cases} \tag{4}$$

$$S = \begin{cases} 0, & \text{if max = 0} \\ \dfrac{\text{Max}-\text{Min}}{\text{Max}} & \text{otherwise} \end{cases} \tag{5}$$

$$V = \text{Max} \tag{6}$$

4   Automatic GrabCut Using FCM

The original GrabCut technique developed by Rother et al. [6] is considered as one of the state-of-the-art semi-automatic techniques for image segmentation. It is a powerful extension of the graph cut algorithm [7] to segment color images iteratively. The first main step to initialize the segmentation process of the GrabCut algorithm requires a degree of user interaction. This user interaction is implemented by simply dragging a rectangle around the desired object to be segmented as shown in Fig. 1. Accordingly, the image is separated into initial foreground and background regions. Both the foreground and background regions are modeled as Gaussian Mixture Models (GMMs). The GrabCut algorithm learns the color distributions by giving each pixel a probability to belong to the most feasible Gaussian component in one of the foreground or background GMMs. A graph is built from the image, and the final segmentation is performed using the iterative minimization algorithm of the graph cut to get a new classification of foreground and background pixels. This process is repeated until classification converges.

Fig. 1. Example of original GrabCut segmentation. (left) GrabCut allows the user to drag a rectangle around the object of interest to be segmented. (right) The segmented object.

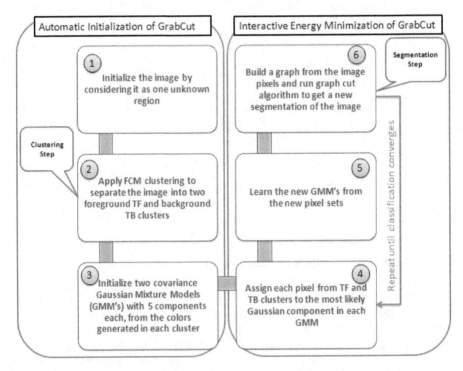

Fig. 2. Flowchart of the automatic GrabCut using FCM clustering for initialization

The initial user interaction is one main disadvantage of GrabCut. It allows GrabCut to be most appropriate for binary-label segmentation and classifies it as a semi-automatic segmentation technique. The modified technique developed by Khattab et al. [8] tried to avoid the previous limitation by modifying GrabCut into an automatic version, where the image can be segmented into proper segments without any user guidance. The novel contribution consists of replacing the semi-automatic/supervised step of GrabCut initialization with a completely automatic/unsupervised one. The modification included the use of the unsupervised image clustering technique of SOFM [9-10] for the GrabCut initialization. Fig. 2 illustrates the flowchart of the automatic GrabCut algorithm as proposed in [8] with the use of FCM as replacement of SOFM clustering technique. The main difference between the automatic and semi-automatic GrabCut occurs mainly in the initialization phase. In this phase, the user selection to create initial foreground and background regions is replaced with the clustering step (Fig. 2, steps 1 – 3). The interactive energy minimization phase of the automatic GrabCut runs exactly as the original semi-automatic GrabCut (Fig. 2, steps 4-6).

Fig. 3. The dataset of images

5   Results and Discussions

A dataset of 30 images, shown in Fig. 3, is used for the evaluation of the proposed automatic GrabCut. The images of the dataset include examples from the benchmark of the Berkeley's database [31] and others from the free images on the internet. These images are selected by certain criteria that include their fitting to the class of binary-label segmentation i.e. to include one object as foreground. Other criteria include having good visual separation in the color regions between the foreground and background.

Two measures; the error rate and the overlap score rate are used for calculating the segmentation accuracy. The error rate is calculated as the fraction of pixels with wrong segmentations (compared to ground truth) divided by the total number of pixels in the image. The overlap score rate is given by $y_1 \cap y_2 / y_1 \cup y_2$, where y_1 and y_2 are any two binary segmentations representing the ground truth and the generated segmentation result respectively.

The automatic GrabCut, which is initialized using FCM, is applied to the dataset images in different color space models, including RGB, XYZ, CMY, YUV and HSV. The features that identify each image pixel are only the values of its three components in the selected color space. The final segmentation results are obtained for all used images. For a quantitative comparison, Table 1 shows the accuracy rates generated for the whole dataset, while Fig. 4 summarizes the average accuracy rates in graph plots for all different color spaces. The results in Table 1 are presented in ascending order from left to right in terms of the total number of good image segmentation results and the average error rates. We can observe from Table 1 that the YUV representation generates better results for most of the images and outperforms the other color space models in terms of the average error rate of 6.2%. The overlap score rates of YUV and RGB are almost identical with 82.27% and 82.79% respectively.

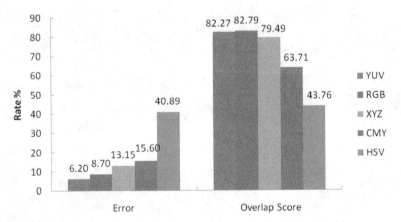

Fig. 4. Average accuracy measures for applying automatic GrabCut using FCM on different color space models

Table 1. Error and overlap score rates for different color space models

Image	Error rate %					Overlap Score rate %				
	YUV	**RGB**	**XYZ**	**CMY**	**HSV**	**YUV**	**RGB**	**XYZ**	**CMY**	**HSV**
1	17.46	17.53	18.76	10.23	52.10	62.22	62.04	59.15	79.22	43.36
2	4.25	4.22	4.23	3.67	66.35	94.05	94.15	94.12	95.13	33.65
3	4.91	4.95	3.90	3.76	40.89	91.48	91.21	94.73	95.09	45.39
4	6.13	5.05	7.41	49.58	11.48	86.51	90.00	82.01	26.72	72.30
5	3.51	2.80	2.84	24.47	43.11	89.31	93.70	93.69	11.94	21.13
6	0.97	0.86	0.88	9.69	70.07	96.41	97.16	97.07	44.37	17.49
7	3.14	61.00	61.07	3.07	19.53	96.88	38.40	38.38	96.36	68.83
8	6.47	6.27	24.97	6.86	17.42	94.81	95.00	67.15	94.19	81.07
9	4.25	3.13	3.21	4.26	4.19	91.62	94.53	94.34	93.15	91.83
10	9.45	2.40	2.32	1.21	71.36	0.01	69.00	69.86	84.30	6.20
11	2.96	2.99	97.04	2.98	2.73	55.70	55.30	2.96	54.94	58.56
12	0.72	1.92	1.08	2.43	69.12	94.62	82.98	90.75	75.40	10.01
13	14.29	2.15	7.78	36.14	46.31	79.41	97.35	88.17	52.77	52.51
14	23.96	2.16	2.17	15.21	20.87	45.45	94.75	94.77	60.61	52.57
15	2.57	2.56	5.12	2.76	76.94	93.95	93.78	83.78	93.15	21.97
16	3.08	24.92	38.96	4.01	53.59	95.89	43.97	38.94	91.67	31.80
17	26.69	36.54	36.94	28.44	56.77	0.27	29.13	28.84	0.33	16.70
18	4.18	4.15	4.15	4.22	22.36	94.56	94.26	94.31	94.19	55.62
19	2.05	2.17	2.16	40.82	2.60	97.03	96.71	96.76	8.23	95.81
20	6.97	4.92	5.08	38.74	70.67	82.31	89.35	88.54	18.02	25.83
21	3.61	2.33	2.38	42.92	68.24	90.95	95.53	95.47	38.01	17.75
22	3.10	2.89	2.86	5.28	38.51	93.41	94.05	94.09	87.19	18.40
23	3.00	2.98	2.93	8.78	38.09	94.58	94.26	94.55	79.26	21.27
24	3.86	3.88	3.87	36.48	73.61	91.05	90.98	90.91	34.25	19.81
25	2.88	37.10	37.14	5.51	59.28	93.48	38.35	38.31	83.63	27.65
26	4.06	2.49	3.11	4.18	8.44	93.05	96.59	95.20	93.03	83.70
27	1.81	1.47	1.46	26.19	65.69	95.11	96.33	96.32	0.21	21.33
28	1.28	1.28	1.31	1.57	38.17	94.58	94.61	94.47	93.01	27.96
29	3.47	3.17	3.29	5.34	4.80	93.01	93.71	93.46	88.51	89.78
30	10.83	10.70	6.03	39.25	13.48	86.48	86.67	93.55	44.51	82.54
Avg.	6.20	8.70	13.15	15.60	40.89	82.27	82.79	79.49	63.71	43.76

Fig. 5 displays visual comparisons of the generated segmentation results for some images having high variance of error rate among the different color models. It can be noticed that the automatic GrabCut using FCM almost failed to get accurate segmentations with the HSV model. One of the interesting results is the segmentation achieved for image no. 11 in the XYZ color space. In this image the GrabCut failed completely to segment the image considering the whole image as one segment. This explains the large error rate of 97.04% and poor overlap score rate of 2.96%. Fig. 6 shows the XYZ conversion of image no. 11 (left) and the initial clustering generated using the FCM (right) before running the segmentation part.

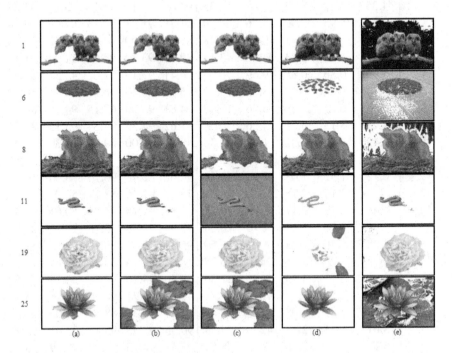

Fig. 5. Samples of image segmentations applied to different color space models, (a) YUV, (b) RGB, (c) XYZ, (d) CMY and (e) HSV

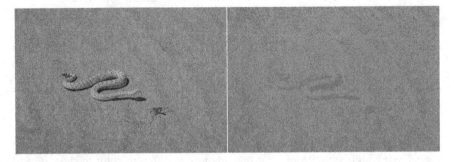

Fig. 6. Image no. 11 in the XYZ color space representation. (left) Original image and (right) initial clustering using FCM.

6   Conclusions and Future Work

User interaction is considered a disadvantage of the GrabCut technique for image segmentation. It allows the segmentation to be more susceptible to errors and the technique to be more appropriate for binary-label segmentation. Automatic GrabCut is used to eliminate the need for initial user interaction. In this paper, the unsupervised clustering of Fuzzy-C means (FCM) is used for GrabCut automation as a replacement of the initial user interaction. As no specific color space is recommended for every segmentation problem, the performance of the FCM-based automatic GrabCut was evaluated on different color space models including RGB, CMY, XYZ, YUV and HSV. Based on a dataset of different 30 images, the YUV representation generated the best segmentation accuracy and outperformed other color spaces. It provided an average error rate of 6.2% and overlap score rate of 82.27%. The future work aims to evaluate the automatic GrabCut on different color spaces for the problem of multi-label image segmentation, where the image can be segmented into more than two segments.

References

1. Karthik, K., Hrushikesh, P.: Image segmentation of homogeneous intensity regions using wavelets based level set. International Journal of Emerging Technology and Advanced Engineering. **3**(10), 215–219 (2013)
2. Lalitha, M., Kiruthiga, M., Loganathan, C.: A survey on image segmentation through clustering algorithm. International Journal of Science and Research (IJSR). **2**(2), 348–358 (2013)
3. Gonzalez, R.C., Woods, R.E.: Digital Image Processing. 3rd ed., Prentice-Hall, Inc. (2006)
4. Sharma, N., Mishra, M., Shrivastava, M.: Colour image segmentation techniques and issues: an approach. International Journal of Scientific & Technology Research. **1**(4), 9–12 (2012)
5. Busin, L., Vandenbroucke, N., Macaire, L.: Color spaces and image segmentation. Advances in imaging and electron physics **151**(1), 1 (2008)
6. Rother, C., Kolmogorov, V., Blake, A.: GrabCut": interactive foreground extraction using iterated graph cuts. ACM Trans. Graph. **23**(3), 309–314 (2004)
7. Boykov, Y., Jolly, M.-P.: Interactive Graph Cuts for Optimal Boundary and Region Segmentation of Objects in N-D Images. In: 8th IEEE International Conference on Computer Vision (ICCV), vol. 1, pp. 105–112 (2001)
8. Khattab, D., Ebied, H.M., Hussein, A.S., Tolba, M.F.: Automatic GrabCut for bi-label image segmentation using SOFM. In: Intelligent Systems' 2014, pp. 579–592. Springer (2015)
9. Kohonen, T., Oja, E., Simula, O., Visa, A., Kangas, J.: Engineering applications of the self-organizing map. Proceedings of the IEEE **84**(10), 1358–1384 (1996)
10. Haykin, S.S.: Neural Networks and Learning Machines, vol. 3. Prentice Hall, New York (2009)
11. Bezdek, J.C.: Pattern Recognition With Fuzzy Objective Function Algorithms. Kluwer Academic Publishers (1981)

12. Dunn, J.C.: A fuzzy relative of the ISODATA process and its use in detecting compact well-separated clusters. Journal of Cybernetics. **3**(3), 32–57 (1973)
13. Gulshan, V., Lempitsky, V.S., Zisserman, A.: Humanising GrabCut: Learning to segment humans using the Kinect. In: IEEE International Conference on Computer Vision (ICCV Workshops), pp. 1127–1133 (2011)
14. Hernández, A., Reyes, M., Escalera, S., Radeva, P.: Spatio-Temporal GrabCut human segmentation for face and pose recovery. In: IEEE Computer Society Conference on Computer Vision and Pattern Recognition Workshops (CVPRW), pp. 33–40 (2010)
15. Hu, Y., Human Body Region Extraction from Photos. In: MVA, pp. 473–476 (2007)
16. Corrigan, D., Robinson, S., Kokaram, A.: Video matting using motion extended GrabCut. In: IET European Conference on Visual Media Production (CVMP), pp. 3–3. London, UK (2008)
17. Göring, C., Fröhlich, B., Denzler, J.: Semantic segmentation using GrabCut. In: Proceedings of the International Conference on Computer Vision Theory and Applications (VISAPP), pp. 597–602 (2012)
18. Ramírez, J., Temoche, P., Carmona, R.: A volume segmentation approach based on GrabCut. CLEI Electronic Journal **16**(2) (2013)
19. Naz, S., Majeed, H., Irshad, H.: Image segmentation using fuzzy clustering: a survey. In: 6th International Conference on Emerging Technologies (ICET), pp. 181–186 (2010)
20. Krinidis, S., Chatzis, V.: A robust fuzzy local information C-means clustering algorithm. IEEE Transactions on Image Processing **19**(5), 1328–1337 (2010)
21. Beevi, S.Z., Sathik, M.M., Senthamaraikannan, K.: A robust fuzzy clustering technique with spatial neighborhood information for effective medical image segmentation. International Journal of Computer Science and Information Security (IJCSIS) **7**(3), 132–138 (2010)
22. Kannan, S., Ramathilagam, S., Pandiyarajan, R., Sathya, A.: Fuzzy clustering Approach in segmentation of T1-T2 brain MRI. Aceee International Journal on signal & Image Processing **1**(2), 43 (2010)
23. Beevi, Z., Sathik, M.: A Robust Segmentation Approach for Noisy Medical Images Using Fuzzy Clustering With Spatial Probability. The International Arab Journal of Information Technology **29**(37), 74–83 (2012)
24. Alata, O., Quintard, L.: Is there a best color space for color image characterization or representation based on Multivariate Gaussian Mixture Model? Computer Vision and Image Understanding **113**(8), 867–877 (2009)
25. Pagola, M., Ortiz, R., Irigoyen, I., Bustince, H., Barrenechea, E., Aparicio-Tejo, P., Lamsfus, C., Lasa, B.: New method to assess barley nitrogen nutrition status based on image colour analysis: Comparison with SPAD-502. Computers and electronics in agriculture **65**(2), 213–218 (2009)
26. Jurio, A., Pagola, M., Galar, M., Lopez-Molina, C., Paternain, D.: A comparison study of different color spaces in clustering based image segmentation. In: Information Processing and Management of Uncertainty in Knowledge-Based Systems. Applications, pp. 532–541. Springer (2010)
27. Chaves-González, J.M., Vega-Rodríguez, M.A., Gómez-Pulido, J.A., Sánchez-Pérez, J.M.: Detecting skin in face recognition systems: A colour spaces study. Digital Signal Processing **20**(3), 806–823 (2010)
28. Du, C.-J., Sun, D.-W.: Comparison of three methods for classification of pizza topping using different colour space transformations. Journal of food engineering **68**(3), 277–287 (2005)

29. Ruiz-Ruiz, G., Gómez-Gil, J., Navas-Gracia, L.: Testing different color spaces based on hue for the environmentally adaptive segmentation algorithm (EASA). Computers and electronics in agriculture **68**(1), 88–96 (2009)
30. Vandenbroucke, N., Macaire, L., Postaire, J.-G.: Color image segmentation by pixel classification in an adapted hybrid color space. Application to soccer image analysis. Computer Vision and Image Understanding **90**(2), 190–216 (2003)
31. Martin, D., Fowlkes, C., Tal, D., Malik, J.: A database of human segmented natural images and its application to evaluating segmentation algorithms and measuring ecological statistics. In: 8th IEEE International Conference on Computer Vision (ICCV), vol. 2, pp. 416–423 (2001)

A Novel Approach to the Weighted Laplacian Formulation Applied to 2D Delaunay Triangulations

Sanderson L. Gonzaga de Oliveira$^{(\boxtimes)}$, Frederico Santos de Oliveira, and Guilherme Oliveira Chagas

Universidade Federal de Lavras, Lavras, MG, Brazil
sanderson@dcc.ufla.br,
{fred.santos.oliveira,guilherme.o.chagas}@gmail.com

Abstract. In this work, a novel smoothing method based on weighted Laplacian formulation is applied to resolve the heat conduction equation by finite-volume discretizations with Voronoi diagram. When a minimum number of vertices is obtained, the mesh is smoothed by means of a new approach to the weighted Laplacian formulation. The combination of techniques allows to solve the resulting linear system by the Conjugate Gradient Method. The new approach to the weighted Laplacian formulation within the set of techniques is compared to other 4 approaches to the weighted Laplacian formulation. Comparative analysis of the results shows that the proposed approach allows to maintain the approximation and presents smaller number of vertices than any of the other 4 approaches. Thus, the computational cost of the resolution is lower when using the proposed approach than when applying any of the other approaches and it is also lower than using only Delaunay refinements.

Keywords: Weighted laplacian formulation · Delaunay triangulation · Heat conduction equation · Voronoi diagram · Laplacian smoothing

1 Introduction

Partial differential equations that model physical phenomena often have large variations in the solutions, or involve domains with complex geometries. One strategy used to improve the numerical solution is to refine the mesh. However, a high computational cost is required to achieve the numerical solution when using a fine uniform mesh throughout the computational domain. Numerical solutions with low computational costs can be reached by increasing the number of vertices into regions of the domain with large variation in the solution. Two types of techniques to adaptively refine a mesh are adaptive mesh refinement and movement of existing vertices of the mesh, that is, the strategies consist of inserting or moving vertices into regions of the domain with large variation in

© Springer International Publishing Switzerland 2015
O. Gervasi et al. (Eds.): ICCSA 2015, Part I, LNCS 9155, pp. 502–515, 2015.
DOI: 10.1007/978-3-319-21404-7_37

the solution. In both cases, the objective is to achieve a low computational-cost numerical solution, maintaining or improving the accuracy of the solution.

A novel approach to the weighted Laplacian formulation is presented in this work. The new approach is easy to be implemented and is associated with Delaunay refinement algorithms in the discretization of the heat conduction equation by finite volumes using Voronoi diagrams [22]. Moreover, the Ruppert [15] and the *off-center* [20,21] Delaunay refinement algorithms are applied [6].

This paper is organized as follows. In Section 2, the set of techniques applied in this work is described. In Section 3, results of the experiments are presented. Finally, in Section 4, conclusions are drawn and future work are discussed.

2 Description of the Set of Techniques for the Solution of the Heat Conduction Equation and Initial Settings in the Experimental Tests

As described, the Ruppert [15] and then the *off-center* [21] algorithms are executed in each refinement step. Afterwards, the finite volume method is applied. Improving locality avoids a large number of cache misses. To reduce the computational cost of the Conjugate Gradient Method [8,11], the reverse Cuthill-McKee method [5] and the George-Liu algorithm [4] are executed. The resulting linear system is then solved by the Conjugate Gradient Method with precision ϵ.

Related to moving mesh methods, the reader is referred to [9,14] for an introduction in this subject. On the other hand, a simple and easy manner to implement a moving mesh scheme can be based on the weighted Laplacian formulation [19]. Thus, a scheme based on the weighted Laplacian formulation was designed. In the new scheme, a local movement is directly proportional to the local variation of the solution. The novel scheme is associate with Delaunay refinement algorithms for discretization of second-order partial differential equations by finite volumes with low computational cost.

The initial mesh is a Delaunay triangulation [3] generated by the Green-Sibson algorithm [7], whose input is a planar straight-line graph (PSLG). Moreover, the Lawson algorithm [12] is used to maintain a Delaunay triangulation when inserting new vertices with the Green-Sibson algorithm. The movement of vertices is carried out between each time step of the numerical solution. This movement is performed while a geometric mesh quality condition is satisfied. When this condition is satisfied, the *off-center* algorithm [21] is executed in order to improve the mesh quality.

The new approach to the weighted Laplacian formulation applied to a moving mesh scheme is called here as $\mathfrak{N}$. One can consider a mesh $M = (M_V, M_A)$, and M_V and M_A are the mesh vertices and edges, respectively. Using the $\mathfrak{N}$ function, the largest difference value of the current solution between adjacent vertices in the entire mesh is considered. The $\mathfrak{N}$ function determines the new position S'_p of vertex $p \in M_V$, such that $S'_p = S_p + \frac{\beta}{(\Delta\phi)_{max}} \sum_{\{i,p\}\in M_A} (S_i - S_p)|\phi_i - \phi_p|$, and S_p is the current position of vertex $p \in M_V$, β is a parameter to control the movement intensity and is within range $(0,1]$, ϕ_i and ϕ_p are values stored in vertices i and p, respectively, and $(\Delta\phi)_{max} = \max\limits_{\forall\{u,w\}\in M_A} |\phi_u - \phi_w|$. The movement of vertices is directly proportional to the local variation of the solution when using this new approach to the weighted Laplacian formulation. The $\mathfrak{N}$ function is compared to other four approaches to the weighted Laplacian formulation $S'_p = S_p + \beta\frac{\sum_{i=1}^n (S_i - S_p)\omega_{p,i}}{\sum_{i=1}^n \omega_{p,i}}$, and $\omega_{p,i}$ is defined according to the following four functions.

- A common approach to the weighted Laplacian formulation, where $\omega_{p,i} = |\phi_i - \phi_p|$, called here as $\mathfrak{L}$.
- The $\lambda|\mu$ function proposed by Taubin [16,17], where $\omega_{p,i} = |\phi_i - \phi_p|$ and combines two successive movements, such that $S'_p = S_p + \lambda\Delta S_p$ and $S''_p = S'_p - \mu\Delta S'_p$, where $\Delta S_p = \frac{\sum_{i=1}^m \omega_{p,i}(S_i^n - S_p^n)}{\sum_{i=1}^m \omega_{p,i}}$ and the relaxation parameters satisfy $0 < \lambda < \mu$. See also Taubin et al. [18] for details about this function.
- The Taubin's function [16,17] set with $\lambda = \mu$, proposed by Kobbelt et al. [10], called here as $\mathfrak{B}$.
- An approach shown by Thompson et al. [19], such that $\omega_{p,i} = \iota\frac{|\phi_i - \phi_p|}{|\phi_i + \phi_p|}$, for $0 < \iota \le 1$, called here as $\mathfrak{T}$.

Functions $\lambda|\mu$ and $\mathfrak{B}$ are variations of the Laplacian smooth and, consequently, both have been used for improving mesh quality. We did not find any work using these functions in movements of vertices with the aim of obtaining better approximations of solutions of partial differential equations.

The procedure for moving the vertices is repeated while the mesh shows a user-defined geometric quality. The metric Shape Regularity Quality [1,2], called here as v, is used. Moreover, a parameter η is used, for $0 < \eta < 1$, so that the procedure for moving the vertices is repeated while $v_{min} \ge \eta$ and v_{min} is the smallest value of the metric Shape Regularity Quality found in the mesh. Then the mesh is examined to verify if the movement has produced a mesh tangling. Afterwards, if the mesh is valid, the *off-center* algorithm is executed in order to generate a mesh with the user-defined quality.

The computational domain was set as a square with dimensions $c \times c$. Specifically, $c = 50$ was used. Boundary conditions in the east, west and south faces were set as 10, and on the north face, as zero. In relation to the parameters shown in Table 1, time steps were set as $t_{end} = 10$ and $\Delta t = 0.1$. The minimum angle was established as $\alpha = 30°$, establishing $\rho_\alpha = 1.0$. This choice for α is due to be near the practical limits of the Delaunay refinement.

The method that generates the initial mesh with at least χ vertices is shown in algorithm 1. The parameter related to insertion of vertices was set as $\theta = 1e-10$ (see line 11 in algorithm 1). The numerical accuracy of the Conjugate Gradient Method was established as $\epsilon = 1e - 10$.

The movement of vertices is shown in algorithm 2. It is repeated while $v_{min} \ge \eta$ (see line 11 in algorithm 2). Thus, $\eta = 0.6$ was established and both v_{min} and η denote values of the metric Shape Regularity Quality [1,2]. This value of η was chosen because of the limit of the off-center algorithm [21], which generates meshes with minimum angle near $32°$.

Table 1. Description of the input parameters of the algorithms 1 and 2. Most parameters are specified by the user.

Parameter	Algorithm(s)	Description
$G = (L_V, L_E)$	1	PSLG: list of vertices L_V and edges L_E of the initial Delaunay triangulation.
θ	1	Quantifying constant of the selection criterion for refinement.
χ	1	Minimum number of vertices in the mesh.
M	2	Initial mesh generated by the algorithm 1.
β	2	Parameter to control intensity in the movement of vertices.
η	2	Tolerance value for the Shape Regularity Quality metric to continue moving the vertices.
t_{end}	2	Maximum number of time steps.
Δt	2	Time variation used in the discretization of the heat conduction equation by the finite volume method.
α	1 and 2	Minimum angle in the mesh.
ϵ	1 and 2	Numerical precision used in the Conjugate Gradient Method.

The input to the five approaches, PSLG $G = (L_V, L_E)$, was established as $L_V = \{v_0, v_1, \cdots, v_4\}$, where $v_0 = (0,0), v_1 = (c,0), v_2 = (0,c), v_3 = (c,c), v_4 = (\lfloor \frac{c}{2} \rfloor, \lfloor \frac{c}{2} \rfloor)$, and $L_E = \{\{v_0, v_1\}, \{v_0, v_2\}, \{v_3, v_2\}, \{v_3, v_1\}, \{v_0, v_4\}, \{v_1, v_4\}, \{v_2, v_4\}, \{v_3, v_4\}\}$.

In simulations with the $\mathfrak{T}$ function, $\iota = 1$ was used. In simulations with the $\lambda|\mu$ function, $\lambda = \beta$ and $\mu = \beta + 0.0001$ were set. In simulations with the $\mathfrak{B}$ function, $\lambda = \mu = \beta$ was established. These settings are summarized in Table 2. To differentiate the values of λ and μ in the $\lambda|\mu$ and $\mathfrak{B}$ functions, these parameters are called as $\lambda_{\lambda|\mu}$, $\lambda_{\mathfrak{B}}$, $\mu_{\lambda|\mu}$ and $\mu_{\mathfrak{B}}$ in Table 2.

Table 2. Values of input parameters for algorithms 1 and 2

Parameter	Algorithm(s)	Value(s) used
L_V	1	$v_0 = (0,0), v_1 = (c,0), v_2 = (0,c),$ $v_3 = (c,c), v_4 = (\lfloor \frac{c}{2} \rfloor, \lfloor \frac{c}{2} \rfloor)$
L_A	1	$\{v_0, v_1\}, \{v_0, v_2\}, \{v_3, v_2\}, \{v_3, v_1\},$ $\{v_0, v_4\}, \{v_1, v_4\}, \{v_2, v_4\}, \{v_3, v_4\}$
θ	1	1e-10
χ	1	15,000; 30,000; 60,000 and 100,000
$\beta, \lambda_{\mathfrak{B}}, \lambda_{\lambda|\mu}, \mu_{\mathfrak{B}}$	2	0.1; 0.2; 0.3; 0.4; 0.5; 0.6; 0.9; 1
$\mu_{\lambda|\mu}$	2	0.1001; 0.2001; 0.3001; 0.4001; 0.5001; 0.6001; 0.9001; 1.0001
ι	2	1
η	2	0.6
t_{end}	2	10
Δt	2	0.1
α	1 and 2	30°
ϵ	1 and 2	1e-10

Input: PSLG $G = (L_V, L_E)$, L_V is a list of vertices and L_A is a list of edges of the initial triangulation, α, θ, ϵ and χ. `// see Table 1`

Output: mesh and initial solution M. `// mesh with angles between` α `and` $\pi - 2\alpha$`, refined in regions of large variation and initial solution stored in the vertices of` M`.`

```
 1 begin
 2 |   σ ← 0;
   |   // generation of mesh M, from PSLG G
 3 |   M ← GreenSibsonAlgorithm(G); // with zero as initial estimative
 4 |   repeat
   |       // generation of linear system by the finite volume method
 5 |       Ab ← FiniteVolumeMethod(M, 0); // initial time step
 6 |       ν ← GeorgeLiuAlgorithm(Ab, M); // pseudo-peripheral vertex
 7 |       Ab ← ReverseCuthillMcKee(Ab, M, ν);
   |       // values of the partial differential equation in the mesh M
 8 |       M ← ConjugateGradientMethod(Ab, ε); // are updated
   |       // γ_max and γ_min are respectively the largest and the lowest
 9 |       γ_max ← calcula_γ_max(M); // values of gradient γ of the mesh
10 |       γ_min ← calcula_γ_min(M);
   |       // σ is a parameter used in the criterion of selection of
11 |       σ ← σ + θ(γ_max − γ_min); // triangles to be refined
   |       // triangles with γ > σ are refined
12 |       M ← RuppertAlgorithm(σ, M);
   |       // triangles with angles lower than α are refined
13 |       M ← Off-centers(α, M);
14 |   until (M.numeroVertices ≥ χ) ;
   |   // values of the partial differential equation are stored in the
15 |   return (M); // vertices of M
16 end
```

Algorithm 1. Mesh generation and refinement

An initial mesh with 3,841 vertices, created with the Green-Sibson algorithm [7], is shown in Figure 1. Considering this figure, examples of meshes after being smoothed with application of the $\mathfrak{N}$, $\mathfrak{L}$, $\lambda|\mu$, $\mathfrak{B}$ and $\mathfrak{T}$ functions are shown in Figures 2, 3, 4, 5 and 6, respectively. After generating the initial mesh, 10 time steps were performed, moving the vertices as shown in algorithm 2. Moreover, the procedure to move the vertices and perform the refinement of triangles with $\angle < \alpha$ is shown in algorithm 2. The inputs of this algorithm are shown in Table 1. Detailed descriptions of this scheme is shown in Oliveira [13].

3   Experimental Tests and Analysis of Results

In this section, comparative results between the $\mathfrak{N}$, $\mathfrak{L}$, $\lambda|\mu$, $\mathfrak{B}$ and $\mathfrak{T}$ functions are presented. To test these approaches to the weighted Laplacian formulation, meshes were generated from $15,788$ to $149,079$ vertices by establishing the minimum number of vertices with $\chi = 15,000; 30,000; 60,000; 100,000$. These meshes present initial quality metrics $\rho_{max} = 1.0$ and $\upsilon_{min} = 0.6$.

Input: initial mesh M, α, β, ϵ, η, Δt and t_{end}. // see Table 1
Output: discretization of the heat conduction equation by finite volumes with
    Voronoi diagram.

1 **begin**
2 | $t \leftarrow 1$; // see algorithm 1 for $t=0$
3 | **while** $(t \leq t_{end})$ **do**
4 | | **repeat**
   | | | // generation of linear system in time t
5 | | | $Ab \leftarrow FiniteVolumeMethod(M, t \times \Delta t)$;
6 | | | $\nu \leftarrow GeorgeLiuAlgorithm(Ab, M)$; // pseudo-peripheral vertex
7 | | | $Ab \leftarrow ReverseCuthillMcKee(Ab, M, \nu)$;
   | | | // values of the partial differential equation
   | | | // in the mesh M are updated in time t
8 | | | $M \leftarrow ConjugateGradientMethod(Ab, \epsilon)$;
9 | | | $Movement \leftarrow Smooth(\beta, M)$; // movement of vertices
   | | | // the mesh quality is checked
10 | | | $v_{min} \leftarrow ShapeRegularityQuality(M)$;
11 | | **until** $((Movement = false) \vee (v_{min} < \eta))$;
   | | // if the mesh is not a Delaunay triangulation anymore,
   | | // the off-center algorithm refines triangles with $\angle < \alpha$
   | | // and generates a new Delaunay triangulation
12 | | **if** $(v_{min} < \eta)$ **then** $Off\text{-}centers(\alpha, M)$;
13 | | $t \leftarrow t + 1$; // a new time step
14 **end**

Algorithm 2. Movement of vertices

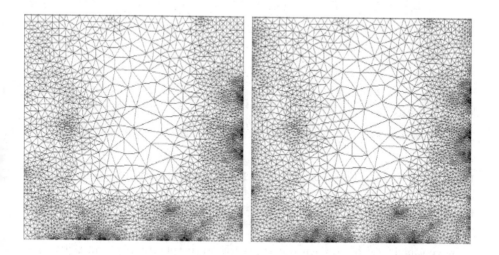

Fig. 1. Initial mesh with 3,841 vertices

Fig. 2. Mesh smoothed by applying the $\mathfrak{R}$ function

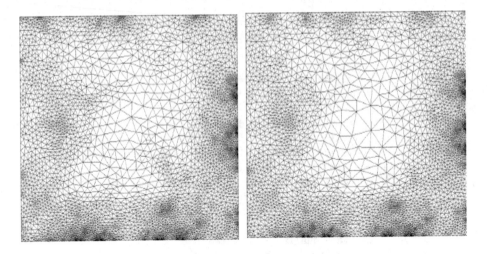

Fig. 3. Mesh smoothed by applying the $\mathfrak{L}$ function

Fig. 4. Mesh smoothed by applying the $\lambda|\mu$ function

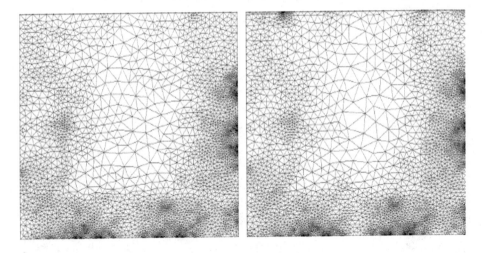

Fig. 5. Mesh smoothed by applying the $\mathfrak{B}$ function

Fig. 6. Mesh smoothed by applying the $\mathfrak{T}$ function

The computational code was developed in the C++ programming language. The experiments were performed using the Ubuntu operation system 12:04 LTS 64-bit and an Intel® processor i3 3.10GHz, 16GB of main memory and 3072KB of L3 cache. In order to obtain solutions in a timely manner and that would meet possible accuracy in accordance with the memory available in every step of the implementation, 512 bits of precision were set when using the library *GNU Multiple Precision Floating-point computations with correct-Rounding* (MPFR).

In the five approaches compared, a parameter β is used to control the movement intensity of vertices. This parameter was studied in the range $0.1 \leq \beta \leq 1$. One obtains the total movement provided by the function with $\beta = 1$ and no movement is obtained with $\beta = 0$. A sample of the results in simulations with β from 0.4 to 0.9 is shown in Table 3: total number of vertices at the end of the simulation, clocks of the experiment and to move the mesh, number of iterations of the repeat loop in algorithm 2, γ_{max} and γ_{med} at the end of the experiments, as shown in algorithm 1. In column "No-Move1" in Table 3, results obtained with 10 time steps for each of the initial meshes are presented. These simulations were performed without moving the vertices and with the input values shown in Table 2. In column "No-move2" in Table 3, results obtained with 10 time steps are also presented. These simulations were carried out without moving the vertices and using meshes with: i) number of vertices larger than the corresponding mesh shown in column 'No-move1"; ii) gradient similar to the simulations with functions to move the vertices. These simulations were performed to obtain γ_{max} and γ_{med} larger than in simulations shown in "No-move1" or similar to simulations with movement of vertices.

In simulations with $\beta < 0.6$, in most cases, the best results were obtained in simulations with the common approach based on the weighted Laplacian formulation, called here as $\mathfrak{L}$. In a few cases, the best results were obtained in simulations with the approach shown by Thompson et al. [19], called here as $\mathfrak{T}$. On the other hand, the lowest final number of vertices in the simulations was obtained with the proposed $\mathfrak{N}$ function. This means that by using this function one obtains good quality meshes and the off-center algorithm inserts less vertices to generate a mesh with the user-defined quality than using any of the other four functions. However, one observes that with $\beta < 0.6$ no advantage is obtained when using the movement of vertices with any of the five tested functions compared to using only Delaunay refinement. This difference in computational cost is reduced with larger meshes.

In simulations with $\beta = 0.6$ and $\chi = 100,000$, the $\mathfrak{N}$ function resulted in the lowest computational cost compared to the other four functions. The higher movement intensity set in the $\mathfrak{N}$ function the fewer iterations in such loop. Hence, simulations with low computational costs were obtained. Moreover, one observes in Table 3 that the computational cost of the experiments with the $\mathfrak{N}$ function and $\beta = 0.6$ and $\beta = 0.9$ obtained lower computational cost than computational costs of the corresponding experiments shown in "No-move2".

The $\mathfrak{N}$ function moves the mesh vertices in a globally normalized manner, and directly proportional to the local variation of the solution. Consequently, the $\mathfrak{N}$ function maintains a reasonable quality of the mesh polygons and fewer vertices are inserted in the mesh by the *off-center* algorithm to keep the user-defined mesh quality. Also, the computational cost of the movement given by the $\mathfrak{N}$ function increases much less than the total time of the experiment when setting β nearer to 1. This can be seen in line % of Table 4, which is (clocks to move the meshes / clocks in the experiment) * 100 in experiments with $\chi = 100,000$. More specifically, the numbers of iterations of the repeat loop in algorithm 2 when

Table 3. Results of the simulations with moving mesh approaches and Delaunay refinement

β	χ	Result	𝔑	𝔏	λ|μ	𝔅	𝔗	No-move1	No-move2
0,4	15.000	Number of vertices	16,701	17,579	16,837	16,821	17,355	15,788	32,515
		Clocks in the experiment	1,62e10	1,27e10	1,55e10	1,56e10	1,32e10	1,33e09	4,05e09
		Clocks in the movement	1,14e09	6,37e08	1,22e09	1,28e09	7,26e08	-	-
		Loop repeat in algorithm 2	30	11	26	25	13	-	-
		γ_{max}	388,50	58,95	26,29	26,28	211,07	26,13	36,13
		γ_{med}	1,14899	0,96986	0,99102	0,99077	1,01422	0,98318	1,03107
	30.000	Number of vertices	33,315	34,980	34,391	34,381	34,746	32,515	67,963
		Clocks in the experiment	4,27e10	3,54e10	4,47e10	4,50e10	3,69e10	4,05e09	1,53e10
		Clocks in the movement	1,88e09	1,13e09	2,67e09	2,51e09	1,29e09	-	-
		Loop repeat in algorithm 2	21	10	29	29	11	-	-
		γ_{max}	224,84	65,94	50,86	50,85	215,36	36,13	52,37
		γ_{med}	1,11159	1,04457	1,04929	1,04969	1,05862	1,03107	1,06202
	60.000	Number of vertices	68,846	71,937	71,170	71,125	72,224	67,963	142,253
		Clocks in the experiment	1,22e11	1,06e11	1,18e11	1,17e11	1,07e11	1,53e10	5,30e10
		Clocks in the movement	4,39e09	2,50e09	4,39e09	4,41e09	2,63e09	-	-
		Loop repeat in algorithm 2	24	11	21	20	11	-	-
		γ_{max}	227,53	106,15	52,68	52,64	365,12	52,37	72,37
		γ_{med}	1,10886	1,09097	1,09448	1,09445	1,10247	1,06202	1,11495
	100.000	Number of vertices	143,306	149,079	147,478	147,508	148,898	142,253	298,574
		Clocks in the experiment	3,48e11	3,33e11	3,42e11	3,41e11	3,20e11	5,30e10	2,18e11
		Clocks in the movement	7,48e09	5,87e09	7,93e09	8,20e09	5,07e09	-	-
		Loop repeat in algorithm 2	19	12	16	16	10	-	-
		γ_{max}	298,82	387,05	100,34	100,35	212,82	72,37	104,80
		γ_{med}	1,12429	1,12151	1,11912	1,11908	1,12472	1,11495	1,13468
0,5	15.000	Number of vertices	16,917	18,196	17,172	17,281	17,933	15,788	32,515
		Clocks in the experiment	1,50e10	1,29e10	1,44e10	1,46e10	1,30e10	1,33e09	4,05e09
		Clocks in the movement	4,24e08	5,98e08	9,41e08	9,67e08	6,16e08	-	-
		Loop repeat in algorithm 2	10	10	19	20	11	-	-
		γ_{max}	67,98	50,23	26,38	26,40	447,48	26,13	36,13
		γ_{med}	1,04517	0,95807	0,98742	0,98068	1,04418	0,98318	1,03107
	30.000	Number of vertices	33,852	35,659	34,639	34,640	35,215	32,515	67,963
		Clocks in the experiment	4,17e10	3,78e10	4,16e10	4,13e10	3,59e10	4,05e09	1,53e10
		Clocks in the movement	1,60e09	1,23e09	2,20e09	2,04e09	1,09e09	-	-
		Loop repeat in algorithm 2	16	11	20	20	10	-	-
		γ_{max}	219,21	48,97	51,01	51,02	374,23	36,13	52,37
		γ_{med}	1,09169	1,03820	1,05415	1,05422	1,05477	1,03107	1,06202
	60.000	Number of vertices	69,059	72,892	71,639	71,622	74,041	67,963	142,253
		Clocks in the experiment	1,18e11	1,06e11	1,10e11	1,10e11	1,11e11	1,53e10	5,30e10
		Clocks in the movement	4,07e09	2,38e09	3,32e09	3,34e09	3,01e09	-	-
		Loop repeat in algorithm 2	25	10	14	14	12	-	-
		γ_{max}	417,86	104,08	67,41	67,42	758,03	52,37	72,37
		γ_{med}	1,12834	1,08956	1,09577	1,09572	1,11319	1,06202	1,11495
	100.000	Number of vertices	143,497	150,672	148,841	148,791	151,400	142,253	298,574
		Clocks in the experiment	3,36e11	3,25e11	3,28e11	3,29e11	3,30e11	5,30e10	2,18e11
		Clocks in the movement	6,49e09	5,05e09	6,02e09	6,00e09	5,51e09	-	-
		Loop repeat in algorithm 2	17	10	12	12	10	-	-
		γ_{max}	202,91	99,86	100,82	100,83	723,86	72,37	104,80
		γ_{med}	1,12883	1,11895	1,11961	1,11971	1,13135	1,11495	1,13468
0,6	100.000	Number of vertices	143,637	157,280	162,631	162,660	157,798	142,253	298,574
		Clocks in the experiment	7,62e10	3,54e11	3,61e11	3,62e11	3,59e11	5,30e10	2,18e11
		Clocks in the movement	6,35e08	5,46e09	5,36e09	5,18e09	5,04e09	-	-
		Loop repeat in algorithm 2	12	10	10	10	10	-	-
		γ_{max}	179,18	353,98	105,56	105,57	223,65	72,37	104,80
		γ_{med}	1,22377	1,11889	1,11625	1,11611	1,14206	1,11495	1,13468
0,9	100.000	Number of vertices	144,959	-	-	-	-	142,253	298,574
		Clocks in the experiment	8,73e10	-	-	-	-	5,30e10	2,18e11
		Clocks in the movement	8,98e08	-	-	-	-	-	-
		Loop repeat in algorithm 2	10	-	-	-	-	-	-
		γ_{max}	275,61	-	-	-	-	72,37	104,80
		γ_{med}	1,30997	-	-	-	-	1,11495	1,13468

Table 4. Number (N) of iterations in the repeat loop in algorithm 2 and % = (*clocks* in movement / *clocks* in the experiment) * 100 in simulations with the $\mathfrak{N}$ function and $\chi = 100.000$

β	0.1	0.2	0.3	0.4	0.5	0.6	0.9
N	71	45	30	19	17	12	10
%	3.8	3.1	2.4	2.2	1.9	0.8	1.0

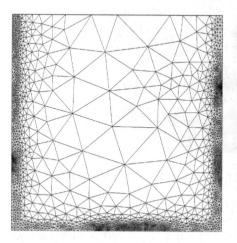

Fig. 7. Mesh with 2005 vertices resulted from a simulation with $\mathfrak{N}$ function set with $\beta = 0.6$, $\chi = 2000$ and $\theta = 0.003$

Fig. 8. Mesh with 10034 vertices resulted from a simulation set with $\mathfrak{N}$ function set with $\beta = 0.6$, $\chi = 10000$ and $\theta = 0.003$

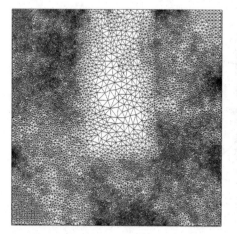

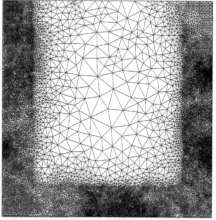

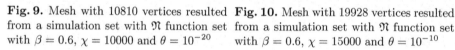

Fig. 9. Mesh with 10810 vertices resulted from a simulation set with $\mathfrak{N}$ function set with $\beta = 0.6$, $\chi = 10000$ and $\theta = 10^{-20}$

Fig. 10. Mesh with 19928 vertices resulted from a simulation set with $\mathfrak{N}$ function set with $\beta = 0.6$, $\chi = 15000$ and $\theta = 10^{-10}$

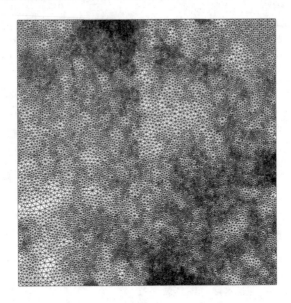

Fig. 11. Mesh with 23417 vertices resulted from a simulation set with $\mathfrak{N}$ function set with $\beta = 0.6$, $\chi = 20000$ and $\theta = 10^{-30}$

applying the $\mathfrak{N}$ function with $\beta = 0, 1; 0, 2; 0, 3; 0, 4; 0, 5; 0, 6; 0, 9$ are shown in Table 4. In Figures 7 to 11, one observes meshes smoothed which have a slightly different topology than the meshes shown in Figures 1 to 6, i.e. meshes with different varied, configuration and density.

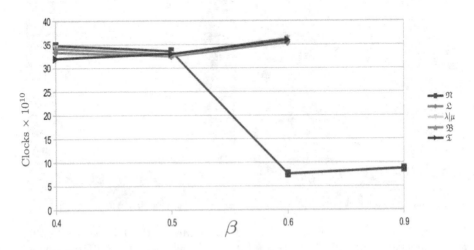

Fig. 12. Clocks in the experiments of the smoothing functions related to tests with $\chi = 100,000$ showed in Table 3

In experiments with $\beta = 0.9$, the simulation with the $\mathfrak{N}$ function was the only one that did not produce mesh tangling. Figure 12 shows the clocks in the experiments of the smoothing functions related to tests with $\chi = 100,000$ showed in Table 3.

Intensity in the movement of vertices with the $\mathfrak{T}$ function [19] may be too high because the value for normalizing the function is less than 1, for $\iota = 1$. Thus, low quality mesh may be obtained with the $\mathfrak{T}$ function when setting $\beta \cong 1$. In addition, vertices must be inserted in the mesh with β quite less than 1 to keep a high quality mesh. One way to make the $\mathfrak{T}$ function [19] competitive is to adjust the parameter ι. However, this may be a complicating factor because, for each problem, one should discover the best value for this parameter. In particular, the normalization in the $\mathfrak{N}$ function avoids setting another parameter such as it is needed in the $\mathfrak{T}$ function.

Experiments with the $\mathfrak{N}$ function with $\beta = 0.6$ and $\beta = 0.9$ showed lower computational cost than only using algorithms for Delaunay refinement with similar average gradient values. This suggests that one may reduce the computational cost of a simulation by using $\beta \cong 1$, that is, a high movement intensity of the vertices with the $\mathfrak{N}$ function resulted in a reduced computational cost compared to only using algorithms for Delaunay refinement. In the experiment with $\beta = 1$, the $\mathfrak{N}$ function also produced mesh tangling.

Tests with the $\mathfrak{N}$ function were also performed in a way that, after the repeat loop in algorithm 2 and before generating the Delaunay triangulation with the *off-center* algorithm, vertices in poor-quality triangles were moved again in the direction of its original position. More specifically, this return in the movement was similar to the second movement performed by the $\lambda|\mu$ and $\mathfrak{B}$ functions, using μ with 0.1, 0.3, 0.5. With this second movement, the expectation was to recover degenerated triangles and, therefore, few vertices would be inserted with *off-centers*. However, the total computational cost of these experiments were higher than the computational cost of the experiments shown in Table 3 with the $\mathfrak{N}$ function and β set as 0.6 or 0.9.

4   Conclusions

In this work, a new approach to the weighted Laplacian formulation applied to the 2D discretization of partial differential equations by finite volumes with Voronoi diagram is proposed. The proposed Laplacian smoothing was called here as $\mathfrak{N}$. When using this function, the movement of vertices is directly proportional to the local variation of the solution, that is, one obtains higher movement in regions with large variation and lower movement in regions with small variation of the solution. This occurs because the function is globally normalized. On the other hand, for example, the movement of vertices is approximately uniform when using a common approach to the weighted Laplacian formulation ($\mathfrak{L}$ function) because it is locally normalized.

The simulation results with the $\mathfrak{N}$ function were compared to results of other four approaches to the weighted Laplacian formulation. Those experiments showed that using the $\mathfrak{N}$ function one can obtain simulations with lower computational costs than using any of the other four functions or using only Delaunay refinement.

A posteriori error estimation shall be implemented in a future work. In addition, simulations with the heat conduction equation with source term and other boundary conditions in complex geometries shall be performed. Additionally, the set of techniques applied to other partial differential equations and application of higher-order schemes shall be studied. This set of techniques for solving partial differential equations in a parallel scheme shall be also implemented.

Acknowledgments. This work was undertaken with the support of the CNPq - Conselho Nacional de Desenvolvimento Científico e Tecnológico (National Council for Scientific and Technological Development) and FAPEMIG - Fundação de Amparo à Pesquisa do Estado de Minas Gerais (Minas Gerais Research Support Foundation).

References

1. Bank, R.E., Smith, R.K.: Mesh smoothing using a posteriori error estimates. SIAM Journal on Numerical Analysis **34**(3), 979–997 (1997)
2. Bank, R.E., Xu, J.: An algorithm for coarsening unstructured meshes. Numerische Mathematik **73**(1), 1–36 (1996)
3. Delaunay, B.: Sur la sphère vide. Izvestia Akademii Nauk SSSR. Otdelenie Matematicheskikh i Estestvennykh Nauk **7**, 793–800 (1934)
4. George, A., Liu, J.W.H.: An implementation of a pseudoperipheral node finder. ACM Transactions on Mathematical Software **5**(3), 284–295 (1979)
5. George, J.A.: Computer implementation of the finite element method. PhD thesis, Computer Science Department, Stanford University, CA (1971)
6. Gonzaga de Oliveira, S.L.: A review on delaunay refinement techniques. In: Murgante, B., Gervasi, O., Misra, S., Nedjah, N., Rocha, A.M.A.C., Taniar, D., Apduhan, B.O. (eds.) ICCSA 2012, Part I. LNCS, vol. 7333, pp. 172–187. Springer, Heidelberg (2012)
7. Green, P.J., Sibson, R.: Computing Dirichlet tessellations in the plane. The Computer Journal **21**(2), 168–173 (1978)
8. Hestenes, M.R., Stiefel, E.: Methods of conjugate gradients for solving linear systems. Journal of Research of the National Bureau of Standards **49**(36), 409–436 (1952)
9. Huang, W., Russell, R.: Adaptive moving mesh methods, 1st edn. Applied mathematical sciences. Springer, New York (2011)
10. Kobbelt, L., Campagna, S., Vorsatz, J., Seidel, H.-P.: Interactive multi-resolution modeling on arbitrary meshes. In: Proceedings of the 25th Annual Conference on Computer Graphics and Interactive Techniques, ACM SIGGRAPH 1998, pp. 105–114. ACM, New York (1998)
11. Lanczos, C.: Solutions of systems of linear equations by minimized iterations. Journal of Research of the National Bureau of Standards **49**(3), 33–53 (1952)
12. Lawson, C.L.: Software for C^1 surface interpolation. In: Rice, J.R. (ed.) Matematical Software III, pp. 161–194. Academic Press, Orlando (1977)

13. Oliveira, F.S.: Numerical solutions of partial differential equations with discretization by finite volume and adaptively refined and moving meshes (in Portuguese). Master's thesis, Departamento de Ciência da Computação - Universidade Federal de Lavras, Lavras, Brazil (2014)

14. Oliveira, F.S., de Oliveira, S.L.G., Kischinhevsky, M., Tavares, J.M.R.S.: Moving mesh methods for numerical solution of partial differential equations (in Portuguese). Revista de Sistemas de Informação da Faculdade Salesiana Maria Auxiliadora-FSMA, 11:11–16 (June 2013)

15. Ruppert, J.: A Delaunay refinement algorithm for quality 2-dimensional mesh generation. Journal of Algorithms 18(3), 548–585 (1995)

16. Taubin, G.: Curve and surface smoothing without shrinkage. In: Proceedings of the Fifth International Conference on Computer Vision, ICCV 1995, vol. 5, pp. 852–857. IEEE Computer Society, Washington, DC (1995)

17. Taubin, G.: A signal processing approach to fair surface design. In: Proceedings of the 22nd Annual Conference on Computer Graphics and Interactive Techniques, SIGGRAPH 1995, pp. 351–358. ACM, New York (1995)

18. Taubin, G., Zhang, T., Golub, G.: Optimal surface smoothing as filter design. In: Buxton, B.F., Cipolla, R. (eds.) ECCV 1996. LNCS, vol. 1064, pp. 283–292. Springer, Heidelberg (1996)

19. Thompson, J.F., Soni, B.K., Weatherhill, N.P.: Handbook of Grid Generation. CRC Press, New York (1999)

20. Üngör, A.: Off-Centers: a new type of steiner points for computing size-optimal quality-guaranteed delaunay triangulations. In: Farach-Colton, M. (ed.) LATIN 2004. LNCS, vol. 2976, pp. 152–161. Springer, Heidelberg (2004)

21. Üngör, A.: Off-centers: A new type of Steiner points for computing size-optimal quality-guaranteed Delaunay triangulations. Computational Geometry 42(2), 109–118 (2009)

22. Voronoi, G.: Nouvelles applications des paramètres continus à la théorie des formes quadratiques. Deuxième mémoire. Recherches sur les parallélloèdres primitifs. Journal für die reine und angewandte Mathematik (Crelles Journal) 1908(134), 198–287 (1908)

Adaptive Clustering-Based Change Prediction for Refreshing Web Repository

Bundit Manaskasemsak[(✉)], Petchpoom Pumjang, and Arnon Rungsawang

Massive Information and Knowledge Engineering Laboratory,
Department of Computer Engineering, Faculty of Engineering,
Kasetsart University, Bangkok 10900, Thailand
{un,arnon}@mikelab.net, g5514552624@ku.ac.th

Abstract. Resource constraints, such as time and network bandwidth, hinder modern search engine providers to keep local database completely synchronize with the Web. In this paper, we propose an adaptive clustering based change prediction approach to refresh the local web repository. Especially, we first group the existing web pages in the current repository into web clusters based on their similar change characteristics. We then sample and examine some pages in each cluster to estimate their change patterns. Selected cluster of web pages with higher change probability will be later downloaded to update the current repository. Finally, the effectiveness of the current download cycle will be examined; either auxiliary (non-downloaded), reward (correct change prediction), or penalty (wrong change prediction) score will be assigned to a web page. This score will later be used to reinforce the consecutive web clustering as well as the change prediction processes. To evaluate the performance of the proposed approach, we run extensive experiments on snapshots of real Web dataset of about 282,000 distinct URLs which are belonging to more than 12,500 websites. The results clearly show that the proposed approach outperforms the existing state-of-the-art on clustering-based web crawling policy in that it can provide fresher local web repository with limited resource.

Keywords: Web change prediction · Refresh policy · Web crawler · Search engine · Sampling · Clustering · Adaptive learning

1 Introduction

The World Wide Web is a global repository storing text documents, images, multimedia, and many other information things. Since the Web is very huge, a web search engine has then been invented as a tool to serve users' information needs. Most search engines are designed to manipulate on their web collections gathered by an important component, called a *web crawler* [4].

One of the crucial challenges of search engines has been identified as an issue of maintaining versions between a local repository and the global one. The Web, in fact, is highly dynamic; a portion of the local copy can possibly get

© Springer International Publishing Switzerland 2015
O. Gervasi et al. (Eds.): ICCSA 2015, Part I, LNCS 9155, pp. 516–528, 2015.
DOI: 10.1007/978-3-319-21404-7_38

outdated. Whenever this situation occurs, it clearly makes the retrieval results do not match the actual data source. This may adversely affect the quality and credibility of the search engine itself. Thus, the web crawler is needed to perform the synchronization process as frequently as web pages are changed, in order to preserve *freshness* of the local repository. Nevertheless, as the size of the Web grows rapidly, it becomes more difficult task to replicate the whole Web, or a large significant portion. In most typical scenarios in which the resource is limited, a crawler is only allowed to periodically download a fixed number of web pages (or bandwidth). Therefore, an effective refresh policy is investigated such that, in each crawl cycle, the crawler must decide which web pages should be downloaded.

Existing studies have proven that a clustering-with-sampling-based strategy can efficiently predict the update pattern of web pages as of the whole cluster [7,10]. A cluster with higher probability of change will be ranked in higher order so that pages belonging to that cluster will be downloaded first. In this work, we also concentrate on this paradigm. Three kinds of feature set are characterized and used to first cluster the entire set of web pages. That is, pages with similar update pattern may share some characteristics and should be grouped together. Afterwards, a small number of web pages are sampled from each cluster and then individually examined their change. As a result, the update pattern of web pages in the whole cluster will be predicted. The key difference to the existing studies is that a cluster of web pages can be re-organized after each crawl. More specifically, we introduce a self-adaptive clustering for the next crawl cycle by given different feedback values as reinforcement features to a web page when it (say, in the current crawl) is either really updated, not changed, or never downloaded for a long time. Furthermore, these feedbacks are also employed in calculating page change probability.

The remainder of this paper is organized as follows. Section 2 briefly mentions to some related work. Section 3 provides detail of the proposed clustering-based web refreshing algorithm. Section 4 reports performance evaluations. Finally, Section 5 concludes the paper.

2   Related Work

Several techniques have been proposed to estimate the change of web pages [3,5,12]. Related to our work, Cho and Ntoulas [7] consider a cluster of web pages as their website. A small number of pages are sampled from each website, and used to estimate change. Hence, a website containing more changed web pages will be allocated with more crawling resources. However, web pages with similar update patterns may distribute across different websites [10].

Tan and Mitra [15] provide the state-of-the-art in clustering based web crawling policy. They first identify static and dynamic features of web pages that are correlated to their change frequency. They then divide the web pages into clusters, and check the sample of web pages from each cluster whether the pages have changed since their last download. If a significant number of web pages have changed, the whole web pages in that cluster are downloaded.

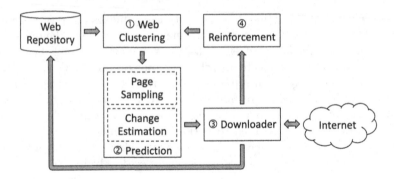

Fig. 1. The adaptive clustering-based web crawling architecture

3   Proposed Clustering-Based Web Refreshing

In this section, we propose a design model of the Adaptive Clustering-based Web Crawler ($ACWC$), attempting to predict the probability of a web page changing in order to determine refresh policies. The corresponding architecture, concluded in Fig. 1, is composed of four main modules: web clustering, change prediction, page downloader, and reinforcement learning, respectively. At the first step, web pages existing in the repository are grouped together based on their similar change characteristics. Next, to estimate the overall update pattern of each cluster, a number of web pages are sampled and their changes are examined. A list of web pages contained in the cluster with higher probability to be updated will be submitted to the page downloader first. Finally, a reinforcement learning is proceeded by feeding back the effectiveness of the current crawl cycle to improve the clustering as well as the change prediction processes again. We further mention to these modules in more detail as follows.

3.1   Clustering Web Pages

Previous studies have found that many characteristic features of web pages—both visible and invisible—are highly correlated to their changes. For instance, Douglis *et al.* [9] observe and show that frequently updated web pages are often large in size and contain many images. Fetterly *et al.* [10] illustrate that the change patterns of web pages are highly associated with their registering top-level domains. That is, a web page residing in a more popular one (e.g., .com, .net, etc.) is possibly updated more frequently. Moreover, Ali and Williams [1] have pointed out a relationship of the past updated content in web pages that can affect their changes in the future.

Relying on those findings, we believe that web pages with similar update patterns will share some characteristics, and can be grouped together. Thus, in the first step of the clustering process, the entire web pages in the local repository are extracted and represented by their feature vectors. We introduce here three types of features: static, dynamic, and reinforcement features, respectively, in which some of the first two types are compiled from [15].

The *static features* are defined as characteristics appeared in the current version of web pages in the repository. There are 20 features in total which are classified into three groups:

- 6 *content features* include words in page title, content type (e.g., text/html, text/plain, etc.), content length (i.e., size of the entity-body), number of words (i.e., word count excluding the HTML tags), number of images, and number of tables.
- 12 *URL features* include depth of the web page (i.e., number of slashes in the URL path), nine dimensions of the top-level domain name (i.e., a binary-value vector indicating which one of the corresponding .com, .edu, .gov, .org, .net, .mil, .co.th, .in.th, or other domain, the web page belongs to), URL length (i.e., character count excluding the protocol), and number of words in URL separated by non-alphanumeric characters.
- 2 *link features* include number of outgoing links and number of web pages in the same site (retrieved from the repository).

The *dynamic features* are defined as how characteristics of web pages are changed. We determine these differences, applying the Dice coefficient [8,14], based on the comparison between two consecutive versions of web pages in the repository, i.e., the version of the current crawl cycle and that of the previous one. Therefore, $\mathcal{D}_{p,i}$ denotes the difference ratio of a web page p based on the ith crawl cycle, defined as:

$$\mathcal{D}_{p,i} = 1 - \frac{2|I_i \cap I_{i-1}|}{|I_i| + |I_{i-1}|}, \tag{1}$$

where I_i and I_{i-1} are sets of feature items appeared in p at the version of ith and $(i-1)^{th}$ crawl cycles, respectively. Notice that the dynamic features can be calculated after finishing the second crawl cycle; therefore, for the first crawl, they are all ignored. Obviously, web pages currently not selected to be downloaded will not have any difference. There are 6 dynamic features in total which are classified into two groups:

- 5 *dynamic content features* include two dimensions of difference in content (i.e., considered by lines and by words), difference in images, difference in tables, and page change frequency (i.e., number of page changes divided by the number of total crawl cycles).
- 1 *dynamic link feature* includes a difference in outgoing links.

The *reinforcement features* are referred to feedbacks on the prediction performance given to improve the clustering process. Similarly, these features can only be calculated after finishing the first crawl cycle; therefore, they are all ignored at the first-time clustering cycle. There are 3 reinforcement features in total. First, the *number of correct predictions* is assigned to a web page as it is downloaded and really changed. The value of this feature will be increased by one when the web page is correctly predicted as change in the consecutive crawls; however, if it is not selected to be downloaded or downloaded but not changed, the value of the feature will be reset to zero. Second, the *number of incorrect predictions* is assigned to a web page as it is downloaded but not changed. In a similar way, the value of this feature will be either increased by one when the web page is still wrong predicted or, otherwise, reset to zero. Last, the *number of stayed times* is assigned to a web page as it is not selected to be downloaded. The value of this feature will be either increased by one when the web page is still the same version in the consecutive crawls or, otherwise, reset to zero.

As an individual web page is represented by a feature vector, in the next step, we afterwards proceed to the clustering process. Since we do not have prior information about the number of clusters, we therefore choose to apply the *X-means* clustering algorithm [13], which can produce a number of proper clusters without information supplied from any user, to construct clusters of web pages based on their similar characteristics.

3.2 Predicting Change Patterns

After finishing the web clustering, the change prediction module attempts to estimate an average change pattern of each cluster as a whole for web pages containing in. However, the size of a cluster may be large; to reduce the computational time in considering the entire web pages, a sampling technique is first suggested in [7]. In our case, we proceed at the cluster level relying on the assumption that web pages residing in the same cluster will have similar change pattern.

Considering a web cluster, the cluster centroid is first determined. Then, a number of web pages (say, 65% in our experiments) which are nearest to their cluster centroid are selected as candidates to represent that cluster. We measure the proximity between a web page and its cluster centroid using the cosine similarity function [2] which returns a floating-point value in range [0,1]. The larger the value is, the closer the web page is to its cluster centroid. These sampled web pages are used to predict change patterns of the cluster.

In practice, the change pattern of a web page can be estimated based on its change history. It is simply to assume that a web page will be updated on its own by a *Poisson process* [11], without being dependent on the others. As this assumption may not be exactly true, it has been proven to be a good approximate model to describe the real Web [3,6]. For each web page p, let T be the time that the next change occurs in a Poisson process with a change rate λ_p. The probability density function for T is

$$f_T(t) = \begin{cases} \lambda_p e^{-\lambda_p t} & \text{if } t > 0, \\ 0 & \text{otherwise.} \end{cases}$$

Hence, it can be known that the probability that p will be changed in the time interval $(0, t]$ is

$$Pr\{T \le t\} = \int_0^t f_T(t)dt = \int_0^t \lambda_p e^{-\lambda_p t} dt = 1 - e^{-\lambda_p t}.$$

In this work, a *download probability* (φ) of a web page p is achieved by two fractions: the *average change probability* ($\mathcal{P}$) with $t = 1$ meaning for one crawl cycle, and the *average reinforcement value* ($\mathcal{R}$) referred to feedbacks obtained after each crawl cycle. Therefore, the download probability of p used for prediction in the next n^{th} crawl cycle is formulated as:

$$\varphi_p = \alpha \mathcal{P}_p + (1 - \alpha)\mathcal{R}_p,$$
$$= \alpha \sum_{i=n-k+1}^{n} w_i(1 - e^{-\lambda_{p,i}}) + (1 - \alpha) \sum_{i=n-k+1}^{n} w_i r_{p,i}, \tag{2}$$

where $\alpha \in [0, 1]$ is a reinforcement coefficient determining the effect of feedbacks.

$\mathcal{P}_p$ is defined as the average of p's change probabilities looking back k crawl cycles (i.e., sliding window from $(n - k + 1)^{th}$ till n^{th}) with each change rate $\lambda_{p,i}$ calculated based on the difference of p between two consecutive versions:

$$\lambda_{p,i} = \beta \mathcal{D}_{p,i}, \tag{3}$$

where $\mathcal{D}_{p,i}$ denotes the *difference ratio* in content (considering the individual lines) of p as previously given in (1), and $\beta \in \mathbb{R}^+$ is a pre-defined constant parameter for determining the effect of this change.

w_i (in the term $\mathcal{P}_p$) is the weight determining an importance effect on change probability associated with each historical crawl cycle with $\sum_{\forall i} w_i = 1$. The more recent crawl cycle can be typically considered as more important, for instance. For this, we also follow with four distributions suggested in [15]:

1) *Non-adaptive* – A uniform weight is assigned to every cycle, meaning that all change probabilities have equal importance: $w_{n-k+1} = w_{n-k+2} = \cdots = w_n = \frac{1}{k}$.

2) *Shortsighted adaptive* – A weight is assigned to the most recent cycle, meaning that the crawler concerns only the current change: $w_{n-k+1} = w_{n-k+2} = \cdots = w_{n-1} = 0, w_n = 1$.

3) *Arithmetically adaptive* – A normalized weight is assigned an arithmetically increasing value to more recent cycle, meaning that the crawler prefers more recent changes: $w_i = \frac{i}{\sum_{i=n-k+1}^{n} i}$.

4) *Geometrically adaptive* – A normalized weight is assigned a geometrically increasing value to more recent cycle, meaning that the crawler prefers more recent changes as well, but in the different distribution: $w_i = \frac{2^{i-1}}{\sum_{i=n-k+1}^{n} 2^{i-1}}$.

Similarly, $\mathcal{R}_p$ is defined as the average of p's feedback values over k historical crawl cycles. We will explain in more details of each feedback $r_{p,i}$ later in Section 3.4. For w_i, the weight is used to determine an importance of effective feedbacks associated with each historical crawl cycle. The more recent cycle can be set to be more important, for instance. Here, we also employ the four distributions as described above.

Finally, a download probability of a cluster C is determined by average download probability over the entire sampled web pages of that cluster:

$$\varphi_C = \frac{\sum_{p \in S_C} \varphi_p}{|S_C|}, \tag{4}$$

where S_C is a set of sampled web pages from C.

3.3 Refreshing Repository

With respect to the resource constraint (i.e., bandwidth), the refreshing module of a crawler, in general, must decide which web pages should be downloaded. However, in our case of maintaining the repository freshness, the criterion (i.e., a refresh policy) setting to our crawler is that web pages residing in a cluster with higher download probability, referred to (4), will have a higher opportunity to be downloaded first.

Let $\{C_1, C_2, \ldots, C_m\}$ be a list of m clusters ranked in the descending order of their download probabilities. For the first trail, the crawler will randomly select web pages with the probability φ_{C_1} from the first order cluster C_1 to be downloaded. Then, if the download resources has still remained, the second trial will proceed by randomly selecting web pages again with the probability φ_{C_2} from the second order cluster C_2, and so on. Finally, the web refreshing process of that crawl cycle will stop when the download resource is exhausted.

3.4 Reinforcing Learning

When the web refreshing process has been stopped, this reinforcing module aims to evaluate how much performance the crawler can be achieved. By examining page changes, the reinforcement module will calculate a result corresponding to an individual web page, called later *reinforcement value*, and then feeds it back to the previous clustering and change prediction modules for further enhancement.

There are three cases of reinforcement which can be occurred for a given web page, i.e., either when the page is not downloaded, downloaded and changed, or downloaded but not changed. Hence, in this subsection, we will describe in detail about how to determine the reinforcement value.

As previously mentioned in (2), the reinforcement $r_{p,i}$ for a web page p examined after the ith crawl cycle can be here formulated as:

$$r_{p,i} = \begin{cases} e^{-\frac{\gamma}{a^2}} & \text{if } p \text{ is not downloaded at } i^{th} \text{ cycle,} \\ e^{-\frac{1}{b^2}} & \text{if } p \text{ is downloaded at } i^{th} \text{ cycle, and really changed,} \\ 0 & \text{otherwise,} \end{cases} \tag{5}$$

where $\gamma \in \mathbb{R}^+$ is a pre-defined constant parameter, a is the number of consecutive crawl cycles that p is not downloaded, and b is the number of consecutive crawl cycles that p is downloaded and also changed.

Table 1. Statistic of changed web pages

	May.	Jun.	Jul.	Aug.	Sep.	Oct.	Nov.	Dec.
% changes	33.68	36.51	37.60	34.65	38.03	37.67	37.76	38.38

As it can be seen for a limitation of the crawler using *the higher probability the higher chance* refresh policy, some web pages with a low download probability may have no chance to be downloaded due to resource constraints. To avoid this frozen page problem, a small *auxiliary score* is reinforced the probability of that web page (see, the first case in (5)). In addition, the parameter γ is set to determine the effect of this attempt. For the second case, a *reward score* is reinforced as the web page is correctly predicted. And last, a *penalty score* is rather applied as the web page is wrong predicted.

4 Experiments

4.1 Dataset

We archived 9 snapshots of a real Web subset by monthly crawling the web pages from April till December, 2014. By excluding new-born and deleted web pages, we obtained approximate 282,000 distinct and overlapping URLs of all snapshots, belonging to more than 12,500 websites. For a web page in the dataset, we compare its content between two consecutive snapshots. A change is defined if the web page has a difference ratio (by (1)) greater than 5%. Table 1 shows the number of changed web pages in each snapshot. Notice that we use web pages in the first snapshot (i.e., April) as the basis for analysis only.

4.2 Evaluation Metrics

To compare the performance of various refresh policies, we use two potential evaluation metrics: *change ratio* [7,9] and *freshness* [5], described below.

Change Ratio. To measure the ability of a refresh policy to predict changes, the change ratio is defined as the number of downloaded and really changed web pages over the total number of downloaded web pages within a crawl cycle. Let D_i be a set of web pages downloaded in the ith crawl cycle. Then, the change ratio (C_i) is formally defined as:

$$C_i = \sum_{p \in D_i} \omega_p \, \mathbf{I}(p), \tag{6}$$

where ω_p is a normalized weight given by different values to the web page p, so that $\sum_{p \in D_i} \omega_p = 1$. However, in our case, we assign a uniform weight (i.e., $\frac{1}{|D_i|}$)

to all downloaded web pages. And, $\mathbf{I}(p)$ is a change indicator function for p:

$$\mathbf{I}(p) = \begin{cases} 1 & \text{if } p \text{ is changed,} \\ 0 & \text{otherwise.} \end{cases}$$

Freshness. To measure how fresh the repository is after finishing a crawl cycle, the freshness metric represents the fraction of web pages in the repository that are up-to-date. From [5], the freshness of a web page p after the ith crawl cycle is defined as:

$$F(p; i) = \begin{cases} 1 & \text{if } p \text{ is up-to-date after } i^{th} \text{ crawl cycle,} \\ 0 & \text{otherwise.} \end{cases}$$

Here, up-to-date means that the local version of p stored in the repository is exactly the same as illustrating at the source.

Let U_i be the set of all web pages stored in the repository after the ith crawl cycle. Then, the freshness over the entire web pages is

$$F(U_i; i) = \frac{1}{|U_i|} \sum_{p \in U_i} F(p; i). \tag{7}$$

That is, if we maintain 100 web pages and if 70 web pages are up-to-date after the ith crawl cycle, the freshness is 0.7, for instance.

4.3 Results

We conducted all experiments on the archived Web data, as described in Section 4.1, for the fairness comparison reason. That is, all refresh policies assigned to a crawler do not need to download web pages on the real Internet, but within the local web collection instead. To simulate the real crawling situation, the crawler will use a data snapshot in April as the starting point to analyze and predict web pages which should be further downloaded in the next crawl cycle in May, and so on.

Our Adaptive Clustering-based Web Crawler are configured using four different settings for w_i discussed in Section 3.2, named *ACWC-nad* (non-adaptive), *ACWC-sad* (shortsighted adaptive), *ACWC-aad* (arithmetically adaptive), and *ACWC-gad* (geometrically adaptive), respectively. For the parameters setting in all experiments, the reinforcement coefficient α in (2) is set to 0.6, the constant parameters β in (3) and γ in (5) are set to 4 and 8, respective.

We compared the performance of our approach with that of the existing state-of-the-art *CLS* algorithm proposed in [15], using four various settings (i.e., *nad*, *sad*, *aad*, and *gad*), as well. For this set of experiments, the sliding window size k is set to maximum 5 crawl cycles and the download resource is set to $90K$ web pages, for both algorithms.

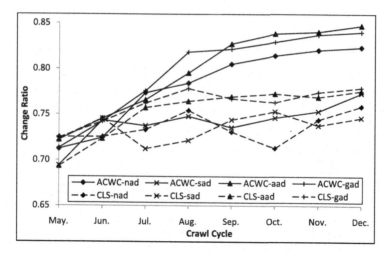

Fig. 2. Change ratio comparison between $ACWC$ and CLS under four adaptive policies

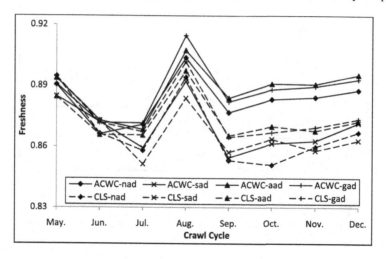

Fig. 3. Freshness comparison between $ACWC$ and CLS under four adaptive policies

Figures 2 and 3 show the comparison results under the change ratio and freshness metrics. As it can be seen that all policies of $ACWC$, except $ACWC$-sad, outperform ones of CLS with higher change ratio and freshness. ACWC provides slightly higher change ratios at the beginning, and then produces significantly higher ratios in the long run, indicating that the reinforcement mechanism indeed affects on the change prediction performance. The reason why the performance of $ACWC$-sad drops is that the reinforcement mechanism will work well under the support of a number of historical reinforcing results while the sad policy adopts the most recent one only. Furthermore, in Fig. 3, the reason why the freshness values are sharply increasing at the 4th crawl cycle (in August) is

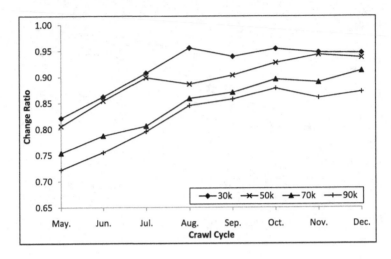

Fig. 4. Change ratio comparison for *ACWC-aad* under different download resources

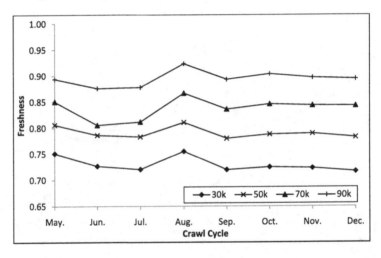

Fig. 5. Freshness comparison for *ACWC-aad* under different download resources

due to the fact that the number of web pages changed is lower than the others (see, Table 1). In other words, a large number of web pages in August are still up-to-date.

We further studied the effect of download resources on the performance of our *ACWC*. The experiments were conducted based on varying amount of download resources to $30K$, $50K$, $70K$, and $90K$ web pages, respectively. However, we only reported the results of *ACWC-aad* since the others also show the similar trend. Figures 4 and 5 illustrate the change ratio and freshness obtained under the same parameter settings $\alpha = 0.6$, $\beta = 4$, $\gamma = 8$, and $k = 5$, respectively.

As it can be seen in Fig. 4, if the resources are increased, the change ratio will be dropped. Since when the download resource has still remained, the crawler will choose web pages with lower download probability so that the crawler will have a higher chance of wrong decision. In contrast to the results depicted in Fig. 5, it is clearly seen that the more the download resources are provided, the fresher the local repository is obtained.

5 Conclusion

In this paper, we investigate the problem of how to maintain freshness of the local repository for the modern search engine providers. Due to the resource constraints, we have designed an effective download policy assigned to a crawler for deciding which web pages should be downloaded first. Relying on an assumption that web pages having similar change patterns will share some similar characteristics, we have shown that the download policy can be determined by change patterns of groups of web pages. In addition, we have proposed a reinforcement mechanism by feeding back some auxiliary, reward, and penalty scores calculated from the past crawls to enhance the next crawl performance. Experimental results show that our approach outperforms the existing clustering-based web crawling policy in many cases. However, we here have chosen to apply only one clustering algorithm from many existing ones in our experiments. We then expect to explore different clustering algorithms and study their impacts on the crawl performance.

References

1. Ali, H., Williams, H.E.: What's changed? measuring document change in web crawling for search engines. In: Nascimento, M.A., de Moura, E.S., Oliveira, A.L. (eds.) SPIRE 2003. LNCS, vol. 2857, pp. 28–42. Springer, Heidelberg (2003)
2. Baeza-Yates, R.A., Ribeiro-Neto, B.A.: Modern Information Retrieval. Addison Wesley, England (1999)
3. Brewington, B.E., Cybenko, G.: Keeping up with the changing web. Computer **33**(5), 52–58 (2000)
4. Burner, M.: Crawling towards eternity: Building an archive of the world wide web. Web Techniques Magazine **2**(5), 37–40 (1997)
5. Cho, J., Garcia-Molina, H.: Synchronizing a database to improve freshness. In: Proceedings of the ACM SIGMOD International Conference on Management of Data, pp. 117–128 (2000)
6. Cho, J., Garcia-Molina, H.: Effective page refresh policies for web crawlers. ACM Transactions on Database Systems **28**(4), 390–426 (2003)
7. Cho, J., Ntoulas, A.: Effective change detection using sampling. In: Proceedings of the 28th International Conference on Very Large Data Bases, pp. 514–525 (2002)
8. Dice, L.R.: Measures of the amount of ecologic association between species. Ecology **26**(3), 297–320 (1945)
9. Douglis, F., Feldmann, A., Krishnamurthy, B., Mogul, J.: Rate of change and other metrics: a live study of the world wide web. In: Proceedings of the USENIX Symposium on Internet Technologies and Systems (1997)

10. Fetterly, D., Manasse, M., Najork, M., Wiener, J.: A large-scale study of the evolution of web pages. In: Proceedings of the 12th International Conference on World Wide Web, pp. 669–678 (2004)

11. Grimmett, G.R., Stirzaker, D.R.: Probability and Random Processes, 3rd edn. Oxford University Press, England (2001)

12. Ntoulas, A., Cho, J., Olston, C.: What's new on the web?: the evolution of the web from a search engine perspective. In: Proceedings of the 13th International Conference on World Wide Web, pp. 1–12 (2004)

13. Pelleg, D., Moore, A.W.: X-means: Extending k-means with efficient estimation of the number of clusters. In: Proceedings of the 17th International Conference on Machine Learning, pp. 727–734 (2000)

14. Sørensen, T.: A method of establishing groups of equal amplitude in plant sociology based on similarity of species and its application to analyses of the vegetation on danish commons. Biologiske Skrifter $5(4)$, 1–34 (1948)

15. Tan, Q., Mitra, P.: Clustering-based incremental web crawling. ACM Transactions on Information Systems $28(4)$, 17:1–17:27 (2010)

A Framework for End-to-End Ontology Management System

Anusha Indika Walisadeera[1,3(✉)], Athula Ginige[2], Gihan Nilendra Wikramanayake[3], A.L. Pamuditha Madushanka[3], and A.A. Shanika Udeshini[3]

[1] Department of Computer Science, University of Ruhuna, Matara, Sri Lanka
waindika@cc.ruh.ac.lk
[2] School of Computing, Engineering and Mathematics, University of Western Sydney, Penrith NSW 2751, Australia
a.ginige@uws.edu.au
[3] University of Colombo School of Computing, Colombo 07, Sri Lanka
gnw@ucsc.cmb.ac.lk, {pammadushanka,udeshini14}@gmail.com

Abstract. An Ontology once developed needs to be kept up-to-date preferably as a collaborative process which will require web based tools. We have developed a large user centered ontology for Sri Lankan agriculture domain to represent agricultural information and relevant knowledge that can be queried in user context. We have generalized our design approach. In doing so we have identified various processes that are required to manage an ontology as a collaborative process. Based on these processes we developed an ontology management system to manage the ontology life cycle. The main processes such as modify, extend and prune the ontology components as required are included. It also has facilities to capture users' information needs in context for modifications, search domain information, reuse and share the ontological knowledge. This is a semi-automatic ontology management system that helps to develop and manage complex real-world applications based ontologies collaboratively.

Keywords: Ontology · Ontology development · Ontology management systems

1 Introduction

The motivation for this work was a problem identified among the people working in the agriculture domain in Sri Lanka. They did not have an agricultural knowledge repository that is consistent, well-defined, and provide a representation of the agricultural information and relevant knowledge needed by the people in agriculture domain within their own context. When providing agricultural information within the context of their specific needs, such information could make a greater impact on their decision-making process [1]. Since farmers are the main stakeholders in this domain we identified the context specific to the farmers in Sri Lanka such as *farm environment*, *types of farmers*, *farmers' preferences*, and *farming stages* (i.e. user context model) [2]. Next we have developed user centered ontology for Sri Lankan farmers to represent the necessary agricultural information and knowledge that can be queried based on the identified context [3]. We then generalized the specific design approach [3] and expanded the ontology for other stakeholders in the agriculture domain in Sri Lanka [4].

© Springer International Publishing Switzerland 2015
O. Gervasi et al. (Eds.): ICCSA 2015, Part I, LNCS 9155, pp. 529–544, 2015.
DOI: 10.1007/978-3-319-21404-7_39

After developing the ontology we had to devise a method to maintain it by giving the facilities to update the structure of the ontology and the knowledge base based on the initial (developed) ontology, search the domain information in user context as required, and gather the user information needs and the related context to extend the ontology further.

This work was carried out as a part of the Social Life Networks for the Middle of the Pyramid (SLN4MOP) (www.sln4mop.org) project. SLN4MOP is an International Collaborative research project aiming to develop a mobile based information system to support livelihood activities of people in developing countries [5]. This research work is aiming to provide agricultural information and relevant knowledge in context to people working in the agriculture domain [6]. The mobile based information system is developed and deployed for farmers that make use of this ontology to provide the needed knowledge (http://webe11.scem.uws.edu.au/slnfarmer/).

If the developed ontology is not up-to-date or the annotation of knowledge resources is inconsistent, redundant or incomplete, then the reliability, accuracy, and effectiveness of the ontology based systems decrease significantly [7][8]. Ontology building is a significant challenge for a number of reasons, for example it takes a considerable amount of time and effort to construct an ontology, it requires a sophisticated understanding of the subject domain, and also it is even greater challenge if the ontology developer or engineer is not familiar with the domain of interest. Due to the increase in volume of information, capturing the information, maintaining it and making it usable are challenges. Gargouri, Lefebvre, and Meunier [9] also have stressed that necessity of an ontology maintenance process due to the complexity of the changes. To enhance the use and maintenance of an ontology, a proper ontology life cycle management system is required. Therefore it is very important to be able to practically maintain the developed ontology by updating the content of the ontology in a timely manner, for example, extending the ontological structure by improving coverage and modifying the instances (individuals) in the knowledge base.

Thus Ontology Management System (OMS) can be very useful for the current age of information system. Through the end-to-end solution the system will provide all the facilities to meet the users' requirements. Thus we developed this as End-to-End Ontology Management System.

In this paper, we have proposed a framework to manage a developed ontology as a collaborative process by introducing a semi-automatic structure for populating the ontology, modifying the ontology content with human interaction (require ontology developers and a team of domain experts), searching the domain information in context, and capturing users' information needs in context. We use a web based application to deploy the proposed framework. With the help of this web based ontology management system, the people with little knowledge about the ontology can help to modify the ontology, and use the ontological information and knowledge for their needs.

The remainder of the paper is organized as follows. Section 2 presents related research in this field. Section 3 describes our framework for an end-to-end OMS. Section 4 summarizes the architecture of proposed OMS. Section 5 concludes the paper and describes the future directions.

2   Related Work

There are several designs, architectures, and tools for OMSs. Here we have only cited some of the studies in this field that help us to clearly identify the challenges and issues in this field at present.

Ding has proposed an OMS which contains three (3) major components as shown in Fig. 1 [10]. This study mainly helps us to identify the basic components of an OMS.

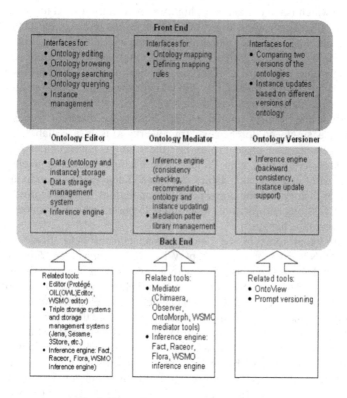

Fig. 1. Three Components of OMS [10]

Ontology Editor facilitates editing an ontology including creating and updating ontology and its instances. The front end is a user friendly interface that helps users to easily create and update ontology and its instances. The back end is the data storage management systems that normally are databases.

Ontology Mediator handles the ontology mapping. It provides the interface for ontology mapping and defining ontology rules. The front end user interface is used for semi-automatic ontology mapping such as drag and drop, help for defining the mapping rules, etc. Inferring new knowledge from the ontology happens in the backend.

Ontology Versioner handles different versions of the ontologies with backward consistency support. It also looks after the versioning of related instances. The front end is used to show the difference of the two versions of the ontologies and back end provides backward consistency among different versions of the ontologies.

Harrison and Chan [11] have developed the Distributed Ontology Management System (DOMS) as a framework for handling the storage, versioning, sharing, and security of ontologies in a repository. However, they do not consider the main components of the OMS such as updating the ontology (add, delete, change the ontology components) and instance management.

The proposed methodology in [9] supports user through discovery of terms and relations which are potentially useful for the ontology maintenance. One dimension of the problem which is the extraction of highly semantically related terms has been only considered (other dimensions are: extraction emergent terms in the texts that are related to the ontology, the integration of new terms and relations to current ontology). This methodology also has two serious limitations such as the proposed model can only handle stable corpora (i.e. if the text changes then all the process must be reworked) and the results produced by the classifications are occasionally problematic in the absence of the linguistic interpretation [9].

The system discussed in reference [12] supports the different modes of ontology evolution in a single comprehensive framework. This is a collaborative approach implemented using two protégé plug-ins: the *Change management plugin* and the *PROMT plugin* combining with Protégé. The Change management plugin provides access to a list of changes and enables users to add annotations to individual changes, see a concept history and the corresponding annotations. PROMT provides comparisons of two versions and facilities to examine the users who performed changes, and accept or reject the changes.

There are other tools in the ontology management field. Some examples are OKBC, KAON2, SymOntoX and pOWL. *Open Knowledge Base Connectivity (OKBC)* provides a set of operations with a generic interface underlying knowledge representation systems (KRSs). This interface isolates an application from many of the idiosyncrasies of a specific KRS and enables the development of tools (e.g. graphical browsers, frame editors, analysis and inference tools) that operate on many KRSs [13]. *KAON2* is a successor to the KAON [14]. It is an infrastructure for managing OWL-DL, SWRL, and F-Logic ontologies. This tool supports ontology management by providing an editor, a Web interface, an API and inference engine based on a deductive database approach. *SymOntoX* is a Web-based ontology editing environment for the e-business domain. This is an open source software system that supports collaborative and distributed ontology authoring activities [15]. *pOWL* is also a Web-based collaborative semantic web ontology editing environment [16].

Based on the literature review we can see that there is no consensus on a method, guideline or tool for the ontology management process. We can also see that the above studies mainly focus on one or two aspects of the ontology maintenance. The main objective of the ontology management systems is to provide holistic control over management activities of the ontological information by externalizing them from the application programs [17]. A collaborative approach to ontology design and development is a joint effort reflecting the experiences and viewpoints of people who intentionally cooperate to produce it [18].

In our application we need to provide the facilities to maintain a large scale and complex ontology based on a real-world application scenario. Moreover, through our ontology development process, we have experienced, when we have large number of instance data, the ontology population using ontology development tools such as Protégé is a tedious and very time consuming task. We also have experienced when we use the ontology management tools we should have some knowledge about the tools such as what are the functions available, how to use them, and so on. To address some of the above shortcomings we have developed a framework for an end-to-end ontology management system to manage the developed ontology in collaborative and holistic manner. Through this framework we overcome the below listed challenges and issues that we have faced in ontology maintenance;

- *way to fully populate a knowledge base based on the ontology by getting the support from user community (e.g. agriculture community) who have no knowledge about the ontology (or ontology development tools)?*
- *way to modify a developed ontology by getting the user requirements from the domain users?*
- *way to reuse, share, and search the information and knowledge in an ontology?*

In section 3 we describe our framework in detail. Based on the framework we also present a Web-based system that we developed to address the above challenges.

3   End-to-End Ontology Management System

We have identified a framework for end-to-end ontology management system through our ontology development process. Fig. 2 shows this framework.

By capturing agricultural information and knowledge from the literature and domain users, for example; domain experts and farmers, we first identified the domain knowledge and user requirements to develop the ontology. Next we designed our ontology based on the design framework which is explained in [3] and implemented the ontology using protégé ontology development environment based on OWL 2-DL [19]. By populating the ontology we have created a knowledge base based on the developed (or initial) ontology. Finally, we have developed an OMS and its framework is shown in Fig. 2. This system has mainly two processes for community based ontology development and maintenance. It also has two facilities for the domain users and these facilities will help the above processes.

- A.   Populate the ontology (Process 1)
- B.   Modify the structure of the ontology (Process 2)
- C.   Search domain information in context (Facility)
- D.   Capture users' information needs and related user contexts (Facility)

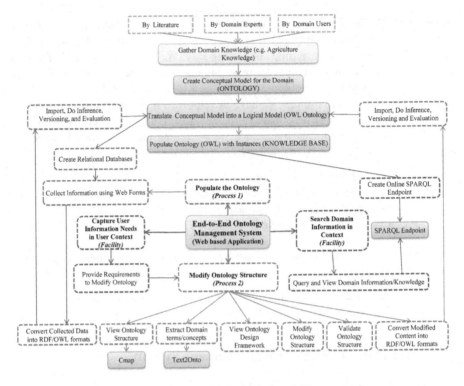

Fig. 2. A Framework for End-to-End Ontology Management System

This framework provides the essential facilities to manage the ontology life cycle by supporting the identified processes. Using the process *"populate the ontology"* we can get the support from the agriculture community to fully populate the knowledge base in the long term. The process to *"modify the structure"* helps us to extend and prune the ontology based on the changing and/or expanding user requirements and related user contexts. This process can be performed by agriculture domain experts and ontology developers. To get the benefits from the knowledge base for all the stakeholders in the domain by finding the right information based on their context we have included facility *"search domain information in context"*. The facility *"capture the user information needs and related user contexts"* collects the information required to extend the ontology further. Therefore this OMS supports a community actively maintain the ontology. Each process/facility is explained below.

A. Populate the Ontology

Agriculture related data such as crops, crop varieties, new hybrids, pest and diseases are constantly changing and need to be updated regularly. Since we have large number of instances, inserting instances directly to the OWL Ontology (using protégé) is a very time consuming and tedious task. The big challenge/issue is to populate the ontologies used in real-word, large-size applications. We therefore implemented a system as a solution to this through our framework by giving the facility to populate the ontology efficiently. To populate, we specially get the involvement of the people in the domain, for example, domain experts such as agricultural instructors, information specialist and researchers in agriculture community. To fully populate the ontology with the real data, we develop a semi-automated system to capture this information using simple web forms that can be easily understood by domain experts. For this we had to research and find solutions to following; how to create relational databases to represent ontology model for data capturing, how to capture instance values through the web forms, how to convert the collected data into OWL or RDF formats, and finally how to integrate collected instances into the existing (or initial) ontology (refer the steps in Fig. 2 with respect to the process to populate the ontology).

Step 1: Create Relational Database

Here we do not need to fully convert the ontology structure into the relational database schema, because we need this only for the ontology population. When creating a relational database, the tables were created for each concept defined in the ontology. The data properties identified in the ontology were defined as attributes of the tables. For example, Crop Length, Crop Weight, Color, etc. are the attributes of the Variety table. In the database structure we do not need to create the inverse object properties, is-a relationships, and inferred relationships in the ontology. Since the data properties of super class can be inherited by subclass of it, we define those data properties related to the subclasses for data gathering. For example, AgroZone is a subclass of the Zone class and the Zone class has data properties such as ZoneType, MaximumRainfall, and MinimumRainfall. Because of the inheritance we need to add these attributes (data properties) to the AgroZone class as well. The object properties of the super class can be inherited by subclass of it, for that we created relationships separately for the super class as well as for subclass. For example since Variety is a subclass of Crop, the growsIn object property which belongs to Crop is introduced as Crop growsIn Location and Variety growsIn Location.

Based on the above basis, we developed a new algorithm to convert an ontology into a relational database schema that can be used to populate the instance values and relationships;

1. Create a table for each concept in the ontology
2. Specify clearly the ownership of the instances
3. Define data properties (including negation data property) of a concept as attributes of specified table of a concept
4. Define object properties (including negation object property) as
   a. If there is an object property (relation) between two concepts then define the tables based on the basics of the creating database schema (i.e. according to the types of the relations; one-to-one, one-to-many, and many-to-many)
   b. If there is an object property between two concepts and one of these two concepts defined as a range is a super concept, then replace this concept (super concept) by all sub concepts of it
5. If a concept is a sub concept of a concept (super concept) then

    a.   Sub concept can inherit all data properties of super concept and use these inherited properties as own; then repeat step 3 for these properties

    b.   Sub concept can inherit all object properties of super concept and use these properties as own; then repeat step 4 for inherited object properties

6.   Identify a primary key of a table based on the basics of the creating database schema

Step 2: Collect Information Using Forms

To collect the information with respect to the relational database schema (defined in step 1), we have used a framework called "CBEADs": **C**omponent **B**ased **E**business **A**pplication **D**evelopment and **D**eployment **S**hell [20] as a data capturing application. This framework which is created using PHP and MySQL has the potential to evolve with changing requirements. Forms as shown in Fig. 3 were created to gather required data to the tables using the CBEAD application.

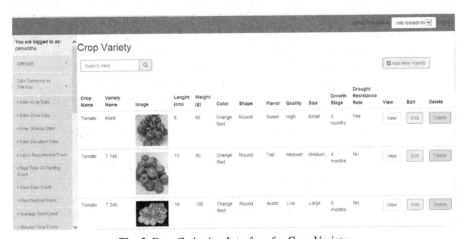

Fig. 3. Data Gathering Interface for Crop Variety

Step 3: Convert Collected Data into RDF/OWL Format

Then we applied the mapping procedure to convert these collected data into the Resource Description Framework (RDF) (http://www.w3.org/RDF/) or OWL format. For this purpose we need especially four syntaxes such as to define the instances, define the ownership of the instance, define related data properties to specified instance, and define object properties (and its negation) to specified instance. In order to generate the text file (OWL format) according to the given structure we developed a PHP code snippet.

Step 4: Integrate Instances into Existing Ontology

Next we upload the generated file (in step 3) into the logical ontology (i.e. OWL Ontology) and then checked for the consistency. The implicit knowledge is derived from the ontology through inference and reasoning procedures attached to the ontology. We finally handle different versions of the knowledge bases using versioning (how to maintain the changes in ontologies – see step 7 in section B).

B.   Modify the Ontology

Since the process to modify the structure of the ontology is complex we need to manage this process carefully. This has three processes; insertion, deletion, and updating (change). Each process has three main activities. For example in insertion process it needs to consider Inserting Concepts, Inserting Data properties and Inserting Object properties. In the same manner these activities can be seen in deletion and updating processes. The following steps have been proposed to modify the ontology. The steps below are referred with the steps in Fig. 2 related to the process Modify Ontology Structure (under process 2).

Step 1: Especially this facility (modify ontology) is provided for the ontology developers and domain experts such as agricultural experts, agricultural instructors and researchers who have domain knowledge to modify and validate the ontology. First they need to identify the requirements to modify the ontology. Through this system we can capture the user requirements explained in section D. We have provided the facility to view the structure of the initial (or developed) ontology. Concept map (Cmap) is a graphical tool for organizing and representing knowledge [21]. We used the Cmap tool to view the graphical representation of the ontology for better user understanding. An extension of CmapTool called COE (Collaborative Ontology Environment) was used to view, navigate, share, and criticize knowledge model. Based on the ontology structure, the people who are responsible for this process can identify what is the missing information, what information they want based on the identified requirements to modify the ontology, etc. Then they can decide the modifications based on the requirements and/or cover (or extend) the domain knowledge accurately.

Step 2: In first step if they identify the missing information or information they want to modify in the ontology structure then we provide the facility to extract the related terms, concepts, and basic hierarchies (taxonomies) based on the knowledge sources (e.g. agriculture leaflets in plain text files or pdf files). Moeng, Ayalew, and Mosweunyane [22] have done an experimental evaluation of five ontology construction tools (OntoGen, TextToOnto, GATE, Terminate and Semantic Vectors) in relation to developing a HIV/AIDS FAQ retrieval system. All the tools are freely available on the Web, semi-automatic, and also support the English language. They have assessed the tools based on the quality of the resulting ontologies and other features provided by the tools such as graphical visualization. Text2Onto (the latest version of Text-To-Onto) is found to be most appropriate and will be used to construct the HIV/AIDS ontology for FAQ retrieval. We therefore decided to select this tool (Text2Onto) to extract the domain terms, concepts, and basic hierarchies as it is capable of processing natural language text which helps to create the ontology. Through this tool we can identify the domain-specific and domain-independent basic ontology components based on the input sources from domain texts (natural language texts). From this process we identify the specific piece of knowledge related to the inputs. However we need to verify this knowledge further with the background knowledge of the input text.

Step 3: We next give the facility to view steps of the ontology design framework to represent the information in user context [3]. Since farmers are the main stakeholders in the domain and other stakeholders are willing to help farmers in various manners, we have identified the users' context specific to the farmers in Sri Lanka [2]. Using the design framework, they can extend the ontology for different scenario problems. This will help them to clearly identify the ontology components further.

Step 4: In our end-to-end framework we then provide the facility to modify the structure based on the identified ontology components in previous steps. Metadata is often called data about data or information about information [23]. The term metadata is used differently in different communities. To modify the structure of the ontology we use the metadata. The metadata provides the information to the users how to modify the structure of the ontology, for example, how to insert the concepts, how to delete the data properties or object properties, and how to change a concept definition or modify the data properties. For example, Table 1, Table 2, and Table 3 show the metadata to insert a new concept to ontology, delete a concept from the ontology, and update an object property in the ontology respectively. For example, the domain experts first need to look at the ontology structure by viewing the Cmap. Then they need to identify whether there are any mismatch, unnecessary, or incorrect concepts. If they are identified, then the identified concepts need to be deleted. Next we need to carefully identify the sub-concepts, data properties, and object properties which belong to the concept (identified). If the identified concept is a super concept then we need to identify the sub-concepts of it. If sub-concepts are engaged with the other concepts of the ontology then we cannot delete these sub-concepts. If it is not, the sub-concepts also need be deleted (see Table 2). In this manner we can delete any concepts in the existing ontology as described in Table 2.

Table 1. Metadata to Insert a New Concept

Property	Description
Concept	Name of the new concept
Concept Definition	Description of the new concept
Super-concept of	If it is a super-concept, specify concepts (in the existing ontology) that to be sub-concepts of this concept (new concept)
Sub-concept of	If it is a sub-concept, specify concepts (in the existing ontology) that to be supper-concepts of this concept (new concept)
Constraints	Conditions (e.g. disjoint with, conjunctions, etc.) or any other assumptions
Comment	Comments by explaining the necessity of this new concept
Note	Specify the meaningful concept names related to the application

Table 2. Metadata to Delete a Concept

Property	Description
Concept	Name of the concept (to be deleted)
Sub-concept (to be deleted)	If this concept is a super-concept of the existing ontology, then name all the sub-concepts of it
Data Property (to be deleted)	If this concept has data properties then name all these data properties
Object Property (to be deleted)	If this concept has object properties then name all these object properties
Comment	Comments by explaining this action (e.g. why do you need to delete this concept)
Note	When deleting a concept, all the sub concepts, data properties and object properties of it, if it is not engaged with other concepts need to be deleted. If it is engaged with other concepts, then all the sub concepts, data properties and object properties of the deleted concept do not need to be deleted.

Table 3. Metadata to Update an Object Property

Property	Description
Object Property	Name of the object property (to be edited)
New object property	Edited object property
Object Property Definition (Annotation)	Description of the object Property (edited) (if changed)
Domain	The domain of the property (if changed)
Range	The range of the property (if changed)
Constraints	Conditions (e.g. Functional, Inverse Functional, Transitive, Symmetric, etc.) (if changed)
Comment	Comments by explaining the necessity of this edition

Step 5: Next we store the modified data in the databases for validation (verification). The content is validated by domain experts and the ontology construction is done by ontology developers. Without proper validation there is no assurance for the content and quality of the ontology. We have used the Delphi method [24] and the Modified Delphi method (or The Nominal Group Technique) [25] to validate our initial ontology. The Delphi method gives a systematic way for gathering experts' feedbacks and critiques on ontology components to validate the ontology. Using the web based tool we are providing a facility to domain experts who are in different places to see the modified ontology components and write their own comments and feedbacks using the web forms. Then the developers and administrators can look at their comments and feedbacks, discuss those and finally get convergent.

Step 6: After validation of the modified data we apply the mapping procedure to convert these data into the RDF or OWL format. For this we need syntaxes to define a concept, data property, object property, domain and range of object property, etc. We automatically generate this OWL file as described in section A (step 3 in section A).

Step 7: The final step is to import the generated file (in step 6) into initial ontology for information integration. After integrating we need to do inferences using reasoners plug-in with Protégé, and internal evaluation (by evaluating the added values) against the user requirements using Description Logic (DL) queries and SPARQL queries already available in Protégé environment. When the ontology is modified, we need to handle the different versions of the ontologies. We managed the different versions of the ontology using a versioning system. It also supports the ontology evolution. We therefore handle the different versions of the ontologies with backward consistencies. For versioning we used the ontology versioning attributes that help us to track version numbers and to see whether the new ontology is compatible with the previous version. Such attributes are versionInfo, priorVersion, incompatibleWith, and backwardsCompatibleWith, where versionInfo is the current version number of the ontology, priorVersion is the previous version number of the ontology, incompatibleWith is the previous version of the ontology that the current version is not compatible with, and backwardsCompatibleWith is the previous version of the ontology that the current version is compatible with. We can take a set of individuals and/or ontology structures as a criterion to decide after modify the ontology whether it is compatible with the previous version of the ontology. For example we checked the compatibility using a same data set (individuals) with different ontology structures and a same ontology structure with different data sets.

C.   Search Domain Information in Context

Here we discuss how to query and view the domain information and knowledge in user context. Through this system under search information we have provided these two facilities (http://webe2.scem.uws.edu.au/oms/searchInformation.php). Especially, farmers can view the domain information in their context and other stakeholders in agriculture domain can retrieve domain information based on their interest. For the farmers, we have provided specific answers to their questions in their context using a natural language (in English, Sinhala and Tamil). Fig. 4 shows a user friendly interface for searching information. For example, from this, farmers can get the correct answers to their questions based on their location, soil type, and other characteristics related to their context.

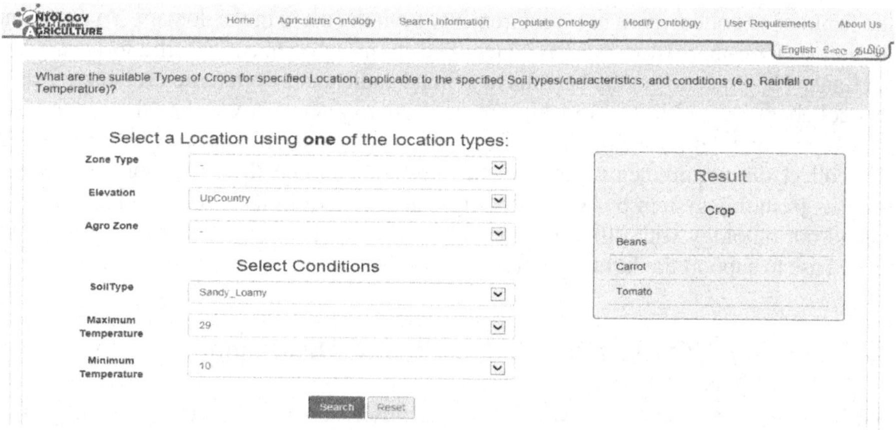

What are the suitable Types of Crops for specified Location, applicable to the specified Soil types/characteristics, and conditions (e.g. Rainfall or Temperature)?

Fig. 4. Information in Context

For other users in agriculture domain we have developed the online knowledge base (SPARQL endpoint) based on the ontology. We first have created the completed RDF model (i.e. both Asserted and Inferred models) to represent and store information on the Web. We first created the SPARQL endpoint using ARC2 (appmosphere RDF classes) Semantic Web toolkit [26] to query the domain knowledge using SPARQL queries (refer http://webe2.scem.uws.edu.au/oms/index.php). This application is also useful to reuse and share domain knowledge through the use of the graph database via the Web. The detail of creating the online knowledge base is described in [3]. For instance, using SPARQL endpoint (in Information Retrieval of Search Information) we can query suitable control methods to control Damping-off disease which affects Tomato and the required pesticide quantities. The related output is shown in Fig. 5.

ControlMethod	Pesticide	Quantity	Unit
Pre-pare_nursery_beds_in_well_drained_virgin_soil_or_subsoil			
Nursery_sterilization	Pad-dy_husk_and_straw		
Use_recommended_fungicides	Captan	6	grams_per_square_meter
Use_recommended_fungicides	Thiram	7	grams_per_square_meter

Fig. 5. Output of the above Query

D. Capture the Users' Information Needs in Context

Our ontology creation begins with the definition of a set of users' information needs. We take these information needs as a main motivation scenario of our application to represent the information in context using ontologies. For example, what are the suitable crops to grow, how to solve the problems of pest, etc. Through this system we provide the facilities to collect more user requirements from the users in the domain. To represent the information in context, we have identified the users' context specific to farmers in Sri Lanka [2]. Since to get the benefits to a broad audience is even more challenging task, this collaborative end-to-end ontology management system via web based interface has now been expanded to include their requirements and related context. For instance we can collect the requirements from all the stakeholders, and then stakeholders can get benefits from this system by finding the right information in their context. Then we can extend our ontology with different motivation scenarios that provide even richer knowledge base to support the farming industry.

4   Architecture of Proposed OMS and Discussion

We have explained our framework using the processes/facilities used to manage the ontology life cycle. Here we discuss the architecture of the proposed framework for end-to-end OMS. Since this is a collaborative approach, the system mostly relies on the users of the domain, their participation to keep the information current, and developers' and administrators' skills in overseeing the collaborative processes. In our system (refer http://webe2.scem.uws.edu.au/oms/index.php), there are three main user categories (e.g. domain experts, normal users, and ontology developers) with different access rights. Domain experts and ontology developers need to be logged into the system for populating and modifying the ontology. Domain experts and ontology developers can change or extend the ontology by getting the requirements and user constraints from the system. There are processes to capture user information needs in user context from the users to represent domain information in context. Domain experts also can involve in populating the ontology by capturing instance data through forms. Through this system all the stakeholders of the domain can search information using user friendly interfaces (for normal users such as farmers) and/or querying the SPARQL endpoint in context (for advanced users).

The quality of the final ontology in terms of correctness and relevancy is checked using modified Delphi method and mobile based application developed based on this ontology (http://webe11.scem.uws.edu.au/slnfarmer/). In Delphi method, 11 Agricultural Instructors gathered and discussed the criteria relevant to the fertilizer application, growing problems, etc. Based on their responses, comments, and suggestions we made judgments for the design criteria. We also received the user feedbacks through questionnaire from 30 farmers in Dambulla and Polonnaruwa area to the ontology knowledge using the mobile based application. The ontology has been refined based on these feedbacks and comments.

For the implementation we have used the Web Services. The OMS has frontend and backend server components. The backend is implemented as an application within the CBEADS. It provides CRUD services to manage data for the OMS. The frontend makes web service calls to the backend and uses the retrieved data to generate views for the client. To develop an attractive and responsive web system easily, we have used Bootstrap framework (http://getbootstrap.com/) which can create responsive web content easily. It also consists of predefined content and we can customize those according to our requirements. We developed this application in three languages as Sri Lankan people mainly use their native languages such as Sinhala and Tamil.

5   Conclusions

User centered ontology for Sri Lankan farmers is a large, complex, and real-world application. To effectively manage this type of ontology applications we need a proper ontology management system. The facilities to manage the structures of the developed ontologies as well as the knowledge bases based on the initial ontologies are needed. In this study we propose a framework shown in Fig. 2 for an end-to-end ontology management system via web based interface for large-scale development and maintenance purposes. Through this system we have given the facilities to the domain users and ontology developers to populate the ontology, modify the structure of the ontology, search and/or retrieve domain information in context, share and reuse ontological knowledge, and finally capture the users' information needs and related user context. To further populate the knowledge base easily, we developed a semi-automatic system to capture agricultural information via simple web forms and imported it into the initial ontology. We have defined the steps based on the metadata to modify the structure of the ontology. This is a collaborative, semi-automatic ontology management system that helps us to develop and manage the ontology in the long term. This is a new unique combination of existing methods and tools to support ontology maintenance; we can evaluate this framework with respect to the knowledge base and the ontology structure.

Acknowledgment. We acknowledge the financial assistance provided to carry out this research work by the HRD Program of the HETC project of the Ministry of Higher Education, Sri Lanka and the valuable assistance from other researchers working on the Social Life Network project. Assistance from the National Science Foundation to carry out the field visits is also acknowledged.

References

1. Glendenning, C.J., Babu, S., Asenso-Okyere, K.: Review of Agriculture Extension in India Are Farmers' Information Needs Being Met? International Food Policy Research Institute (2010)
2. Walisadeera, A.I., Wikramanayake, G.N., Ginige, A.: An ontological approach to meet information needs of farmers in Sri Lanka. In: Murgante, B., Misra, S., Carlini, M., Torre, C.M., Nguyen, H.-Q., Taniar, D., Apduhan, B.O., Gervasi, O. (eds.) ICCSA 2013, Part I. LNCS, vol. 7971, pp. 228–240. Springer, Heidelberg (2013)
3. Walisadeera, A.I., Ginige, A., Wikramanayake, G.N.: User centered ontology for Sri Lankan farmers. Ecological Informatics 26(2), 140–150 (2015)
4. Walisadeera, A.I., Ginige, A., Wikramanayake, G.N.: User centered ontology for Sri Lankan agriculture domain. In: 14th International Conference on Advances in ICT for Emerging Regions (ICTer 2014), Colombo, Sri Lanka, pp. 149–155 (2014)
5. Ginige, A.: Social Life Networks for the Middle of the Pyramid. http://www.sln4mop.org//index.php/sln/articles/index/1/3
6. Ginige, A., De Silva, L.N.C., Ginige, T., Giovanni, P.D., Walisadeera, A.I., Mathai, M., Goonetillake, J.S., Wikramanayake, G.N., Vitiello, G., Sebillo, M., Tortora, G., Richards, D., Jain, R.: Towards an agriculture knowledge ecosystem: a social life network for farmers in Sri Lanka. In: 9th Conference of the Asian Federation for Information Technology in Agriculture (AFITA 2014): ICT's for future Economic and Sustainable Agricultural Systems, Perth, Australia, pp. 170–179 (2014)

7. Stojanovic, L., Maedche, A., Motik, B., Stojanovic, N.: User-driven ontology evolution management. In: Gómez-Pérez, A., Benjamins, V. (eds.) EKAW 2002. LNCS (LNAI), vol. 2473, pp. 285–300. Springer, Heidelberg (2002)
8. Ding, Y., Fensel, D.: Ontology library systems: the key to successful ontology reuse. In: SWWS, Citeseer (2001)
9. Gargouri, Y., Lefebvre, B., Meunier, J.G.: Ontology maintenance using textual analysis. In: The Seventh World Multi-Conference on Systemics, Cybernetics and Informatics (SCI 2003) (2003)
10. Ding, Y.: Ontology Management System. http://sw-portal.deri.at/papers/deliverables/d17_v01.pdf
11. Harrison R., Chan, C.W.: Distributed ontology management system. In: 18th Annual Canadian Conference on Electrical and Computer Engineering (CCECE 2005), IEEE, Saskatoon (2005)
12. Noy, N.F., Chugh, A., Liu, W., Musen, M.A.: A framework for ontology evolution in collaborative environments. In: Cruz, I., Decker, S., Allemang, D., Preist, C., Schwabe, D., Mika, P., Uschold, M., Aroyo, L.M. (eds.) ISWC 2006. LNCS, vol. 4273, pp. 544–558. Springer, Heidelberg (2006)
13. OKBC. http://www.ai.sri.com/~okbc/
14. KAON2. http://kaon2.semanticweb.org/
15. Missikoff, M., Taglino, F.: SymOntoX: A web-ontology tool for ebusiness domains. In: The 14th International Conference on Web Information Systems Engineering (WISE 2003), Rome, Italy, pp. 10–12 (2003)
16. Sören, A.: pOWL – A Web Based Platform for Collaborative Semantic Web Development. http://powl.sourceforge.net/overview.php
17. Lee, J., Goodwin, R.: Ontology management for large-scale enterprise systems. Journal of Electronic Commerce Research and Applications 5(1), 2–15 (2006)
18. Holsapple, C.W., Joshi, K.D.: A collaborative approach to ontology design. Communication of ACM 45(2), 42–47 (2002)
19. Knublauch, H., Fergerson, R.W., Noy, N.F., Musen, M.A.: The protégé OWL plugin: an open development environment for semantic web applications. In: McIlraith, S.A., Plexousakis, D., van Harmelen, F. (eds.) ISWC 2004. LNCS, vol. 3298, pp. 229–243. Springer, Heidelberg (2004)
20. Ginige, A., De Silva, B.: CBEAD: a framework to support meta-design paradigm. In: Stephanidis, C. (ed.) HCI 2007. LNCS, vol. 4554, pp. 107–116. Springer, Heidelberg (2007)
21. Novak, J., Cañas, A.: The Theory Underlying Concept Maps and How to Construct Them. Technical Report IHMC CmapTools 2006-01, Florida Institute for Human and Machine Cognition. http://cmap.ihmc.us/Publications/ResearchPapers/TheoryUnderlyingConceptMaps.pdf
22. Moeng, B., Ayalew, Y., Mosweunyane, G.: Experimental Evaluation of HIV/AIDS Ontology Construction Tools. In: The Second IASTED African Conference on Health Informatics (AfricaHI 2012), pp. 339–346 (2012)
23. Understanding Metadata, National Information Standards Organization. http://www.niso.org/publications/press/UnderstandingMetadata.pdf
24. Mattingley-Scott, M.: Delphi Method. http://www.12manage.com/methods_helmer_delphi_method.html
25. NOMINAL GROUP TECHNIQUE1, HANDOUT: The Skilled Group Leader. http://www2.ca.uky.edu/agpsd/nominal.pdf
26. PHP and Semantic Web. http://blog.m1k.info/category/semantic-web-2/?lang=en&lang=en

Towards a Cloud Ontology Clustering Mechanism to Enhance IaaS Service Discovery and Selection

Toshihiro Uchibayashi[1], Bernady Apduhan[1(✉)], and Norio Shiratori[2]

[1] Faculty of Information Science, Kyushu Sangyo University, Fukuoka, Japan
{uchibayashi,bob}@is.kyusan-u.ac.jp
[2] Research Institute of Electrical Communication, Tohoku University, Sendai, Japan
norio@shiratori.riec.tohoku.ac.jp

Abstract. The continuing advances in cloud computing technology, infrastructures, applications, and hybrid cloud have led to provide solutions to challenges in big data and high performance computing applications. The increasing number of cloud service providers offering cloud services with non-uniform descriptions has made it time consuming to find the best match service with the user's requirements.

This paper is an effort to speed up the service discovery and selection of IaaS cloud services which is „best-match" to the user requirements. Preliminary experiments provided promising results which demonstrates the viability of the approach.

Keywords: Ontology · Agents · Clustering

1 Introduction

In recent years, due to the reduced cost of high speed network infrastructure and server equipments, cloud computing has been rapidly expanding. Cloud computing has a wide range of services XaaS (SaaS, IaaS, PaaS, etc.). The combination of cloud computing and other existing technologies has produced new technology research areas, like Social Cloud[1] and Hybrid Cloud[2,3] which represents novel usage of cloud computing technology. In Social Cloud, the cloud acts as a social media which facilitates the exchange of information. By directionally coupling all information relating to life through the cloud, it aims to support towards an energy saving comfortable social life. Likewise, hybrid cloud is a technology that combines the private and public clouds. The public cloud provides services to unspecified number of personal clients or private companies. Whereas, a private cloud provides in-house services limited to groups or corporate departments of a specific company. Aside from high confidentiality of important data in private clouds, the front-end part of the system provides easier access to use in public clouds. This feature provides a crucial role since easy and secure access to public cloud services are highly desired to meet the user's changing demand of computing resources in private clouds which in turn provides cost reduction while ensuring data security.

© Springer International Publishing Switzerland 2015
O. Gervasi et al. (Eds.): ICCSA 2015, Part I, LNCS 9155, pp. 545–556, 2015.
DOI: 10.1007/978-3-319-21404-7_40

In this paper, we focus on hybrid cloud. Many researches on private and public clouds have been conducted or on-going to ensure security of its distributed resources. A hybrid cloud comprises a private cloud and one or many different public clouds. To use a hybrid cloud with many public clouds, a common cloud which can act as the common interface to different public cloud resources is sought. The notion of common cloud has attracted much attention in inter-cloud research area which acts as a common gateway to access resources of public clouds. The inter-cloud is a combination of different public clouds[4,5], and proposed a common organization in the architecture and application levels.

To this, it is envisioned that by placing the cloud ontology in public clouds in the foreseeable future, a unified representation between different public clouds can be carried out. Ontology is defined as a systematic classification which exist in the world of interest, and is obtained by explicitly stating the formal relationships [6]. Through the use of ontology, more precise cloud services can be obtained that matches or nearest-match to the user requirements. As cloud computing technology advances and cloud service providers increase, the cloud ontology likewise evolves and the number of ontology elements and relationships will significantly increase making it longer to search for the required cloud service(s).

In this paper, we tackle the construction of an efficient cloud service discovery system employing ontology on a hybrid cloud computing environment. We then proposed the use of heuristic clustering methodology to enhance the discovery and selection of cloud services. Preliminary experiments provided promising results to pursue our efforts.

In the following, Section 2 describes some related work, while section 3 describes the cloud service discovery system. The cloud clustering mechanism is described in Section 4. Section 5 described some preliminary experiments and discussed its results. Lastly, Section 6 provides our concluding remarks and some directions for future research.

2   Related Works

SC Punitha [7] introduced a Hybrid Scheme for Text Clustering (HSTC) Method and evaluated its application on Text Clustering with Feature Selection (TCFS) Method. In TCFS, the group division is performed by using the semantic weight. That is, elements with closer meaning are grouped together in a cluster, which is a similar procedure we adapted in our   study. However, since our study is aimed to speed-up the search process, some specifics in forming the cluster differs.

Latifur Khan [8], uses clustering to perform an automatic construction of ontology structure. A parent serves as one cluster and the ontology structure is built by connecting these clusters. The criteria that defines the similarity differs and automatic clustering may include unintended elements. In our study, the pre-structure of the cluster is pre-determined and this exclude the possibility that unintended elements will be included in the cluster.

3   The Cloud Service Discovery System

In [9], the following, we describe a discovery system that collects information of public cloud services using agents and ontology on a hybrid cloud computing environment which constitutes a private cloud, public clouds, and a broker server, as illustrated in Figure 1. The broker server is the core of the cloud service discovery system. All cloud services information is aggregated in the broker server, and the user performs a search of cloud services by connecting to the broker server. The public cloud owns a database that stores information such as the usage and the rates of instance type cloud service which it provides. The private cloud may or may not have the same type of cloud resources as the public clouds.

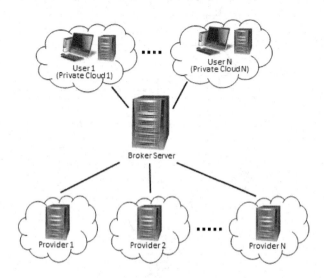

Fig. 1. A hybrid cloud computing environment

Figure 2 shows the architecture of the cloud service discovery system. The core of the cloud service discovery system consists of "Search" cloud information and "Collect" cloud information. In "Collect" cloud information, the cloud service discovery system is connected to the cloud via the Internet, and the collected information on cloud services is stored in the cloud. In "Search" cloud information, it searches for information on cloud services that are to be collected. The user can specify the algorithm to be used to perform a search of cloud services as the conditions may require.

Figure 3 shows the organization of "Collect" cloud information module which monitors the service information stored in the cloud using agents which detects any changes that occurs, and then collect the update information. The public cloud has a database that stores information, such as the instance type and price of cloud services that it offers. By linking the information in the database and ontology, we can have a common meaning of the stored information in the broker server. When a user

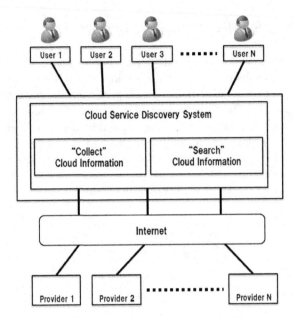

Fig. 2. Cloud Service Discovery System Architecture

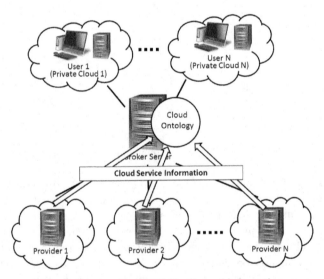

Fig. 3. Construction of "Collect" cloud information

searches for a cloud service, it is possible to do the search in the broker server with the desired specifications. The key feature in using cloud ontology in service discovery is that it can perform a commonality of instance type which is expressed differently by public clouds. The advantage of using cloud ontology for shared information is that it can utilize the existing data. By defining systematically the terms by ontology, even using different representations for each public cloud service, it is then possible to perform a unified representation without changing the original data.

The following describes a structural example of the IaaS cloud service elements and its values. In this scenario, the cloud ontology has 647 elements. Root element is "IaaS" and has five elemetnts; namely, "CaaS", "DaaS", "Cloud Service", "Infra Software", and "Computing Resource", shown in Figure 4.

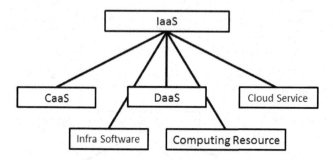

Fig. 4. The IaaS cloud service and its elements

The "Computing Resource" has seven resource elements whose presence and values are dependent on the service type offered by the provider; that is, "CPU (Hz)", "CPU (Core)", "Storage", "OS", "SLA", "Memory", and "Monthly Cost", shown in Figure 5.

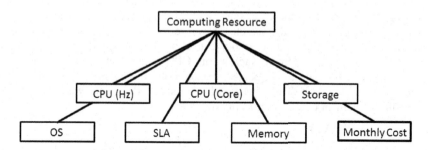

Fig. 5. Computing Resource element and sub-elements

The resource name element have values as provided by the service type, Figure 6.

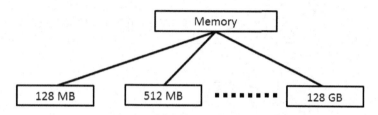

Fig. 6. Memory element and offered values

The "Cloud Service" gives the names of the cloud service providers, which in this study, we have 20 providers, Figure 7.

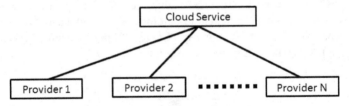

Fig. 7. Cloud Service and its elements

Furthermore, each provider has different service types and its corresponding values, as in Figure 6.

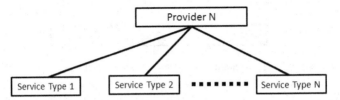

Fig. 8. A cloud service provider and its service types

4   Cloud Service Clustering Mechanism

Clustering is performed to speed up the search process. The computing resource is used as a unit in performing the search. Therefore to create a cluster, it should only contain the same type of computing resource at search time to speed up the search.

In cloud ontology, the element such as the CPU is an element of computing resource where it is assumed to be located. Furthermore, the possible value of the element is identified below it. The values with its associated service type and provider name forms an element of a cluster, as in Figure 9.

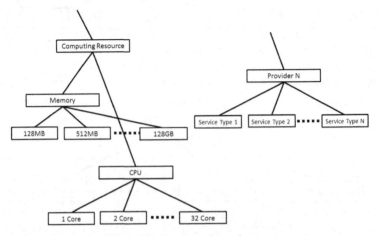

Fig. 9. The cloud ontology elements

Broadly, if the sub-clusters belong to the same provider, then this can be integrated into a cluster, Figure 10. This integration is repeated until the number of clusters equals to the number of computing resources. After clustering, the configuration will be provider, service type, and value.

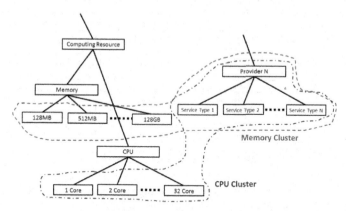

Fig. 10. Cloud ontology elements with cluster

Figure 11 illustrates a portion of the cloud ontology. The provider offers six types of services. Each service type has a set of computer resources such as CPU, memory and storage.

For example, consider the service type 1 in which the CPU is 2 Core, memory is 128MB, and storage is 12TB, and create a cluster of the cloud ontology in Figure 12.

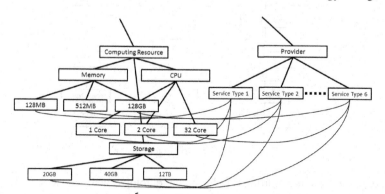

Fig. 11. A portion of the cloud ontology

Based on a heuristic clustering rule, the lowest elements of a computing resource with each corresponding service type are formed as sub-clusters. Nine sub-clusters are formed as in Figure 12. This is on the assumption that the sub-cluster is composed of the provider, service type and value.

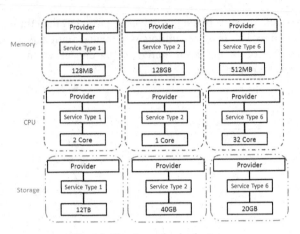

Fig. 12. The formed sub-clusters

Then sub-clusters with the same computing resource are formed into bigger clusters reducing the number of cluster counts, as in Figure 13.

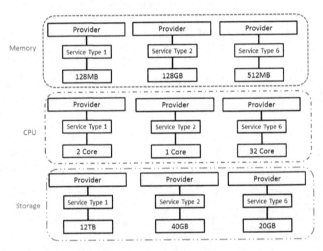

Fig. 13. formed into bigger clusters

Figure 14 shows the portion of the cloud ontology with clusters of the three computing resources, i.e., CPU, memory, and storage.

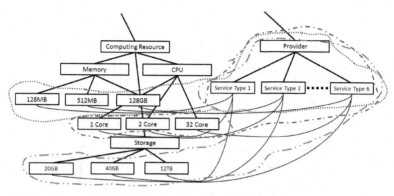

Fig. 14. The cloud ontology with clusters

Figure 15 illustrates the extraction of elements sub-ontology from the base cloud ontology based on heuristic rules to form its corresponding cluster. This process makes the description of provided services simpler and then stored in the cloud service provider database.

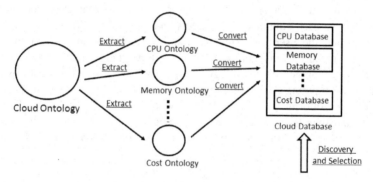

Fig. 15. The IaaS cloud discovery architecture

Figure 16 shows the architecture when cloud ontology update occurs. When the base cloud ontology is updated, the clustering process is not repeated. Instead, the difference cloud ontology is merge/added to the previously extracted sub-ontology.

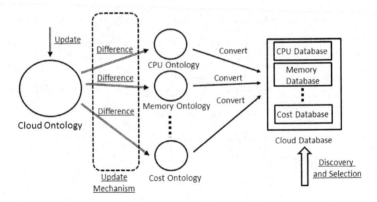

Fig. 16. The cloud ontology update architecture

5   Experiments and Evaluations

We compare the time to search for the sub-ontology and sub-ontology information which is stored in the database with or without using clustering technique. The cloud ontology that we used contains 20 providers and 194 service types. We measure the time from when the user requested the required resources until the return of the results. The resource information that the user can specify are "CPU(Hz)", "CPU(Core)", "Storage", "OS", "SLA", "Memory", and "Monthly Cost", which are elements of „Computing Resource".

The resource information and its values were specified at random and measured 100 times, and then the average times were compared, as shown in Figure 17. The average time to search for ontology without clustering was 10152ms. By adapting

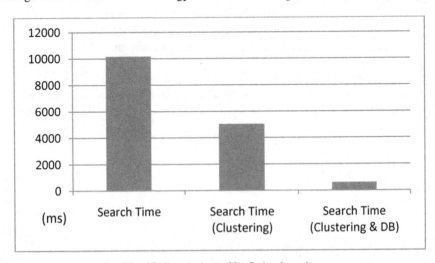

Fig. 17. Search time of IaaS cloud service

clustering, the average time to search for sub-ontology from the generated sub-ontology clusters is 5021ms, which is 50% faster. The clusters can be segregated based on the sub-ontology of the resources which can eliminate unnecessary search operation. Then the extracted sub-ontology are stored in the database, and the average search time was 601ms which is 88% faster compared with searching without segregation of sub-ontologies. Furthermore, it's 94% faster as compared to generic ontology search.

Typically, the ontology search traverses on the resources tree, but in this paper, the search is focused on the required resource created by clustering. Therefore, from the above, we can observe that by clustering, it can speed-up the search of the required cloud ontology.

6   Concluding Remarks and Future Work

Ontology can be used to discover and select the user's required IaaS cloud service. However, ontology is evolving over time and the number of elements and relationships significantly increase making it longer to search for the required ontology.

In this paper, we used clustering to speed up the search of cloud ontology. After clustering, the sub-ontologies are extracted from the clusters to create the minimum required elements of the ontology search. The relationships between elements in the sub-ontology becomes very simple making it easy to store in the database. We then compared the cloud ontology search time using the generic method (traversing the ontology tree), using sub-ontology, and using database. The measured search time after clustering and using the database exhibited a much shorter time, which exhibits promising results.

This study is still in its infancy and much work has still to be done. In the future, we envisioned to develop an algorithm which can also be applied to other cloud service models aside from the IaaS cloud ontology.

Acknowledgment. This research was supported in part by the Japan Society for the Promotion of Science Grants-in-Aid for Scientific Research 24500100.

References

1. Chard, K., Caton, S., Rana, O., Bubendorfe, K.: Social cloud cloud computing in social networks. In: 2010 IEEE 3rd International Conference on Cloud Computing (CLOUD), pp. 99–106 (2010)
2. Palwe, R., Kulkarni, G., Dongare, A.: A New Approach to Hybrid Cloud. International Journal of Computer Science and Engineering Research and Development (IJCSERD) 2(1), 01-06 (2012)
3. Gupta, A.K., Gupta, M.K.:. A New Era of Cloud Computing in Private and Public Sector Organization. International Archive of Applied Sciences and Technology 3, 80–85 (2012)

4. Buyya, R., Ranjan, R., Calheiros, R N.: InterCloud: utility-oriented federation of cloud computing environments for scaling of application services. In: The 10th International Conference on Algorithms and Architectures for Parallel Processing, pp.13–31 (2010)
5. Demchenko, Y., Ngo, C., de Laat, C., Makkes, M.X., Strijkers, R: Intercloud Architecture for Multi-Provider Cloud based Infrastructure Services Provisioning and Management. International Journal of Next-generation Conputing, 4(2) (2013)
6. Yoo, H., Hur, C., Kim, S., Kim, Y.: An ontology-based resource selection service on science cloud. In: Ślęzak, D., Kim, Tai-hoon, Yau, S.S., Gervasi, O., Kang, B.-H. (eds.) GDC 2009. CCIS, vol. 63, pp. 221–228. Springer, Heidelberg (2009)
7. Punitha, S.C., Punithavalli, M.: Performance evaluation of semantic based and ontology based text document clustering techniques. In: International Conference on Communication Technology and System Design, pp.100–106 (2011)
8. Khan, L., Luo, F., Yen, I.-L.: Automatic Ontology Derivation fromDocuments. http://ipm.lviv.ua/library/0/004/004.8/Automatic%20ontology%20derivation%20from%20documents.pdf
9. Uchibayashi, T., Apduhan, B., Shiratori, N: Towards a resilient hybrid iaas cloud with ontology and agents. In: The 14th International Conference on Computational Science and its Applications, pp. 70-73 (2014)

Improving Reliability and Availability of IaaS Services in Hybrid Clouds

Bernady Apduhan[1](✉), Muhammad Younas[2], and Toshihiro Uchibayashi[1]

[1] Faculty of Information Science, Kyushu Sangyo University, Fukuoka, Japan
{bob,uchibayashi}@is.kyusan-u.ac.jp
[2] Department of Computing and Communication Technologies,
Oxford Brookes University, Oxford, UK
m.younas@brookes.ac.uk

Abstract. This paper investigates into IaaS service provisioning in hybrid cloud which comprises private and public clouds. It proposes a hybrid cloud framework in order to improve reliability and availability of IaaS services by taking into account alternative services which are available through public clouds. However, provisioning of alternative services in hybrid cloud involves complex processing, intelligent decision making and reliability and consistency issues. In the proposed framework, we develop an agent-based system using cloud ontology in order to identify and rank alternative cloud services which users can acquire in the event of failures or unavailability of desired services. The proposed framework also exploits transactional techniques in order to ensure the reliability and consistency of the service acquisition process. The proposed framework is evaluated through various experiments which show that it improves service availability and reliability in hybrid cloud.

Keywords: Component · Hybrid cloud · IaaS · Availability · Reliability

1 Introduction

In hybrid cloud environment, organizations and companies can provide some of its services internally using private cloud, while other services are provided externally using public cloud [1]. For instance, mission critical applications and data can be deployed at private cloud while other part of applications can be delegated to public cloud. While hybrid clouds provide immense benefits and promising returns, they are still vulnerable to various kinds of failures which can be attributed to the unavailability of required services, complexity of the underlying cloud infrastructure, large-scale, and widely distributed computing resources.

In this paper we propose a hybrid cloud framework in order to improve reliability and availability of IaaS (Infrastructure-as-a-Service) services (e.g., CPU, disk storage, memory, etc.) by taking into account alternative services which are available through public clouds. However, the process of provisioning and acquisition of alternative services in hybrid cloud is not trivial. First, it involves complex processing and intelligent decision making in order to decide on which alternative services should be

© Springer International Publishing Switzerland 2015
O. Gervasi et al. (Eds.): ICCSA 2015, Part I, LNCS 9155, pp. 557–568, 2015.
DOI: 10.1007/978-3-319-21404-7_41

provided to cloud users in the case of unavailability of desired services. Second, the required services should be acquired in a reliable way such that they are consistent and correct even in the event of failures such as system failures or network communication failures.

In the proposed framework, we develop an agent-based system using cloud ontology [4] in order to identify and rank alternative cloud services which users can acquire in the event of failures or unavailability of desired services. It includes a broker server within the private cloud that collects information about availability and status of services from different public cloud providers. Individual agents are stationed on public clouds that monitor and transmit (in real time) the availability and service status changes to the broker server. Using a heuristic algorithm the broker server ranks available services at public cloud in an order that best match specification of user's requests. Once services are ranked the broker server then starts acquiring the services. It exploits transactional techniques in service acquisition in order to maintain the correctness and consistency of user's requests and the required services. Transactions have been employed in various cloud services [5]. However, they have not been exploited in the acquisition of services in hybrid cloud environment. We believe that employing transactions in hybrid cloud service acquisition can guarantee correctness and consistency of applications and services despite failures of computing nodes, or network failures.

The rest of the paper is organized as follows. Section 2 discusses related work. Section 3 presents architecture and an example scenario of IaaS service provisioning in a hybrid cloud. Section 4 describes the selection and ranking process of IaaS services. Section 5 presents service acquisition. Section 6 presents empirical evaluation and results. Section 7 describes the concluding remarks.

2   Related Work

The work in [2] introduces an agent-based cloud service search engine for cloud resource management. However, SLA and network latency were not considered. Authors in [3] identified the characteristics of IaaS and used the analytic hierarchy process (AHP) as the multi-criteria decision-making technique to compare IaaS providers. In our case, we used ontology and semantics to find the provider services that best-fit the user requirements. While we share similar goals with the above efforts, we focus on the infrastructure level and used ontology and agents for discovery and selection of IaaS services to provide resiliency support in hybrid cloud computing. Our work does not intend to replace existing or classic cloud computing resiliency techniques, but can be considered as a complement. Authors in [5] study various alternative architectures of cloud computing for transaction processing in web and database applications. Based on alternative architectures, this work investigates into performance and cost of running such applications on different commercial cloud services such as Google and Amazon AWS. The findings are that all major cloud service providers have adopted a different architecture for their cloud service provisioning. Thus depending on the workload, there are significant differences in the cost and performance of cloud services. Paper [7] proposed a scalable transaction management approach which is based on snapshot isolation. The objective is to achieve high scalability by decoupling transaction management functions from storage systems and integrating them with application-level processes. Paper [6] presents a Deuteronomy system that ensures ACID properties in data anywhere or cloud environment. This work also separates transactions from underlying data such that the former are managed by transactional component and the latter by data component. Such separation is claimed to result in improved performance while still maintaining ACID properties. The aforementioned approaches do not exploit transactions in service acquisition in hybrid cloud. Instead, they are concerned with transaction processing in cloud data management.

3   The Hybrid Cloud Framework

A generic architecture of the proposed hybrid cloud framework is depicted in Figure 1. It is comprised of a private cloud, broker server and a number of public clouds. Users can acquire services from a private cloud. If required services are not available from the private cloud then they can be acquired from one or more public clouds. In this paper, we focus on the IaaS services such as CPU, memory, disk space, and so on.

Consider a scenario where a highway control office wish to run a cloud application in order to collect large volume of data (e.g., cameras, sensors, traffic vehicles, etc) about current traffic conditions on roads; and process such data in order to assess the impact of various events such as traffic jams, accidents, etc. Based on the assessment and analysis of data the highway control office can then provide highway operation staff with useful information so they can respond to events in a timely manner and ensure smooth traffic flow on roads. In order to process such a huge data in a timely manner, such application will require various IaaS services such as CPUs, memory, disk space, and so on. Highway control office may use private cloud to run such an application. But if a private cloud does not provide all the required services then this application will not be able to run and will eventually fail. In this situation, hybrid cloud provides alternative provisioning of services from public clouds. That is, in the case of unavailability of some services, the highway control office can acquire services from public clouds in order to run their application.

In the proposed framework, a broker server is deployed (on the private cloud side) which facilities the alternative service provisioning. Broker server provides various functionalities which we classify in the following classes:

Service Selection and Ranking: As shown in Figure 1, there are various public clouds which provide similar alternative services but with varying level of quality. Thus it is important to identify (search), rank and select the alternative services. The proposed broker-server is built around multi-agents and cloud ontology in order to carry out searching, ranking and selection of services.

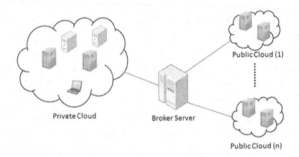

Fig. 1. Architecture of the Hybrid Cloud

Service Acquisition: In order for a cloud application (e.g., highway application) to run, it is important to guarantee that all required services are acquired in a consistent and reliable manner. For a cloud application, it would not be acceptable, if a CPU service is acquired but not a memory or a disk space. The broker server uses transactions in order to avoid such situation.

4   Service Selection and Ranking

This section describes an agent-based system which is used to collect and monitor the availability and status of services provided by the selected public cloud IaaS service providers. The system comprises a broker server and various agents — i.e., each public cloud has an agent deployed on its side. Such agents send to the broker server all updated information about the user-required services of public clouds. In other words, all updated services of each cloud provider are in place in the broker server database. This helps to instantly process user's requests (or transactions) which are explained in the subsequent section. As an example we consider the following two cases:

All Services in one Public Cloud Provider
The broker server, based on the requirements from the user application (e.g., load spikes), will conduct a first search of all member public clouds providing a first-list of public clouds which meets the minimum requirements demanded by the user application. With the first-list and using a heuristic algorithm, the broker server will conduct an individual search of each cloud service (e.g., OS, Cost, CPU, memory, storage, SLA) and rank the providers according to the level of services availability (i.e., the more memory the provider offers, the many points it gets) that it can provide based on the user requirements.

The provider having the most number of high service availability or having the most cumulative points of its services will be ranked top and others follow accordingly. The rational idea is to have all required cloud services to be provisioned by the same provider so as to minimize communication overhead.

Some Services in other Public Clouds
Considering that other clients or customers are also vying to avail of the same public cloud service(s), there's a possibility that one or more requested public cloud services

in the top-ranked provider may not be available by the time the broker server made its selection decision. The broker server will then proceed to the next-ranked cloud provider, and so on.

However, if the set of services cannot be provisioned by one cloud provider, the broker server will then search in the next rank provider the unavailable requested service(s). If the broker server cannot find a match, the same procedure is repeated to the next lower providers until the first-list is used up. Up to this point when no match is found, the broker server repeats the same procedure from the beginning, i.e., looking up to its updated database.

Results and Observations

Tables 1 and 2 show the first-list of public cloud providers (top five) which have met the requirements of two users, and ranked according to the procedures described above. Each cloud provider in the list offers all the required services. The tables include the SLA (System Level Agreement) guarantee offered by each public cloud provider and slightly differ from each other. These SLAs are used to break up whenever two or more providers have the same accumulated scores for ranking. If a tie score persists, then a random selection is employed.

Since some users may be requesting the same service(s) concurrently on the same provider, one or some services may not be available by the time the broker server made its final acquisition decision. The broker server will then tap the lacking services from other cloud providers.

A possible dilemma is that if one or more services will be provided by a different public cloud, such operation may incur a large communication overhead (latency) which may be detrimental to the application performance. The broker server may send a ping message to the public cloud providers to determine the latency which can be considered as a parameter to deal with so as not to compromise the service transaction process.

To this, the user will now have a full knowledge on the best-fit public cloud providers ranked accordingly. It is now up to the user to select the desired provider for optimum benefits.

5   Service Acquisition

This section describe hybrid cloud transactions which are executed as part of users' requests in order to correctly and consistently acquire the cloud services which are ranked by the agent-based system.

To run an application, user makes a request (or initiate a transaction) in order to acquire required services from a hybrid cloud, for example, 4_Core CPU, 2GB memory, 200GB disk space, Ubuntu OS. All these four services are necessary for a user to run an application. If some of these resources (e.g., disk space) are not available from one cloud provider then it can be acquired from an alternative public cloud provider.

We define hybrid cloud transaction as an execution of a user's request which can be divided into well defined units (or component transactions) that provide correctness and consistency of cloud services. A hybrid cloud transaction can be considered as a unit of different component transactions each of which is used to acquire service — e.g. one component transaction can be used to acquire CPU service, another to acquire memory, and so on.

In order to ensure that cloud resources are consistently acquired we define the following rules (properties) [8] which must be followed by hybrid cloud transactions.

Semantic atomicity requires that a hybrid cloud transaction should execute as an atomic unit of work. That is, its component transactions should be run completely such that they can acquire those services which are essential for running a user's application. If any of the essential resource (e.g., CPU) is not available then other resources (e.g. OS) are not required.

Consistency: This requires that a hybrid cloud transaction should maintain the consistency of information related to the availability of cloud resources and user's application requirements.

Durability requires that effects of a completed hybrid cloud transaction must be made permanent in their respective data sources, in order to ensure recovery in the event of failures.

Resiliency is the ability to complete a hybrid cloud transaction in spite of failures or service unavailability. Resiliency is achieved by providing alternative services from public clouds. For example, if a disk space cannot be acquired from one cloud provider then it can be acquired from another one as there exist various alternative cloud providers.

5.1   Hybrid Cloud Transaction Protocol

In the following, we describe a protocol that enforces the above rules in hybrid cloud transactions. The proposed protocol is implemented as one Hybrid Cloud Coordinator (HCC) and several Component Transaction Coordinators (CTC). HCC is deployed as part of the broker server while each CTC is deployed on individual public cloud (as in Figure 1). The main functionality of the HCC is to coordinate the execution of the overall hybrid cloud transaction in relation to the component transactions. Each CTC is responsible for coordinating its individual component transaction (e.g., acquiring CPU or disk space). HCC and each CTC maintain log files in order to record the required information about the execution of hybrid cloud transaction.

In order to acquire required resources for user's requests, HCC needs to execute hybrid cloud transaction. During the execution of hybrid cloud transaction, HCC communicates various messages with CTCc. Transaction execution is accomplished through the following steps:

1. A new hybrid cloud transaction is assigned to a HCC, which records the start of the transaction in a log file, and send 'start' message to CTCi to initiate component service transaction.

2. CTCi records the start of a component service transaction in the log file. After processing, CTC sends either a 'local success' or 'local fail' decision to HCC. 'local success' decision is made if component service transaction can acquire the required cloud resource (e.g., CPU, etc). 'local fail' decision is made if it cannot acquire the required cloud resource.

In case of 'local fail' decision, CTC records the failure of component service transaction, and declares it to be locally failed. Otherwise, CTC writes a 'local success' decision, and awaits from HCC final decision.

3. HCC receives all decisions from CTCc. If all the decisions are 'success' (i.e., all resources, CPU, disk space, memory are acquired), then HCC records 'global success' decision in the log file and inform all CTCs of the success decision. The CTCs commits, by forcibly writing the commit decision and then terminates the transaction and starts the processing of a new one.

Table 1. User1 Requirements

Table 1. User1 Requirements:				cpu	memory	storage	cost	os	
				2_Core	2GB	200GB	10000JPY	Ubuntu	
rank	point	provider	service	cpu(core)	memory(GB)	storage(GB)	SLA(%)	cost(JPY)	OS
1	3	Softbank_Telecom	Type_Dual	2	2	100	99	15750	WindowsSever2012 Ubuntu CentOS
2	3	NIFTY	Spec_Type6	2	2	30	99.99	25410	WindowsServer2012 Ubuntu RedHatEnterpriseLinux CentOS
3	2	GOGRID	Large	4	25	200	100	13140	Ubuntu Debian RedHatEnterpriseLinux WindowsServer2012 CentOS
4	2	NTTCommunications	Plan_v2	2	4	10	99.99	7560	WindowsSever2012 Ubuntu CentOS
5	2	IDC_Frontier	Type_M2	2	8	15	99.999	21000	WindowsServer2012 Ubuntu RedHatEnterpriseLinux CentOS

Table 2. User2 Requirements

Table 2. User 2 Requirements:				cpu	memory	storage	cost	os	
				4_Core	2GB	200GB	10000JPY	Ubuntu	
rank	point	provider	service	cpu(core)	memory(GB)	storage(GB)	SLA(%)	cost(JPY)	OS
1	3	GOGRID	Large	4	25	200	100	13140	Ubuntu Debian RedHatEnterpriseLinux WindowsServer2012 CentOS
2	2	Softbank_Telecom	Type_Dual	2	2	100	99	15750	WindowsSever2012 Ubuntu CentOS
3	2	NTTCommunications	Plan_v4	4	8	10	99.99	15120	WindowsSever2012 Ubuntu CentOS
4	2	IDC_Frontier	Type_L	4	4	15	99.999	29400	WindowsServer2012 Ubuntu RedHatEnterpriseLinux CentOS
5	2	NIFTY	Spec_Type3	1	2	30	99.99	18144	WindowsServer2012 Ubuntu RedHatEnterpriseLinux CentOS

4. If any 'failed' component service transaction is replaceable (having alternative service from public cloud), MTC initiates the alternative component service transaction and awaits the CTC's decision regarding 'success' or 'fail' of the alternative component service. If component service transaction is not replaceable HCC records 'global fail', and informs all CTCs about its decision. In this case, the overall hybrid cloud transaction has failed and cloud resources cannot be acquired.

5. After receiving a global decision, CTC writes the global success of component service transaction. If component service transaction is local-success and CTC receives a 'global fail' from HCC, then CTC must cancel that service. This situation occurs when one component service transaction acquires a resource (e.g. disk space) and another does not acquire (e.g. CPU). In this case the acquired resource needs to be cancelled as user will need both resources (CPU and disk). This is the constraint set by the semantic atomicity rule (as described above).

6   Empirical Evaluation

This section describes the evaluation of the proposed framework in terms of failure resilience and communication overhead. One of the main objectives of the proposed framework is to enhance the resiliency of cloud services through alternative public cloud services. We therefore evaluate the resiliency aspect of the framework. However, such resiliency may incur performance delay mainly due to the network communication delays. We therefore evaluate the overhead caused through the alternative services provisioning from the public clouds.

A. Failure Resiliency

Our evaluation criteria for failure resiliency are based on probability theory and are simulated through a prototype tool. In order to simulate the success/failure rate of a hybrid cloud transaction, we define the following set of probabilities:

private cloud success (PriCS): This refers to the probability that user can acquire required services from a private cloud by executing a hybrid cloud transaction.

private cloud failure (PriCF): This refers to the probability that user cannot acquire required resources from a private cloud by executing a hybrid cloud transaction. In other words, if the private cloud cannot meet the minimum requirement of user's request then it is considered as a failure of private cloud. In this case, the user's requests should be sent to the broker server (as described in the subsequent section).

alternative services availability (AltSA): This refers to the probability that there exist alternative services which user can acquire from public clouds. In other words, this depicts the situation where users cannot acquire first choice services from private cloud. But they have the opportunity to acquire alternative services from public cloud. As described above, we use a heuristic algorithm to rank different public cloud providers.

public cloud success (PubCS): This refers to the probability that user can acquire required services from a public cloud (first-list – see below) by executing a hybrid cloud transaction.

public cloud failure (PubCF): This refers to the probability that user cannot acquire required resources from a public cloud by executing a hybrid cloud transaction.

total success rate (TSR): This refers to the probability that a user can acquire required services from private and one or more public clouds.

total failure rate (TFR): This refers to the probability that user cannot acquire required services from private or any of the public clouds.

Based on the above probabilities, we design various mathematical expressions in order to calculate the total success rate (TSR) and total failure rate (TFR) of a hybrid cloud transaction. TSR and TFR are calculated using the probabilities PriCS, PriCF, AltSA, PubCS and PubCF.

We consider multiple alternative cloud services from public clouds. For example, if required storage is not available from a private cloud then it can be acquired from public clouds (see Section 4).

Table 3. Total Success and Failure Rates

	Case 1	Case 2	Case 3	Case 4	Case 5
PriCS	0.9	0.8	0.7	0.6	0.5
PriCF	0.1	0.2	0.3	0.4	0.5
TFR1	0.028	0.072	0.132	0.208	0.3
TFR2	0.0208	0.0464	0.0816	0.1312	0.2
TSR1	0.972	0.928	0.868	0.792	0.7
TSR2	0.9792	0.9536	0.9184	0.8688	0.8

As shown in Table 3 we conduct various experiments to evaluate the resiliency of the proposed framework. We assume that the probability of alternative services availability (AltSA) is 0.9 — showing that there exist 90% chance of acquiring alternative services from public clouds. In each case, different values of the PriCS and PriCF probabilities are also used in conjunction with AltSA. Various situations are considered. For example, scenarios where the failure probability is high, and also scenarios where the success probability is high.

The total success and failure rates of hybrid cloud transactions are calculated in Table 3 and are graphically represented in Figures 2 and 3. The graphs clearly indicate that the proposed framework increases the resiliency of the service provisioning by successfully executing hybrid cloud transactions and acquiring the cloud services. For instance, the success rate of hybrid cloud transaction in Figure 2 is higher when there are services available from public clouds (TSR1 and TSR2). Similarly, Figure 3 shows that failure rate of TFR1 and TFR2 is lower than the PriCF when public cloud services are available.

B. Processing and Communication Delays

The improvement in failure resiliency may result in performance degradation. If one or more services are acquired from public clouds, then it may incur a communication overhead (latency) which may be detrimental to the application performance. In this section, we evaluate such overhead by taking into account some of the important parameters such as the network communication delay, the number of messages communicated between Hybrid Cloud Coordinator and several Component Transaction Coordinators (see Section 5-A), processing time of each component transaction spent on acquiring services from public cloud, the total number of component transactions acquiring services from public cloud, and the probability that a particular service is acquired from a public cloud.

The average time taken to process a hybrid cloud transaction, T_{proc}, can be calculated as follows:

$$T_{proc} = \sum_{i=0}^{n} CTPR_i + \sum_{j=0}^{m} CTPUB_j + NET_d$$

T_{proc} is the sum of processing time of its component transactions ($CTPR_i$) acquiring resources from a private cloud plus component transactions ($CTPUB_i$) acquiring resources from a public cloud plus the network communication delay (NET_d).

Processing time of a component transaction is the time it takes to acquire a cloud service (e.g., disk storage, memory, etc).

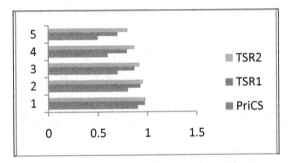

Fig. 2. Success Rate of Hybrid Cloud Transactions

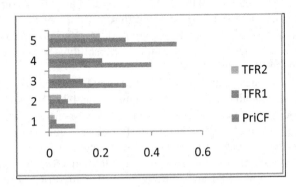

Fig. 3. Failure rate of Hybrid Cloud Transactions

NET_d is a network delay which is calculated as follows:

$$NET_d = (M_{pr} \times CD_{pr} \times N_{pr}) + ((M_{pub} \times CD_{pub} \times N_{pub}) \times P_{pub})$$

M_{pr} and M_{pub} are the number of messages communicated between hybrid cloud systems in order to respectively acquire services from private cloud and public cloud(s). CD_{pr} and CD_{pub} are the message delays for private and public clouds. N_{pr} and N_{pub} represent the number of component transactions respectively acquiring services from private cloud and public cloud. P_{pub} is the probability that one or more cloud services can be acquired from a public cloud. $P_{pub} = 0$, if all services can be acquired from a private cloud. In that case, there is no extra communication overhead of acquiring services from public clouds. $P_{pub} = 1$, if any of the service is acquired from a public cloud.

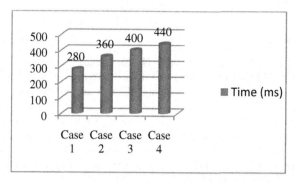

Fig. 4. Average Processing Time

We simulate an average time taken to process a hybrid cloud transaction, T_{proc}, by taking into account four different cases. We consider that as part of user request, a hybrid cloud transaction has to acquire four services by executing four component transactions.

The average time for four different cases is graphically represented in Figure 4. Case 1 represents the situation where all services are acquired from private cloud and thus average processing time, T_{proc}, is less than other cases. Case 4 represents the opposite that all services are acquired from public cloud showing steep increase in the processing time. Case 2 shows the situation where half of the services are acquired from private cloud and half from public cloud. Case 3 shows that three services are acquired from public cloud and one from private cloud.

7   Conclusion

We presented a hybrid cloud framework in order to improve reliability and availability of IaaS services through a synergetic approach of agents, ontology and transactions. The framework employed agents and ontology to gather current status of IaaS services from each member public clouds. The service information status is consolidated on a broker server which can provide, using a heuristic selection algorithm, the first-list public cloud providers which meet the user requirements through the private cloud at runtime. Transactional techniques are then used in order to ensure the reliability and consistency of the service acquisition process. The proposed framework can be used by cloud users and providers in order to make intelligent decision in service consumption/provisioning in hybrid cloud by taking into account various crucial factors such as (i) ranking and selection of alternative cloud services from public cloud(s) (ii) enhanced failure resiliency and (iii) the performance and communication delay caused when services are acquired from private and/or public clouds.

Acknowledgment. This research was supported in part by the Japan Society for the Promotion of Science Grants-in-Aid for Scientific Research 24500100. Part of this work was completed during Prof. Apduhan's visit to Oxford Brookes University, Oxford, UK.

References

1. Uchibayashi, T., Apduhan, B.O., Shiratori, N.: An ontology update mechanism in iaas service discovery system. Int. Journal of Web Information Systems **9**(4), 330–343 (2013)
2. Sim, K.W.: Agent-Based Cloud Computing. IEEE Trans. on Services Computing **5**(4), 564–577 (2012)
3. Lee, S., Seo, K-K.: A Multi-Criteria Decision-making Model for an IaaS Provider Selection Problem. Int. Journal of Advancements in Computing Technology **5**(12) (2013)
4. Androcec, D., Vrcek, N., Seva, J.: Cloud Computing ontologies: a systematic overview. In: Proc. of the 3rd Int. Conf. on Models and Ontology-based Design of Protocols, Architectures and Services, pp. 9–14 (2012)
5. Kossmann, D., Kraska, T., Loesing, S.: An Evaluation of alternative architectures for transaction processing in the Cloud. In: Proc. of SIGMOD, pp. 579–590 (2010)
6. Levandoski, J.J., Lomet, D.B., Mokbel, M.F., Zhao, K.: Deuteronomy: transaction support for cloud data. In: 5th Biennial Conference on Innovative Data Systems Research, January 9-12, Asilomar, California, USA, pp. 123–133 (2011)
7. Padhye, V., Tripathi, A.: Scalable transaction management with snapshot isolation on cloud data management systems. In: Proc. of IEEE 5th Intl. Conference on Cloud Computing, pp. 542–549 (2012)
8. Younas, M., Eagelstone, B., Holton, R.: A review of multidatabase transactions on the web: from the ACID to the SACReD. In: Jeffery, K., Lings, B. (eds.) BNCOD 2000. LNCS, vol. 1832, pp. 140–152. Springer, Heidelberg (2000)

Mixed Reality for Improving Tele-rehabilitation Practices

Osvaldo Gervasi[1]([✉]), Riccardo Magni[2], and Matteo Riganelli[1]

[1] Department of Mathematics and Computer Science,
University of Perugia, Perugia, Italy
osvaldo.gervasi@gmail.com
[2] Pragma Engineering SrL, Perugia, Italy

Abstract. In the present work we propose a methodology for improving tele-rehabilitation practices adopting mixed reality techniques. The implemented system analyzes the scene of a tele-rehabilitation practice acquired by a RGB-D optical sensor, and detects the objects present in the scene, identifying the type of object, its position and rotation. Furthermore we adopted Mixed Reality techniques to implement more complex rehabilitation exercises.

After an initial training period, in which the set of objects are classified by the system, the method analyzes the acquired images in real time and the identified objects (which are included in the set of preliminarily identified objects) are evidenced with a rectangle, the size and location of which are variable. All regions of the image are analyzed and objects of different shape and size are identified. In addition in a file associated to the object, the most relevant object features are stored using XML format.

The implemented method is able to identify a vast set of objects used regularly in tele-rehabilitation exercises and allows the therapist to perform the quantitative assessment of the patient practices.

1 Introduction

The importance of tele-rehabilitation is constantly growing due to the limited economical resources available and the increasing need to rationalize the health care costs. The availability of systems able to identify objects, monitoring their movements in the space, may facilitate therapist's work, allowing the distance assessment of tele-rehabilitation practices.

The system proposed in the present paper presents the innovative characteristics of being very fast in computing the objects present in the images acquired via a RGBD camera (in our case we used a Kinect® device, however there are several alternatives available as low cost RGB-D video acquisition systems) in real time (at a frame rate ranging from 15 to 30 frames per second), being able to process multiple objects at the same time, and identify rotated and scaled objects.

The implemented method uses the *integral image*, defined during the initialization phase and optimized to speed up the comparisons made by the system. The method is based on the works of Papageorgiou [6] and Viola-Jones [8], and take advantage of the following characteristics:

© Springer International Publishing Switzerland 2015
O. Gervasi et al. (Eds.): ICCSA 2015, Part I, LNCS 9155, pp. 569–580, 2015.
DOI: 10.1007/978-3-319-21404-7_42

- The definition and subsequent usage of the *integral image* to speed up the object detection process;
- The *Haar-like Features*, portions of pixels which are applied to the image and are able to provide numerical information about the object. It can be seen as a small window used to analyze a small portion of the image to be identified. The window position and size are variable, allowing the detection of objects at various distances and positions from the optical sensor. Once available the integral image the computation of the Haar-like Feature occurs in a constant time, allowing to implement a very fast detector.
- The most relevant characteristics of the object are stored in a XML file, so that after an initial training phase, the *detector* can be able to identify the objects present in the input frames very efficiently.

Only objects which have a integral image and are described in the system by the associated XML file may be recognized during the tele-rehabilitation exercises. Once completed the initialization phase, the *detector* will scan in real-time all images acquired by the optical sensor, trying to identify each object present in the image. The detector communicates with a program conceived to manage the setup of tele-rehabilitation exercises, via a telecommunication protocol. The proposed system can be integrated, in such a way, into tele-rehabilitation systems like the Nu!Reha desk [3,10].

The detector selects [1] a small number of features using the *AdaBoost* algorithm [2]. In particular, the implemented method combines some classifiers in a cascade structure, which speedup drastically the object identification, focusing the attention on the most promising area of the image. The classifier is able to recognize only a single object: for this reason, the cascade structure of the classifiers increase significantly the performance of the system when searching for several objects.

Such a fast engine for the object recognition has several application fields.

In figure 1 is sketched the conceptual scheme of the proposed system. *Nu!Reha creator* is the mixed reality component which allows the creation of tele-rahabilitation practices. In fact, once identified by the *Object recognition system*, the object can be associated to any representation of it and this aspect is really helpful in defining tele-rehabilitation exercises.

2   The Proposed Method

Our approach is based on the extraction of the image features, namely the most meaningful parts of the image, from each of the selected objects; these features will be retrieved and compared from the object database file during object recognition, according to Lowe method [5], shown in the simplified scheme of Figure 2.

The extracted features are used to build the model figure to be identified in each frame. The model figure speeds up significantly the object recognition process, allowing to perform the object identification in real time, on the frames acquired by the RGBD camera.

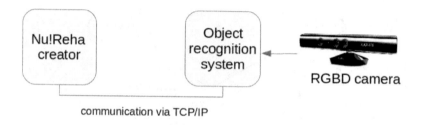

communication via TCP/IP

Fig. 1. Architecture of the proposed system

The quality and the quantity of the features detected in the initial image associated to the object to be identified are crucial for the efficiency and the quality of the identification process. In fact, the variations of color intensity and of the type of light may condition the type of information associated with the features in the model image. Once identified the points of the image in which the information is locally highest (keypoints), an appropriated shape is selected to represent the feature.

Usually the features are much smaller than the original image and may be conveniently represented in a vector. The identification process is much faster in such a way, since analyzing a small set of features is faster than analyzing the whole set of image pixels.

The identification of the features represents a crucial phase of the artificial vision, since an efficient feature extraction allows to properly represent the selected object, taking advantage of a small set of numerical values, which allow to speed up the identification process. Each feature is a function of a set of measures, representing each an object property and a quantitative measure of the presence of such property,

Papageorgiu [6] proposed a framework based one Haar wavelet. The system is trained using a database of positive samples (set of images in which the object to be detected is present) and a set of negative samples (set of images in which the object is not present). To improve the performances of such method, a bootstrapping learning method [7] is used. The learning process is iterative and the set of negative samples is regularly updated.

As proposed by Viola-Jones [8], a Haar-like feature takes into account a rectangle of the detection window, evaluating the rectangle intensity as a sum of the individual pixel intensity, and evaluates the difference between adjacent regions. In such a way the various parts of an image are classified. Even if the method has been originally designed for face detection, it can be used to detect any object.

In Figure 3 is represented the definition of 4 feature inside a detection window in an image. The algorithm defines the following feature typologies:

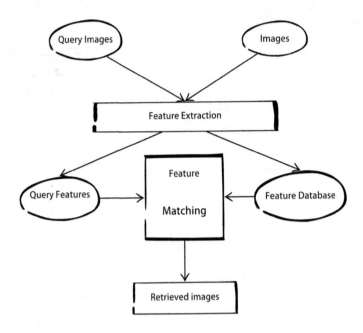

Fig. 2. The systematization of the presented Object Detection system

- two-rectangle-feature(A-B): as the difference between the sums of pixel inside of the two rectangles
- three-rectangle feature(C): as the result of calculation of sum of pixels of external rectangles subtracted to the pixel sum of the internal rectangles
- four-rectangle feature(D): as difference of pixel sums between the diagonal rectangles

Each feature may vary in position and dimension inside the detection window and it takes the value equivalent to the normalized difference between the gray levels of the pixels belonging to the clearer region and those of the darker one. To slide these rectangles in the entire integral image requires a constant time, and this is a great advantage over the other methods.

2.1 The Integral Image

The integral image is represented by a matrix representing for each pixel in (x, y) position the sum of all pixel intensity whose positions are less then or equal to (x, y). To speed-up the computation, an integral image is computed for each frame. After having converted the image as a gray scale, the integral image is generated applying for each point of coordinates (x, y) the following equation:

$$ii(x, y) = \sum_{x' < x, y' < y} i(x', y') \tag{1}$$

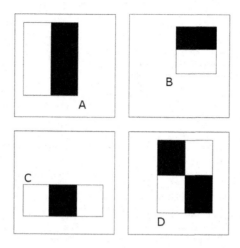

Fig. 3. Definition of 4 Haar-like features

where $ii(x,y)$ represents the element of the integral image and $i(x,y)$ the element of the original image.

Figure 4 shows how the integral image is used to evaluate the Haar-like features. Supposing we want to know the sum of the pixels internal to the rectangle D. It can be evaluated accessing 4 matrix elements. The value of the integral image in point 1 represents the sum of all area A pixels. The value in point 2 represents the sum of the intensity of the pixels of the rectangles A and B. Similarly point 3 represents the sum of pixels of rectangles A and C, while point 4 is equal to the sum of all pixels of rectangles A, B, C and D. The sum of all pixel contained in D is given by:

$$D = 4 + 1 - 2 - 3$$

2.2 Adaboost

In its original formulation, the method described so far is used to improve the classification performances of a learning algorithm. Considering a window of 24x24 pixels, the expected number of features is greater than 45,000. If we consider a larger image, the number of features should be multiplied by the number of sub-windows present in the image. The resulting computational cost will increase significantly. For this reason the system implements a modified version of the AdaBoost learning algorithm, so that only the best features are taken into account, building classifiers based on them.

Adaboost is a *machine learning* meta algorithm formulated by Freund and Schapire [9] based on the creation of some variations of the main classifier. To each sample created is assigned a weight and iteratively are built models dependent from

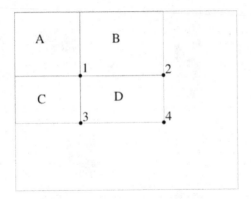

Fig. 4. Sample Haar-like features evaluation

weights. Each iteration AdaBoost updates the weights. The algorithm improve the classifier identifying the wrong samples (higher values of the weight).

2.3   The Cascading Architecture

In each iteration of the Adaboost algorithm a base classifier is created. According to Viola-Jones approach, the best classifiers are organized according to a cascading architecture, based on their complexity, as described in Figure 5. In this way each classifier is trained only on the sub-windows (portions of the given image) in which the previous classifier detected some features. If a sub-window has been discarded by a previous classifier, it is not anymore considered in the subsequent phases, and the system moves to the next sub-window to be analyzed. In this way the analysis is concentrated only on the sub-windows where are present some features, and the analysis of the whole image is carried out very quickly.

The classifiers in the cascading system are arranged in order of increasing complexity. In this way the simple and fast classifiers discard the majority of the image regions, so that the sub-windows without features are discarded rapidly and immediately (in the first levels of the cascading architecture), while the more complex classifiers are mainly used to reduce the false positive sub-windows.

In Figure 6 the features obtained with the above described method are shown. In particular, such features have been obtained in a 4 levels cascading system. As shown in the figure, each feature catch a a different feature of the object to be detected.

Once implemented the cascading system, we implemented the *Detector*, which scales the size of the detected features in order to be able to identify variable dimensions objects.

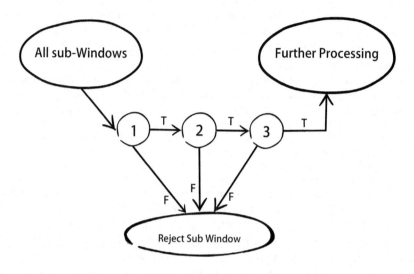

Fig. 5. The cascading architecture in a Haar-like system

Fig. 6. Features detected by a 4-levels cascading classifier

3   Analysis of Performances

We carried out several experiments to prove that the Viola-Jones method is the fastest method in order to identify in real time the objects in a series of frames acquired by a RGBD camera at a frame rate in the range 15-30 frames per second, even if the training period and the cascading model set-up are time consuming and complex processes.

We considered Viola-Jones and other 3 methods, we implemented on the basis of the available OpenCV APIs. All the methods were implemented on the target platform (referred to the same hardware and same RGBD camera characteristics).

3.1   Color Matching

This method has been implemented in the initial phase of the study to detect object made by only one color. It uses the OpenCV APIs and has been applied to analyze the frames acquired by the RGBD camera. The method is really fast, however we observed a very high variation in the efficiency of the detection process as a function of the light intensity and of the light sources.

The method has been improved introducing some image correction methods in conjunction with some methods for identifying the shape of the objects present in the scene, and comparing the results with some known shapes (rectangle, triangle, circle).

3.2   Shape Detection

The method analyzes the frames coming from the RGBD camera computing the simple shapes contained in the image (circle, triangle, square, rectangle). The method is fast and robust when detecting simple shapes. However also in such case the light intensity and the type of light source influence dramatically the results. Furthermore, complex shapes cannot be detected.

3.3   SURF

This algorithm [4] is based on the detection of the *keypoints*, and is able to detect complex shapes. However it is able to detect objects only when they are just a few and cannot be utilized when an object has several faces, like a book. If the method has to detect the keypoints for every face of a box or any other real object it becomes unusable. Also in such a case the light intensity may influence the evaluation of key points.

3.4   Performance Tests

In table 1 the mean time required to identify some objects are shown. The values expressed in terms of the milliseconds required to identify a moving object (during the simulated tele-rahabilitation practice) shown in the table have been obtained on a computer Intel Core i5-3210M, 64 bits, CPU 2.50GHz, 4,00 GB RAM.

As discussed previously, even if the Color matching and Shape detection algorithms are much faster that the others 2 methods, they are unusable in real settings, such as the one we are interested in the tele-rehabilitation applications. In fact, the light intensity and the type of light sources influence dramatically the efficiency of the detector.

Table 1. Measures related to object recognition made using different methods

n.	Algorithm used	time (ms)
1	Color Matching	instant
2	Shape Detection	instant
3	SURF	120-140ms
4	Viola-Jones	15-30ms

4   Discussion of Results

The reported implementation proposes a complex work to classify the object visual features, through the processing of a huge number of images (photos with and without the object) taken in different light conditions and angles. This process is time consuming and should be repeated for each different object the patient is asked to handle. Anyway, considering the limited number of different objects needed for tele-therapy exercises (estimated in 10) and the provision of a *standard kit* to be brought at patient home, the classification complexity has been considered bearable with respect to the reliability and velocity of the recognition of the object in the scene.

The proposed system is based on the present NU!Reha platform implementation. The hardware elements are constituted by:

- Single Board Computer: 4,00Gb RAM, Intel Core I5, 2,50 Ghz, Microsoft Windows 7 - 64 bit, USB 2, HDMI interface
- Wide LCD TV (30-5"): simulating patient home TV
- Kinect 1.0 (or RGBD equivalent camera): 640x48
- Real objects to be handle by the patient: plastic jar, glass, tetrapack brick, etc.

The software running on the platform is mainly consisting in:

- Exercise manager (Nu!Reha Manager): application that is synchronizing the exercise and their data for configuration and results

Fig. 7. An example of object recognition made on the images acquired from a RGBD camera

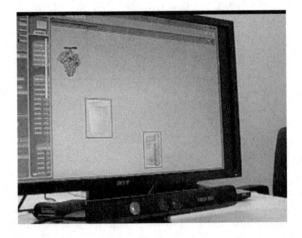

Fig. 8. The occupational therapy exercise

- Exercise engine (Nu!Reha Creator): a Smalltalk-Squeak based application which is performing the execution of each programmed exercise, recording variables linked to patient behavior. It allows the interaction via mouse or other compatible HCI and some external sensors (as they should be useful to interact with patient considering his/her specific conditions).

In Figure 7 is shown the graphical rendering (given for development/debugging purposes) of the Viola-Jones detector applied in a complex scene.

The system is able to detect rotated objects (in our tests the XML models have been created rotating the objects for 25 degree at the maximum) and several issues have been considered in the model, resulting a very stable and robust system for the object detection.

Using such a system we implemented an occupational therapy exercise with the following characteristics:

- the objects the patient handles (which have to be identified by the system) are a bag of milk and a glass.
- the system has to detect the objects in real time in the frames sent to the system by the RGBD camera, identifying the detected objects with a colored rectangle (each object will be identified with a different color).
- the exercise platform generates the virtual representation of a bunch of grapes or a cow which appear from the top of the screen, falling slowly.
- the patient has to pick up the right object (glass or bag of milk) and move it towards the object falling to the bottom part of the screen.
- the system will count the right and wrong moves respectively and at the end sends the outcome to the therapist.

A sample image related to the occupational therapy exercise is shown in Figure 8.

5 Conclusions

We proposed an object recognition platform able to identify a limited set of predefined objects in a frame sequence acquired by a RGBD camera, which allows to implement occupational therapy exercises adopting a Mixed Reality approach.

The system is based on the present NU!Reha platform implementation, which is used since longtime as a tele-rehabilitation desk suitable for neurologic patients (Stroke, TBI, MS) living at home.

Considering the actual implementation, this work explore the chance to transform the RGBD camera into another "complex sensor", extracting from the image the position of some reference objects to be used during the execution of the specific exercise. The performance of such system during the occupational therapy exercise fulfills the requirements of NU!Reha system in terms of responsiveness and reliability.

The adopted Mixed reality approach enables the therapist to deploy several exercises at a different level of difficulty and this is really useful for improving the tele-rehabilitation practices.

References

1. Vapnik, V.N., Boser, B.E., Guyon, I.M.: A training algorithm for optimal margin classifiers, pp. 144–152
2. Freund, Y., Schapire, R., Abe, N.: A short introduction to boosting. Journal-Japanese Society For Artificial Intelligence 14(771–780), 1612 (1999)
3. Magni, R., Zampolini, M., Gervasi, O.: Nu!reavr: Virtual reality for neuro tele-rehabilitation for patients with traumatic brain injury. Virtal Reality 14, 131–141 (2010)
4. Bay, H., Tuytelaars, T., Van Gool, L.: SURF: speeded up robust features. In: Leonardis, A., Bischof, H., Pinz, A. (eds.) ECCV 2006, Part I. LNCS, vol. 3951, pp. 404–417. Springer, Heidelberg (2006)
5. Lowe, D.G.: Object recognition from local scale-invariant features, vol. 2, pp. 1150–1157, September 1999

6. Poggio, T., Papageorgiou, C.P., Oren, M.: A general framework for object detection, pp. 555–562, January 1998
7. Lienhart, R., Kuranov, A., Pisarevsky, V.: Empirical analysis of detection cascades of boosted classifiers for rapid object detection. In: Michaelis, B., Krell, G. (eds.) DAGM 2003. LNCS, vol. 2781, pp. 297–304. Springer, Heidelberg (2003)
8. Jones, M., Viola, P.: Rapid object detection using a boosted cascade of simple features, vol. 1, pp. I-511 - I-518 (2001)
9. Schapire, R.E., Freund, Y.: A decision-theoretic generalization of on-line learning and an application to boosting, **55**, 119–139, August 1997
10. Zampolini, M., Magni, R., Gervasi, O.: An x3d approach to neuro-rehabilitation. In: Gervasi, O., Murgante, B., Laganà, A., Taniar, D., Mun, Y., Gavrilova, M.L. (eds.) ICCSA 2008, Part II. LNCS, vol. 5073, pp. 78–90. Springer, Heidelberg (2008)

Author Index

Printed in the United States
By Bookmasters